ARTIFICIAL CHEMISTRIES

ARTIFICIAL CHEMISTRIES

Wolfgang Banzhaf and Lidia Yamamoto

The MIT Press
Cambridge, Massachusetts
London, England

This book was set in New Times Roman using the LaTeX— Memoir class.

Library of Congress Cataloging-in-Publication Data

Banzhaf, Wolfgang, 1955—

Artificial chemistries / Wolfgang Banzhaf and Lidia Yamamoto.

Includes bibliographical references and index.

ISBN 978-0-262-02943-8 (hardcover : alk. paper)
ISBN 978-0-262-55152-6 (paperback) 1. Biochemistry. 2. Molecular evolution. 3. Chemistry, Physical and theoretical. 4. Evolution (Biology) 5. Life—Origin. I. Yamamoto, Lidia. II. Title.

QD415.B24 2015

572—dc23

2014046071

artificial, *a.*, contrived by art rather than nature: "artificial flowers";

http://www.thefreedictionary.com/artificial, accessed 2014

Made or produced by human beings rather than occurring naturally, typically as a copy of something natural.

https://www.google.ca, accessed 2014

chemistry, *n.*, a branch of physical science that studies the composition, structure, properties and change of matter, [from Greek: χημεια],

http://en.wikipedia.org/wiki/Chemistry, accessed 2014

The investigation of substances, their properties and reactions, and the use of such reactions to form new substances.

http://www.oxforddictionaries.com/definition/english/chemistry, accessed 2014

CONTENTS

The field of Artificial Life (ALife) is now firmly established in the scientific world. A number of conference series have been founded, books and edited volumes have been written, journals are being published, and software has taken up the philosophy of Artificial Life. One of the original goals of the field of Artificial Life, the understanding of the emergence of life on Earth, however, has not (yet) been achieved. It has been argued that, before life could emerge on the planet, more primitive forms of self-organization might have been necessary. So just as life likely was preceded by a prebiotic phase of the organization of macromolecules, perhaps Artificial Life would gain from the introduction of models explicitly studying chemical reaction systems. It is from this motivation that artificial chemistries were conceived.

Artificial chemistries (ACs) have been around for more than two decades. Perhaps sometimes in disguise, they have been a fascinating topic for scientist of numerous disciplines, mathematicians and computer scientists foremost among them. From the term itself we can glean that they are not primarily meant to produce new materials, but rather new interactions. The interactions between molecules are to be man-made, and the molecules themselves might be artificial, or even virtual. As a consequence, the results of artificial chemistries can be found in the virtual world, e.g. in certain multiagent systems, or in the real world, in the form of new (artificial) reaction systems.

Book Overview

This book will provide an introduction and broad overview of this still relatively young field of artificial chemistries (ACs). The field is highly interdisciplinary, touching chemistry, biology, computer science, mathematics, engineering, and even physics and social sciences to a certain extent. Therefore, we have tried to make the book as accessible as possible to people from these different disciplines. We assume that the reader has some basic knowledge of mathematics and computer algorithms, and we provide a general introduction to the concepts of chemistry and biology that are most often encountered in ACs.

The book will start with a gentle introduction, discuss different examples of artificial chemistries from different areas proposed in the literature, and then delve into a deeper discussion of both theoretical aspects of ACs and their application in different fields of science and technology. An appendix will offer a hands-on example of an AC toolkit ready for the reader to implement his or her own artificial chemistry on a computer.

This book is divided into the following parts:

- Part I holds introductory chapters, with chapter 1 discussing the relation to Artificial Life and other roots of the field, chapter 2 offering a more general methodological introduction,

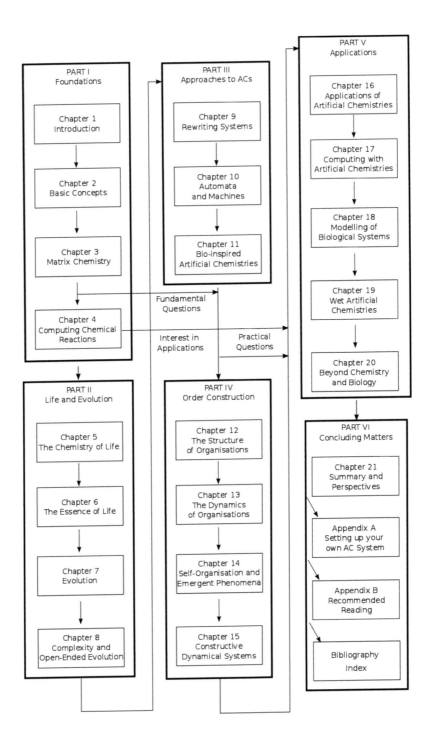

and chapter 3 exposing, by way of example, key problems that will reappear throughout the book. Chapter 4 gives an introduction to the algorithms applied to chemical reactions, which are also used in ACs.

- Part II, on life and evolution, then steps back for some time to briefly review important aspects of Biology: Chapter 5 discusses the relation between life and chemistry, chapter 6 delves into the essence of life and its origins, chapter 7 reviews the basic mechanisms of evolution and their relation to ACs, and chapter 8 discusses the issue of complexity growth and openness of evolution, a topic that attracts much interest from AC and ALife researchers.

- Part III, Approaches to Artificial Chemistries, is dedicated to a discussion of different Artificial Chemistries introduced in the literature of the last two decades and classifies them into groups: Rewriting Systems (chapter 9) include ACs where the transformation of molecules during "chemical" reactions is dictated by rules that rewrite the contents of these molecules. Automata and Machines (chapter 10) cover ACs that operate in a similar way as a computer with a processor, a memory, and some registers. Bio-inspired ACs (chapter 11) take inspiration from various biological processes, such as the way enzymes operate on DNA molecules.

- Part IV, on order construction, comprises chapters 12 to 15, which expose fundamental questions of how order can emerge. Central to this part of the book is the concept of *chemical organization*, a closed and self-maintaining set of chemicals. Chemical organization theory formalizes this concept and is the topic of chapters 12 and 13. Chapter 12 (co-authored with Pietro Speroni di Fenizio) elaborates on how the structure of organizations can be formalized. Chapter 13 undertakes first steps into the dynamics of organizations. Chapter 14 discusses self-organization and emergent phenomena, which play a key role in ACs. Chapter 15 wraps up with an exposure of constructive dynamical systems, systems that are able to produce novelty and thereby to create new organizations.

- Part V, on applications, returns to reality and is dedicated to a discussion of a broad range of AC applications: Chapter 16 introduces some selected applications; chapter 17 looks at how artificial chemistries can be used for computing; chapter 18 discusses how ACs can be used for the modeling of systems in biology; chapter 19 reports on synthetic life and wet chemical computers; and chapter 20 discusses implementations in different media ranging from mechanical to social systems.

- Part VI concludes the book with a summary chapter (chapter 21), followed by some further suggested reading, and an appendix that introduces a Python toolkit for implementing ACs.

Given the interdisciplinary nature of this book, readers with different backgrounds might find interest in different parts of the book. For instance, readers familiar with chemistry and biology might prefer to skip chapter 5. Readers interested in more fundamental questions will certainly enjoy part IV of the book, while a more practically oriented readership might wish to focus on the application areas in part V. At any time readers should feel free to set up their own AC simulation system. In order to help with that process, we have provided an easy-to-follow guide in the appendix, with code written in Python.

Personal Remarks by Wolfgang Banzhaf

I first became involved with artificial chemistries when I discovered an interest in self-replicating structures. Von Neumann's proof that self-replicating machines are possible, together with the "Game of Life" rules for cellular automata, exerted an enormous influence on me. While those ideas had been established in the 1950s and 1970s, respectively, Artificial Life as a concept took another decade to arrive. At that time I was a freshly minted theoretical physicist, with lots of ideas and no way to realize them.

What is the essence of life? How is it that the molecules in our bodies are replaced every so often, yet we consider ourselves the same, regardless of which molecules are doing their heavy lifting within our bodies? Surely, it could not be the material constitution of our bodies that would make us alive? It was in this context that I began reading books and articles about organization, self-organization, and the processes of life. I realized that below the wonderful phenomena of real life there was a huge machinery of chemical and physical processes, all the time providing the raw material and energy gradients, the driving forces, and the constraints necessary for life to exist.

Following the lead of others, and together with my students, I started to develop notions of artificial chemistry, which at the same time were reinforced and further developed by a whole group of researchers fascinated by Life.

When MIT Press approached me and asked whether I would like to write a book about the topic of artificial chemistries, I was hesitant at first. In a field so new and inhomogeneous as this, how could one hope to find a thread and to provide a concise discussion of the major ideas without sacrificing many of the potentially valuable avenues that are being explored as one writes them down? Fascination with the topic, however, prevailed over caution, and it is in this exploring spirit that we hope you receive this book. It is not intended to discourage any effort in this field — just the opposite, we hope to convey some of this fascination, and to draw as many of you as possible into this field, if you have not yet been dedicated to the cause: to understand life, in all its forms, and therefore ultimately, to understand where we come from.

Personal Remarks by Lidia Yamamoto

When I was invited to help write this book, I felt of course very honored and excited, but at the same time I also felt very scared by the scale of the challenge ahead of me. My enthusiasm finally won over my fear, and I decided to leap at this chance.

Writing this book was a big learning experience for me. Our collaboration was a great pleasure and a source of inspiration for me. This book represents about four years of our joint effort. As I had feared, I was quickly overwhelmed by the huge amount of information that I had to synthesize into a coherent and mature overview of this still burgeoning field. My goal was to make this book as accessible as possible, remembering the person I was when I first came into artificial chemistries, all the questions I had and for which I couldn't find easy answers, and all the questions that I even didn't know how to formulate. One of my past advisors once said that I am a walking question mark. Indeed.

In this book we tried to bring answers to those questions, even if most of the answers consist in saying that no firm answer is currently known, at least we tried to point the reader to the places where people are looking for the answers. We hope that this will stimulate many more people into finding the answers, and then contribute to the progress in the field.

It quickly became apparent that we had to achieve much more than just a big survey of the field of artificial chemistries. This was an opportunity to finally put together the fundamental concepts needed to understand and work with artificial chemistries, the main motivations and rationale behind the work done so far, and where we should be heading from now on. This was a daunting possibility, and perhaps the main value of the book. I don't know if we have attained this ambitious goal; the readers will tell us. However, I think the book at least made a big leap in this direction, although we might not be fully there yet.

Acknowledgments

We are indebted to many people: Bob Prior, executive editor at MIT Press, first planted the idea to consider the plan for a book on artificial chemistries seriously. Susan Buckley, associate acquisition editor at MIT Press, regularly (and patiently) checked in to ask about the status of the book, Christopher Eyer took over from her and helped in the later stages of getting the book draft ready. Katherine Almeida accompanied the production of the book with unfailing diligence.

Over the years, we worked with many individuals (students, colleagues and collaborators) on various aspects of artificial chemistries: Jens Busch, Pierre Collet, Peter Dittrich, Burkard Eller, Mark Hatcher, Christian Lasarczyk, Bas Straatman, Thomas Meyer, Daniele Miorandi, Hilmar Rauhe, Pietro Speroni di Fenizio, Christian Tschudin, Roger White, and Jens Ziegler. We also had many discussions with colleagues in the community who in various ways helped to shape our thinking about Artificial Chemistries or read parts of the book and gave feedback: Mark Bedau, Ernesto Costa, Rene Doursat, Harold Fellermann, Christoph Flamm, Walter Fontana, James Foster, Martin Hanczyc, Stuart Kauffman, Jan Kim, Krzysztof Krawiec, Julian Miller, Una-May O'Reilly, Steen Rasmussen, Hiroki Sayama, Peter Schuster, Lee Spector, Peter Stadler, Brian Staveley, Susan Stepney, Yasuhiro Suzuki, and Hiroshi Tanaka. We gratefully acknowledge the rich and inspiring interactions we had with them all.

Many of our colleagues were so friendly to allow us to reproduce figures from their own work. We have acknowledged these permissions in the appropriate places, but want to express our gratitude here as well to Erwan Bigan, Aaron Clauset, Mark Hatcher, Martin Hanczyc, Simon Hickinbotham, Sandeep Krishna, Thomas Miconi, Robert Pennock, Hiroki Sayama, Peter Stadler, and Sébastien Verel.

This book was typeset in LaTeX's Memoir class and counted upon various tools from the free software and open source communities, including Python, Gnuplot, Inkscape, OpenOffice and many more. We would like to express our sincere gratitude to these developer communities.

W.B. would like to cordially thank his wife Pia for her interest and continuous encouragement in the process of writing this book. L.Y. wishes to thank her husband Pascal for his unconditional support, encouragement, and sacrifices that were essential to her successful contribution to this extensive book writing effort. Both of us are indebted to Pierre Collet from the University of Strasbourg, who kindly provided us with the academic infrastructure needed to pursue part of this project. Finally, we both owe many apologies to our families and friends who were too often neglected on the way toward completion of this project.

Wolfgang Banzhaf, St. John's, Newfoundland, Canada
Lidia Yamamoto, Brussels, Belgium
February 25, 2015

Part I

Foundations

Information should be generated from information just as organisms from organisms.

STANISLAW LEM, SUMMA TECHNOLOGIAE, 1964

CHAPTER **1**

INTRODUCTION

The study of Artificial Chemistries (ACs) originated from the efforts to understand the origins of life. Similarly to how Artificial Life studies models of "life as it could be," artificial chemistries propose and study various models of chemical reactions as they could be imagined. In real chemistry, reactions happen when a set of molecules (the reactants) come into contact in such a way that some of the bonds between their atoms are broken and other bonds are formed, rearranging the atoms into the final products of the reaction. There are a vast number of different ways to rearrange atoms into valid molecular structures. In artificial chemistries, this combinatorial space of possible structures is extended beyond pure chemistry: one is allowed to "play god" and invent his or her own rules to determine how objects interact to produce other objects. In an abstracted sense, we mean by chemical reaction in an artificial chemistry the transformation and sometimes the new generation of objects in combinatorial spaces. In this general notion, artificial chemistries can be considered tools that study the dynamics of systems of interacting objects in these combinatorial spaces. A surprisingly broad set of natural and man-made systems and phenomena fall under this class, from chemistries to even economies. As such, artificial chemistries have become a field of interest to a number of different research and engineering disciplines.

The strength of artificial chemistries draws from the fact that they offer tools for an examination of both quantitative and qualitative phenomena. The concentration or frequency of occurrence of some particular agents (chemical types) offers a window into the quantitative world, while the particular types of agents being under consideration allow a qualitative picture to develop, e.g. a new type of agents appears for the first time, or some types start cooperating (e.g., by supporting reactions that produce each other). In the same way that real chemistry is concerned with both qualitative and quantitative aspects, by measuring concentrations of reacting substances and by analyzing newly developed substances, artificial chemistry is concerned with observing the concentrations of collections of objects and with the analysis of their internal properties.

Stanislaw Lem, the famous Polish philosopher, literary critic and author, has wonderfully summarized the key idea of this field by suggesting the symbols of mathematics as the basic objects of consideration. In one of his most influential books, *Summa Technologiae*, meant as a summary (both review and outlook) on the technology humans have developed, much as Thomas Aquinas summarized the status and perspectives of theology in the Middle Ages in his famous *Summa Theologiae*, Lem stated:

> "Information should be generated from information just as organisms from organisms. The pieces should fertilize each other, they should be crossed over, they should be mutated, that is, varied to a small degree, but also to a larger degree by radical changes not known in genetics. This could perhaps happen in some vessels where reactions between "information-carrying molecules" take place, molecules that carry information in a similar way as chromosomes carry the features of organisms."

So the problem Lem envisioned at the core of this new discipline was, how to set up a system akin to a chemistry, with the interacting objects being symbols like numbers. And Lem already correctly identified the need for a combinatorial approach, by envisioning molecules — today one would say macromolecules — akin to DNA chromosomes. Finally, another aspect of his view on these reactions of information-carrying molecules is an evolutionary approach, something that became well known in the meantime under the term "evolutionary algorithm" or "genetic algorithm." We shall later come back to this notion, but will now concentrate on the proposition of mathematical symbols as reactants, as it is one of the representations for artificial chemistries. Instead of mathematical operations being executed on symbols without connection to a material world, however, in an artificial chemistry, operations include a notion of material. For example, two symbols from the set of the natural numbers, say "5" and "6" as being part of an addition, $5 + 6$, would actually be two representatives of type "5" and "6" respectively, that would not form a symbolic equation

$$5 + 6 = 11, \tag{1.1}$$

but instead would perform a quasi-chemical reaction that leads to a new product, "11," in the following form

$$5 + 6 \rightarrow 11. \tag{1.2}$$

As such, a new symbol, "11" would be produced by this reaction, and the arguments of the addition operation would disappear due to them now being contained in the new product.[1] In a sense, these particular representatives of the symbol types "5" and "6" would be consumed by the reaction. This is indeed a very materialistic understanding of the symbols, with an assumption of conservation of matter behind the consumption of the original symbols.

In real chemistry, of course, there is a backward reaction as well, which would translate into a possible decay of "11" into the two constituents "5" and "6":

$$11 \rightarrow 5 + 6. \tag{1.3}$$

Both forward and backward reactions in chemistry are governed by the kinetics of a particular reaction process. The kinetics determines the speed of the reactions based on the concentrations

---

[1] Note that this understanding of an operation is much closer to computer operations than the structural understanding of mathematics.

of reactants and on rate coefficients stemming from energy considerations. For our toy arithmetic chemistry above, it is an interesting question whether it would be possible to formulate a mathematics based on such processes. It would probably entail to assume an infinite resource of symbols of any particular type, and perhaps the assumption that the kinetic constants for all possible reactions are exactly equal. If we would relax the latter assumption, would we end up with a different mathematics?

To recapitulate: The gist of artificial chemistry is that a set of (possibly abstract) objects is allowed to interact and produce other (sometimes new) objects. The observation of those systems would then give a hint at their potential for producing regularity, fault-tolerance, and innovation. What we will need to define for an artificial chemistry is therefore objects, rules for their interactions, and an "experimental" system where we can let them react.

In real chemistry, both the objects and their rules of interaction are given by nature, the former being atoms of chemical elements or collections of atoms of elements in the form of assemblages called molecules, the latter in being nature's laws of chemical reactions. Variability can be introduced in choosing the conditions for the reaction vessel, its geometry and environmental parameters (pressure, temperature, flow, etc.). The reaction laws cannot be tampered with, however, and in most cases the objects are known beforehand.

Contrast that to artificial chemistries: Everything is free to be varied, certainly the reaction vessel and its parameters, but then also the laws of interaction between objects and even the objects themselves. For instance, the objects could be shares of companies, held by a trader on the New York Stock Exchange. Reactions could be the buying and selling of shares through the trader, based on an exchange of money. And the reaction vessel could be the market, with its particular conditions, e.g., price level, amount of shares on offer, economic climate, and so on.

One key feature of both artificial and real chemistry is that not all reactions between objects are equally likely to happen or even possible. Some are simply not allowed on energetic reasons, for instance (if there is a notion of energy), some others are not possible based on other constraints of the system under consideration. If there is no acceptable prize, shares will simply not be sold or bought by the trader in the example mentioned before, despite a hypothetical offer. So no reaction takes place, what is sometimes called "elastic collision," from the notion that objects have come close, even to the point where they encountered each other, but the constraints didn't allow a reaction to happen. The objects are non-reactive, and will not form compounds, or produce anything new.

In general, artificial chemistries follow their natural counterpart in studying interactions between two objects. If there are more objects participating in an overall reaction, a separation into a number of two-object reactions is often used to build the overall reaction. This follows from a principle of physics: A collision between two particles is much more likely than a simultaneous collision of several particles together. This simplest form of interaction is also able to capture reactions with more partners, since two-body interactions can be concatenated with other two-body interactions to provide an overall result.

Real chemistry is, of course, about more than just single two-body interactions. A huge number of reactions are possible in principle, and only energetically disfavored. This produces a patterning of objects, with some being more abundant than others. Thus the study of abundances and the energy kinetics of reactions is a key element of chemistry, as it determines these patterns. In artificial chemistries, energy considerations have only started to enter studies recently. In principle, abstract interactions can be formulated without recurrence to energy, a freedom that many researchers in artificial chemistries have made use of. The observation of abundance patterns, however, is a central part of all artificial chemistries, since it provides the information necessary to judge which interactions produce which objects in the system under study.

Although very often energy does not play a constraining role in artificial chemistries, constraints in general are considered and easily implementable. They are called filter conditions, and are included as one step in the process of calculating a reaction product in the algorithm that implements an artificial chemistry. Thus, an artificial chemistry might allow a reaction product to be generated, but not to be released into the pool of objects. It is filtered out before it enters the pool. An example might help to explain this point: Suppose again, we have an artificial chemistry of natural numbers. This time, however, reactions are chosen to be the division of numbers. So two elements of the pool of objects, say an "8" and a "4," might collide and react. In this case, we have

$$8 : 4 \rightarrow 2 \tag{1.4}$$

The first complication here arises from the fact that, mathematically speaking, the arguments of the division operator are not commutable, i.e., it makes a difference whether we divide "8" by "4" or "4" by "8." This asymmetry in the application of the division produces a conceptual difficulty for artificial chemistries, since chemical reactions are symmetrical. Two reactants "A" and "B" will react to the same product, regardless of their order of writing in the equation. We can evade this difficulty for now by simply saying that always the larger of the two numbers will be divided by the smaller. So, the operation ":" that we wrote in equation 1.4 really is not just a division, but a sorting of the numbers, followed by a division. Suppose now that we are interested in a particular feature of numbers, namely their divisibility, and that we want to restrict reactions to those that produce only natural numbers. This is the aforementioned filter condition, in that we determine that if a number "a" does not divide a number "b," and produces a natural number, then this reaction is not allowed. We say, the reaction is an elastic collision. Technically, we can still let the division happen, but then need to prohibit the result from getting back into the pool of objects. For instance,

$$8 : 5 \rightarrow \varnothing \tag{1.5}$$

despite the fact that, in principle, "8" divided by "5" could be calculated as "1.6." This filter condition is not the same as trying to make sure that the operation of division stays within the allowed set of natural numbers. Such a strategy would, for instance, be represented by the division with subsequent choice of the next integer. In our example above, the result would be "2" since "1.6" is closer to "2" than to "1."

As we can see from this example, performing a reaction can be more than a simple mathematical operation, it can be an entire program or algorithm being executed prior to the production of the result, and another program being executed after the production of the result, to check whether the result is indeed allowed by the constraints of the system. In principle, even a certain degree of randomness can be injected into this process, for example, by allowing a result to enter the pool only based on the cast of a dice.

In this sense, the expressive power of artificial chemistries is larger than that of their natural counterparts, since they allow arbitrary constraints to be implemented, not to speak of the objects and interactions which also can be chosen with relative freedom.

Suppose now that we have determined the types of objects and the rules of interaction of such a system of objects, together with the general method for producing reactions. We already said that by observing the development of abundances of objects over time, we can start to discern one dynamics from another and find ways to characterize the objects and their interactions. If we had a full classification of objects and looked at each of them separately, what could we learn from these? We can ask questions like: Which of the objects is stable, i.e., will not be replaced by others through reactions? Which subset of objects is stable in its appearance, and is produced in large numbers? Which interactions produce which patterns of abundances? What is the influence of initial conditions on the outcome? Generally speaking, we can ask questions relating to the nondynamic ("static") character of the artificial chemistry under study. In simple cases, all reactions can be exhaustively listed and put into a reaction matrix. But we can also ask questions relating to the dynamic character of the AC, e.g., how abundances develop under which conditions, etc. Many of the simpler Artificial Chemistries being studied will have a notion of equilibrium, just as it happens in real chemistry. In other words, after some time the dynamics will reach a steady state, upon which not very much will happen, or no qualitative changes will occur any more in the behavior of the system.

While we can learn a lot from these kinds of studies, artificial chemistries really come into their own when we start discussing the production of new molecules, new entities that never before had appeared in the reaction vessel ("pool"), whose production happens for the first time in a single event. For that to happen, it is important that the set of objects is not fully known to us from the outset, which means that reactions can only be defined implicitly, or that at least the number of knowable objects is much greater than the number of realized objects. We can see from here how important it is to consider combinatorial spaces, spaces that are very large, yet not necessarily populated to the full extent. So only a subgroup of artificial chemistries will be able to exhibit such events, namely those based on combinatorial spaces, because only there is the number of possible different objects much greater than the number of existing objects.

Once a new object appears, it has the chance to participate in subsequent interactions. If these interactions, now having a new participant, produce new results again, we might even get a whole slew of new objects in a very short time, some of which might be produced repeatedly, some others of which only intermittently, soon to disappear again. Yet through this process, the system as a whole might move to a different equilibrium state. So while the emergence of a new object has suddenly thrown the system out of balance, a new, different balance might be established through the integration and with the help of the new object.

The innovation potential, therefore, that can be part of an artificial chemistry is extremely interesting in that the establishment of an artificial chemistry is one of the very few methods with which we can conceptualize and study the emergence of novelty and of qualitative changes to a system in a simulation environment. For a long time, novelty has eluded the grasp of exact science and could be treated only qualitatively. Here, finally, is a method that allows to study the phenomenon in a model, and to determine the conditions under which innovation thrives.

It will be no surprise that, together with novelty, the question of complexity growth will play an important role in artificial chemistries. Is it possible to set up systems in a way that they are open to grow in complexity? Where does this complexity growth ultimately come from, and what are the conditions under which it thrives? How does complexity growth relate to novelty in these systems? These fascinating questions are suddenly open for a closer examination using simulation tools in an artificial chemistry. Admittedly, simulated systems might look artificial, yet there is a chance to find valuable answers that can be transferred to more realistic, perhaps even to real systems.

The condition that needs to be fulfilled for innovation to be a constitutive part of a system is that the number of possible different types of objects, N, in a system is much larger than the number of existing objects (of any type), M, or

$$N \gg M.$$

(1.6)

This condition will ensure that there is always enough possibility for innovation. How can such a condition become a constituting part of a system? There might be systems where this is just a natural condition that can be fulfilled under most any circumstances. Or it might be that this is a particular corner of the system space in which these conditions prevail. We are of the opinion that nature can be considered to realize systems that most naturally allow innovation to occur. The very fact that macromolecular chemistry, and carbon-based macromolecular chemistry in particular, allows the assembly of molecules like beads on a strand virtually guarantees a combinatorial explosion of system spaces and thus conditions proliferating innovation.

Even simple systems such as those that can be constructed from using sequences or strings of binary numbers "0" and "1," the most basic of all number representations, already possess the power to produce novelty in the above sense. Let's just explore for a moment how easy that is. Suppose we consider the memory system of a modern computer with a storage capacity of a terabyte as our reaction system. Suppose now that we fill this storage with strings of binary numbers of length 100, i.e., each of our strings is 100 bits long. You can't put a lot of information into 100 bits, as we all know. Given that each packet of data traversing the Internet holds at least 160 bits, such a string could not even hold a single e-mail packet. Yet there are $2^{100} \simeq 10^{30}$ different strings of length 100 bits. Filling the entire terabyte of storage with strings will allow us to store 8×10^{10} strings of this length. Since there are, however, that number of different strings available, even if we would allow every string to be only held once, only one out of every 10^{19} possible strings would be in the storage. Thus, the combinatorial space of binary strings is essentially empty. Had we required that strings need to be there in multiple copies, an even smaller proportion of all possible strings could have been realized, providing tremendous potential for novelty. Would there be a mechanism in place which allowed strings to be replaced by other strings, with an occasional variation incorporated in the process, we would naturally expect the emergence of strings never seen before.

One of the key driving questions for the field of Artificial Life has been how life emerged in the first place. Origin-of-life research is closely related to the earlier question: How did qualitative change happen and a nonliving system suddenly transform itself into a living one? We have argued previously, that while this is an important question, perhaps one stands to learn a lot from studying artificial chemistries first, before asking about the origin of life. Indeed, there exist many "origin" questions for which Artificial Chemistries might have lessons to teach. Origin of language, origin of consciousness, origin of human society, origin of, you name it, are questions that target a very special, innovative event in their respective systems. These are fascinating questions as they tie the dynamics of their underlying system to qualitative changes these systems undergo at certain key junctions in their development.

Beyond the simulation realm, with the progress in natural and material sciences, it now becomes feasible to set up chemistries that can be called artificial, i.e., man-made chemistries, that heretofore were not possible or realized in nature. Simple examples close to the border with biochemistry are systems where additional amino acids are used to produce "artificial proteins," or where more than four nucleotides are used to create an "artificial genetic code." Many similar ideas are presently being examined, which can be subsumed under the general term "wet artificial chemistries." This term stems from the idea that most of the artificial chemistries are used to model lifelike systems, which in reality are systems in a wet environment. Note that innovation in these wet artificial chemistries is based on the exploration of novel combinations of atoms leading to new chemical compositions that were not known before, but which still obey the same universal laws as natural chemistry. In contrast to virtual or "in silico" artificial chemistries, the laws of chemistry here cannot be tampered with. Nonetheless, the potential of such new chemistries is enormous: from new materials to synthetic biology, enabling perhaps totally new forms of "wet artificial life" based on a different genetic code or fueled by different metabolic processes. A new technology based on such engineered living entities opens up a whole new range of applications, and of course also raises numerous ethical issues.

Applications of artificial chemistries range from astrophysics to nanoscience, with intermediate stops at social and economic systems. At a certain level of abstraction human societies can be considered artificial chemistries. Models have been built that encapsulate these ideas and predict collective behavior of groups of individuals. On a more elementary level we can look at social behavior and study social interactions in AC-type models. Observation and imitation would be important aspects of such studies, since they allow social norms to develop and to be reinforced.

It remains to be seen how far the paradigm is able to carry. For the time being, many systems can be looked at from this point of view, and potentially new insights might be gained by a thorough modeling and treatment as an artificial chemistry. In the next chapter, we will look at a very simple embodiment of an AC, but first we need to understand some principles of modeling and simulation in general.

Computer models and simulation attempt to achieve the ultimate goals of science and thus represent the ultimate extension of science.

CHARLES BLILIE, THE PROMISE AND LIMITS OF COMPUTER MODELING, 2007

CHAPTER **2**

BASIC CONCEPTS OF ARTIFICIAL CHEMISTRIES

Before we can discuss the general structure of an artificial chemistry, we have to set the stage for this discussion by addressing two topics that together constitute the pillars of any artificial chemistry: the first one is a topic of very broad relevance to all of science and engineering, a topic fraught with misunderstandings: modeling and simulation. The second pillar is the topic of real chemistry, which is used as a source of inspiration for artificial chemistries, or as the target system to be modeled. In the first section of this chapter, we will therefore address the principles behind modeling and simulation, and try to argue for their value in the current world, dominated by computers. In section 2.2 we will introduce some basic concepts of chemistry that are often used in ACs. Note that in contrast to chemistry textbooks, we will present these concepts from a computational perspective. Section 2.3 will then introduce us to the general structure of an artificial chemistry. From the vantage point of modeling and simulation, we will then see that the general structure of an artificial chemistry is nothing but a specialization of a simulation tool, geared toward the interactions possible in (artificial) chemistries. Section 2.4 supplies a few important distinctions between AC approaches. Section 2.5 then introduces a few simple examples of ACs, a discussion that will be beefed up by later chapters. A number of frequently used AC techniques are then addressed in section 2.6, among them the notion of space brought about by the possibility to define a topology among reacting objects. Because we are still in the introductory part of the book, we try to stay as high-level as possible in this chapter.

2.1 Modeling and Simulation

As we try to understand the world around us, humankind has adopted a number of successful strategies, some of which have now been formalized. Epistemology is the branch of philosophy that concerns itself with how humans acquire and process knowledge. Empirical strategies of acquiring knowledge center on experience and how it can be used to learn about the world. The emergence of science is closely connected to the recognition that if the acquisition of knowledge

Table 2.1 Empirical Hierarchy: From Data to Understanding

LEVEL	FROM	TO
Elements	Data	Relations between facts
Tools	m Experimental trials	n Hypotheses
Methodology	Experiment	Theory
Result	Observation	Explanation / Understanding

is subjected to certain constraints often termed the scientific method, this will bear fruit in the form of an increased understanding of the world around us, but also in the form of allowing us to predict phenomena and to apply them to satisfy human needs.

At the core of the scientific method is the interaction between observation and explanation/understanding. We say that we understand a phenomenon observed either through our senses or by using scientific instruments if we can reproduce the phenomenon using our understanding, and if we further can predict the outcome of hitherto unobserved phenomena. In order to reach this goal, science works with two complementary tools: well-defined experiments and theories, the former to observe phenomena and the latter to explain them in a logically consistent way.

On the experimental side, many independent trials need to be conducted, possibly under different initial or boundary conditions. On the theoretical side, many hypotheses need to be formulated as well, in order to test different aspects of the phenomenon under study and to gradually refine our understanding and achieve a level of explanatory power that merits the term "theory" for a consistent set of hypotheses. In a sense, this is an evolutionary process involving some trial and error, that will ultimately lead to a better reflection of reality in our theories.

Experiments under constrained conditions produce data that need to be analyzed and checked against the predictions of hypotheses. The hypotheses themselves are trying to uncover relations among these data. This is why correlation analysis plays such a fundamental role for scientists in trying to discover simple hypothesis about data: The simplest relations, i.e., linear relations, can be easily detected by this tool.

Where in all this do we locate models and simulations? The short answer is: at the core, right in between theory and experiment. A simulation produces data (and thus stands in for an experiment) based on a model (standing in for a theory) formulated as a computer program.

The longer answer is this: With the advent of computers, the traditional scientific method has been augmented and refined to produce an even more efficient interaction between theory and experiment, provided the theory can be translated into a computer model, which is an algorithm incorporating the relationship and transition rules between factual objects. This computer algorithm can then be executed as a program to run a simulation which could be termed a model in motion [122]. The typical work cycle of a scientist would therefore read: Theory — Computer Model (algorithm/program) — Simulation — Experiment. Simulation sometimes replaces, sometimes only precedes an experiment. It allows science to progress more efficiently in that not all experimental trials actually need to be carried out, potentially saving time, materials, lives of laboratory animals, and so on.

Simulations are built upon a model of the system under study. A model is the next step from a theory or hypothesis toward an experiment, in that it is a theoretical construct attempting to replicate a part of reality. Bunge defines a model to be a general theory plus a special description of an object or a system [144]. The term "model" is often used synonymous with that of "theory," but one should avoid this use. As we can see from Bunge's above definition, a theory is more ambitious than a model in explaining phenomena, since it strives to be more general.

Blilie states [122] that computer models and simulation "attempt to achieve the ultimate goals of science and thus represent the ultimate extension of science."

In the preceding discussion, we have completely glossed over a very important point: in what sense models reflect, or represent reality. This is a deep question that we cannot address in our short discussion here. Interested readers are referred to the literature on epistemology [671]. However, one point is too important not to be mentioned: Models always are caricatures of the real system they intend to represent. They are abstractions that reduce the complexity of the real system by projecting systems into subspaces, e.g., by neglecting certain interactions between the elements of the original system or by insulating it from its environment. Models have a purpose, which dictates the kind of simplification of the original system or the perspective from which they are considered.

Models come in different flavors, physical, imaginary, and conceptual. The following discussion will only be concerned with conceptual models. Computer models are conceptual models, based on the formalism of algorithms. There are other well-known conceptual models, e.g., based on logic or on mathematics, but algorithms are a broader category, and we believe they include the latter two. Algorithmic conceptual models of systems have the advantage that they can be simulated, i.e., executed on computers. This execution produces behavior that can be observed and compared to the behavior of the original system under study. Depending on the input of the computer model, various types of behavior might be observable, and the range of these behaviors might be similar to or completely different from the range of behaviors the original system exhibits. Based on such comparisons, computer models can be refined. Different inputs produce predictions of behavior that can be confirmed or falsified by observing the real system.

It is an interesting question whether mathematical and algorithmic models are equivalent to each other. Clearly, mathematical descriptions like differential equations have their counterpart in algorithms. However, do algorithmic implementations always have mathematical descriptions? The issue we raised in the previous chapter, whether one could formulate a mathematical framework based on chemistry, in which instances of symbols "react" with each other to produce instances of new symbols is closely connected to that. If, as we stated, algorithmic formalism contains mathematical formalisms but is not equivalent, the question is nontrivial. Perhaps one type of formalism can be considered a model of the other, with all the connotations we previously mentioned about models (abstraction, simplification, projection)?

Once a mathematical or algorithmic model of the system under study has been constructed, a computer simulation of the model can then be implemented. Typically, the simulation algorithm will try to mimic the behavior of the real system in time, by a set of transition rules that define how the system moves from an initial state to future states. What happens if we let the simulation algorithm execute? It will show a particular behavior, based on its input and initial conditions. The behavior can be followed by observing its state changes and collecting this information to compare it with different starting conditions and inputs, and the real system it models.

However, there is more to it than simply behavior of known components. In fact, the observation of system behavior needs to be preceded by the determination of which state variables (or collections of state variables) should be observed at all. In principle, a simulation of a system can produce an infinite amount of data. We therefore have to restrict ourselves to the ones we believe will be relevant. This might not be determined in one go; rather, an iterative process of simulation and observation might be necessary to focus on relevant quantities. This has to do with the constructive nature of a simulation. Rasmussen and Barrett [690] propose to formulate an entire theory of simulation, based on the power of computer systems to execute algorithms and produce phenomena.

Before we can understand this latter aspect, a few words are in order about the goal of simulations. Again, following Blilie [122], we can discern three goals of computer models and their dynamic execution: (1) a unified representation of physical phenomena, (2) a unified representation of human society and history, and (3) construction and presentation of different, seemingly realistic alternate worlds. Blilie further divides the latter into "simulated realities" useful for, e.g., training purposes and "fictional realities" as they appear in gaming and entertainment applications. The playful aspect of the latter is something we need to come back to.

All three goals have in common that the elements of a computer model, when set in motion and interacting with each other in a simulation, will show a coherent dynamic behavior, that can only be captured when we observe the system at some higher level, e.g., above the level of simple elements. We shall see an example of this characteristic feature in the next chapter. For the moment, we refer the reader to the famous "Game of Life" simulation [111] as an illustration (see chapter 10). One of the features of this game is to exhibit coherent patterns on a spatially extended grid of cellular automata, such as gliders and blinkers. One might then look at those "blinkers" and "gliders" and other coherent patterns, count their numbers, measure the speed with which they traverse the universe, or observe some other quantities related to them. This way, computer simulations can and usually do expose higher level regularities in the underlying simulated system, a source of complex (emergent) behavior from possibly simple objects and interaction rules.

In general, simulations will — by virtue of being virtual — allow to explore different scenarios and possibilities for a system, depending on the input and initial conditions of the simulation. We are interested here in event-driven simulations, that is simulations whose natural timing is determined by discrete elementary moments, events like the collision of two atoms. Simulations of this type are called discrete time simulations, in contrast to continuous-time simulations in which time is considered a continuous quantity.

According to [690] in order for a simulation to be possible, four systems must be assumed to exist: (1) The real system Σ_R to be simulated, (2) a simulation system Σ_S with an update rule U, (3) subsystems Σ_{S_i} with $S_i \in M$, M being the set of models including their rules of interaction, and finally (4) a formal computing machine Σ_C on which the simulation is run. These requirements implicitly assume that we are interested in studying the emergent phenomena that come about by simulating the interaction of the subsystems S_i, which is a valid assumption in many contexts.

In practical terms, we are here mainly concerned with (2) and (3), since those are the core of the modeling and simulation process. The setup of a simulation system proceeds through the following steps [122]: system identification, determination of boundary conditions, determination of system components and their relation, determination of component properties and quantifiable state variables, setting up of transition rules between states (of subsystems) and their relation to each other, determination of the simulation management, implementation, verification, and validation. All of this has to happen on the background of a clearly defined modeling purpose.

It is the modeling purpose that dictates the perspective of the modeler, and ultimately, the expectations that we have for the results of the simulation. Do we expect to find data that mimic the original system Σ_R as close as possible? Do we want to show that the behavior of both systems, Σ_R and Σ_S is qualitatively the same? Do we only want to make statistical statements of the type, that the class of behavior in both systems is comparable? Finally, do we hope to find unexpected phenomena in Σ_S that we can use to predict behavior of Σ_R, under certain circumstances?

Turning our attention now to artificial chemistries, a preliminary comment is in order. A frequently expressed criticism about AC approaches is raised by scientists in the following form: If ACs wish to study real chemistry, they need to become much more sophisticated and realistic than is the case at present. They are currently weak representations of real systems. If, however, ACs aim at exploring the realm of the possibility of chemistries, without a necessary connection to reality, what is the point? They look like mere entertainment and idle play. It must be answered that the goals of ACs are various, and goals cannot be fixed in general. Each AC, by virtue of being a model, looks at real chemistry from a different perspective. As this book will discuss in later chapters, the spectrum is indeed very wide. There are very realistic artificial chemistries, wet ACs, that involve the interaction of matter. Next are ACs that aim at simulating real chemistry. At the other end of the spectrum there are playful variants not aimed at current chemistries. In a sense, these make use of the formal nature of the simulation process that allows them to probe virtual spaces not bound to reality. Some are aimed at visualization and demonstration of AC concepts. The same concern has been raised against mathematics exploring universes of possible mathematical worlds, without being bound by reality. In the case of mathematics, this playful exploration of the realm of the possible has in many cases yielded unexpected insights and connections to reality. It is our strong belief that artificial chemistries, based on firm principles of modeling and simulation, can achieve the same: Unexpected bridges into reality and real systems that we do not even dream of now. Without an exploration of the possible, we shall never understand the real.

2.2 Chemistry Concepts

Before we can discuss in more detail how to build an artificial chemistry in a computer, we need to understand a few basic concepts of how chemistry can be modeled and simulated. Thus, this section will now introduce some of the notions used to treat chemistry computationally.

In a chemical reaction vessel, a number of objects (molecules) interact via chemical reactions that transform these molecules according to the rules of chemistry. A molecular *species* is the family of substances with the same molecular composition and shape. Substances may be in a solid, liquid, or gas phase. The liquid phase plays a prominent role in artificial chemistries, since it enables the complex interactions at the origins of life. Some artificial chemistries are also based on gases, and a few consider the combination of multiple phases and phase transitions.

Two types of vessels can be distinguished: well-stirred and spatial. They are depicted in figure 2.1. In a *well-stirred* (or well-mixed) vessel, the actual position of each molecule in space has no impact on the overall outcome of the reactions taking place inside. In terms of an artificial chemistry, this means that the probability that a group of molecules react is independent of their position. This is in contrast to *spatial reactors* such as Petri dishes, where chemicals form patterns in space. These patterns depend on the way chemicals move and interact with other nearby chemicals. This second type of reactor will be covered later in this chapter and throughout the book.

For now, we focus on the well-stirred case. In this case, the state of the system is fully described by the concentration of each species present in the reactor. The *concentration* measures

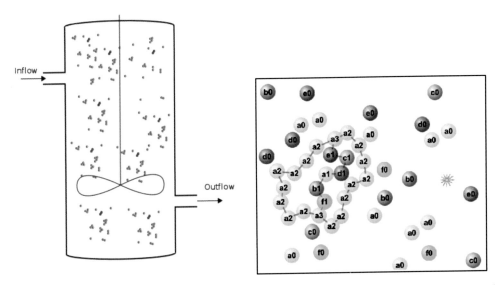

Figure 2.1 Examples of reaction vessels with molecules inside. Left: Well-stirred tank reactor in which all molecules are equally likely to collide with each other and react. Right: Screenshot of the Organic Builder [413], a 2D spatial artificial chemistry akin to a virtual Petri dish, where nearby molecules can collide and react (in this example: yellow 'a'-type molecules bond together to form a cell-like structure). Screenshot from the running application downloaded from organicbuilder.sourceforge.net.

the number of (moles of) molecules per unit of volume in a vessel. For a given molecular species S, the concentration of S is denoted by $[S]$ and is measured as:

$$[S] = \frac{n_S}{N_A V} \qquad (2.1)$$

where:

- n_S is the number of molecules of species S in the vessel;

- N_A is the Avogadro constant: $N_A = 6.02214 \times 10^{23}$ molecules;

- V is the volume of the vessel.

A chemical reaction is a mass-conserving transformation from a multiset of "input" molecules (educts, or reactants) to a multiset of "output" molecules (products). A *multiset* is a collection of objects where elements may occur more than once. More about multisets will appear in Section 2.3.3. In the following chemical reaction, one molecule of nitrogen and three molecules of hydrogen react to form two molecules of ammonia:

$$N_2 + 3H_2 \xrightarrow{k} 2NH_3 \qquad (2.2)$$

The educt multiset is therefore $\mathcal{M}_e = \{N_2, H_2, H_2, H_2\}$, and the product multiset is $\mathcal{M}_p = \{NH_3, NH_3\}$. The coefficient k is the kinetic coefficient or rate coefficient of the reaction, and it measures the speed of the reaction when one mol of each needed molecule is present (in this case, one mol of N_2 and three of H_2).

The inflow and outflow of substances in and out of a vessel can also be represented as chemical reactions, in this case, with addition or removal of mass. For instance, the following reaction represents the injection of molecules of species A at constant rate k:

$$\xrightarrow{k} A \tag{2.3}$$

The reaction below represents the removal of B:

$$B \xrightarrow{k} \tag{2.4}$$

Removal reactions are also often used to model the degradation of substances into some inert chemical (e.g. a waste product) that is not considered for the system under study.

Reactions are often reversible, with different coefficients on the forward and backward reaction. This is represented as:

$$A + B \underset{k_r}{\overset{k_f}{\rightleftharpoons}} C \tag{2.5}$$

In the preceding example, the forward reaction consumes A and B to produce C with a rate coefficient of k_f, and the reverse reaction produces A and B with coefficient k_r.

2.2.1 Chemical Reaction Networks

In a chemical reactor, multiple substances react in different ways, leading to a network of chemical reactions: a node in such a network represents either a reaction or a chemical species, while an edge linking two nodes represent the flow of substances participating in the reactions. The graph representing this network is *bipartite*, since chemicals always connect to reactions and not directly to other chemicals (similarly, reactions connect to chemicals and not directly to other reactions). For example, consider the following reactions:

$$R_1: \quad A \quad \rightarrow B + C \tag{2.6}$$
$$R_2: \quad C \quad \rightarrow D \tag{2.7}$$
$$R_3: \quad B + D \quad \rightarrow E \tag{2.8}$$
$$R_4: \quad C + D \quad \rightarrow A + B \tag{2.9}$$

The corresponding chemical reaction network is shown in figure 2.2, left. Often, chemical reaction networks are depicted with substances as nodes and reactions as edges, as in figure 2.2, right. This is simpler and more intuitive, though in some instances this simplification may be misleading.

Reaction networks have a great importance in chemistry and also in biology. The processes taking place in cells form complex reaction networks that system biologists strive to analyze. Some basic analytical methods will be introduced in this chapter. The three major categories of cellular networks are: metabolic networks, genetic regulatory networks, and cell signaling networks. Artificial chemistries have been extensively used to either model such biological networks or to find alternative representations leading to analogous lifelike phenomena, such as artificial metabolic networks in which the chemical substances are represented by strings of letters from a given alphabet. Several examples of such chemistries will appear throughout this book.

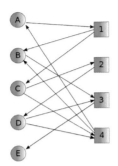 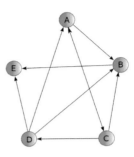

Figure 2.2 Left: Bipartite reaction graph of reactions symbolized by equations 2.6 — 2.9; circles are substances, squares are reactions. Right: Reaction graph with edges representing the reactions, nodes representing the substances participating.

2.2.2 Law of Mass Action

We have seen that reactions can be accompanied by a kinetic coefficient k linked to the speed of the reaction. The most common way to specify the speed of the reaction given its coefficient k is through the *law of mass action*. This law governs the way the kinetic coefficient k and the educt concentrations influence the speed v of the reaction: it states that the speed (rate) of a chemical reaction in a well-stirred reactor is proportional to the product of the concentrations of the reactants present. According to this law, the rate of reaction 2.2 becomes:

$$v = k[N_2][H_2]^3 \tag{2.10}$$

The law of mass action stems from counting the number of possible collisions between the molecules involved in the reaction, and is only valid for large numbers of molecules, and for reactions that proceed in one single step. Many reactions in nature are composed of several intermediate steps. Moreover, in practice, the likelihood of a simultaneous collision of more than two molecules is negligible. Hence, this law has to be taken with care, since it is a simplification of the reality. However, it is sufficient and useful in many cases of real and artificial chemistries.

More generally, for a chemical reaction r with n reactants:

$$r : \alpha_1 S_1 + \alpha_2 S_2 + \ldots \alpha_n S_n \xrightarrow{k} \beta_1 P_1 + \ldots + \beta_m P_m \tag{2.11}$$

The speed of the above reaction according to the law of mass action is:

$$v = k \prod_{i=1}^{n} [S_i]^{\alpha_i} \tag{2.12}$$

where:

- $[S_i]$ is the concentration of the i-th reactant $S_i, i = 1..n$

- α_i is the stoichiometric coefficient of S_i in r, i.e. the number of molecules of S_i required in one chemical reaction r involving S_i.

- α_i, β_j are the stoichiometric coefficients of species S_i and P_j in the reaction, respectively.

Stoichiometric coefficients tell how many molecules of S_i (respectively P_j) are required (resp. produced) in one chemical reaction r. For example, in reaction 2.2, $\alpha_1 = 1$, $\alpha_2 = 3$ and $\beta_1 = 2$. Note that the speed of the reaction only depends on the reactant quantities, never on the quantity of products.

The law of mass action is not the only way to characterize the speed of a chemical reaction. Other laws are very common, such as Michaelis-Menten for enzyme kinetics, or Hill kinetics for genetic regulatory networks (see chapter 18) which are mostly derived from mass action. We will now see how this law can be used to understand the dynamics of networks of chemical reactions.

2.2.3 Concentration Dynamics

In a well-stirred reactor, a large number of molecules is usually present, frantically colliding and reacting with each other. In this case, the concentration dynamics of the molecules in the vessel can be described by a system of ordinary differential equations (ODE), one equation per molecular species. Each equation quantifies the net rate of production of a given chemical, by adding all reactions that produce the chemical, and subtracting those that consume it.

An example for these reactions would be:

$$r_1: \quad 2A + 3B \xrightarrow{k_1} C \tag{2.13}$$

$$r_2: \quad 2C \xrightarrow{k_2} \tag{2.14}$$

Using the law of mass action, the corresponding differential equations are:

$$\frac{d[A]}{dt} = -2k_1[A]^2[B]^3 \tag{2.15}$$

$$\frac{d[B]}{dt} = -3k_1[A]^2[B]^3 \tag{2.16}$$

$$\frac{d[C]}{dt} = +k_1[A]^2[B]^3 - 2k_2[C] \tag{2.17}$$

For instance, take equation (2.17): Reaction r_1 produces one molecule of C at rate $k_1[A]^2[B]^3$, represented by the first term of equation (2.17). Reaction r_2 consumes two molecules of C at rate $k_2[C]$; therefore, the second term multiplies this rate by two. The same procedure is applied to the other substances to obtain the complete ODE in a straightforward manner.

More generally, for a system of n substances $S_1,...,S_n$ and m reactions $r_1,...,r_m$ the equations read:[1]

$$r_1: \quad \alpha_{1,1}S_1 + \alpha_{2,1}S_2 + ...\alpha_{n,1}S_n \xrightarrow{k_1} \beta_{1,1}S_1 + ... + \beta_{n,1}S_n \tag{2.18}$$

$$...$$

$$r_m: \quad \alpha_{1,m}S_1 + \alpha_{2,m}S_2 + ...\alpha_{n,m}S_n \xrightarrow{k_m} \beta_{1,m}S_1 + ... + \beta_{n,m}S_n \tag{2.19}$$

---

[1] Note that the stoichiometric coefficients $\alpha_{i,j}$ and $\beta_{i,j}$ are set to zero if a species S_i does not participate in the reaction r_j ($\alpha_{i,j} = 0$ if S_i is not a reactant and $\beta_{i,j} = 0$ if it is not a product of reaction r_j).

The change in the overall concentration of S_i is obtained by summing over all reactions r_j that possibly contribute, and can be then written as:

$$\frac{d[S_i]}{dt} = \sum_{j=1}^{m} \left[k_j (\beta_{i,j} - \alpha_{i,j}) \prod_{l=1}^{n} [S_l]^{\alpha_{l,j}} \right] \quad , \quad i = 1, \dots, n \tag{2.20}$$

This system of differential equations can be expressed in a more compact form using a matrix notation:

$$\frac{d\vec{s}}{dt} = \mathbf{M} \vec{v} \tag{2.21}$$

where

- $d\vec{s}/dt = [ds_1/dt \quad \dots \quad ds_i/dt \quad \dots \quad ds_n/dt]^\top$ is the vector of differential equations for each of the species S_i in the system and $^\top$ is the sign for transposition.

- $\mathbf{M}$ is the *stoichiometric matrix* for the reaction, which quantifies the net production of chemical S_i in each single reaction r_j. An element of $\mathbf{M}$ is given by:

$$M_{i,j} = \beta_{i,j} - \alpha_{i,j} \tag{2.22}$$

- $\vec{v} = [v_1 \quad \dots \quad v_j \quad \dots \quad v_m]^\top$ is the vector containing the speeds of each reaction r_j:

$$v_j = k_j \prod_{l=1}^{n} [S_l]^{\alpha_{l,j}} \tag{2.23}$$

Such ODE systems are common in the modeling of biological systems and in some forms of biochemical computation (discussed in chapters 17 and 19), in which the concentration of substances represents values to be computed by a chemical reaction network.

The fact that chemical reactions can be translated into a set of deterministic differential equations following precise laws, such as the law of mass action, means that many mathematical methods and tools available for the analysis of dynamical systems can also be applied to chemistry. This is one of the basic premises behind systems chemistry and biology. By extension, an artificial chemistry designed to follow similar precise laws can also benefit from such analytical tools, one of the main advantages of the ODE approach.

ODEs have many limitations however: they are only sufficiently accurate for very large numbers of molecules participating in the reaction. Therefore, they are not appropriate to model reactions involving only a few molecules, such as those taking place inside living cells. Moreover, they can only model systems where the number of possible molecular species and reactions is known in advance, showing limited innovation potential. These are important shortcomings from an Artificial Chemistry perspective. We will come back to this topic in section 2.3.3.

2.2.4 Chemical Equilibrium

Although the changes in concentration hold a lot of information about the behavior of a system, it is often interesting to consider such a system when it reaches an equilibrium state. At equilibrium, the concentrations of all substances in the system do not change:

$$\frac{d\vec{s}}{dt} = \vec{0} \tag{2.24}$$

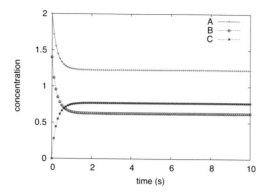

Figure 2.3 Changes in concentrations of A, B, and $C = A.B$ in a reversible dimerization reaction: for $k_f = k_r = 1$, starting from initial concentrations $A(0) = 2.0$, $B(0) = 1.4$, $C(0) = 0$, the system approaches a chemical equilibrium where the concentrations no longer change.

Consider, for instance, a reversible dimerization reaction:

$$A + B \underset{k_r}{\overset{k_f}{\rightleftharpoons}} A.B \tag{2.25}$$

In this reaction, molecules A and B are monomers that combine to form a dimer $A.B$. This process is reversible, so the dimer can dissociate back into its constituent monomers.

At equilibrium:

$$\frac{d[A]}{dt} = \frac{d[B]}{dt} = k_r[A.B] - k_f[A][B] = 0 \tag{2.26}$$

$$\Rightarrow \frac{[A.B]}{[A][B]} = \frac{k_f}{k_r} \equiv K \tag{2.27}$$

K is called the equilibrium constant of the reversible reaction. Figure 2.3 depicts what happens with this system as it converges to the equilibrium concentrations.

Throughout this book we will see many examples of chemistries that either aim to settle to an equilibrium point or, on the contrary, strive to stay away from it by a permanent inflow and outflow of substances.

2.2.5 Molecular Structure and Chemical Bonds

So far we have not looked much inside the structure of the molecules considered in our examples. However, in both real and artificial chemistries, we will often be very interested in the internal structures of molecules and how these structures affect their function.

In real chemistry, most of the atomic elements (except noble gases such as helium) react with other atoms to form molecules, and some of these molecules may in turn react to form other molecules. Chemical bonds are formed and broken during reactions. There are basically two types of chemical bonds: *ionic* and *covalent*. Ionic bonds are formed by electron transfer from an electron donor (typically a metal) to an electron acceptor. The donor atom becomes positively charged (cation), the receptor atom becomes negatively charged (anion), and these two oppositely charged ions attract each other to form a ionic bond. An example is the well-known reaction $2\,Na + Cl_2 \rightarrow 2\,Na^+ + 2\,Cl^- \rightarrow 2\,NaCl$ that forms sodium chloride, the common table salt. Covalent bonds are formed by sharing electron pairs between two nonmetal atoms. For instance, water (H_2O), carbon dioxide (CO_2), and a large number of organic compounds are built from covalent bonds.

Since there is no limit to the size of a molecule, there are nearly infinitely many ways to recombine atoms into valid molecular structures, especially in organic chemistry, where macromolecules such as proteins or DNA may contain thousands or even millions of atoms. As discussed in chapter 1, it is precisely this combinatorial nature of chemistry that ACs seek to explore in alternative ways, not necessarily bound by the laws of real chemistry.

In artificial chemistries, bonds are not always explicitly represented. Molecules in ACs are often represented as strings of symbols (such as letters or numbers), where each symbol represents an atom. For instance, in [233] molecules are binary strings, whereas in [638] they are strings made of the 8 letters from a to h. Among the ACs that explicitly represent bonds, we can mention [103, 384]. In [384] molecules are made of digits that represent the amount of bonds that they form, for instance, "1-2-3=2" is a valid molecule in this chemistry. In [103] molecules are graphs that are closer to the molecular structures found in real chemistry, and their reactions are a simplified model of real chemical reactions. These and other chemistries will be discussed in more detail throughout the book.

2.2.6 Energy and Thermodynamics

Energy plays a key role in the execution of chemical reactions. Within a reactor vessel, molecules collide and react, releasing or absorbing energy in the process.

The first law of thermodynamics is the law of conservation of energy: it states that the internal energy of an isolated system remains constant. In reality, however, most systems are not isolated: They lose energy to the environment, for example in the form of dissipated heat, and must then rely on an external source of energy to keep operating.

The second law of thermodynamics is the law of increasing entropy : The entropy of an isolated system that is not in equilibrium will tend to increase over time, approaching a maximum value at equilibrium.

The total energy E of a molecule in the reactor is the sum of its kinetic and potential energies:

$$E = E_k + E_p \tag{2.28}$$

The kinetic energy E_k of a molecule is given by its speed inside the reaction vessel. The potential energy E_p of a molecule is given by its composition, bonds between atoms and shape (spatial conformation of atoms and their bonds).

Two colliding molecules may react only if the sum of their kinetic energies is high enough to overcome the activation energy *barrier* (E_a) for the reaction, which corresponds to the minimum amount of kinetic energy that is needed for the reaction to occur. In this case, the collision is said to be effective. If the kinetic energy of the colliding molecules is lower than E_a, the reaction does not occur, and the collision is called elastic. Effective and elastic collisions are very commonly modeled in artificial chemistries, as we will see in section 2.3.

The formation of chemical bonds leads to a more stable molecule, with a lower potential energy. Therefore breaking these bonds requires an input of energy. This energy is provided by the kinetic energy of the movement of the colliding molecules. The collision results in broken bonds,

sending the molecules to a higher potential energy state that is highly unstable. This intermediate cluster of atoms is called *transition state*. As new bonds are formed, the transition state
decays almost instantly into a lower energy state of one or more stable molecules, constituting
the products of the reaction. The height of the energy barrier E_a corresponds to the difference in
potential energy between the transition state and the reactant state.

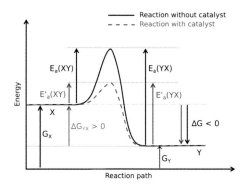

Figure 2.4 Energy changes during catalyzed (black curve) and uncatalyzed (red dashed curve) chemical
reactions. The forward reaction proceeds from left to right, converting reactants X to products Y;
conversely, the reverse reaction (Y back to X) can be read from right to left. The vertical axis indicates
the Gibbs free energy (G). On both directions, the reaction only occurs when the reactants are able
to overcome the activation energy barrier E_a. In this example, E_a is lower for the forward reaction
$E_a(XY)$ than for the reverse reaction $E_a(YX)$; therefore the forward direction is favored ($\Delta G < 0$).
Arrows pointing upward indicate positive energy values, while those pointing downward refer to negative
energy values. Modified from the original image by Brian Kell, http://en.wikipedia.org/wiki/
Activation_energy, available for unrestricted use.

A common way to depict the energy changes during a reaction is through an *energy diagram*,
such as the one shown in figure 2.4. For the moment we ignore the red parts of this figure, which
will be the topic of section 2.2.7. Since it is difficult to observe and measure the behavior of
individual molecules, energy diagrams usually depict macroscopic quantities at the scale of mols
of molecules. Such macroscopic values reflect the average behavior of individual reactions.

The horizontal axis of the plot in figure 2.4 is called *reaction coordinate*, and can be read in
two directions: from left to right, it shows the progression of the forward reaction from reactants
X to products Y; symmetrically, the corresponding reverse reaction (from Y back to X) can be
read from right to left. The vertical axis corresponds to the Gibbs free energy (G) of the substances
involved, a quantity that takes entropy into account:

$$G \quad = \quad H - TS \qquad\qquad (2.29)$$
$$H \quad = \quad U + pV \qquad\qquad (2.30)$$

H is the enthalpy , T is the temperature, S is the entropy, p is the pressure, V is the volume, and U
is the internal energy of the system, which includes the potential energy.

In figure 2.4, the X substances have total free energy G_x, and the Y substances have G_y. The
difference in Gibbs free energy before and after the reaction is denoted by $\Delta G = G_y - G_x$. If $\Delta G > 0$
then the reaction is *endergonic*, i.e., it absorbs energy from its surroundings, while if $\Delta G < 0$ it
is *exergonic*, releasing energy. The reaction of figure 2.4 is exergonic in the forward direction

($\Delta G < 0$) and endergonic in the reverse direction ($\Delta G_{yx} > 0$). The energy barrier from X to Y is $E_a(XY)$, and from Y to X it is $E_a(YX) = E_a(XY) - \Delta G$. Because it faces a smaller energy barrier $E_a(XY) < E_a(YX)$, the forward direction is easier to occur than the reverse reaction; therefore, it is said to be *spontaneous*. Let us refer to the difference in activation energy before and after the reaction as $\Delta E_a = Ea(YX) - E_a(XY)$. Note that for any given reversible reaction, the fact that $\Delta E_a > 0$ implies that $\Delta G < 0$:

$$\Delta E_a = E_a(YX) - E_a(XY) > 0 \implies E_a(XY) < E_a(YX) \implies E_a(XY) < E_a(XY) - \Delta G \implies \Delta G < 0$$
(2.31)

Hence, reactions with a negative Gibbs free energy change ($\Delta G < 0$) are often referred to as spontaneous, thermodynamically favored, or "*downhill*" (since they go from a higher to a lower energy landscape). Conversely, because they see a higher E_a barrier than their reverse counterpart, reactions with $\Delta G > 0$ are referred to as nonspontaneous, thermodynamically unfavored, or "*uphill*."

All reactions are in principle reversible: in figure 2.4, the reaction can flow from left to right or from right to left, as long as the colliding molecules acquire sufficient kinetic energy to jump over the E_a barrier. The speeds with which the reactions proceed in the forward and reverse directions are determined by the E_a height observed at each side of the barrier: the higher the barrier, the slower the reaction. The relation between E_a and the speed of a chemical reaction is given by the Arrhenius equation [42]:

$$k = Ae^{\frac{-E_a}{RT}}$$
(2.32)

where:

- k is the kinetic coefficient for the reaction (recall that k is proportional to the speed of the reaction);

- A is the so-called *preexponential factor* of the reaction;

- R is the gas constant: $R = k_B N_A$, where k_B is the Boltzmann constant ($k_B = 1.38065 \times 10^{-23}\ JK^{-1}$), and N_A is the Avogadro constant (thus $R = 8.31451\ JK^{-1}mol^{-1}$);

- T is the absolute temperature inside the reactor.

Irreversible reactions are reactions in which the participating substances fall into a considerably lower energy state, such that the height of E_a barrier seen at the opposite side is so great that it can no longer be overcome in practice: the substances remain "trapped" at the other side.

Uphill reactions require an input of energy to drive them in the forward direction, such their equilibrium can be shifted to yield more products. The relationship between ΔG and the equilibrium constant K of a reversible reaction can be derived from the Arrhenius equation:

$$K = \frac{k_f}{k_r} = \frac{Ae^{-E_a/(RT)}}{Ae^{-(E_a-\Delta G)/(RT)}} = e^{\frac{-\Delta G}{RT}} \implies ln(K) = \frac{-\Delta G}{RT} \implies$$
$$\Delta G = -RT\, ln(K)$$
(2.33)

Hence, for $k_f > k_r$ (forward reaction favored) we have $K > 1$ which results in $\Delta G < 0$; conversely, for $k_f < k_r$ (reverse reaction favored), $K < 1$ and $\Delta G > 0$. For $\Delta G = 0$ both sides of the reaction occur with equal probability hence their average speed is the same ($k_f = k_r$).

In summary, ΔG (or equivalently ΔE_a) determines the equilibrium properties of the reactions, whereas E_a determines its dynamic (speed-related) properties.

2.2.7 Catalysis

A catalyst is a substance that participates in a chemical reaction by accelerating it without being consumed in the process. Its effect is to lower the reaction's activation energy peak, thereby accelerating the reaction, while leaving the initial and final states unchanged. The decrease in E_a is depicted by the red/dashed curve in figure 2.4. Following the Arrhenius equation (Eq. 2.32), the kinetic coefficient k for the reaction increases exponentially with a decrease in E_a. Hence, decreasing E_a has a dramatic effect on k, which is proportional to the speed of the reaction. This explains the role of enzymes in accelerating reactions.

It is important to highlight that the enzyme activity does not shift the equilibrium of a reaction: because it accelerates both the forward and the reverse reactions in the same proportion, ΔG does not change, therefore according to equation 2.33, the equilibrium constant K also remains the same.

Catalysts play a fundamental role in both natural and artificial chemistries. Enzymes are proteins that act as biological catalysts. They are essential to numerous cellular processes, such as metabolism, gene expression, and cell replication. Many artificial chemistries rely on catalysts to emulate phenomena related to life, such as autocatalytic cycles, the formation of organizations, self-replication, and Darwinian evolution.

After this brief excursion to chemistry, we now turn to the general structure of an artificial chemistry.

2.3 General Structure of an Artificial Chemistry

This section now introduces the basic concepts of artificial chemistries. Let us start with a broad definition, taken from our previous review of the state-of-the-art of ACs [236]: *An* artificial chemistry *is a man-made system that is similar to a real chemical system.* By man-made we mean a model system that can either be executed as a simulation (computer model), or as a physical system, i.e., where real physical/chemical processes play a role that do not appear combined naturally in our environment. This definition has been kept as general as possible in order not to exclude any relevant work.

Formally, an *artificial chemistry* can be defined by a triple (S, R, A), where S is the set of all possible molecules, R is a set of collision rules representing the interactions among the molecules, and A is an algorithm describing the reaction vessel or domain and how the rules are applied to the molecules. Both, the set of molecules S and the set of reaction rules R, can be defined explicitly by stating which molecule(s) reacts (react) to which other molecule or implicitly by giving an algorithm or a mathematical operation for what will happen.

If we recall briefly what we stated in the first section of this chapter, we have the central elements of any simulation at hand: The components or subsystems, here the molecules S (previously ΣS_i), their possible interactions, here R, and the update rule, here algorithm A (previously U).

2.3.1 The Set of Molecules S – Objects

Just as in the general approach to simulation, we start with the identification of components or subsystems of the system to be simulated. Those are the objects or molecules of an artificial chemistry.

The set of molecules $S = \{s_1, \ldots, s_i, \ldots, s_n\}$, with n possibly infinite, describes all different valid molecules that might appear in the AC. A vast variety of molecule definitions can be found in different ACs. A molecule's representation is often referred to as its *structure*, which needs to be kept apart from its *function*, which is given by the reaction rules R. The description of the valid molecules and their structure is usually the first step in the definition of an artificial chemistry. This is analogous to the part of chemistry that describes what kind of atomic configurations form stable molecules and how these molecules appear.

Note that the molecules as components of an AC do not need to be elementary; rather, they could already be subsystems. This is a very important and interesting possibility that we shall later have to elaborate on in detail.

2.3.2   The Set of Rules R – Interactions

The set of reaction rules R describes the interactions between molecules $s_i \in S$. A rule $r \in R$ can be written according to the chemical notation of reaction rules in the form (see eq. 2.11):

$$s_1 + s_2 + \cdots + s_n \rightarrow s_1' + s_2' + \cdots + s_m' \tag{2.34}$$

A reaction rule determines the n components (objects, molecules) on the left hand side that can react and subsequently be replaced by the m components on the right hand side. In the general case, n is called the *order* of the reaction , but there are exceptions.

A rule is applicable only if certain conditions are fulfilled. The major condition is that all components written on the left hand side of the rule are available. This condition can be broadened easily to include other parameters such as neighborhood, rate constants, probability for a reaction, or energy consumption. In these cases a reaction rule would need to hold additional information or further parameters. Whether or not these additional predicates are taken into account depends on the objectives of the AC. If it is meant to simulate real chemistry as accurately as possible, it is necessary to integrate these parameters into the simulation system. If the goal is to build an abstract model, many of these parameters might be left out.

2.3.3   Reactor Algorithm A – Dynamics

The reactor algorithm A is in charge of simulating the dynamics of a reaction vessel. It determines how the set R of rules is applied to a collection of molecules P, called *reactor, soup, reaction vessel*, or *population*. Note that P cannot be identical to S, since some molecules might be present in many exemplars, while others might not be present at all. So while we can speak of S as being a set, P is correctly termed a multiset . The index of S indicates a type of molecule, while the index of P indicates a place in the population which could be occupied by any type of molecule.

Because multisets are a concept used in many artificial chemistries, we'll take some time now to define and understand it. A *multiset* is a generalization of a set allowing identical elements to be members. Thus, in order to characterize a multiset, the identity of each element must be known, as well as how many of these elements are contained in the multiset. For instance, $M = \{a, b, b, c, c, c\}$ is a multiset, while $S = \{a, b, c\}$ would be a set. Multisets are used in computer science, sometimes also called "bags." The order of elements appearing in a multiset is not important, such that $\{c, a, b, c, b, c\} = M$ is the same multiset as the one above. The *cardinality* of the multiset is the total number of elements, i.e., $card(M) = 6$. The *multiplicity* of an element is the number of times it appears in the multiset. For M, the multiplicity of element a is 1, for b it is 2, and for c it is 3.

From this it can be easily seen that an alternative definition of a multiset is that it is a 2-tuple (S, m) where S is a set and m is a function from S into the natural numbers $m : S \rightarrow \mathbb{N}$. m counts the number of occurrences of an element $s \in S$, i.e. its multiplicity. For the above example we have again, $m(a) = 1$; $m(b) = 2$; $m(c) = 3$. The multiset M can therefore also be written as the tuple: $(\{a, b, c\}, \{1, 2, 3\})$.

As we can see here, the state of a chemistry (in terms of the number of molecules of a particular element) can be represented as multiset, where all the molecular species correspond to elements of S and the molecule count corresponds to their respective multiplicity. It is, of course, not so interesting to see only a snapshot of a chemistry, but rather its changes over time. As chemical reactions happen, new molecules appear and previously existing molecules are consumed. So it is really the changes of the counts of molecules (and possibly) the increase or decrease in the cardinality of the multiset (and the underlying set) that is of interest when looking at chemical reactions.

Two classes of reactor algorithms can be distinguished: microscopic and macroscopic. The former takes into account individual molecular collisions, and is inherently stochastic. The latter approximates a system containing a vast amount of molecules, and consists mostly in numerical methods for the deterministic integration of systems of differential equations such as equations. (2.20) and (2.21). We will briefly describe a few reaction algorithms in this section, while their detailed explanation is reserved for chapter 4.

Although the algorithm A is conceptually independent of the representation of R and P, its performance and algorithmic complexity will often be influenced by the way R and P are represented as data structures in the computer. This is especially true for P, which may represent huge numbers of molecules. An analysis of the algorithmic complexity of a few popular reaction algorithms will be discussed in chapter 4.

(1) Stochastic molecular collisions: In this approach, every molecule is explicitly simulated and the population is represented as a multiset P. A typical algorithm draws a random sample of molecules from the population P, assumes a collision of these molecules, and checks whether any rule $r \in R$ can be applied. If the conditions of a rule r are fulfilled, the molecules are replaced by the molecules on the right-hand side of r. In the event that more than one rule can apply, a decision is made which rule to employ. If no rule can be applied, a new random set of molecules is drawn. The actual algorithm, though, must not be that simple. Further parameters such as rate constants, energy, spatial information, or temperature can be introduced into the rules for the chemistry to become more realistic.

The following example is an algorithm used for an AC with second-order reactions only:

```
while ¬terminate() do
  s₁ := draw(P);
  P := remove(P, s₁);
  s₂ := draw(P);
    if ∃ (s₁ + s₂ → s'₁ + s'₂ + ⋯ + s'ₘ) ∈ R
      then
        P := remove(P, s₂);
        P := insert(P, s'₁, s'₂, ..., s'ₘ);
    else
        P := insert(P, s₁);
      fi
      od
```

The function draw returns a randomly chosen molecule from P (without removing it from P). The probability that a specific type is returned is proportional to the concentration of this type in P. If a reaction rule that applies to the selected molecules is found, the corresponding reaction products are injected into P. In this case, we say that the collision between the two molecules was effective. If no applicable reaction rule can be found, the drawing of the two molecules does not produce any change to the reactor vessel. We say that the collision between the two molecules was elastic. The collision is simulated by the drawing of molecules, whereas the reaction is given by the reaction rule. Note that the first molecule (s_1) is removed such that it cannot be drawn twice in the same reaction, as in chemistry; if the collision is elastic, s_1 is returned to the vessel in order to restore its original state before the collision.

The previous algorithm simulates a closed system in which no molecules flow in and out of the vessel. Most systems of interest, however, are open systems relying on an inflow of raw material and/or energy, and an outflow of product molecules or elimination of waste. To begin with, a dilution flux of molecules can be easily added to the algorithm. In chemistry, dilution means the addition of a solvent such as water: the overall volume of the vessel increases, and the concentration of each molecular species that it contains decreases accordingly. In an artificial chemistry, the notion of a dilution flux is frequently associated to a vessel that overflows, such that any excess molecules are removed. This is the notion that we are going to use for the next algorithm. In a well-stirred reactor, all molecules have an equal probability of dropping out of the vessel by excess dilution. Therefore, for a given molecular species, the probability of being chosen for removal is proportional to its concentration. So the following lines could be added to the while loop:

```
s := draw(P);
P := remove(P, s);
```

We have now assumed that molecules s_1, s_2 are consumed by the reaction, molecules $s'_1, s'_2, ..., s'_m$ are newly produced and one arbitrary molecule s is removed from the reaction vessel.

The simulation of every molecular collision is realistic and circumvents some difficulties generated by collective descriptions we shall discuss. There are, however, certain disadvantages to an explicit simulation of every collision. If reaction probabilities of molecular species differ by several orders of magnitude this low level simulation is not efficient. Also, if the total number of different molecular species is low or the population is large, or the number of different reaction rules is large, such an explicit simulation will be slow.

(2) Stochastic simulation of chemical reactions: The stochastic molecular collision algorithm presented before correctly mimics the stochastic dynamics of reactions triggered by individual molecular collisions. However, it has a number of shortcomings. First, the algorithm is inefficient when most collisions turn out to be elastic. Moreover, it is only applicable to the case of bimolecular reactions. It is difficult to extend it to arbitrary reactions with different kinetic coefficients.

In order to simulate the stochastic behavior of a chemical reaction system in an accurate and efficient way, a number of specialized algorithms have been proposed. Perhaps the most well-known one among these is Gillespie's stochastic simulation algorithm (SSA) [328]: at every iteration, it decides which reaction will occur and when. Therefore, only effective collisions are computed and elastic collisions do not play a role in this algorithm, which can save substantially on simulation time. The reaction chosen is drawn from the set R with a probability proportional to its propensity, which is proportional to the rate of the reaction. So faster reactions will occur more often, as expected. The time increment when the reaction should occur is drawn from an exponential distribution with mean inversely proportional to the sum of propensities of all reactions in R. This means that in a system with a large number of fast reactions, the interval between two reactions is small; conversely, a slow system with few reactions will display a long time interval between reactions, on average; which is again the expected behavior. Gillespie proved that his algorithm accurately simulates the stochastic behavior of chemical reactions in a well-mixed vessel. Gillespie's algorithm will be described in more detail in chapter 4, along with other algorithms in this category.

(3) Continuous differential or discrete difference equations: A further step away from individual reaction events is often more appropriate to describe the behavior of a chemical system. A typical example is the simulation of large numbers of molecules, like in real chemistry. In such a scenario, simulating individual molecular collisions would be inefficient; a collective approach based on deterministic simulation of the macroscopic behavior of the system is sufficiently accurate. For this purpose the dynamics of the chemical system is described by differential rate equations reflecting the development of the concentrations of molecular species, as explained in section 2.2.3. The index carried by a molecule now characterizes its species (S_i), not its place in the population.

The method to simulate the chemical reaction system now consists in numerically integrating the corresponding ODE: given the stoichiometric matrix and the rate vectors for the system, and starting from the initial concentrations of all substances S_i, a numerical integrator computes the changes in concentration for each substance in time according to equation (2.21), and outputs a set of vectors containing the concentration timeline for each substance. Several numerical methods are readily available for this purpose, in well-known scientific software packages. The simplest one is to compute the concentration changes in discrete time steps Δt, and then add the resulting ΔS_i value to the absolute concentration S_i. This is called the explicit Euler method.

Although simplistic and inaccurate, it can offer sufficiently good approximations provided that the time step Δt can be made small enough, and that the integration process does not last too long (since this method accumulates errors with time).

Rate equations are a continuous model for a discrete situation. While concentrations might approach zero continuously, this is not the case for the real system: If the last molecule of a particular species disappears in the volume under consideration, there is a discrete change in concentration of this species from a finite quantity to zero. This causes a number of anomalies, and the modeler should be aware of the limitations of this model. Using differential equations is an approximation for large numbers of molecules of the same species in the volume under consideration.

One of the disadvantages of this approach deserves mentioning: Concentration values are normally represented in a computer by floating point numbers. They may be so small as to be below the threshold equivalent of a single molecule in the reactor. For example, if the assumed population size is 1,000 and $[s_i]$ is below 0.001 , then every further calculation of reactions using such a small concentration value s_i would cause unnecessary computational effort because species s_i is effectively no longer present in the reactor. In fact, these calculations would be misleading as they would still pretend that a possibility exists for a reaction in which s_i might participate, when in reality s_i has disappeared from the reactor. In order for the differential equation approach to be useful, the number of different molecular species N should be small relative to the total number of molecules, M,

$$N \ll M. \tag{2.35}$$

One can see that this condition contradicts the condition for the existence of novelty in a system, see eq. 1.6. We conclude that for a treatment of systems with novelty differential equations don't seem to be an appropriate approach. We shall discuss this in more detail later.

(4) Metadynamics: The metadynamics approach acknowledges the limitations of the differential equation model because it assumes that the number of species and therefore the number of differential equations may change over time [49,50]. The equations at a given time only represent dominant species, i.e., species with a concentration value above a certain threshold. All other species are considered to be part of the background (or might not even exist). As concentration values change, the set of dominant species may change as well. Thus the differential equation system is modified by adding and/or removing equations. One can distinguish between deterministic and stochastic metadynamics. In deterministic metadynamics, there are basically no surprises. The set of dominant species follows the development of its differential equations, until the concentration threshold causes a change in the set. The sequence of sets resulting from this dynamic explores the potential reactions embodied in the differential equation system but is purely deterministic. All pathways are internal to the system. As a result, a final state, the metadynamic fixed point, or pinned state, is reached.

Stochastic metadynamics assumes the existence of random events that change the background of dominant species. Every so often, a species from the background is allowed to interact as if it were above threshold. This will potentially result in a chain reaction causing the metadynamic fixed point to change. If that happens, a new pinned state can be reached and the system has shown some sort of innovation. There is a second type of stochastic metadynamics where, instead of bringing in a random species from the background, the set of reactions is changed through a random event [425,426]. This can also result in a change in the fixed point behavior of the underlying system. We will need to discuss this type of approach in more detail in chapter 15 (section 15.2.2).

If applied in the right way, metadynamics can combine the speed of a differential equation approach with the physical accuracy of a stochastic molecular collision simulation.

There is one further variant on this approach: If the differential equation system (2.20) is sufficiently simple, symbolic integration of the equations becomes possible. This way, the steady state behavior of the system (fixed point, limit cycle, chaotic behavior ...) can be derived analyt-

Table 2.2 Comparison of reactor algorithms

ENTITIES	FORMALISM	INTERACTION
Individual molecules	Stochastic molecular collisions	Collisions and rule-based reactions
Molecular species	Stochastic simulation of chemical reactions	Possible reaction channels
Molecular species	Concentration/rate differential equations	Coupling between equations
Dominant molecular species and background	Metadynamics (ODEs and random events)	Coupling between equations and random changes to couplings
Dominant molecular species and single molecules	Explicit simulation of molecules and ODE of dominant species	Coupling between equations and collision-based reactions of a small number of molecules

ically. This technique can be applied in the metadynamics approach to calculate the dynamical fixed point of the differential equations in a very compact way.

In most cases, however, the differential equation system will contain nonlinear coupling terms and, as a result, will be difficult to solve analytically. Thus, we should keep in mind that the analytical approach is feasible only in a very small subset of ACs. In most cases, numerical integration methods have to be applied.

(5) Hybrid approaches: There are also approaches where single macromolecules are simulated explicitly and a small number of molecules are represented by their concentrations (see, for instance [952, 953]).

Table 2.2 compares the different approaches to modeling ACs through the reactor algorithm, all based on the assumption of a well-stirred chemical reactor.

2.4 A Few Important Distinctions

Each of the elements $\{S, R, A\}$ that characterize an artificial chemistry can be defined in an explicit or implicit way. These differences will have an important impact on the behavior of the chemistry; therefore, we will often use this distinction to classify or compare the various ACs introduced later in the book.

2.4.1 Definition of Molecules: Explicit or Implicit

In an artificial chemistry $\{S, R, A\}$ the set S represents the interacting objects or molecules. S can be defined explicitly or implicitly.

Molecules are *explicitly* defined if the set S is given as an enumeration of symbols, e.g., $S = \{r, g, b\}$. An *implicit* definition is a description of how to construct a molecule from more basic elements. This description may be a grammar, an algorithm, or a mathematical operation.

Examples for implicit definitions are: $S = \{1, 2, 3, \ldots\}$, the set of natural numbers; or $S = \{0, 1\}^*$, the set of all bit strings.

To build constructive dynamical systems (see chapter 15) it is convenient to define molecules implicitly. Typical implicitly defined molecules are character sequences (e.g., *abbaab*), mathematical objects (e.g., numbers), or compound objects which consist of different elements. Compound objects can be represented by data structures [562]. We will refer to the representation of a molecule as its *structure*.

In some artificial chemistries, the structure of a molecule is not a priori defined. Instead, the arrival of molecules is an *emergent* phenomenon and interpreting a structure as a molecule is possible only a posteriori. In these cases, it is not possible to define the molecules because they are emergent to the process. However, ACs of this type will possess lower level elements that can be defined.

2.4.2   Definition of Interactions/Reaction Laws: Explicit or Implicit

Once we have defined the molecules of our AC, it is time to specify their interactions. These interactions can be defined, again, in two different ways.

Explicit: The definition of the interaction between molecules is independent of the molecular structure. Molecules are represented by abstract interchangeable symbols and the total number of possible elements of the AC is fixed. Reaction rules can be enumerated and explicitly given. All possible elements and their behavior in reactions is known a priori. Their interaction rules do not change during the experiment.

Implicit: The definition of the interaction between molecules depends on their structure. Examples for implicit reaction rules are concatenation or cleavage reactions of polymers. From the number division example we discussed in chapter 1 we remember that particular numbers under division behave differently. We can say that this is the result of their structure. An artificial chemistry with an implicit reaction scheme allows to derive the outcome of a collision from the structure of its participating molecules. Therefore, there is no a priori way of knowing or enumerating the molecules beforehand. The number of possible molecules can be infinite, even. Implicitly defined interactions are commonly used for constructive artificial chemistries.

2.4.3   Definition of the Algorithm/Dynamics: Explicit or Implicit

Finally, the algorithm A of the artificial chemistry $\{S, R, A\}$ producing the dynamics of the system needs to be defined. The *explicit* definition of the system's dynamics has already been given in section 2.3.3. This definition is necessary from a computational point of view, because in a computer only the effective execution of a series of formally defined interactions causes the system to change and to show any kind of dynamic behavior. An explicit definition of the dynamics is based on the interactions determined through R and may include various parameters, ranging from temperature, pressure, pH-value to field effects of secondary or higher level molecule structures, resulting in an artificial chemistry that can be used in the field of computational chemistry [182]. It is shown in section 2.5, however, that even a random execution of interactions with no additional parameters causes interesting dynamical phenomena.

Sometimes, the definition of interactions makes the additional definition of the *A* unnecessary. This *implicit* definition is used, for example, in a cellular automata chemistry, where the dynamics is caused by the synchronous or asynchronous update of lattice automata sites.

2.5 Two Examples

Let us now look at two examples of very simple artificial chemistries, in order to get a feeling how one would go about setting up these kinds of systems.

2.5.1 A Nonconstructive Explicit Chemistry

The first example is intended to demonstrate the relation of an explicit simulation of molecules to their differential equation model. We call this a nonconstructive AC, since all molecular objects are known beforehand and can be explicitly taken care of in a simulation model. The example is inspired by a problem posed in the "Puzzled" column of a computer science magazine [922]. The puzzle goes like this:

> A colony of chameleons includes 20 red, 18 blue, and 16 green individuals. Whenever two chameleons of different color meet, each changes to the third color. Some time passes during which no chameleons are born or die, nor do any enter or leave the colony. Is it possible that at the end of this period all 54 chameleons are the same color?

The answer is "no" for this particular distribution of chameleons at the outset; see [921]. Yet there are many other configurations in which this is possible, namely those where the difference between the number of chameleons of different colors is a multiple of 3. However, any such outcome is extremely unlikely, because in a world where probabilities reign, encounters of chameleons are dominated by the most frequent ones, just the opposite of the very low number required for the mathematical problem posed here. Be that as it may, we are not so much interested in the answer to this concrete question (more on it later), but to take the idea of an interaction of chameleons and model it as an artificial chemistry. In the realm of chemistry and ACs as well, encounters are also governed by probabilities, i.e., the most frequent molecules bump into each other most frequently.

Objects/molecules: In order to model this problem as an AC, we need to first determine the interacting objects. There are three types of objects here, red *r*, green *g*, and blue *b* chameleons, each in different quantities.

Interactions/reactions: Their interaction is determined by the following rules:

$$r + g \rightarrow b + b \tag{2.36}$$

$$r + b \rightarrow g + g \tag{2.37}$$

$$b + g \rightarrow r + r \tag{2.38}$$

Rules obey symmetry in that an encounter of r and g has the same effect as an encounter of g and r.[2] All other encounters between chameleons ("collisions") are elastic, i.e., do not lead to changes in color. These reaction equations can be compactly written as

$$i + j \rightarrow k + k \quad \forall \ i \neq j \neq k \tag{2.39}$$

where i stands for any chameleon with color i and indices i, j, k run through r, g, b colors.

Dynamics/algorithm: The algorithm is very simple in this case. Since there is no additional production of objects, we don't need a flow term, the total population will always remain of the same size. A flow term would guarantee a constant population size by flushing out an appropriate number of objects, if that were needed. The algorithm is really a simple adaptation of the algorithm introduced before:

```
while ¬terminate() do
  s₁ := draw(P);
  s₂ := draw(P);
    if (s₁ ≠ s₂)
        then
        P := remove(P, s₁, s₂);
        P := insert(P, s₃, s₃);
        fi
        od
```

The only variation we have made here is that the `insert`-operation inserts the new objects into the same places where the reactants were found. One iteration of the while-loop simulates one collision of two molecules and we can count the number of each molecular type over the course of iterations. In appendix A we have included the code for this example in the Python programming language. There, you will also find a small simulation environment for allowing you to set up ACs conveniently. You can download this and later examples from `www.artificial-chemistries.org`.

Before we now look at the behavior of this AC, a few notation techniques need to be mentioned. We can write the reaction rules in the form of a reaction table:

rules	reactants r	g	b	products r	g	b
r_1	1	1				2
r_2	1		1		2	
r_3		1	1	2		

or, shortly,

	r	g	b
r	$\emptyset$	$2b$	$2g$
g	$2b$	$\emptyset$	$2r$
b	$2g$	$2r$	$\emptyset$

where $\emptyset$ means that no objects are produced by the interaction listed. These tables can then be translated into reactions graphs of the sorts depicted in figure 2.2.

If we start with approximately the same number of chameleons, we shall not see much development, except that in many places, color changes due to a random encounter. However, if we start with chameleons of only two colors, we will see initial development toward an equilibrium of colors, see figure 2.5.

---

[2]While this is the natural case in chemistry due to the physical features of space, it is not required for artificial chemistries and must be explicitly considered.

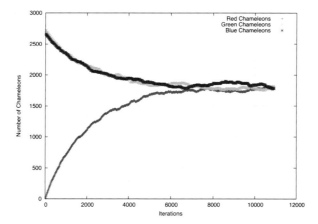

Figure 2.5 The first two generations (equivalent to 10,800 iterations) of an explicit simulation of encounters between chameleons changing colors. A total of 5,400 chameleons is simulated, starting from a condition where only blue and read chameleons are present, in virtually equal numbers, due to randomized initial condition. Red chameleons quickly pick up in numbers, until only fluctuations are left between numbers.

We can see that dynamics at work more clearly if we now formulate the differential equation system for the chameleon example and integrate these equations numerically.

The model with differential equations: As said previously, an alternative to a molecular simulation of each and every encounter between molecules is to write down rate equations for the reaction system. We start by representing each species by a concentration vector $\vec{x}(t) = (x_r(t), x_g(t), x_b(t))$, where $x_r(t)$, $x_g(t)$ and $x_b(t)$ denote the concentration of chameleon types r, g, b at time t, respectively. The reactions are set up such that the total number of chameleons is constant $0 \leq x_{r,g,b} \leq 1$ and their sum is the total concentration, $\sum_{i=r,g,b} x_i(t) = 1$, summing to 1. The corresponding differential equations read:

$$\frac{dx_r}{dt} = 2x_g x_b - x_r x_b - x_r x_g \tag{2.40}$$

$$\frac{dx_g}{dt} = 2x_r x_b - x_g x_b - x_g x_r \tag{2.41}$$

$$\frac{dx_b}{dt} = 2x_r x_g - x_b x_r - x_b x_g \tag{2.42}$$

A dilution flux $\Phi(t)$ is not necessary here because as a result of each collision two objects are created and two are removed.

Comparison of the two descriptions: Figure 2.6 shows the simulation of an explicit molecular system with a system of differential rate equations. Initial state has been prepared carefully so as to be identical in both approaches. The time scale of the differential equation simulation has to be carefully chosen, such that a number of single molecular reactions is equivalent to one step of the numerical integration is executed. Notably, precaution has to be taken about

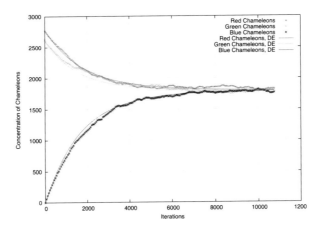

Figure 2.6 A comparison of two generations (equivalent to 10,800 iterations) of an explicit simulation of encounters with an integration of rate equations 2.40 — 2.42. Again, a total of 5,400 chameleons is simulated. Here, colors are distributed differently, but equilibrate after some time. The differential equation (DE) curves show good agreement with the explicit simulation.

the fact that there are a number of elastic collisions in the explicit simulation model. Those elastic collisions correspond to a change in the time scale for the update rule of rate equations. In our particular example, elastic collisions happen when two chameleons of the same color encounter each other, i.e., with a probability $p_e = \sum_{i=r,g,b} x_i^2$. In other words, only in $1 - p_e$ cases a reaction takes place, and this is the quantity with which the time scale of the integration has to be corrected.

As we increase the number of chameleons, the fluctuations in a stochastic explicit simulation average out and the description approaches the rate equation model, which assumes an infinite population of each type of reactants.

2.5.2 A Constructive Implicit Chemistry

As example for a constructive, implicitly defined artificial chemistry we explain the number-division chemistry examined previously in [72, 113]. In chapter 1 we alluded to the fact that we can imagine numbers reacting with each other to produce new numbers. The arithmetic operations of multiplication and division of natural numbers are particularly fruitful examples for demonstration purposes. Here we shall focus on the division operation as our implicit reaction, which can be defined over the infinite set of natural numbers. We shall, however, exclude the numbers "0" and "1" from our artificial chemistry, since they behave rather exceptionally under division. Otherwise, a potentially infinite number of molecular types are present. So let's again start with the objects of our system and work our way to the dynamics from there.

Molecules: The set of molecules S are all natural numbers greater than one: $S = \{2, 3, 4, \ldots\}$. This set is infinite, but we do not expect to have an infinite reaction vessel. Thus, only a finite subset of numbers will be realized in our actual reaction system at any given time.

Reactions: Mathematical division is used to calculate the reaction product. If two molecules $s_1, s_2 \in S$ collide and s_1 can divide s_2 without remainder, s_2 is transformed into s_2/s_1. Thus s_1 acts as a catalyst. The set of reactions can also be written implicitly as

$$R = \{s_1 + s_2 \rightarrow s_1 + s_3 : s_1, s_2, s_3 \in S \wedge s_3 = s_2/s_1\} \qquad (2.43)$$

If two molecules collide and there is no corresponding reaction rule in R, the collision is elastic.

Dynamics: The algorithms draws randomly two molecules from the population and replaces the larger number provided it can be divided by the smaller reactant. In case they are equal, or cannot be divided the collision is elastic. Written in pseudo-code, the algorithm is:

```
while ¬terminate() do
    s₁ := draw(P);
    s₂ := draw(P);
    if (s₂ > s₁) ∧ (s₂ mod s₁ = 0)
        then
            s₃ := s₂/s₁;
            P := remove(P, s₂);
            P := insert(P, s₃);
    else
            if (s₂ < s₁) ∧ (s₁ mod s₂ = 0)
            s₃ := s₁/s₂;
            P := remove(P, s₁);
            P := insert(P, s₃);
        fi
    fi
od
```

Despite the simple algorithm, the dynamic behavior of the system is fairly interesting. After repeated application of the reaction rule, prime numbers emerge as stable (nonreactive) elements and the whole population tends to eliminate nonprime numbers (figure 2.7). Elsewhere we have shown that only prime numbers remain after some time. This is surprising at first, because nothing in the experimental setup (neither in S nor in R) specifies any condition regarding primes. Upon further analysis, however, one realizes that the reaction rule precisely determines primes as nonreactive.

We have now seen the implicit character of the reaction system prohibiting the formulation of a reaction matrix, but it remains to be seen in what sense this system is constructive. After all, initializing a population with all numbers between 2 and 100 doesn't produce new types of numbers.

The previous simulation can quickly be turned into a constructive system, through the initialization with 100 numbers out of a larger range of integers. For instance, if we initialize in the interval $I = [2, 1000]$, we will see new numbers, which have not appeared before. If again a population of $M = 100$ numbers is initialized (this time through a random draw from the interval I), the condition $N \gg M$ is fulfilled. An analysis of the systems under these new conditions must be done carefully, since we don't know from the outset which number types will appear. Figures 2.8 and 2.9 show one run of the reaction system with new numbers being generated.

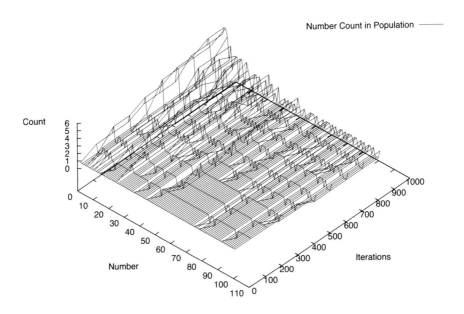

Figure 2.7 Development of (prime) number count over time in iterations. Population size $M = 100$. Population initialized with each number from $S = \{2,3,\ldots,101\}$. Prime numbers emerge as dominant, since they become stable in the population.

From the initial status of the system, gradually more smaller numbers and primes are produced. The smaller numbers further react until all numbers are divided into primes. The primes cannot be divided any more and thus accumulate in the reactor. The concentration of primes shows a typical sigmoidal growth pattern and finally reaches a maximum at prime concentration $c_{primes} = 1.0$. Note, however, that this AC does not produce the prime factorization of the initial set of numbers, because divisors act as catalysts in the reaction system of equation 2.43, i.e., there are no copies produced in the division operation. This explains the fact that we do not see more than a limited number of copies of primes in the final state of the reaction vessel.

2.6   Frequently Used Techniques in ACs

A number of techniques often used in ACs can be highlighted: for instance, sometimes it is interesting to measure time in terms of the number of collisions that happened in the system, instead of an absolute time unit. Pattern matching to emulate molecule recognition and binding (mimicking the way biological enzymes recognize their protein or DNA targets) is another commonly used technique in ACs. A third technique is the use of a spatial topology as opposed to a well-mixed vessel that was discussed in section 2.2. An overview of these techniques can be found below, and more about them will appear throughout the book as they are discussed in the context of the various ACs covered.

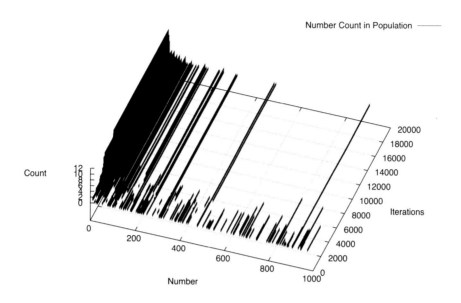

Figure 2.8 Development of (prime) number count over time in iterations. Population size $M = 100$, population initialized with random integers taken from the interval $S = \{2,3,\ldots,1000\}$. Some numbers appear and disappear. Again, primes emerge and stay.

2.6.1 Measuring Time

One step of the algorithm in section 2.3.3 can be interpreted as a collision of two (or more) molecules. Simulated time is proportional to the number of collisions divided by reactor size M. It is common to measure the simulated time in generations, where one *generation* consists of M collisions, independent of whether a collision causes a reaction or not. Using M collisions (a generation) as a unit of time is appropriate because otherwise an increase of the reactor size M would result in a slowdown of development speed. If, for example, in a reactor with a population size of 10^5, 10^{10} collisions were executed, this would mean that — statistically speaking — every molecule of the population did participate twice in a reaction. If the same number of collisions were executed in a reactor of size 10^{20}, only half of all molecules would have taken part in a reaction. If time would not be scaled by reactor size, this would result in a development of the reactor at a quarter of the speed as the previous one.

In continuously modeled ACs the progress of time depends on the integration step size of the numerical ODE solver. For a standard solver, a fixed step length of $h = 10^{-2}$ is a common setting. Thus, 100 evaluations of the ODE are needed to get a unit length time progress. If one wants to compare a stochastic simulation with a ODE numerical model, care must be taken to adapt the time measurements for both approaches. This will have to take into account the probability of elastic collisions as we have exemplified in the chameleon example.

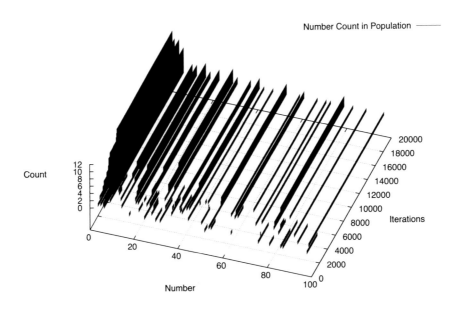

Figure 2.9 Closer look at development of (prime) number count over time in iterations, only interval [2, 100] shown. New numbers appearing can be clearly identified, as they come and go.

2.6.2   Pattern Matching

Pattern matching is a method widely used in artificial chemistries and other artificial life systems. A pattern can be regarded as information about (or as a means to identify) the semantics of a subcomponent or location. It allows to refer to parts of a system in an associative way independent of the absolute position of these parts. For instance, Koza used the shape of trees to address subtrees and to select reaction partners [472]. In the field of DNA computing, pattern matching is a central mechanism in the alignment of two DNA strands. There, a pattern is a sequence of nucleotides such as $CGATTGAGGGA...$ In the Tierra artificial life platform [701], a pattern is given by a sequence of NOP0 and NOP1 operations in the genome of an individual and is used to direct jumps of the instruction pointer, which points to the next instruction the program will execute. We shall encounter many other examples throughout the book.

The accuracy of a match can be used to calculate the probability of a reaction or rate constants, as suggested in, e.g., [49]. In order to compute the accuracy, a distance measure between patterns is needed, that allows to compute a "similarity" between two molecules by assigning a numerical value. One convenient distance measure for binary strings of the same length is the *Hamming distance*, whereas for variable length strings one can use the *string edit distance*. The Hamming distance is the number of bits that differ between the two binary strings. The edit distance between two strings is the minimum number of editing operations (insertions, deletions and replacements of single characters in a string) necessary to convert one of the strings into the other.

2.6.3 Spatial Topology

In many artificial chemistries, the reactor is modeled as a well-stirred tank reactor, often with a continuous inflow and outflow of substances. In a well-stirred tank reactor the probability for a molecule s_i to participate in a reaction r is independent of its position in the reactor.

In a reactor with topology, however, the reaction probability depends on the *neighborhood* of a molecule s_i. This neighborhood may be determined by the vicinity of s_i, for example, as measured by a Euclidian metric. This space will usually have a two- or three-dimensional extension. Alternatively, the neighborhood may be defined as the neighborhood in a cellular automaton (e.g., [530, 881]) or as a self-organizing associative space ([232]). In some cases, specific features, like diffusion, can be present in a reaction system [939].

All these different spatial structures have one feature in common: the topology/spatial metric has some influence on the selection of reactants by the algorithm A. While the analysis of ACs becomes more complicated by adopting an additional spatial topology, some properties can be observed only with a spatial structure and will not occur at all otherwise.

2.6.4 Limited Resources

Limited resources often determine which models can be possibly simulated and examined and which cannot. Therefore, it is important to consider the resource implications of certain decisions made in setting up a simulation model. A comparison of computational costs is first among those. Here, we shall look at a comparison of the two basic approaches, differential rate equation simulations and models using discrete molecular collisions. In particular, we want make more quantitative statements about when it would be useful to apply one of these methods versus the other.

In order to arrive at those statements, we shall need to estimate the computational costs of computing explicit molecular collisions vs. numerically solving the equivalent ODE system. We shall assume that $\langle n \rangle$ is the mean value of the number of molecules participating in each reaction r in R, and N is the number of different species in a population P of size M. First we need to estimate computational costs of performing a single step of both major approaches.

In an explicitly performed simulation, a reaction requires $O(\langle n \rangle)$ `draw`-operations, followed by $O(R)$ lookups in the reaction rule set R. Thereafter $O(\langle n \rangle)$ `insert`-operations are performed for an intermediate update. This leads to the estimation of $O(\langle n \rangle) + O(R)$ costs for each reaction, which can be approximated by $O(R)$, because in almost every (artificial) chemistry the total number of different reactions is much larger than the average number of reactants of a single reaction. A full update of the population P thus requires $O(R) \cdot O(M) = O(R \cdot M)$ operations.[3]

An integration of the system of ODE equations (eq. 2.20) requires N differential equations, each of which is of order $\langle n \rangle$. Assuming an integration step size h, an update of the concentration vector $\vec{s}$ with a numerical integration method thus has computational costs of $O(\frac{1}{h} \cdot N \cdot \langle n \rangle)$ function evaluations, because $\frac{1}{h}$ evaluations of the system with N equations of order $\langle n \rangle$ are needed to perform a single step. So $O(\frac{1}{h} \cdot N \cdot \langle n \rangle)$ computations have to be performed to have the same progress as M reactions would generate in an explicit simulation with a population P of size M. If a very common integration method like the standard Runge-Kutta method is used,

---

[3]These computational costs will reduce to $O(M)$ if the lookup of a reaction can be done in $O(1)$ which is, e.g., the case for reactions implicitly defined by the structure of a molecule. However, calculation of implicit reactions will then replace the lookup, and might carry even higher costs. So the comparison of costs in table 2.3 assumes $O(R)$ lookup costs.

Table 2.3 Comparison of the computational costs (in instructions). Fixed step size $h = 0.01$, assumed average reaction order $\langle n \rangle = 2$: White rows: Numerical integration is less expensive; Blue rows: Atomic collision simulation less expensive; Pink rows: Costs are equal.

PARAMETERS			COSTS	
Pop. size M	Species N	Reactions R	Collisions	Integration
10^3	10	10	10^4	10^3
10^3	10	10^2	10^5	10^3
10^3	10	10^4	10^7	10^3
10^3	10^4	10	10^4	10^6
10^3	10^4	10^2	10^5	10^6
10^3	10^4	10^4	10^7	10^6
10^3	10^8	10	10^4	10^{10}
10^3	10^8	10^2	10^5	10^{10}
10^3	10^8	10^4	10^7	10^{10}
10^6	10	10	10^7	10^3
10^6	10	10^2	10^8	10^3
10^6	10	10^4	10^{10}	10^3
10^6	10^4	10	10^7	10^6
10^6	10^4	10^2	10^8	10^6
10^6	10^4	10^4	10^{10}	10^6
10^6	10^8	10	10^7	10^{10}
10^6	10^8	10^2	10^8	10^{10}
10^6	10^8	10^4	10^{10}	10^{10}
10^9	10	10	10^{10}	10^3
10^9	10	10^2	10^{11}	10^3
10^9	10	10^4	10^{13}	10^3
10^9	10^4	10	10^{10}	10^6
10^9	10^4	10^2	10^{11}	10^6
10^9	10^4	10^4	10^{13}	10^6
10^9	10^8	10	10^{10}	10^{10}
10^9	10^8	10^2	10^{11}	10^{10}
10^9	10^8	10^4	10^{13}	10^{10}

an additional factor of 4 (because of the intermediate steps) needs to be weighed in, so that the overall costs sum up to $O(4 \cdot \frac{1}{h} \cdot N \cdot \langle n \rangle)$, which can be approximated by $O(\frac{1}{h} \cdot N)$. Note that this simple Runga-Kutta method uses a fixed step size. If the solver employs a variable step size, parameter h will change accordingly.[4]

Though constant factors were omitted in the assumption of runtime behavior expressed in $O(\ldots)$ terminology, they may play an important role in the calculation of the actual computa-

---

[4]Variable step size solvers might be necessary if the rate constants of the reactions differ by several orders of magnitude, which in turn makes the system become stiff).

tional effort.[5] Results of the comparison of the two approaches are shown in table 2.3. In general, integration is advantageous if the population is large and the number of species is small. In the opposite, case an explicit simulation of molecular collisions shows better efficiency.

Even without taking the different costs of a function evaluation into consideration (which is of course much more expensive compared to accessing an element of the population) it is clear that both approaches have advantages and drawbacks.

Computational costs as defined here are a narrow term that translates mainly into "time of computation." However, space of computation is another resource, and stems from memory requirements of parameters that need to be stored. There are different demands on memory for both methods as well. For instance, in order to integrate ODEs the reaction parameters need to be stored in matrices prior to integration. This requires $O(N^{\langle n \rangle})$ storage locations. In a molecular simulation with explicit reactions, requirements are similar. In a molecular simulation with implicit reactions, however, reactions can be stored in a lookup table that grows as the number of new reactions that never had been encountered before grows. In this way, a more efficient use of space can be achieved. Keep in mind that — as in other algorithms — one can sometimes trade time for memory or the other way around, so the entire picture needs to take into account both requirements when dealing with resource limitations.

2.7 Summary

In this chapter we have discussed general principles of chemistry and how chemical reactions can be modeled mathematically and simulated in a computer. We have also seen how complex chemical reactions networks can be formed from simple reactions. We have learned about the general structure of an artificial chemistry and realized that it reflects the needs of all modeling and simulation approaches. We have then looked at two examples of very simple ACs to demonstrate the idea. Finally, we have discussed a number of frequently used terms and techniques in AC. The discussion in this chapter tried to summarize on an introductory level some of the bare essentials necessary to understand the rest of the book.

---

[5]This comparison does not consider other effects such as creating the reaction table, or the creation and deletion of new species during the evolution of a system. It also neglects the memory requirement of both approaches, which causes additional computational effort. Nevertheless, this comparison gives as a result a coarse evaluation of the approaches.

matrix, *n. pl.* **matrices** or **matrixes**
A situation or surrounding substance within which something else originates,
develops, or is contained.

THE AMERICAN HERITAGE DICTIONARY OF THE ENGLISH LANGUAGE,
UPDATED 2009

CHAPTER

3

THE MATRIX CHEMISTRY AS AN EXAMPLE

This chapter discusses in depth a more complex example of an artificial chemistry, an algorithmic chemistry. In fact, it is an entire class of systems that can be treated in a similar way. The matrix chemistry is a system that was devised some time ago [63] to explore fundamental aspects of artificial chemistries in a simple abstract system. Despite its simplicity, the matrix chemistry provides sufficient richness in interactions to exemplify and discuss a number of important AC features. Its investigation was originally inspired by the application of genetic algorithms [703], [395] to problems of combinatorial optimization. It studies a different sort of algorithm, however, using strings of binary number that have a closer similarity to entities playing a role in prebiotic evolution.

As will be discussed in more detail in chapter 6, it is now believed that before DNA and proteins started to cooperate to provide the molecular machinery of life, a more primitive system, based on RNA, had developed. This idea, known as the RNA world [327], has become increasingly plausible, with the discovery of RNA acting in both a DNA-like information storing role and a protein-like chemical operator role. What has long been the subject of speculation was finally confirmed in 1981 [167]: The existence of this simpler RNA system, capable of both, information storage *and* enzyme-like operations. This system is visible in higher organisms even today, when it has been fully integrated by performing many useful auxiliary functions for the DNA-protein system.

RNA has, like DNA, a natural tendency to fold itself into a secondary and tertiary structure, depending on the sequence of nucleotides in the strand, the main difference being that a ribose sugar is substituted for a deoxyribose sugar. In a folded form RNA molecules can occasionally perform operations, either on themselves or on other strands of RNA (these are called "ribozymes"). Hence, the sequence on a strand of RNA can be interpreted as information of the macromolecule, whose three-dimensional shape resulting from attractive and repulsive forces of physics determines its operational counterpart, that is its function.

In modern computers, sequences of binary numbers, bit strings, are the simplest form of information storage and representation. Thus the idea is not far-fetched to empower those information strings with some sort of operator-like behavior, capable of transforming other strings, and maybe even themselves. This is the fundamental idea of the matrix chemistry: binary strings capable of assuming an alternative operator form ("matrix") that can act on strings.

In the next section we shall first introduce the basic idea in some more detail before discussing concrete examples in later sections.

3.1   The Basic Matrix Chemistry

The order of discussion will follow the order set up for previous examples in chapter 2. The matrix chemistry has implicitly defined molecules, with implicit reaction rules in an explicit algorithm.

Objects/molecules: Since we want to construct an AC as simple as possible, *binary* strings are our choice for the elements of the system. Binary strings (as any other type of string system) are a combinatorial system because they allow us to freely combine the elements into larger objects (strings here). Let us consider string objects $\vec{s}$, consisting of concatenated numbers "0" and "1", which we shall write in the following form:

$$\vec{s} = (s_1, s_2, ..., s_i, ..., s_N), \qquad s_i \in \{0, 1\}, \qquad 1 \le i \le N. \tag{3.1}$$

For N, the length of the string, a square number should be chosen, at least for the time being.

The restriction for string components s_i to be a binary number is by no means a necessity. It only serves to facilitate our discussion. In principle, real numbers and even strings of symbols may be considered, provided appropriate and natural rules can be found to govern their interaction.

Because we can have any pattern of binary numbers on a string, our molecules are defined implicitly.

Interactions/reactions: Strings react with each other through assuming a second form, provided by a mapping into operators, that allows each string to assume a form able to mathematically operate on another string by matrix multiplication. Thus, the structure of a string determines its operator form and therefore its effect on another string. In this way, interactions are defined implicitly.[1]

For the folding of binary strings into matrix operators we use the following procedure: Strings $\vec{s}$ with N components fold into operators $\mathscr{P}$ represented mathematically by quadratic matrices of size $\sqrt{N} \times \sqrt{N}$ (remember, N should be a square number!), as it is schematically depicted in figure 3.1. At this stage, it is not necessary to fix the exact method of string folding. Any kind of mapping of the topologically one dimensional strings of binary numbers into a two dimensional (quadratic) array of numbers is allowed. Depending on the method of folding, we can expect various transformation paths between strings. Also, the folding must not be deterministic, but can map strings stochastically into different operators.

We now take two strings $\vec{s}_1$ and $\vec{s}_2$ and assume, that one of them is folded into an operator, say $\mathscr{P}_{\vec{s}_1}$. Upon meeting $\vec{s}_2$ another string $\vec{s}_3$ will be generated, according to the rules set up for their interaction (see figure 3.2):

$$\mathscr{P}_{\vec{s}_1} \times \vec{s}_2 \Rightarrow \vec{s}_3, \tag{3.2}$$

---

[1] In nature, the analogous process at work is folding: Based on physical law and on chemical interactions in three-dimensional space, folded secondary and tertiary structures are generated from linear strings of nucleotides and amino-acids.

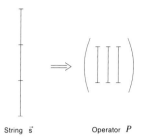

Figure 3.1 A string $\vec{s}$ folds into an operator $\mathscr{P}$

where × symbolizes the interaction.

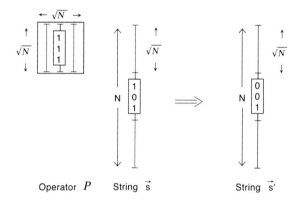

Figure 3.2 An operator $\mathscr{P}$ acts upon a string $\vec{s}$ to produce a new string. A very special case of this interaction is a matrix - vector interaction that can be couched in terms of two matrices interacting. $\vec{s}'$

So, how exactly is that going to work? Consider figure 3.2. We can think of a string $\vec{s}$ as being concatenated from $\sqrt{N}$ fragments with length $\sqrt{N}$ each. The operator $\mathscr{P}$ is able to transform one of these fragments at a time in an imagined semi-local operation. In this way, it moves down the string in steps of size $\sqrt{N}$, until it has finally completed the production of a new string $\vec{s}'$. Then, operator $\mathscr{P}$ unfolds back into its corresponding form as the string corresponding to $\mathscr{P}$ and is released, together with $\vec{s}$ and $\vec{s}'$, into the ensemble of strings, which will be called string population from now on.

More technically, we will use, for the time being, a variant of the mathematical operation of matrix multiplication as the realization of the interaction rule. That will mean calculating the piecewise scalar product of $\mathscr{P}$ with $\vec{s}$:

$$s'_{i+k\sqrt{N}} = \sum_{j=1}^{j=\sqrt{N}} P_{ij} \times s_{j+k\sqrt{N}}, \tag{3.3}$$

with $i = 1, ..., \sqrt{N}$ and $k = 0, ..., \sqrt{N} - 1$, where k counts the steps the operator has taken down the length of the string. This computation, however, will not conserve the binary character of strings, unless we introduce a nonlinearity. Therefore, we shall use the following related computation:

$$s'_{i+k\sqrt{N}} = \sigma \left[\sum_{j=1}^{j=\sqrt{N}} P_{ij} \times s_{j+k\sqrt{N}} - \Theta \right],$$

(3.4)

with $i = 1, ..., \sqrt{N}$, $k = 0, ..., \sqrt{N} - 1$ and $\sigma[x]$ symbolizing the squashing function

$$\sigma[x] = \begin{cases} 1 & \text{for } x \geq 0 \\ 0 & \text{for } x < 0. \end{cases}$$

(3.5)

Θ is an adjustable threshold. Equation 3.3 can be interpreted as a matrix multiplication, whereas eq. 3.4 amounts to a combination of Boolean operations (for the case $\Theta = 0$ discussed later).

This is all very particular to our system, for the sake of concreteness of the example. Other operations are allowed as well, so the following general transformation rule is really all we need:

$$\vec{s}' = \vec{f}(\mathcal{P}, \vec{s}).$$

(3.6)

As in previous examples, we adopt the reaction notation of chemistry and consider an operator $\mathcal{P}$, formed from s_1,[2] which reacts with s_2 to produce a string s_3 under conservation of all reactants. We can write this as a general reaction of the kind

$$s^{(1)} + s^{(2)} \longrightarrow s^{(1)} + s^{(2)} + s^{(3)},$$

(3.7)

where a new string is produced through the interaction of the two strings $s^{(1)}, s^{(2)}$. In the real world, there would need to be some sort of material X added to the left-hand side of this reaction, needed to build the pattern of string s_3, but we do not require this conservation of matter in our current study.

The general reaction of eq. (3.7) can be further examined and reactions can be classified into three different classes: (1) Self-replication reactions, for instance,

$$s^{(1)} + s^{(1)} \longrightarrow s^{(1)} + s^{(1)} + s^{(1)},$$

(3.8)

(2) Replication reactions, for instance

$$s^{(1)} + s^{(2)} \longrightarrow s^{(1)} + s^{(2)} + s^{(1)},$$

(3.9)

where one of the partners replicates with the help of another string, and (3) heterogeneous reactions, like

$$s^{(1)} + s^{(2)} \longrightarrow s^{(1)} + s^{(2)} + s^{(3)},$$

(3.10)

where $s^{(1)} \neq s^{(2)} \neq s^{(3)}$.

Based on the realization of these interaction laws in the matrix chemistry, there is one string, which really destroys a balanced system development if left uncontrolled: The string consisting of "0"s only (called "destructor") is a self-replicating string that is able to replicate with all other strings. So it will quickly gain dominance under any condition. We therefore will have to disable it from reactions by using the special rule that interactions with the "0" string will not produce any string and instead be elastic collisions. However, there is another potentially hazardous string in this system, the one consisting of all "1" entries (called "exploitor"). In addition to being able to replicate itself, it is capable of replicating with many other strings. However, we don't need to prohibit the reaction of this string with others; instead, we can use a softer method to control it.

---

[2] In the future, when discussing string types, we shall omit the vector arrow and only number different string sorts.

Table 3.1 Some low dimensional examples. $\sqrt{N}$: Matrix size in one dimension; N: Length of strings; n_S: Number of different strings, excluding destructor; n_R: Number of possible reactions, excluding self-reactions.

$\sqrt{N}$	1	2	3	4	5	10
N	1	4	9	16	25	100
n_S	1	15	511	65535	$\sim 10^7$	$\sim 10^{30}$
n_R	0	210	$\sim 2.6 \cdot 10^5$	$\sim 4 \cdot 10^9$	$\sim 10^{15}$	$\sim 10^{60}$

Algorithm: To this end, we shall introduce the following general stability criterion for strings: A string may be considered the more stable, the less "1"'s it contains. Its chance to decay, hence, depends on

$$I^{(k)} = \sum_{i=1}^{N} s_i^{(k)}, \qquad k = 1, ..., M. \tag{3.11}$$

$I^{(k)}$ measures the amount of "1"'s in string k and will determine a probability

$$p^{(k)} = (I^{(k)}/N)^n \tag{3.12}$$

with which an encountered string should decay. The parameter n can serve us to tilt probabilities slightly. Note that the exploitor has probability $p = 1$ and must decay upon encounter.

The entire algorithm can now be stated as follows:

```
while ¬terminate() do
    s₁ := draw(P);
    s₂ := draw(P);
    s₄ := draw(P);
    O :=: fold(s₁);
        if ∃ (s₁ + s₂ → s₁ + s₂ + s₃) ∈ R
            then
            P := remove(P, s₄);
            P := insert(P, s₃);
            fi
    od
```

We close this section by giving a short table (table 3.1) showing the impressive amount of possible interactions between strings as we increase their length N. For arbitrary N we have

$$n_s = 2^N - 1 \tag{3.13}$$

strings and

$$n_R = 2^{2N} - 3 \cdot 2^N + 2 \tag{3.14}$$

reactions, excluding reactions with the destructor and self-reactions. The number of potential self-replications is, of course

$$n_{SR} = n_S. \tag{3.15}$$

3.2   The Simplest System, $N = 4$

With these general remarks we have set the stage to discuss the simplest nontrivial system with strings of length $N = 4$.

A still simpler and in fact trivial system is $N = 1$. It has two strings, $s_1 = 0$ and $s_2 = 1$, which coincide with their operators, since there is not much to fold in one-component strings. Using the operations defined by eq. 3.4, we readily observe that both strings are able to self-replicate:[3]

$$0 \cdot 0 = 0 \qquad\qquad 1 \cdot 1 = 1$$

and that the destructor s_1 replicates using s_2:

$$0 \cdot 1 = 0 \qquad\qquad 1 \cdot 0 = 0.$$

Since we have to remove the destructor, nothing is left than one string, prompting us to call this system trivial.

So let us start the examination of the first nontrivial system $N = 4$ by agreeing on a naming convention for strings: We shall use integer numbers that correspond to the binary numbers carried by a string as compact descriptions for a string. Thus, e.g. $\vec{s} = \begin{pmatrix} 1 \\ 0 \\ 1 \\ 0 \end{pmatrix}$ will be called $s^{(5)}$.

Folding strings into 2×2 matrices can take place in various ways. One of these ways will allow us to consider the operations involving scalar products (according to eq. 3.3) with the string acting on itself as ordinary matrix vector multiplications, so we shall call this the canonical folding or folding of the first kind. The arrangement is

$$\vec{s} = \begin{pmatrix} s_1 \\ s_2 \\ s_3 \\ s_4 \end{pmatrix} \rightarrow \mathscr{P}_s = \begin{pmatrix} s_1 & s_2 \\ s_3 & s_4 \end{pmatrix}, \tag{3.16}$$

which can be easily generalized to arbitrary size $\sqrt{N}$.

There are other kinds of folding, for instance the transposed version of (3.16):

$$\mathscr{P}_s' = \begin{pmatrix} s_1 & s_3 \\ s_2 & s_4 \end{pmatrix}, \tag{3.17}$$

but also $\mathscr{P}_s'' = \begin{pmatrix} s_1 & s_2 \\ s_4 & s_3 \end{pmatrix}$ and $\mathscr{P}_s''' = \begin{pmatrix} s_1 & s_4 \\ s_2 & s_3 \end{pmatrix}$. Essentially, the different folding methods amount to a permutation of the correspondence between matrices and strings. They have been listed in more detail in [64]. Table 3.2 gives the resulting operators for the folding method we shall consider here.

We shall now give an example for each class of reactions listed in equations (3.8 to 3.10) in the last section. Using the notation of eq. 3.2 for better readability, we find for the squashed scalar product and the canonical folding, for example

$$\mathscr{P}_{s^{(8)}} \times s^{(8)} \Rightarrow s^{(8)} \tag{3.18}$$

---

[3]It is astonishing that this self-replicating system was around for centuries!

Table 3.2 Results of folding method according to eq. 3.16 for strings $0,...,15$.

	String number						
0	1	2	3	4	5	6	7
$\begin{pmatrix} 0 & 0 \\ 0 & 0 \end{pmatrix}$	$\begin{pmatrix} 1 & 0 \\ 0 & 0 \end{pmatrix}$	$\begin{pmatrix} 0 & 1 \\ 0 & 0 \end{pmatrix}$	$\begin{pmatrix} 1 & 1 \\ 0 & 0 \end{pmatrix}$	$\begin{pmatrix} 0 & 0 \\ 1 & 0 \end{pmatrix}$	$\begin{pmatrix} 1 & 0 \\ 1 & 0 \end{pmatrix}$	$\begin{pmatrix} 0 & 1 \\ 1 & 0 \end{pmatrix}$	$\begin{pmatrix} 1 & 1 \\ 1 & 0 \end{pmatrix}$

	String number						
8	9	10	11	12	13	14	15
$\begin{pmatrix} 0 & 0 \\ 0 & 1 \end{pmatrix}$	$\begin{pmatrix} 1 & 0 \\ 0 & 1 \end{pmatrix}$	$\begin{pmatrix} 0 & 1 \\ 0 & 1 \end{pmatrix}$	$\begin{pmatrix} 1 & 1 \\ 0 & 1 \end{pmatrix}$	$\begin{pmatrix} 0 & 0 \\ 1 & 1 \end{pmatrix}$	$\begin{pmatrix} 1 & 0 \\ 1 & 1 \end{pmatrix}$	$\begin{pmatrix} 0 & 1 \\ 1 & 1 \end{pmatrix}$	$\begin{pmatrix} 1 & 1 \\ 1 & 1 \end{pmatrix}$

Table 3.3 Reactions using computations according to eq. 3.4 with first kind of folding. Four reactions are self-replications, 76 are replications.

Operator	String														
	1	2	3	4	5	6	7	8	9	10	11	12	13	14	15
1	1	0	1	4	5	4	5	0	1	0	1	4	5	4	5
2	0	1	1	0	0	1	1	4	4	5	5	4	4	5	5
3	1	1	1	4	5	5	5	4	5	5	5	4	5	5	5
4	2	0	2	8	10	8	10	0	2	0	2	8	10	8	10
5	3	0	3	12	15	12	15	0	3	0	3	12	15	12	15
6	2	1	3	8	10	9	11	4	6	5	7	12	14	13	15
7	3	1	3	12	15	13	15	4	7	5	7	12	15	13	15
8	0	2	2	0	0	2	2	8	8	10	10	8	8	10	10
9	1	2	3	4	5	6	7	8	9	10	11	12	13	14	15
10	0	3	3	0	0	3	3	12	12	15	15	12	12	15	15
11	1	3	3	4	5	7	7	12	13	15	15	12	13	15	15
12	2	2	2	8	10	10	10	8	10	10	10	8	10	10	10
13	3	2	3	12	15	14	15	8	11	10	11	12	15	14	15
14	2	3	3	8	10	11	11	12	14	15	15	12	14	15	15
15	3	3	3	12	15	15	15	12	15	15	15	12	15	15	15

$$\mathscr{P}_{s^{(1)}} \times s^{(11)} \Rightarrow s^{(1)} \tag{3.19}$$

$$\mathscr{P}_{s^{(1)}} \times s^{(6)} \Rightarrow s^{(4)}, \tag{3.20}$$

where the $\Rightarrow$ sign indicates only the string that was newly produced by the interaction (suppressing the conserved reactants). A list of all reactions for the present case is given in table 3.3. Similar reaction tables can be derived for the other three main folding methods, see [64].

At this point we pause and note again the fact that we are dealing with a system of binary strings, each of which has only four components. This poor material is able to "react" in quite a complicated manner, as a glance at table 3.3 tells us. Hence, already for $N = 4$, we expect rather complicated dynamical behavior. In general, systems of this type exploit the phenomenon of combinatorial explosion shown in table 3.1. Therefore, by studying $N = 4$, we shall have gained only a slight impression of what might be possible in larger systems of this sort. An entire reaction universe is opened, once we consider larger strings with length, say, of order $O(100)$.

In our real world, the smallest virus contains about $3,000$ base pairs, whose elements themselves are certainly able to interact in much more complicated ways than the binary strings we are considering here. This comparison may give us a small hint of how intricate the fundamental mechanisms of life really may be.

3.2.1 Dynamic Behavior

We now want to discuss the dynamic behavior of this system. For later use, we introduce some global quantities that characterize the time development of our system. These are the concentrations x_i of all different string sorts $s^{(i)}$:

$$x_i = m_i / M, \tag{3.21}$$

where m_i is the number of actual appearances of string type $s^{(i)}$ in the soup and M, as before, the constant total number of strings. We have

$$\sum_{i=1}^{n_s} m_i = M \tag{3.22}$$

or

$$\sum_{i=1}^{n_s} x_i = 1. \tag{3.23}$$

Figure 3.3 shows the first 10^6 iterations through the algorithm of two runs with $M = 1,000$ strings. All molecule counts m_i, except for the destructor are shown. Strings are randomly initialized in the population. After some time of growing and falling concentrations, $s^{(1)}$ (left figure) and $s^{(15)}$ (right figure) dominate and finally suppress all other string types in the population. Both $s^{(1)}$ and $s^{(15)}$ are self-replicators and can thus maintain themselves once in a strong position.

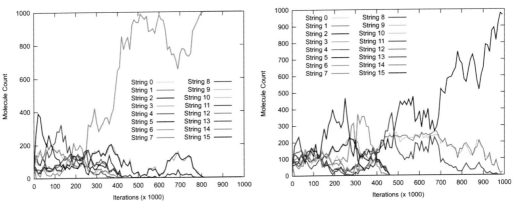

Figure 3.3 Matrix Chemistry System with $N = 4$, first kind of folding. Number of strings: $M = 1,000$. Number of different sorts: $n_s = 15$. Shown are the first 10^6 iterations in two runs. Left: $s^{(1)}$ becomes dominant; Right: $s^{(15)}$ becomes dominant.

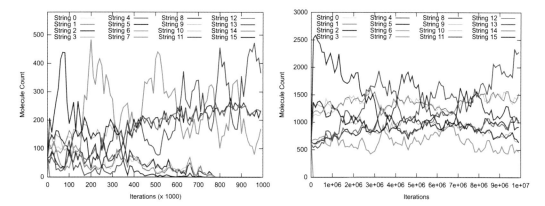

Figure 3.4 Same matrix chemistry system ($N = 4$, first kind of folding, $M = 1,000$, $n_s = 15$). Shown are the first 10^6 (left) and 10^7 (right) iterations of two runs starting from different initial conditions and resulting in two different attractors, made up of a collection of string types. Left: $s^{(1)}, s^{(2)}, s^{(4)}, s^{(8)}$; Right: $s^{(1)}, s^{(2)}, s^{(3)}, s^{(4)}, s^{(5)}, s^{(8)}, s^{(10)}, s^{(12)}, s^{(15)}$.

The system relaxes to a macroscopic attractor state as determined by the self-replicating nature of the winning string, in combination with the initial state of the system and the detailed sequence of reactions that happen during the iterations. Figure 3.4 (left and right) shows two different outcomes of similar runs, again only different in their initial conditions. Here, not one but a number of strings survive, while some others go extinct. While we cannot be sure that with 10^6 or 10^7 iterations we really have reached an attractor, it is interesting to see so many string types survive. The systems under these initial conditions seem to behave more like ecosystems, where some string types support others and are, in exchange, themselves supported.

The fluctuations seem to hide some of the reasons for this behavior. Figure 3.5 shows a system with 100 times more strings, which should allow us to see better what is really going on between all these random events. Here, the behavior of figure 3.4, right, is reproduced. This does happen with most of the initial conditions, so that we can consider the other outcomes as finite size effects of the reaction vessel.

3.2.2 Model Equations

We now want to model matrix chemistry reactions by an ODE approach. The system of coupled differential equations used will be similar to those studied by Eigen and Schuster for the hypercycle [249–251], except of higher order. As in chapter 2, we describe the important aspects of our system by a concentration vector $x_i(t)$ with $i = 1, ..., 15$ under the constraint $0 \le x_i(t) \le 1$.

The rate equations (written with the abbreviation $\dot{x}_i = \frac{dx_i}{dt}$) for this system read:

$$\dot{x}_i(t) = \sum_{j,k \neq i} W_{ijk} x_j(t) x_k(t) - \frac{x_i(t)}{\sum_k x_k(t)} \Phi(t), \tag{3.24}$$

where W_{ijk} are constants from the reaction matrix and $\Phi(t)$ is a flow term used to enact competition between the various string sorts $s^{(i)}$.

Let us discuss in more detail the different contributions to equation (3.24): The first term describes a reaction between operator $\mathscr{P}_{s^{(j)}}$ and string $s^{(k)}$ leading to string $s^{(i)}$. If this reaction

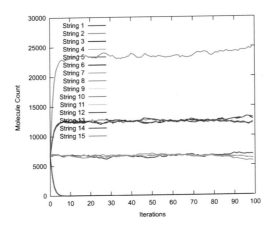

Figure 3.5 System $N = 4$, first kind of folding. Number of strings: $M = 100,000$. Number of different sorts: $n_s = 15$. Shown are the first 100 generations.

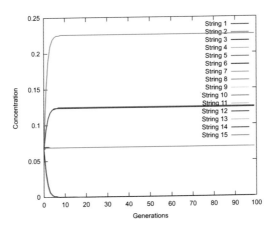

Figure 3.6 Numerical integration of a $N = 4$ system, first kind of folding. Same Initial conditions as in figure 3.5 Number of different sorts: $n_s = 15$. Shown are the first 100 generations.

exists for a given i, j, k, $W_{ijk} = 1$, otherwise we have $W_{ijk} = 0$. The second term, called flow term, assures that the sum of all concentrations is constant over the relaxation process. $\Phi(t)$ is defined as

$$\Phi(t) = \sum_i \dot{x}_i(t). \tag{3.25}$$

By keeping the sum of concentrations constant and normalized $\sum_i x_i = 1$, a strong competition between string sorts is caused. The flow term corresponds to removal operation in the algorithm.

Figure 3.6 shows a run with the couplings W_{ijk} set based on a $N = 4$ binary string system. The simulation of the differential equations (3.24) is started under the same initial conditions as previous stochastic simulations. We can clearly observe that some concentrations merge into the same levels, due to the particular interactions present in the dynamics of the binary string system. Again, we can see that modeling by rate equations shows a deterministic competitive

Table 3.4 Number of replication and self-replication reactions in $N = 9$. Foldings are numbered, with 1 being the folding considered in more detail here.

Folding	Self-replications	Replications
1	14	12,028
2	122	21,310
3	18	11,822
4	94	16,830

system. The comparison between an explicit stochastic simulation (see figure 3.5) and the numerical integration of (3.24) shows very good agreement.

If we want to apply the equations to cases involving larger strings, however, we will have a harder time. For $N = 9$, W_{ijk} would have 1.3×10^8 components due to the 511 string types ($2^9 - 1$) participating, and we will need a proper data structure to handle this case.

Before we move on to that more complex system, one remark is in order in regard to an alternative formulation of the rate equations. Instead of representing all reactions in the reaction matrix $W_{ijk} \forall i, j, k$ one can divide the reactions according to type into self-replications, replication and heterogeneous reactions. This will result in the following equivalent equations:

$$\dot{x}_i(t) = [B_i x_i(t) + \sum_{k \neq i} C_{ik} x_k(t)] x_i(t) + \sum_{j,k \neq i} W_{ijk} x_j(t) x_k(t) - \frac{x_i(t)}{\sum_k x_k(t)} \Phi(t). \qquad (3.26)$$

The parameter B_i is 1 for a string type i capable of self-replication, otherwise $B_i = 0$ and the term does not contribute. If string type i is able to replicate with the help of string type k then $C_{ik} = 1$, otherwise $C_{ik} = 0$.

This concludes the section on $N = 4$. As we move to longer strings, more complicated systems will emerge and soon the combinatorial explosion will prohibit us from completely investigating the state space of strings. This was the reason why we studied the $N = 4$ system in more detail.

3.3 The System $N = 9$

The system $N = 9$ is much more complicated than $N = 4$. Due to combinatorics we now face $n_S = 511$ strings with a total of $n_R = 261,121$ reactions.

We first have to generalize the folding methods given in the last section to the present case. Corresponding to (3.16) we have:

$$\vec{s} = \begin{pmatrix} s_1 \\ s_2 \\ \vdots \\ s_9 \end{pmatrix} \rightarrow \mathscr{P}_s = \begin{pmatrix} s_1 & s_2 & s_3 \\ s_4 & s_5 & s_6 \\ s_7 & s_8 & s_9 \end{pmatrix}. \qquad (3.27)$$

The alternative folding methods can be generalized as well, which has been done in [65]. Here we want to only mention that there is indeed a difference between these foldings in terms of the number of self-replicating and replicating string types available in each reaction universe. Table 3.4 lists the number of resulting replication and self-replication reactions.

Table 3.5 Reactions for $N = 9$ with first kind of folding. Reactions for $N = 9$ with 1st kind of folding. Selected are reactions between operator/string pairs $165, ..., 179$.

Op.	String														
	165	166	167	168	169	170	171	172	173	174	175	176	177	178	179
165	283	287	287	280	281	284	285	283	283	287	287	312	313	316	317
166	347	351	351	344	344	349	349	347	347	351	351	376	376	381	381
167	347	351	351	344	345	349	349	347	347	351	351	376	377	381	381
168	274	278	278	272	274	276	278	274	274	278	278	304	306	308	310
169	275	278	279	280	283	284	287	282	283	286	287	304	307	308	311
170	338	343	343	336	338	341	343	338	338	343	343	376	378	381	383
171	339	343	343	344	347	349	351	346	347	351	351	376	379	381	383
172	283	287	287	280	282	284	286	283	283	287	287	312	314	316	318
173	283	287	287	280	283	284	287	283	283	287	287	312	315	316	319
174	347	351	351	344	346	349	351	347	347	351	351	376	378	381	383
175	347	351	351	344	347	349	351	347	347	351	351	376	379	381	383
176	402	406	406	400	400	406	406	402	402	406	406	432	432	438	438
177	403	406	407	408	409	414	415	410	411	414	415	432	433	438	439
178	466	471	471	464	464	471	471	466	466	471	471	504	504	511	511
179	467	471	471	472	473	479	479	474	475	479	479	504	505	511	511

Table 3.6 Sorts and their concentrations of figure 3.7's simulation, taken at generation 60. Only the nine string sorts with highest concentration are shown. Among them are newly appearing string sorts $s^{(16)}, s^{(18)}, s^{(24)}, s^{(27)}$.

Sort	1	3	9	27	8	24	2	18	16
Concentration	2.4 %	2.1 %	2.0 %	1.7 %	0.5 %	0.5 %	0.4 %	0.4 %	0.1 %

Since it is difficult to visualize all possible reactions, we can only present a qualitative picture by focusing on part of the reaction matrix. Table 3.3 shows a randomly selected part of all reactions. The size of this part is the same as the total table for $N = 4$. We can see that in this region only a limited number of strings is formed from the interactions, so there is a kind of locality in sequence space.

In dynamical terms, the $N = 9$ system is much more complicated, too. Figure 3.7 shows one stochastic molecular simulation. We have selected concentrations of some 15 initial string sorts plus those generated by their interactions. Table 3.6 shows string sorts dominating at generation 60, among them newly generated sorts, like $s^{(27)}$. This points to the constructive character of these systems: If there are enough string sorts previously never present new types appear easily during a run. Other string sorts will disappear during a run, due to the limited size of the population. With a total string population of $M = 1,000$ it is no surprise that many sorts disappear in the course of a simulation due to the competitive process.

One could argue that undersampling a system by restricting the population so severely would lead to false conclusions, since not all string sorts have a chance to survive in sufficient numbers

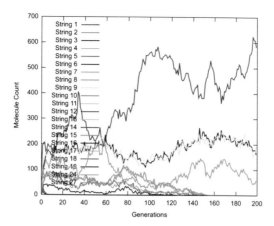

Figure 3.7 System $N = 9$. Number of strings: $M = 1,000$. Number of different sorts: $n_s = 511$. Shown are the first 100 generations of a run started with a nearly equal initial distribution of the first 15 strings $s^{(1)}...s^{(15)}$. Additional string sorts that start appearing frequently are also shown.

to gain resistance against random fluctuations. We have, however, to face this situations sooner or later and there is no way to increase the population by factors of 10^{10} or more as it would be required for larger systems. Therefore, we take the opportunity here, while things are still simple, to consider the implications of a changing mix of strings present. Note, the situation is still not severe, as the number of string types is of the order of the number of strings (511 vs. 1,000).

We can now ask the following question: What will happen during a deterministic simulation of an $N = 9$ system using the previously introduced rate equations, adapted to the particular case of folding? We would start from the same initial distribution of strings. A full-blown simulation of the system is already so tedious that we have to resort to a number of tricks to get the numerical integration running at acceptable speed. What we shall do is to first consider which string types will be present in the system. To this end, we have to look at the full reaction table and extract the subset of strings produced by string sorts present at the outset. We will learn in chapter 12 why this procedure is so relevant to study any type of organization.

For instance, as table 3.7 shows, the strings produced by reactions between $s^{(10)}$ and any of the string types $s^{(1)},...s^{(15)}$ are $s^{(1)}, s^{(2)}, s^{(3)}$ of the present types, but then also $s^{(16)}, s^{(17)}, s^{(17)} s^{(19)}$, which we hitherto have not encountered. So we have to extend the set of strings participating in the reactions by these, and reconsider the enlarged reaction matrix to see whether further strings need to be added. And indeed, further string types $s^{(24)}, s^{(25)}, s^{(26)}, s^{(27)}$ have to be added. This examination has to be done for each string type in the system, until we do not find any more new strings and reaction products. In our case, all string types up to $s^{(27)}$ have to be added, with the exception of $s^{(20)}, s^{(21)}, s^{(22)}, s^{(23)}$. The system of strings is now closed (we shall see in chapter 12 what we mean by this expression in mathematical terms), and we can run the deterministic simulation.

Figure 3.8 shows a run with the same 15 string types initially present in figure 3.7. From a comparison of these figures we can see that there is only a very coarse similarity, due to the sampling error in small populations. For example, we can observe that $s^{(1)}$ dominates in both simulations, and that $s^{(3)}$ and $s^{(9)}$ develop with the same strength. Other than that, there are too many fluctuations in the stochastic molecular simulation to make sure statements. However, it

Table 3.7 Extract from reaction table of a 9-bit matrix-multiplication chemistry. Again, "0" indicates an elastic collision.

	1	2	3	4	5	6	7	8	9	10	11	12	13	14	15	16	17	18	19	20	21	22	23	24	25	26	27
1	1	0	1	0	1	0	1	8	9	8	9	8	9	8	9	0	1	0	1	0	1	0	1	8	9	8	9
2	0	1	1	0	0	1	1	0	0	1	1	0	0	1	1	8	8	9	9	8	8	9	9	8	8	9	9
3	1	1	1	0	1	1	1	8	9	9	9	8	9	9	9	8	9	9	9	8	9	9	9	8	9	9	9
4	0	0	0	1	1	1	1	0	0	0	0	1	1	1	1	0	0	0	0	1	1	1	1	0	0	0	0
5	1	0	1	1	1	1	1	8	9	8	9	9	9	9	9	0	1	0	1	1	1	1	1	8	9	8	9
6	0	1	1	1	1	1	1	0	0	1	1	1	1	1	1	8	8	9	9	9	9	9	9	8	8	9	9
7	1	1	1	1	1	1	1	8	9	9	9	9	9	9	9	8	9	9	9	9	9	9	9	8	9	9	9
8	2	0	2	0	2	0	2	16	18	16	18	16	18	16	18	0	2	0	2	0	2	0	2	16	18	16	18
9	3	0	3	0	3	0	3	24	27	24	27	24	27	24	27	0	3	0	3	0	3	0	3	24	27	24	27
10	2	1	3	0	2	1	3	16	18	17	19	16	18	17	19	8	10	9	11	8	10	9	11	24	26	25	27
11	3	1	3	0	3	1	3	24	27	25	27	24	27	25	27	8	11	9	11	8	11	9	11	24	27	25	27
12	2	0	2	1	3	1	3	16	18	16	18	17	19	17	19	0	2	0	2	1	3	1	3	16	18	16	18
13	3	0	3	1	3	1	3	24	27	24	27	25	27	25	27	0	3	0	3	1	3	1	3	24	27	24	27
14	2	1	3	1	3	1	3	16	18	17	19	17	19	17	19	8	10	9	11	9	11	9	11	24	26	25	27
15	3	1	3	1	3	1	3	24	27	25	27	25	27	25	27	8	11	9	11	9	11	9	11	24	27	25	27
16	0	2	2	0	0	2	2	0	0	2	2	0	0	2	2	16	16	18	18	16	16	18	18	16	16	18	18
17	1	2	3	0	1	2	3	8	9	10	11	8	9	10	11	16	17	18	19	16	17	18	19	24	25	26	27
18	0	3	3	0	0	3	3	0	0	3	3	0	0	3	3	24	24	27	27	24	24	27	27	24	24	27	27
19	1	3	3	0	1	3	3	8	9	11	11	8	9	11	11	24	25	27	27	24	25	27	27	24	25	27	27
20	0	2	2	1	1	3	3	0	0	2	2	1	1	3	3	16	16	18	18	17	17	19	19	16	16	18	18
21	1	2	3	1	1	3	3	8	9	10	11	9	9	11	11	16	17	18	19	17	17	19	19	24	25	26	27
22	0	3	3	1	1	3	3	0	0	3	3	1	1	3	3	24	24	27	27	25	25	27	27	24	24	27	27
23	1	3	3	1	1	3	3	8	9	11	11	9	9	11	11	24	25	27	27	25	25	27	27	24	25	27	27
24	2	2	2	0	2	2	2	16	18	18	18	16	18	18	18	16	18	18	18	16	18	18	18	16	18	18	18
25	3	2	3	0	3	2	3	24	27	26	27	24	27	26	27	16	19	18	19	16	19	18	19	24	27	26	27
26	2	3	3	0	2	3	3	16	18	19	19	16	18	19	19	24	26	27	27	24	26	27	27	24	26	27	27
27	3	3	3	0	3	3	3	24	27	27	27	24	27	27	27	24	27	27	27	24	27	27	27	24	27	27	27

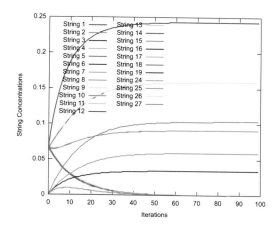

Figure 3.8 Deterministic rate equation simulation of system $N = 9$. Initial string distribution of the first 15 string types, as in figure 3.7. Additional string types as produced through their reactions are also depicted.

is easily seen from the simulation using differential rate equations that new string types appear that are able to survive the competition and remain strong in the population.

In this example, the constructive character of the system has forced us to first analyze the reaction matrix and examine the full set of products accessible to the particular initial condition we have chosen. What, however, happens, if we don't have access to the reaction matrix (in this case of dimension 511×511) in other, possibly even larger systems? In such a case we enter the truly constructive systems, which require a different treatment. This will be the subject of chapter 15.

As a final step in a discussion of the matrix chemistry, we will discuss N-dimensional systems in the next section.

3.4 Systems with Larger N

A further generalization of the folding method to arbitrary $N, \sqrt{N} \in \mathcal{N}$ reads like the following:

$$\vec{s} = \begin{pmatrix} s_1 \\ s_2 \\ \vdots \\ s_N \end{pmatrix} \rightarrow \mathcal{P}_s = \begin{pmatrix} s_1 & \cdots & s_{\sqrt{N}} \\ s_{\sqrt{N}+1} & \cdots & s_{2 \cdot \sqrt{N}} \\ \vdots & \ddots & \vdots \\ s_{(\sqrt{N}-1) \cdot \sqrt{N}+1} & \cdots & s_N \end{pmatrix}. \tag{3.28}$$

If we recall, that in systems with longer strings the number of different sorts grows exponentially, we can imagine that before N reaches 100 the naming of strings becomes impossible. Therefore, for larger systems we have to completely rely on the binary notation for strings.

Another more important change is required in connection with the rapid increase in the number of string sorts: The introduction of a mutation rate becomes unavoidable, because no population of data strings in a computer, whatever size it may assume, can totally explore the state space of a system of this type any more. A local mutation will allow us to explore — at least partially — the neighborhood of populated areas (in terms of string sorts). We can clearly see that if we consider that for $N = 100$ there are approximately $n_S = 1.3 \times 10^{30}$ different string sorts.

So while an exploration of this "sequence" space will not be possible in parallel, i.e., with a large population, it may be possible serially, i.e., by introducing the potential of transformations between all string sorts. This is precisely what a mutation achieves. Otherwise, a system would be closed presumably to those sorts which were achievable in reactions between those string sorts that were present at the outset of the run and their respective products. By the addition of a mutational transformation, new things become possible, at least if we wait long enough. The system will show history dependence and the attractor behavior (settling in) will be interrupted by changes due to mutations allowing new string types to appear.

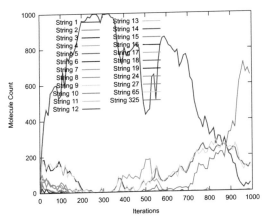

Figure 3.9 System $N = 9$. Number of strings: $M = 1,000$. First $1,000$ generations of a run started with a nearly equal initial distribution of the first 15 strings $s^{(1)}...s^{(15)}$. Mutation probability $p_m = 10^{-3}$. After diversity reduces quickly to the point where self-replicating string type $s^{(1)}$ dominates the population completely, mutations are able to bring about new string types, e.g. $s^{(65)}, s^{(325)}$. There is a ground level of fluctuations always present in this system.

For the introduction of a mutation exist various options, two of which we will discuss here. The first option is to allow for certain errors to appear in the production process of a new string. The assumption would be that the interaction between operator and string cannot be perfectly reproducible, but only to a high degree. Therefore, each newly created component of a string has a very small probability of being erroneous. This probability will be called the elementary error probability p_e. Consequently, in a string with N components, the probability for at least one error in a component is

$$P_1 = N p_e. \tag{3.29}$$

We have to require that $p_e < \frac{1}{N}$ (still better $p_e << \frac{1}{N}$), otherwise the system experiences an error catastrophe. k errors shall occur with a probability of approximately

$$P_k = (N p_e)^k. \tag{3.30}$$

Whereas this type of mutation does occur in nature's RNA or DNA systems, we perceive it to be quite artificial in the binary string system. Since we have defined the production of a new string as a mathematical operation, there is no reason that would force us to attach a mutation just to this operation. After all, mathematics is perfect!

The second option in using a mutation is an occasional change hitting each string with a probability depending on its size. We define p_m to be the probability that one element of a string changes to a randomly selected symbol from the alphabet, here "0" to "1" or vice versa. Since it may hit each element, this is again a size-dependent mutation, and the probability that at least one error occurs in a string is

$$P_1 = N p_m, \tag{3.31}$$

with the provision that $p_m << \frac{1}{N}$. In this way, we add a further step to the general algorithm explained in section 3.1. Note that this mutation depends on the concentration of strings in the reaction vessel and will be more frequent for sorts with larger concentration levels. That is to say, a more successful string sort will spawn more mutations. In nature, at least one instance of this mutation occurs, too, as mutation caused by natural radiation.

Figure 3.9 shows a $N = 9$ system with a mutation probability of $p_m = 0.001$ starting from an initial condition with the first 15 string types present. We can see that after some time, approximately 200 to 300 iterations, the diversity is substantially diminished to the point that one string $(s^{(1)})$ dominates the entire population. But then mutations on that string type produce new and well reproducing string types that are able to win the competition against the originally dominating $s^{(1)}$.

3.5 Summary

So far, we have only discussed string systems of a length that is a quadratic number. This had to do with the type of interaction we chose between strings and operators. However, in no way are we restricted to such systems. Arbitrary length of strings is also possible if we provide for rules that allow interaction of strings and operators of different sizes [66]. It turns out that in such systems a tendency toward changing the length of strings is prevalent. Naturally, that will lead to an overall tendency to increase the average length of strings (as there is a bias towards extension).

In many ways, the matrix chemistry is but one example of a class of artificial chemistries. We have chosen this example merely for illustration purposes. It is deceptively simple in its construction, yet it shows the richness of more complex chemistries. At the same time, it remains a source of inspiration, and we are sure that some of its aspects haven't been uncovered yet.

COMPUTING CHEMICAL REACTIONS

I n chapter 2, the elements of an artificial chemistry were introduced: the set of molecules S, the set of reaction rules R, and the reactor algorithm A. A few reaction algorithms were then briefly described or just enumerated. In this chapter we will examine reactor algorithms in more detail. Before going any deeper in the realm of artificial chemistries, a look at the underlying reality and its description in real chemistry is in order.

Chemical reactions are fundamentally interactions of material objects. Chemistry has, for centuries, heavily relied on the experimental scientific method to learn from chemical reality in our world. Besides the experimental mode, there always was room for theoretical and later mathematical considerations of what atoms could do and how one might best describe them in formal models.

More recently, however, the new aspect of computation has taken hold in chemistry and transformed it into a fully quantitative science. While many chemists still prefer to stand in the laboratory and make experiments, a growing number of numerically inclined chemists sit in front of a computer screen, designing numerical experiments or simulations that should illuminate certain behavior of material chemistry. The field of "computational chemistry" [429] is broad, but one of its interests is the simulation of the details of chemical reactions.

Here we shall focus mainly on stochastic reaction simulation algorithms, first in well-stirred vessels and then in spatial and compartmentalized systems. It turns out that there are sophisticated methods to simulate reactions in stochastic environments. These algorithms might well be applied when simulating artificial chemistries.

4.1 From Macroscopic to Microscopic Chemical Dynamics

In chapters 2 and 3 we have already seen two different ways to describe (artificial) chemical reaction systems. One method is using deterministic rate equations; the other we have termed explicit molecular simulation. We have seen that under appropriate conditions the descriptions at both levels coincide with each other, which demonstrates that the approximation made in those particular examples was valid. However, the different descriptions of how chemical dynamics happens have been intuitive in those examples, and it is time to put them on a firm formal basis.

For many decades, chemists have developed descriptions of reaction kinetics. Because atoms are extremely small, chemistry worked for a long time with the assumption that a large number of molecules is reacting in any given chemical reaction system.[1] This assumption has led to the following consequences: (1) the number of molecules in a reaction can be approximated by a continuous variable (describing the concentration $[X] = X/V$ of the substance with X molecules in a given volume V); and (2) this continuous variable changes deterministically over time, leading to reaction rate equations for chemical species $X_1, X_2, \ldots$

$$\frac{d[X_1]}{dt} = f_1([X_1], [X_2], \ldots) \tag{4.1}$$

$$\frac{d[X_2]}{dt} = f_2([X_1], [X_2], \ldots). \tag{4.2}$$

These ordinary differential equations can be integrated to solve for the resulting concentrations $[X_1](t), [X_2](t), \ldots$, as already discussed briefly in chapter 2. Sometimes it is assumed that the concentrations have assumed equilibrium values and do not change any more, a condition summarized for one variable by

$$\frac{d[X_1]}{dt} = 0 = f_1([X_1], [X_2], \ldots), \tag{4.3}$$

which leads to algebraic equations that can be solved.

However, in small systems, the model assumption of continuous quantities (concentrations) is not certain any more, and we have to take into account molecule counts and consider that under some circumstances some of the reactions cannot even take place since reaction partners might not be available. This might happen, for instance, if we consider small (cellular) compartments, or any kind of systems with highly complex macromolecules that come in small numbers. The assumption of continuous concentrations breaks down at the latest when the concentration falls to a value that would correspond to less than one molecule in the volume under consideration, but often much earlier. In these cases, we have to adopt alternatives to the reaction rate equations from above for describing the behavior of a system. Notably, things will not any longer be deterministic as the random collision of molecules might drive a system into one or the other direction, depending on when it happens and what molecules result from the reaction. A stochastic treatment is necessary.

This is just one step in an entire chain of refinements that can be considered for a chemical system. The next step after a stochastic approach to molecular reactions would be the spatial-stochastic modeling and simulation of chemical reaction systems. Here, molecules diffuse either in discrete/fixed-length steps (also called "on-lattice") or as point particles (called "off-lattice" without volume exclusion) or as billiard balls (called "off-lattice" with volume exclusion).

The next closer look would be to consider single molecules as moving under the laws of Newtonian mechanics. This would then lead to a consideration of speed, momentum, or volume of individual molecules and how those characteristics would impact on possible reaction collisions between molecules. This is part of the field known as molecular dynamics (MD) simulation. It is much more ambitious and realistic than the stochastic simulation mentioned previously, but

---

[1]See, for instance, the Avogadro constant for the number of molecules of a particular sort in a mol of the species: $N_A \approx 6 \times 10^{23} \, mol^{-1}$, a huge number.

feasible only for simple molecules. For large molecules that need to be simulated over extended periods of time the method cannot produce practical results at the current time.

Finally, there is the possibility of considering atoms and molecules as quantum systems and solve for their quantum states. This adds another level of complexity, however, and the approach becomes intractable for what we want to achieve here. But we should always keep in mind that the stochastic treatment discussed here has already stripped away a number of layers of complexity from the treatment of molecules, layers that, for an exact treatment, would be necessary in principle, although they might not be feasibly applied to produce results.

4.2 Stochastic Reaction Algorithms

Gillespie [328] has introduced the now popular version of stochastic chemical kinetics as a way to describe the evolution in time of a chemical system where the discrete and stochastic nature of interactions is taken into account.[2] The simplest approach of this kind is for a well-stirred reactor of chemicals. We will show now how this approach can capture the essentials of a stochastic description and still be practically executable on a computer.[3]

Let us consider n types of chemicals $\{S_1, \ldots, S_n\}$ that interact with a set of m possible reactions $\{R_1, \ldots, R_m\}$. In stochastic chemical kinetics one often speaks of the m different reaction channels, each one of which could "fire," meaning that a reaction takes place in this channel. Further simplifying assumptions are that the volume V of the reactions does not change and that the temperature is constant, i.e. the system is in thermal equilibrium. $X_i(t)$ is the number of molecules of type S_i in the volume at time t. We would like to know the state vector $\mathbf{X}(t) \equiv (X_1(t), \ldots, X_n(t))$, given that the system was in state $\mathbf{X}(t_0) = \mathbf{x_0}$ at an initial time t_0.

We have now characterized the state of our system. How can state changes be described? Two quantities will suffice to characterize the state changes embodied in the m reaction channels. Each reaction R_j, $j = 1, \cdots m$ will be assigned a state change vector $\mathbf{v}_j \equiv (v_{1j}, \ldots v_{Nj})$, with v_{ij} being the change in species S_i brought about through a reaction in channel R_j. If the system is in state $\mathbf{x}$ prior to the reaction, then it will be in state $\mathbf{x} + \mathbf{v}_j$ after the reaction. The other quantity assigned to R_j is a so-called reaction propensity a_j which quantifies the probability p

$$p = a_j(\mathbf{x})dt \tag{4.4}$$

that one reaction R_j will happen in volume V in the time interval $[t, t + dt)$, given that $\mathbf{X}(t) = \mathbf{x}$, in other words

$$\mathbf{X}(t + dt) = \mathbf{x} + \mathbf{v}_j \text{ with probability } p. \tag{4.5}$$

This definition of the probability of a reaction can be derived from theoretical considerations, and is called the fundamental premise of stochastic chemical kinetics. Everything follows from here using the laws of probability. So we may ask: What is the probability to find the system in state $\mathbf{x}$ at time t when it previously was in state $\mathbf{x_0}$ at time t_0? We therefore want to calculate

$$P(\mathbf{x}, t \mid \mathbf{x_0}, t_0) \equiv \text{Prob}\{\mathbf{X}(t) = \mathbf{x}, \text{ given } \mathbf{X}(t_0) = \mathbf{x_0}\}. \tag{4.6}$$

---

[2]There were earlier proposals for similar algorithms and multiple versions of implementation exist.

[3]It should be noted that this type of stochasticity can be considered as intrinsic noise (when does the next reaction happen and which reaction is that) and is distinguished from extrinsic noise. The latter is due to external factors of the chemical reaction system such as variation in temperature or volume over time. These factors can be introduced into the SSA framework by using time-dependent reaction rates.

This requires solving the chemical master equation (CME) which can be set up as the result of using the probability p from equation 4.4. The CME captures the change in probabilities of states, and reads:

$$\frac{\partial P(\mathbf{x}, t \mid \mathbf{x}_0, t_0)}{\partial t} = \sum_{j=1}^{M} [a_j(\mathbf{x} - \mathbf{v}_j) P(\mathbf{x} - \mathbf{v}_j, t \mid \mathbf{x}_0, t_0) - a_j(\mathbf{x}) P(\mathbf{x}, t \mid \mathbf{x}_0, t_0)]. \quad (4.7)$$

The first term in the parenthesis symbolizes the change from a previous state into $\mathbf{x}(t)$, whereas the second term is the flow out of state $\mathbf{x}(t)$. Readers more interested in the details of a derivation of this master equation should consult [330, 879]. Unfortunately, the CME is not easy to solve, as it is a set of coupled partial differential equations (PDEs). It can, however, be approached from another angle. Rather than trying to solve the equation by integrating for $P(\cdot \mid \cdot)$ we can numerically construct realizations of state trajectories, and then average over these trajectories to estimate $P(\cdot \mid \cdot)$.

We now turn to an example of how this works. Suppose we have the following reaction channels

$$R_1 : A + B \xrightarrow{k_1} C \quad (4.8)$$

$$R_2 : B + C \xrightarrow{k_2} D \quad (4.9)$$

$$R_3 : D + E \xrightarrow{k_3} E + F \quad (4.10)$$

$$R_4 : F \xrightarrow{k_4} D + G \quad (4.11)$$

$$R_5 : E + G \xrightarrow{k_5} A. \quad (4.12)$$

We can use these to simulate a specific stochastic reaction sequence as one realization of events unfolding with the preceding reaction channels.[4] Propensities a_j for these reactions now read

$$a_j = k_j \times [\#educt1] \times [\#educt2], \quad (4.13)$$

with the exception of the equation with only one educt.

There are two methods to produce such a sequence of states, either an explicit chemical reaction simulation or stochastic chemical kinetics. In the former case we use single molecules and draw as many as we need for the chemical reactions to take place (in the preceding example, drawing a single molecule of A would not be sufficient since no reaction condition (left-hand side) is fulfilled, but drawing a molecule F would be sufficient). This approach is based on collision theory originally formulated in 1916 [859], but simplified for the simulation. It is the approach many artificial chemistry simulations are using and which we have discussed in earlier chapters.

The latter approach is stochastic chemical kinetics, another special case of discrete event simulation methods, also called kinetic Monte Carlo methods [328, 500, 947]. In this method, the time to a next event is drawn from an appropriate distribution (usually exponential), and the reaction channel is selected according to the probability of reactions. We will discuss it here,

---

[4]Care should always be taken when using kinetic reaction constants. If we are in the deterministic realm and use concentrations of molecular species (concentrations are particle counts over volume), kinetic rate constants are "macroscopic" quantities and are independent of volume. However, in the case discussed here we use a discretized system in which we consider molecule counts as our variables. In this case, rate constants are "mesoscopic" in scale and depend on volume. Thus, volume dependence has been shifted between variables and constants.

since it is extremely popular in chemistry and systems biology. It would be naive, however, to assume that one run of such a stochastic system can provide a reliable picture of what kind of reactions happen at what frequency. Therefore, we have to repeat this calculation multiple times to produce an entire set of runs, the results of which need to be averaged to provide an estimate for P.

How does this approach to stochastic simulation work? In order to generate simulated trajectories $\mathbf{X}(t)$ we have to consider another probability function, $p(\tau, j \mid \mathbf{x}, t)$, which is defined as the probability that the next reaction will occur in the time interval $[t + \tau, t + \tau + d\tau)$ and will be R_j, if the system is in state $\mathbf{X}(t) = \mathbf{x}$. This probability can be derived exactly [328–330] and reads:

$$p(\tau, j \mid \mathbf{x}, t) = a_j(\mathbf{x}) e^{-a_0(\mathbf{x})\tau} \tag{4.14}$$

with

$$a_0(\mathbf{x}) \equiv \sum_{k=1}^{M} a_k(\mathbf{x}). \tag{4.15}$$

This means that τ is a real random variable which is exponentially distributed with mean and standard deviation $1/a_0(\mathbf{x})$ whereas j is an integer random variable with probabilities $a_j(\mathbf{x})/a_0(\mathbf{x})$.

Generating these random variables is the realm of Monte Carlo calculations. Monte Carlo methods in general are used to produce, by computational algorithms, random samples of given distributions [574]. By using these samples, results can be produced with otherwise deterministic algorithms. For our purposes here, there are different Monte Carlo procedures to generate samples according to the distributions we need. The simplest is called the "direct method," which runs like this: Two random numbers r_1, r_2, drawn uniformly from the unit interval $(0, 1)$ are used to determine τ, j:

$$\tau = \frac{1}{a_0(\mathbf{x})} ln\left(\frac{1}{r_1}\right) \tag{4.16}$$

and

$$j = \text{smallest integer number fulfilling} \sum_{k=1}^{j} a_k(\mathbf{x}) > r_2 a_0(\mathbf{x}). \tag{4.17}$$

The direct method of Gillespie's stochastic simulation algorithm (SSA)then can be summarized as follows:

```
while ¬terminate() do
determine a_j(x), a_0(x);
r_1 := uniform(0, 1);
r_2 := uniform(0, 1);
determine τ, j;
t := t + τ;
x := x + v_j;
od
```

with `uniform(.)` returning a random number in the interval, and the determination of quantities according to equations 4.13 — 4.17.

An example might help to understand how this works. For example, if we start the reaction system of equations 4.8 — 4.12 with the particular initial state $\mathbf{x}_0 = (7, 6, 5, 4, 3, 2, 1)$ counting the

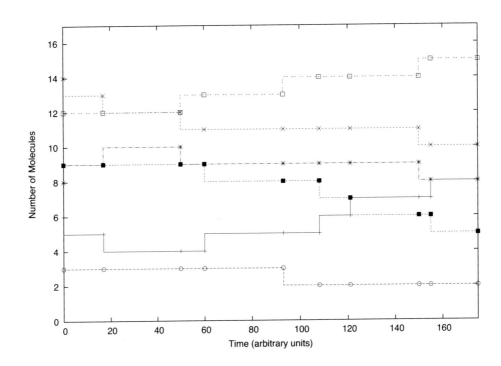

Figure 4.1 Graphical representation of a sample trajectory. Molecule counts (each type with a different marker) are observed over time. Markers are X: A, circle: B, triangle: C, square: D, triangle: E, star: F, line: G. Adapted from [324].

Table 4.1 State representation; the reaction row indicates which reaction has occurred. Adapted from [324].

Time	0	17	50	60	93	108	121	150	155	175
#A	6	5	4	4	5	5	6	7	7	8
#B	14	13	12	11	11	11	11	11	10	10
#C	8	9	10	9	9	9	9	9	8	8
#D	12	12	12	13	13	14	14	14	15	15
#E	9	9	9	9	8	8	7	6	6	5
#F	3	3	3	3	3	2	2	2	2	2
#G	5	5	5	5	4	5	4	3	3	2
Reaction	—	1	1	2	5	4	5	5	2	5

number of molecules $\{A, B, C, D, E, F, G\}$ we might see the following dynamics in a sample run (see figure 4.1 and table 4.1).

This is one realization of the reaction trajectory only. Many realizations of trajectories need to be calculated to get an accurate picture of the underlying chemical dynamics, and therefore simple and efficient implementations of the method are important. With the direct method, we can see that the update to τ will make time a quantity that sometimes changes quickly and sometimes more slowly, rather than ticking away always at the same rate. This can be an advantage of the method under certain circumstances, but it also can be a drag on simulation speed among others.

Note that while the system doesn't have to explicitly wait until τ, the SSA nevertheless simulates the exact timing of reaction events. This is different from the reaction rate equation based on the assumption of infinitesimal dt and continuous concentrations, which is an approximation.

A number of alternative realizations of the stochastic simulation algorithm have been proposed, like the first reaction method also suggested by Gillespie [328]. As well, improvements to the algorithm, like better data structures allowing efficient execution, see e.g., the next reaction method (NRM) by Gibson and Bruck [324], have been examined. We have to refer the reader to the literature to find out more about these variants. One improvement, however, of the classical SSA needs more explanation, since it spawned a number of further works itself, the tau leaping algorithm.

The tau leaping algorithm [331] is not an exact stochastic simulation algorithm but an approximation. It starts from the assumption that during the time $[t, t+\tau)$, the reaction propensities a_j stay constant for all j. This can only be an approximation, of course, since with every reaction the molecule counts will change, which has a bearing on the reaction propensities. But with the assumption that they don't change significantly, a reaction channel R_j will fire with a probability determined from a Poisson distribution $\mathscr{P}(m_j)$ with mean (and variance) $m_j = a_j(\mathbf{x})\tau$. This allows an update of all reaction channels at the same time, leaping from t to $t+\tau$. Thus, we can update the reaction system much faster by taking

$$\mathbf{X}(t+\tau) \approx \mathbf{x} + \sum_{j=1}^{M} \mathscr{P}_j(a_j(\mathbf{x})\tau)\mathbf{v}_j \tag{4.18}$$

with $\mathbf{x} = \mathbf{X}(t)$. This is the basic equation for updating with the tau leaping algorithm.

By further assuming that τ is sufficiently large so that each reaction channel fires repeatedly, i.e.,

$$a_j(\mathbf{x})\tau \gg 1 \text{ for all } j = 1, \ldots, M, \tag{4.19}$$

we can replace the Poisson distribution by a Gaussian distribution $\mathscr{N}_j$, which is easier to handle:

$$\mathbf{X}(t+\tau) \approx \mathbf{x} + \sum_{j=1}^{M} \mathscr{N}_j(a_j(\mathbf{x})\tau, a_j(\mathbf{x})\tau)\mathbf{v}_j. \tag{4.20}$$

Using well known relations for the normal distribution this expression can be further rewritten as :

$$\mathbf{X}(t+\tau) \approx \mathbf{x} + \sum_{j=1}^{M} \mathbf{v}_j a_j(\mathbf{x})\tau + \sum_{j=1}^{M} \mathbf{v}_j \sqrt{a_j(\mathbf{x})} \mathscr{N}_j(0,1)\sqrt{\tau}, \tag{4.21}$$

where $\mathcal{N}_j(0,1)$ are M statistically independent normal distributions with mean 0 and variance 1. If we formulate this equation as a differential equation, we have arrived at the chemical Langevin equation.

It is clear that τ now has a lower and an upper bound (using equation 4.19 and the condition that the propensities don't change significantly). But how exactly to choose τ? A number of different procedures have been proposed [155,331,332], with the latest of these the most refined. We want to choose τ as large as possible such that a relative change of the propensities $\Delta_\tau a_j / a_j$ is smaller than a user-specified parameter ϵ with ($0 < \epsilon \ll 1$). Since we want to avoid to calculate the propensities directly, this relative change is estimated by calculating $\Delta_\tau x_i / x_i$ of each chemical species. To achieve this, we can consider the propensity requirement fulfilled if

$$\langle \Delta_\tau x_i \rangle \approx \sum_{j=1}^{M} v_{ij}(a_j \tau) \leq max\{\epsilon_i x_i, 1\} \tag{4.22}$$

and

$$\sqrt{var[\Delta_\tau x_i]} \approx \sqrt{\sum_{j=1}^{M} v_{ij}^2(a_j \tau)} \leq max\{\epsilon_i x_i, 1\}, \tag{4.23}$$

which allows us to solve for τ.

So the tau leaping simulation algorithm now can be summarized as follows:

```
while ¬terminate() do
    estimate τ;
    determine k_j random numbers from independent Poisson distributions;
    t := t + τ;
    x := x + Σ_j k_j v_j;
od
```

Since we have approximated the SSA using tau-leaping, the question arises whether there is some limit, e.g. $\tau \to 0$ in which this algorithm approaches the exact SSA. This is indeed the case, with the following precautions. One of the problems of tau-leaping is that while most reaction channels work perfectly fine within the numerical boundaries, some reaction channels might consume so many molecules of particular types that those species would be driven to negative counts (theoretically). I.e. the concentration of these chemicals would be below zero, which definitely cannot happen in reality and needs to be taken care of. The solution to this problem is to define a critical number of molecules, n_c, that signals a reaction channel might be in danger of coming close to exhausting its supply of molecules [153]. All reaction channels identified thus are subjected to slower step simulation, by using the direct method of equations 4.16 and 4.17.[5] This method guarantees to fire only one reaction channel per step, thus prohibiting any case of negative population count. All other (noncritical) reaction channels are updated using the tau-leaping algorithm above. We can see that in the extreme case where every reaction channel is critical, the tau-leaping method smoothly morphs into the exact direct method.

There is one last detail we need to discuss in this section, a problem that is frequently encountered both in the deterministic as well as in the stochastic simulation of chemical systems:

---

[5]Should it turn out that τ produced by equation 4.16 is larger than the τ from the tau-leaping method, the smaller one is used.

stiffness. A stiff system is characterized by a separation of time-scales of the underlying dynamical variables (in our case concentrations or population counts). In other words, there are
population counts that develop quickly (those with high propensity for reaction often having
many molecules present in the reaction vessel) and others that develop slowly (those with a low
propensity for reaction, often having only a small number of molecules in the reaction vessel).
As we can immediately see, this is a condition that must happen often in chemical systems, and
one that is very interesting to study in simulations.

Our tau-leaping approximation method discussed earlier would not work well with stiff systems since it would force a very small τ due to the critical number of molecules. The reaction
channels that fire least frequent will hold up the simulation for the entire system. Fortunately,
there are many methods to overcome the difficulty of stiff systems, and readers interested in
more background are advised to consult the corresponding literature [40]. In the context of
stochastic reaction systems, [698] discusses the so-called implicit tau-leaping method. Briefly,
in this method the explicit update formula for $\mathbf{X}(t + \tau)$ of equation 4.20 is replaced by an implicit
formula which contains $t + \tau$ dependency on the right-hand side.

Here, however, we want to explain another approach to stiff systems. This works by separating the slowly developing and the fast developing variables. First, the reaction channels
$R = \{R_1, \cdots, R_M\}$ are separated into fast and slow reaction channels R^f and R^s by applying a
partition criterion that determines what it means to develop "fast." Next, all chemical species
$S = \{S_1, \cdots S_N\}$ are divided into fast and slow species S^f and S^s, with species being changed by
fast reactions R^f being part of S^f, all others being part of S^s. This means that we now have two
different types of variables $\mathbf{X}^f(t)$ and $\mathbf{X}^s(t)$. Third, we define the virtual fast process $\hat{\mathbf{X}}^f(t)$ which
is an approximation to the fast process $\mathbf{X}^f(t)$, but with all slow reactions switched off. This virtual
fast process is next required to go to equilibrium values, i.e. $\hat{\mathbf{X}}^f(t) \to \hat{\mathbf{X}}^f(\infty)$ in a time that is small
compared to the expected time to the next slow reaction, such that we can substitute the values
$\hat{\mathbf{X}}^f(\infty)$ for $\hat{\mathbf{X}}^f(t)$. These are the stiffness conditions that provide a clear measure for the separation of time scales. With these conditions satisfied, we apply the slow-scale approximation: We
ignore the fast reactions and only consider slowly reacting species, simulating them one reaction at a time. To do this, we have to replace the propensity functions for slowly reacting species
by an average over the time-scale of the virtual fast process. If $\hat{P}(\mathbf{y}^f, \infty | \mathbf{x}^f, \mathbf{x}^s)$ is the probability
that $\hat{\mathbf{X}}^f(\infty) = \mathbf{y}^f$, given $\mathbf{X}(t) = (\mathbf{x}^f, \mathbf{x}^s)$, then the averaged propensity function for a slow reaction
channel R_j^s at time t can be calculated by

$$\bar{a}_j^s(\mathbf{x}^f, \mathbf{x}^s) = \sum_{\mathbf{y}^f} \hat{P}(\mathbf{y}^f, \infty | \mathbf{x}^f, \mathbf{x}^s) a_j^s(\mathbf{y}^f, \mathbf{x}^s). \tag{4.24}$$

This way, the slow scale stochastic simulation algorithm (ssSSA) [154] only simulates the
slowly developing variables, approximating the fast developing variables by their equilibrium
values. The simulation steps of the ssSSA, however, are exact in the sense of an SSA, except that
they do not take into account changes in fast-developing variables.

4.3 Spatial and Multicompartmental Algorithms

Up to now we have considered algorithms operating in well-mixed vessels only. An important class of algorithms takes space into account: molecules may be placed in compartments
enclosed by membranes, or in locations with well-defined coordinates in a two- or three-
dimensional space.

Spatial considerations are important because they offer a more realistic model for many systems where the location of objects plays a significant role in the phenomena observed: the interaction among objects is confined within a neighborhood radius, and objects may move from one location to another at different speeds. The combination of motion and local interactions may result in the formation of patterns in space. These patterns can be relevant to understanding the system under study, or they can be interpreted as the outcome of a parallel computation process, implemented by the underlying molecular motions and reactions.

Sometimes it is interesting to place molecules in separate compartments even when the exact location of the compartments in space is not important. A typical example of this case is the membrane computing or P systems formalism, from which more specific chemistries such as ARMS have been derived (see chapter 9). The importance of compartments lies in their ability to scale in complexity by encapsulating functions in a hierarchical way, resembling a biological organization: molecules are encapsulated into organelles, these organelles into cells, cells into multicellular organisms, and so on.

In a spatial chemistry, apart from reacting, molecules may move in space, usually by diffusion, but they may also be dragged by fluid or atmospheric currents or they may be actively transported by various mechanical, electrical or chemical forces. When molecules simply diffuse in space we have a *reaction-diffusion* process already discussed in [868] in 1952 (see also [573]) which is typically described by a system of partial differential equations (PDE):

$$\frac{\partial \mathbf{c}(\vec{p}, t)}{\partial t} = \mathbf{f}(\vec{c}(\vec{p}, t)) + D\nabla^2 \vec{c}(\vec{p}, t) \tag{4.25}$$

The vector $\vec{c}(\vec{p}, t)$ contains the concentrations c_i of each chemical i at time t and position $\vec{p} = (x, y, z)$ in space. The reaction term $\mathbf{f}(\cdot)$ describes the reaction rates at each point in space, and the diffusion term $D\nabla^2 \cdot$ tells how fast each chemical substance diffuses in space. D is a matrix containing the diffusion coefficients, and ∇^2 is the Laplacian operator.

The simplest numerical integration method for PDEs is a straightforward extension of its ODE counterpart: it consists in discretizing t in equation 4.25 into small fixed-sized time steps Δt. For each successive integration time step Δt, the change in concentration $\Delta \vec{c}$ (in one time unit) is calculated using equation 4.25, for each molecular species in the system at each point in space. The concentration vector $\vec{c}$ is then updated accordingly: $\vec{c}(\vec{p}, t + \Delta t) = \vec{c}(\vec{p}, t) + \Delta \vec{c} \Delta t$. Note that space must be discretized too, so the Laplacian operator "∇^2" is implemented as a concentration flow between a point in space and a finite number of its neighboring points, with proper rescaling of the diffusion coefficient to reflect such discretization. More details can be found in [253, 817]. Both the choice of the time step Δt and the granularity of the discretized space have an impact on the accuracy of the results, therefore more sophisticated algorithms are often needed [146, 149].

Algorithms for the stochastic simulation of reaction-diffusion systems also rely on the discretization of space into small lattice sites or larger containers called *subvolumes*. Each subvolume is treated as a well-mixed vessel with a given volume and a given coordinate in space, in which the usual rate laws such as the law of mass action apply. Diffusion is handled as the transport of an integer amount of molecules between neighboring reactors. A diffusion step can be implemented as a unimolecular reaction in which the educt is the molecule in the source compartment, and the product is the molecule in the destination compartment. Based on this idea, a number of spatial extensions of Gillespie's SSA for the stochastic simulation of reaction-diffusion systems have been proposed.

The next subvolume method (NSM) [253] is one of the most well-known algorithms for the stochastic simulation of reaction-diffusion systems. It is a spatial extension of the next reaction method (NRM) [324], in which subvolumes are scheduled by event time, as single reactions would be in NRM. The event time for a subvolume is computed as a function of the total propensity of the reactions (including diffusion as a unimolecular reaction) within the subvolume. At each iteration, one subvolume is removed from the top of the waiting list. The basic Gillespie SSA is applied to this subvolume in order to choose a reaction to fire. The propensities and event times for the concerned subvolumes are updated accordingly, and the algorithm proceeds to the next iteration. Since events are kept in a binary tree, the execution time for one iteration of the NSM algorithm scales logarithmically with the number of subvolumes, and therefore represents a significant advantage over a linear search for the right subvolume, as would result from a naive extension of SSA to a spatial context.

In order to improve the efficiency even further, the Bτ-SSSA algorithm (binomial τ-leap spatial SSA) [544] combines a variant of τ-leap with NSM such that leaps of several reactions can be taken whenever possible, within selected subvolumes.

The multicompartmental Gillespie's algorithm (MGA) [661] and variants thereof [308,718] extend SSA to support multiple compartments in P systems. P system rules originally executed in a maximally parallel way, which hampered their application to systems biology, where more realistic reaction timing is required. This becomes possible with MGA. Like NSM, MGA is also based on NRM: The events occurring in each membrane are ordered by firing time, and at each iteration the algorithm picks the events with lowest time for firing, updating the affected variables accordingly. Here however, the tree data structure comes for free from the existing membrane hierarchy.

An extension of Gillespie for compartments with variable volume is introduced in [528], in order to simulate cellular growth and division. Indeed, the authors show that when the volume changes dynamically, the propensities are affected in a nonstraightforward way, and adaptations to the original SSA are needed in order to accurately reflect this.

All the algorithms above assume that molecules are dimensionless particles moving and colliding randomly. However, in reality, intracellular environments are crowded with big macromolecules such as proteins and nucleic acids, that fold in complex shapes. In such an environment, the law of mass action no longer applies. Simulations of reaction kinetics in such crowded spaces need different algorithms, such as [346,752].

4.4 Summary

The wealth of algorithms discussed in this section is only a small sample of a vast landscape of existing algorithms, with new algorithms probably being proposed as we write these lines. Each algorithm has its strengths and weaknesses, and choosing the right one for a situation can prove challenging.

Recognizing that there is no perfect "one size fits all" algorithm for all possible applications in systems and cell biology, a meta-algorithm was proposed by [841]. It is able to run several potentially different subalgorithms in an integrated way, such as NRM together with numerical ODE integration. Time synchronization across subalgorithms is achieved with an event scheduling policy similar to NRM and NSM.

The modeling and simulation of physical, chemical, and biological systems is still a subject under active research. Since the systems under study tend to grow quite large and complex, the computational cost of such simulations is still considerably high. One can try to lump or abridge chemical reaction systems to systems that perform identically or at least very similarly with respect to the chemical species of interests, but use a smaller number of reactions and/or species. When applied properly, this can yield significant computational savings. For instance, Michaelis and Menten's famous enzyme kinetics is one such attempt to abridge a more complex reaction system - simply by using a more complex kinetic law. More information on abridgment methods is given in [81, 503].

Alternatively, one can try to cope with such high computational load by exploiting the natural parallelism provided by artificial chemistries. Solutions based on parallel hardware such as GP-GPUs (general-purpose graphic processing units) are beginning to emerge, and they are expected to gain more prominence as such hardware becomes increasingly affordable and more easily programmable. We refer the interested reader to our recent survey [940] for more information.

Part II

Life and Evolution

Since everything that happens in life is without exception chemistry (unless one wants to play with words and argue that chemistry is made up of different sciences), the only conclusion is that the appearance of life was the culmination of spontaneous chemistry becoming increasingly complex, ordered, and fast.

CLIVE TROTMAN

CHAPTER

5

THE CHEMISTRY OF LIFE

A rtificial chemistries stand to Artificial Life as chemistry stands to biology. The link between chemistry and biology is also intimately related to the origin of life from inanimate matter. The urge to understand the origin of life and the essence of the phenomenon of life was among the main motivations of Artificial Life that gave birth to Artificial Chemistry. A significant amount of the literature in Artificial Life and Artificial Chemistry can be truly understood only in the context of this long quest for the truth behind the organization of living beings. This is why the present chapter is devoted to an overview of the current knowledge about how life gets organized from basic chemistry.

Understanding life starts with understanding basic chemistry, and from there, organic chemistry and biochemistry. Organic chemistry is the chemistry of carbon-based compounds (which are not necessarily part of living beings, such as plastics and petroleum). Biochemistry, the chemistry of life, studies the molecular structures and chemical reaction processes that occur inside living organisms, including the complex "machinery" inside cells and the chemical interactions among cells.

One of the reasons why biochemistry is so much more complex than its inorganic counterpart is precisely because of the ability of many biomolecules to form new compounds in a combinatorial way, by the concatenation of elementary building blocks (monomers) into long strands (polymers), such as DNA, RNA, and proteins. Although much is known today about all the chemical structures needed for life, there are still significant gaps in our knowledge about their complex interactions, and how these interactions lead to the robust and yet versatile adaptive systems that we call living organisms. Organic compounds in living beings interact in multiple ways, forming large and intricate networks of chemical reactions that still elude analysis. Given an arbitrary protein sequence, how to determine what it does, i.e., what functions it will perform inside the organism? How do the interactions among genes, proteins, and various substances present in the environment influence the development and health state of living beings? How did life arise on Earth, and how exactly did it manage to evolve to the diverse, complex, and intricate web of life forms that we see today? Is it possible for life to arise under totally different conditions, with a totally different chemistry? Can we synthesize alternative forms of life in the laboratory? These and many other questions are still the subject of intense research, and they will be briefly discussed in this chapter.

Two methods can be used to study a given phenomenon: analytic or top-down, and synthetic or bottom-up. The former tries to dissect the object under study in order to understand its various parts. The latter tries to construct individual constituent elements from the object, and assemble them together. Until recently, biology was mostly based on the former method. As such, it was mainly a "reverse-engineering" science: It tried to understand life by dissecting its compounds to study them. Later, biologists gave in to the difficulty in understanding complex organisms by studying their individual parts in isolation because the whole is more than the sum of its parts. It is only possible to understand organisms as complex systems of interacting parts. With the help of modern computers and new mathematical methods, systems biology was born as a science that studies the complex interactions among the different biological elements. Recently, the door to the bottom-up method has opened, with the promise to synthesize (artificial) life forms in laboratory, giving rise to synthetic biology. Most of the work in synthetic biology today still preserves the top-down nature that characterized biology for so long: For instance, synthesized DNA strands have been recently inserted into recipient natural cells (bacteria) [323,782] in order to build semisynthetic artificial cells [535]. But a rising wave of truly bottom-up research is gaining ground, in which various cell components are synthesized individually and composed together within self-assembling artificial containers into a minimum synthetic cell [542,691,787]. The latter research draws heavily on the existing theories about the origin of life on Earth, in order to build *protocells*: minimum elementary cells with lifelike properties. These synthetic cells help to understand how life can emerge, and have many potential applications in medicine, engineering and other fields.

Learning the chemistry of life is important for artificial chemistries for various reasons: First, it helps to understand and model those origin of life phenomena that were the main motivation for artificial chemistries in the first place. Second, it offers a common view on the combinatorial spaces (discussed in chapter 1) and their associated innovation potential. Furthermore, it is a source of inspiration for chemical computing models where algorithms displaying emergent properties are expected to solve problems of increasing complexity by self-organizing into the proper structures, and to display robustness properties similar to those of living beings. Last but not least, it is important to understand and engineer new wet artificial chemistries in the realm of unconventional and natural computation.

The goal of this chapter is therefore not to introduce the chemistry of life as in a biology textbook, but to present it from an AC perspective, in order to situate the context of artificial chemistry research, and provide motivation and inspiration for new AC models.

We start by looking at how life can be defined, and at how the main types of biomolecules look like. We then discuss how these biomolecules interact to compose life forms. It is all about offering choices: from a vast combinatorial explosion of possibilities, only a few lead to viable organisms. Exploring this vast amount of possibilities requires more than pure chance. Evolution by natural selection is a feasible way to explore this space. It is all about making choices: Those combinations leading to better adapted individuals have a higher chance of passing on to the next generation, while those leading to individuals unable to survive will quickly disappear. The dynamics of evolutionary processes can be captured by artificial chemistries. Chapter 7 is therefore dedicated to an overview of Darwinian evolution and evolutionary dynamics.

Evolution explains how life may change once it already exists, but it does not explain how life emerges from scratch. Here again, artificial chemistries can be used to model the self-organizing chemical processes that could possibly lead to early life forms. In this context, the existing theories about the origin of life will be briefly reviewed in Chapter 6, including how life on Earth might have originated from inanimate matter, and how early evolution mechanisms might kick-off from a prebiotic weak selection dynamics.

5.1 What Is Life?

Before we start discussing about the chemistry of life, we need of course to clarify what we mean by "life." Since our observation of life is restricted to life on Earth, any definition of life is necessarily biased. Nevertheless, attempts have been made at a broader definition that would also encompass life outside our planet. Some define life as a *"self-sustaining chemical system capable of undergoing Darwinian evolution"* [434]. Others take a metabolic view of life as an out-of-equilibrium system able to build internal order (thus to resist entropic forces) by taking nutrients from the environment. Others look at life's self-organizing properties and describe it as an *autopoietic* system (self-producing and self-maintaining [534, 558, 881]).

These definitions tell us more about the properties of life than about what life actually is. The definition of life remains to a certain extent controversial [185, 532, 534], but from an operational, pragmatic point of view, a consensus is forming around the cell as the minimal unit of life [691]. A minimal cell would then be the smallest or simplest entity that can still be considered "alive." This minimal cell operates autonomously by extracting energy from its environment (for instance, from food or light sources), and using this energy to maintain itself, usually expelling some waste material in the process. The minimal cell requires three basic subsystems:

- a membrane defining the cell's boundary and enclosing its content;

- an information carrier subsystem containing its genetic material, and determining how it is reliably copied to the next generation;

- and a metabolic subsystem, that keeps the cell alive by turning food materials or energy into the essential building blocks needed for maintaining all the three subsystems together in a functional way.

The operation of the cell is guided by a blueprint deposited in its genetic material. Genetic information may get slightly transformed from one generation to the next, and the resulting cells may be subject to various selection pressures from the environment, promoting more efficient and better adapted cells, at the expense of weaker or ill-formed ones. As a result, populations of these cells may undergo Darwinian evolution, resulting in more complex organisms.

According to the minimal cell definition, a bacterium is alive, but a virus is not: The bacterium has the three subsystems interconnected in a functional way, but the virus has only the genetic subsystem (and sometimes a primitive compartment in the form of a protein coating) and is not autonomous, needing a host cell to replicate itself. Although viruses can reproduce and evolve, their lack of reproductive autonomy relegates them to the border between living and nonliving. Note that according to this standpoint, anything that contains the above three minimum subsystems can be considered alive, whether it ever existed on Earth or not. In particular, a synthetic cell can be classified as alive if it displays the three subsystems, and is able to autopoietically maintain them together and operational. In connection to artificial chemistries and Artificial Life, a potentially controversial case is that of a simulated artificial cell displaying all the properties of a living cell: Would such a simulation be also considered alive? Today it is mostly agreed that being alive also requires embodiment, i.e., a realization in the physical world. According to this view, a virtual simulation is not truly alive, but a philosophical debate about this issue might not be excluded.

5.2   The Building Blocks of Life

All life on Earth is based on the same chemistry. The basic atoms in organic chemistry are carbon (C), hydrogen (H), oxygen (O), nitrogen (N), phosphorus (P) and sulfur (S). Other important elements include calcium (Ca), sodium (Na), potassium (K), magnesium (Mg), iron (Fe), and chlorine (Cl), albeit occurring in smaller quantities. These elements can combine in infinitely many ways to form biomolecules, the basic molecules of life. Many biomolecules take the form of arbitrarily long polymers. They are formed by the concatenation of simple monomers (small molecules that have the ability to concatenate to others of a similar kind) to form oligomers (small chains of monomers), then polymers (long chains of monomers), and macromolecules (especially long polymers). The main classes of biomolecules are: carbohydrates, lipids, proteins, and nucleic acids. These molecules are immersed in water, the universal solvent of life as we know it.

In this section we will outline the role of water and the composition of the basic biomolecules. We will also mention a few recurrent motifs found in organic molecules (the so-called functional groups) and highlight the crucial role of energy and the ways to capture and use it.

5.2.1   Water

All living beings on Earth rely on liquid water, and it is believed that life started in the early oceans and ponds. Water (H_2O) plays a key role in molecular interactions: Some biomolecules are *hydrophilic* (have an affinity with water), others are *hydrophobic* (are repelled from water), and others are *amphiphilic* (have a head that likes water and a tail that hates it).

Hydrophiles form hydrogen bonds with water. Hydrogen bonds are noncovalent interactions between the hydrogen atom of one molecule and an electronegative atom (oxygen, nitrogen, or fluorine) of another molecule. These bonds are weaker than covalent bonds and make the hydrophilic molecules easy to dissolve in water, where they can move freely.

In contrast, hydrophobic molecules are repelled by water and therefore tend to cluster together while moving away from it. An important group of hydrophobic molecules are *hydrocarbons*, organic compounds consisting only of carbons and hydrogens. Hydrocarbons typically form long chains or rings (the ring forms are called aromatic hydrocarbons) and are commonly encountered in the composition of lipids.

Amphiphiles are interesting: Their hydrophilic heads are attracted to the water molecules, while their tails cluster together away from it, forming self-organizing structures such as micelles (like oil droplets) and vesicles (like soap bubbles), which can be regarded as precursors of cell membranes. Amphiphiles, vesicles and micelles will be introduced in Section 5.2.5 below.

The physical properties resulting from the interactions of biomolecules with water contribute to determine their behavior and function, and they play such an important role in the processes of life that many scientists believe that the existence of life is very improbable anywhere in the universe without liquid water.

5.2.2 Functional Groups

Biomolecules are formed by assembling small functional groups together in various ways, like Lego bricks. Functional groups confer molecules some of their key characteristics. Some groups that we will frequently encounter in our discussions below are:

- *Hydroxyl group* $(-OH)$: it forms hydrogen bonds with water, and is therefore hydrophilic; hence, substances containing many hydroxyl groups are very soluble in water, such as common sugars.

- *Carboxyl group* $(-COOH)$: another hydrophilic group; depending on the pH of the solution in which it is immersed, it will tend to lose its hydrogen, forming a negatively charged ion $-COO^{\ominus}$, frequently encountered in amino acids.

- *Amine group*, or *amino group* $(-NH_2)$: a nitrogen-based compound derived from ammonia; depending on the pH, it tends to gain an extra hydrogen atom, forming the ion $-NH_3^{\oplus}$, often found in amino acids.

- *Phosphate group* $(-PO_4^{2-})$: the most common functional group containing phosphorus (P); enters the composition of nucleic acids, where it links nucleotides together to form DNA or RNA chains. The ion PO_4^{3-} alone is called a *phosphate*, and is often represented as P_i.

5.2.3 Energy

All life forms rely on an input of energy to maintain their cellular activities. Energy is first captured by plants (photoautotrophs) from sunlight during photosynthesis, transformed in metabolic reactions, and stored in the form of fat, starch, or sugar. Animals (chemoheterotrophs) then harvest such energy by eating energy-rich cells from other organisms. Some bacteria (chemoautotrophs) are able to extract energy from inorganic compounds.

According to Kauffman [448], all living organisms are examples of autonomous agents. An autonomous agent is a self-reproducing system able to perform at least one thermodynamic work cycle. This perspective puts energy at the core of the operation of living systems.

Most biochemical reactions are catalyzed by biological enzymes. Enzymes are specialized proteins that facilitate chemical reactions by bringing reactants together and binding to their transition state, decreasing the activation energy necessary for the reaction to occur (see chapter 2). Enzymes accelerate biochemical reactions by several orders of magnitude. Without enzymes, these reactions would be too slow to sustain any proper rhythm for life.

In spite of the crucial role of enzymes in accelerating biochemical reactions, enzymes are not sufficient to keep all needed reactions flowing. Many biochemical reactions are endergonic, relying on an input of energy in order to drive them uphill in the energy landscape. Such input of energy occurs mainly through coupled reactions: in a coupled reaction, a downhill reaction is combined with an uphill reaction, such that the former provides energy to the latter. The net reaction is downhill and can therefore occur spontaneously.

There are two main ways in which energy can be transferred in such coupled reactions: One is via ATP hydrolysis, and the other is via coupled redox reactions:

- ATP hydrolysis: ATP (adenosine triphosphate) is a *nucleotide* composed of adenosine (a nucleobase that also enters the composition of DNA), ribose (a sugar), and three phosphate groups. During hydrolysis, ATP releases one phosphate becoming ADP (adenosine diphosphate):

$$ATP + H_2O \longrightarrow ADP + P_i. \tag{5.1}$$

This reaction is highly exergonic, releasing a lot of energy, which can be used to drive endergonic reactions. Other similar reactions occur in the cells, but the ATP-ADP conversion is by far the most common. For this reason, ATP is considered as the "energy currency" of the cells. ATP can also be regarded as a "loaded battery," whereas ADP is its unloaded form. In order to "load" ADP back to ATP, a significant energy input is needed. This occurs in mitochondria and chloroplasts, where proton gradients flowing across their inner membrane are used to drive ATP synthase, a molecular motor that can be found in such membranes: each turn of the motor loads an ADP with a phosphate, converting it to an ATP, ready for reuse.

- Coupled redox reactions: In a *redox* (short for reduction-oxidation) reaction, electrons are transferred from one atom to another: One substance loses (donates) electrons and becomes oxidized (oxidation step), while the other gains them and becomes reduced (reduction step). The oxidation and reduction steps (each of them considered as a "half reaction") are combined into a coupled redox reaction.

The most obvious examples of redox reactions involve the formation of ionic bonds, since these bonds are formed by electron transfer. For instance, the formation of $NaCl$ is a simple example of a redox reaction:

$$\text{oxidation of sodium:} \quad 2\,Na \quad \longrightarrow 2Na^+ + 2e^- \tag{5.2}$$

$$\text{reduction of chlorine:} \quad Cl_2 + 2\,e^- \quad \longrightarrow 2\,Cl^- \tag{5.3}$$

$$\text{combined redox reaction:} \quad 2\,Na + Cl_2 \quad \longrightarrow 2\,NaCl. \tag{5.4}$$

Another well-known example is the oxidation of iron when it rusts: $4\,Fe + 3\,O_2 \rightarrow 2\,Fe_2O_3$.

Redox reactions involving covalent bonds are also widespread. An important example from biology is the following reaction:

$$\text{NADP}^\oplus + H^\ominus \rightleftharpoons \text{NADPH}. \tag{5.5}$$

In this reaction, a carbon atom from NADP$^\oplus$ binds covalently to $H^\ominus$ (hydride ion). NADP (nicotinamide adenine dinucleotide phosphate) is a nucleotide that can occur in oxidized form as NADP$^\oplus$ and in reduced form as NADPH. It acts as a coenzyme in various metabolic reactions. Coenzymes or cofactors are molecules that help in the activity of enzymes. The energy released in the oxidation of NADPH to NADP$^\oplus$ is used to drive metabolic reactions. NADP$^\oplus$ is later "reloaded" to NADPH using energy from photosynthesis in chloroplasts (for plants) and respiration in mitochondria (for animals).

Substances that readily donate electrons (becoming oxidized in the process) are called reductants (because they reduce the acceptor substance) while oxidants readily accept them (becoming reduced). The term "oxidation" comes from the fact that oxygen is itself a strong oxidant, but other redox reactions not involving oxygen are of course perfectly legal. Coupled redox reactions are crucial for metabolism, as will become apparent in section 5.3.2.

Life is also described as a system that is kept constantly away from equilibrium: The products of biochemical reactions are quickly removed and transported to their target locations. When a substance reaches a sufficient quantity, a feedback loop ensures that its production slows down or comes to a halt. Enzymes are tightly regulated to prevent reactions from reaching equilibrium concentrations, which would be mostly fatal for the cell.

5.2.4 Carbohydrates

Carbohydrates are biomolecules made of monomers called monosaccharides. Sugars such as glucose, lactose, and fructose are simple carbohydrates made of one or two of such monomers. Glucose is produced by plants from inorganic materials during the photosynthesis process. It is then converted into other forms of carbohydrates such as starch in plants, or along the food chain into other forms used by animals such as lactose and glycogen.

Sugars can be readily used as fuel source in the metabolism of many organisms. Larger and more complex carbohydrates (polysaccharides) such as starch and glycogen typically serve as longer-term energy storage. In order to retrieve the stored energy, a catabolic process breaks down such large molecules back into their constituent sugars. Two special types of sugar, ribose and deoxyribose, are also components of nucleic acids, and therefore play a role in the genetic storage of information.

Figure 5.1 shows the shapes of some common carbohydrates, in stereochemistry notation, an organic chemistry notation that shows the three-dimensional spatial arrangements of the atoms in the molecule. 3D shapes are very important in biochemistry, because the shape of a molecule has a direct influence on its function: for instance, given two different shapes (isomers) of the same compound (i.e., with the same number of atoms of each type), one of them might be an effective medicine while the other might be a lethal poison. Stereochemistry, in a nutshell, represents a molecule by a graph in which nodes (vertices) indicate atoms, and line segments (edges) indicate covalent bonds between two atoms.

Both glucose and fructose (fruit sugar) are monosaccharides with formula $C_6H_{12}O_6$, and can appear in different shapes, i.e., have many different isomers, such as chains and rings. Their most common isomers are D-glucose (dextrose), and D-fructose, both of which frequently adopt a cyclic shape. These are depicted in figure 5.1. The common table sugar is sucrose (or saccharose, $C_{12}H_{22}O_{11}$), a disaccharide formed by linking together one glucose and one fructose molecule (releasing one molecule of water in the process, in a *condensation* type of reaction). In these molecules we can see several hydroxyl group ($-OH$) sticking ends. Hydroxyl groups form hydrogen bonds with water and are therefore hydrophilic, making these sugars soluble in water. The hydroxyl group is a common building block in carbohydrates and other large organic molecules, as we will see. More complex carbohydrates such as starch and glycogen are large polymers made of 300 to 3,000 or more monosaccharide units aligned in a linear chain or forming branched structures.

5.2.5 Lipids

Lipid is a denomination for a broad class of hydrophobic and amphiphilic substances encompassing common fats and oils, waxes, steroids, and many other compounds [596]. Besides energy storage, lipids also play other important roles, notably as cell membrane constituents and signaling molecules among cells.

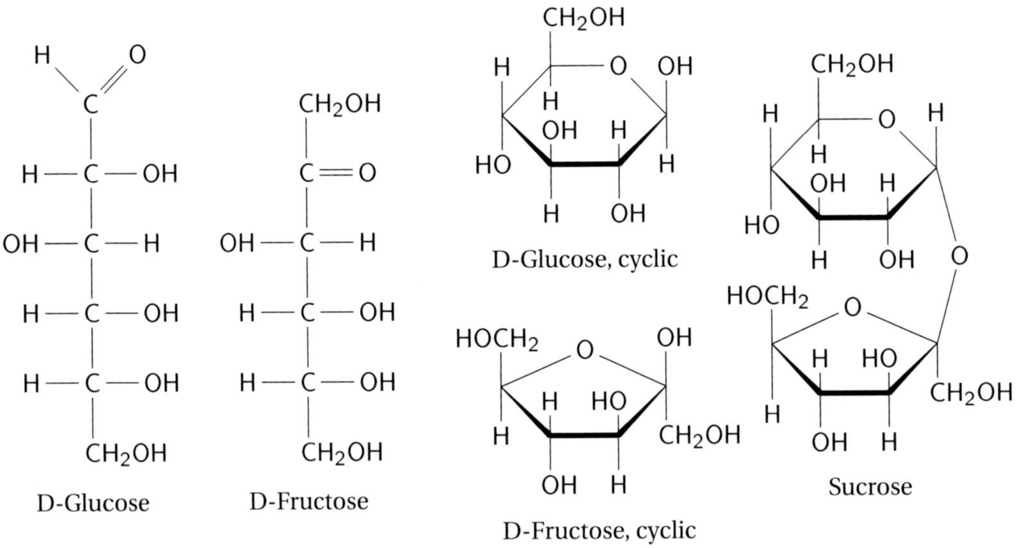

D-Glucose

D-Fructose

D-Glucose, cyclic

D-Fructose, cyclic

Sucrose

Figure 5.1 Some common carbohydrates. D-Glucose, or simply glucose, is a monosaccharide found in the bloodstream and used as energy source. D-Fructose, or fructose is a monosaccharide commonly found in fruits. Both are most frequently found in their cyclic forms. Sucrose is a disaccharide used as table sugar, and results from a condensation reaction between a glucose molecule and a fructose molecule. In chemical notation, carbon and hydrogen atoms are most often represented implicitly: each nonlabeled vertex is assumed to contain one carbon atom, and each carbon atom is assumed to be bound to enough hydrogen atoms such that its total number of bonds is four. A straight line represents an edge that is on the plane of the page; a thick wedged bar represents an edge that sticks out of the page toward the reader; a dashed bar represents an edge that points behind the page (away from the reader).

The simplest kind of lipid is a fatty acid. A *fatty acid* is a pin-shaped molecule made of a carboxylic acid head ($-COOH$) and a tail consisting of a potentially long hydrocarbon chain. *Carboxylic acid* is an organic acid made of at least one *carboxyl group*. Some common fatty acids are depicted in figure 5.2. Fatty acids are amphiphiles, since their head group is hydrophilic, and their tail is hydrophobic. Fatty acids may be *saturated* (with no double bonds between carbons in the carbon chain) or *unsaturated* (with one or more double bonds between carbons). Saturated fatty acids tend to present a straight, linear tail structure, whereas unsaturated ones may bend at some double bond positions.

Common fats and oils are triglycerides, composed of one glycerol and three fatty acids. *Glycerol* (or glycerine) is a type of alcohol composed of three hydroxyl groups. Glycerol and an example of triglyceride can be seen in figure 5.2. Fats are formed from intermediate products of sugar metabolism, during a process called *lipogenesis*. Energy can be efficiently stored in fats. When metabolized, the fatty acids contained in fat molecules release several ATP molecules, which can then be used as energy input in other reactions that require such energy for vital functions such as locomotion and cell structure maintenance.

An important class of lipids is specially relevant to our discussion: Phospholipids (Fig. 5.2) form lipid bilayers, the basic structure of modern cell membranes (see Fig. 5.3). Phospholipids have a hairpin shape, typically composed of a diglyceride (forming a tail with two fatty acids),

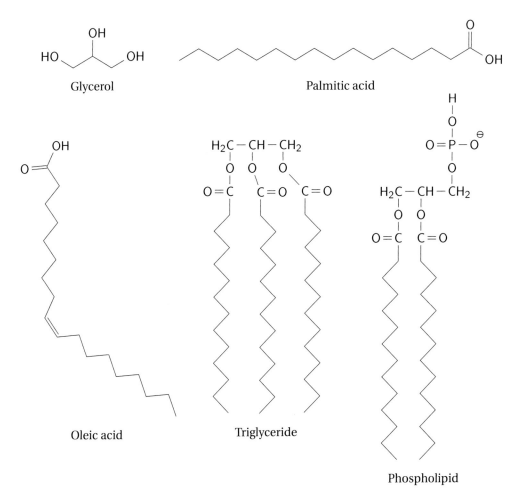

Figure 5.2 Lipids and their components: Glycerol, a component of some lipids; palmitic acid (saturated, common in plant and animal fat) and oleic acid (unsaturated, abundant in olive oil) are examples of fatty acids, the simplest kinds of lipid, with amphiphilic properties; triglyceride, an example of fat molecule, formed by the combination of a glycerol and three fatty acids (the three fatty acids are identical in this example, but in general triglycerides may carry heterogeneous fatty acid mixes); a simple phospholipid (phosphatidate), in which the phosphate group at the head of the molecule is bound to a hydrogen. In more complex phospholipids, the place of this hydrogen is occupied by more complex compounds, ranging from glycerol derivates to various nitrogen-based compounds.

and a head containing a phosphate group bound to a small compound that often contains nitrogen. Fatty acids and phospholipids are amphiphiles, so they tend to self-assemble into micelles and vesicles when thrown in water (see figure 5.3), a useful property when building cell compartments in natural and artificial cells, as will be discussed in section 6.1.1.

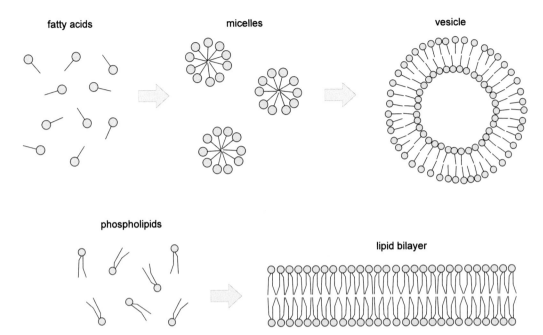

Figure 5.3 Some typical structures made of amphiphiles: Self-assembling fatty acid structures: Micelles and vesicles (top); Phospholipids forming a lipid bilayer (bottom).

5.2.6 Proteins

Proteins are large chains of monomers called amino acids. Smaller chains of amino acids are called *peptides*. Hence proteins are also referred to as polypeptides. An *amino acid* is composed of a main chain and a side chain. The main chain is composed of a carbon, a hydrogen, an amine group, and a carboxylic acid group. The side chain composition varies from one amino acid to another, and is therefore the part that identifies the amino acid.

Figure 5.4(a) shows the structure of an amino acid. Figures 5.4(b) and 5.4(c) show two examples of amino acids commonly found in nature.

Protein sequences are formed by joining two amino acids together by condensation (releasing one water molecule). The amino acids are linked together by *peptide bonds*, which are covalent bonds between the amine group of one amino acid and the carboxylic group of the next. The amino acids linked together in this way are called *residues*. Figure 5.5 shows the process of peptide bond formation.

There are 20 types of amino acids commonly found in existing organisms. These are referred to as α-amino acids. The 20 types of amino acids differ by their side chains. Thus the protein "alphabet" is said to consist of 20 "letters" that can be combined in any order and length to form the protein sequence. Indeed, amino acids are often represented by their single-letter abbreviation, leading to the protein alphabet $\mathscr{A} = \{$ A, R, N, D, C, Q, E, G, H, I, L, K, M, F, P, S, T, W, Y, V $\}$. Table 5.1 lists the 20 common α-amino acids with their names and their three- and single-letter abbreviations.

Amino acids differ in their properties: Some are hydrophilic, others are hydrophobic; some are acids, others are bases. Other relevant properties include their charge and size. Hydrogen

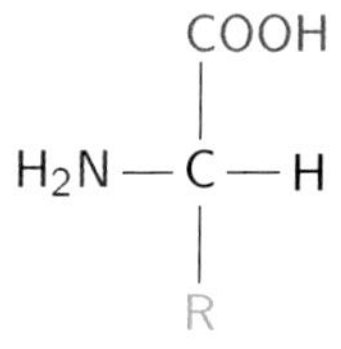

COOH
|
H₂N — C — H
|
R

Amino acid

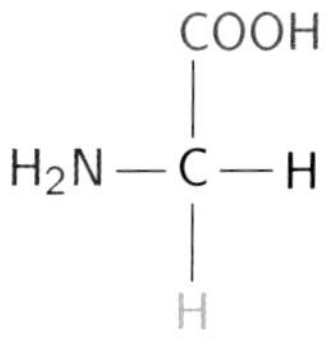

Glycine

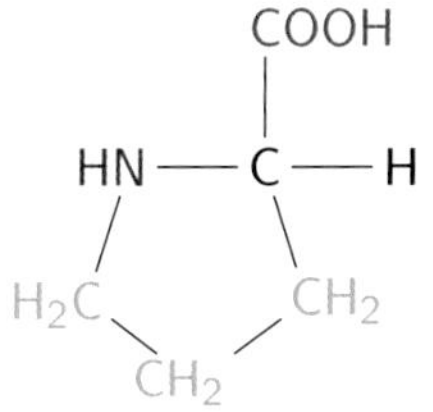

Proline

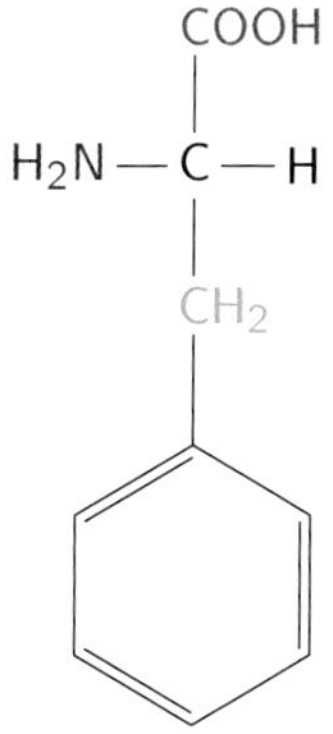

Phenylalanine

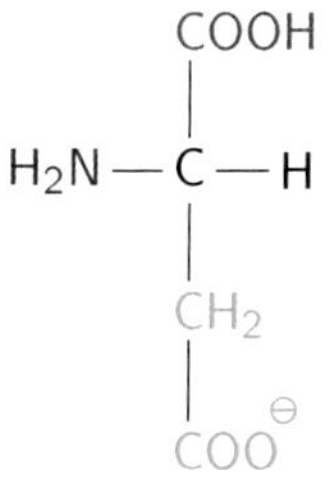

Aspartate

Figure 5.4 Top: Amino acid structure: Carboxyl group (red), amine group (blue), and side chain (R, green). Bottom: Four common examples of amino acids: Glycine, the smallest amino acid; proline, a special amino acid in which the side chain bonds with the amine group forming a ring; phenylalanine, a hydrophobic amino acid; aspartate (or aspartic acid), which carries a negatively-charged side chain.

bonds can form between amino acids in different regions of the protein. All these properties taken together confer proteins the interesting ability to fold into intricate 3D shapes in water.

One can look at the protein structure at multiple levels, from primary to quaternary structure. A protein's *primary structure* is its sequence of amino acid "letters." Its *secondary structure* is the shape observed in segments of the sequence, typically coils (alpha helices) or ribbons (beta sheets), formed due to hydrogen bonds. The protein's *tertiary structure* is its final 3D shape where, roughly, hydrophobic regions tend to cluster together inside the protein, while hydrophilic regions tend to stick out, toward the water molecules. Finally, a *quaternary structure* corresponds to aggregates of multiple proteins forming protein complexes. Two important research topics in computational biology are the prediction of the 3D protein structure from its sequence, and the prediction of the protein's function from its shape. Some typical protein shapes are shown in figure 5.6.

The protein's 3D shape determines its function. Some proteins act as structural components in the organism, such as muscle and hair fibers. Other proteins act as enzymes that catalyze metabolic reactions, help in the duplication and repair of DNA strands, orchestrate the manufacturing of other proteins in the ribosome, and so on. An enzyme may have a "pocket" shape where molecules may "fall into" and react, facilitating their contact and therefore accelerating their reaction. Other enzymes may have a "scissor" shape, able to "cut" other polymers at specific locations. Other protein shapes bind well to certain molecules like a lock and key mechanism,

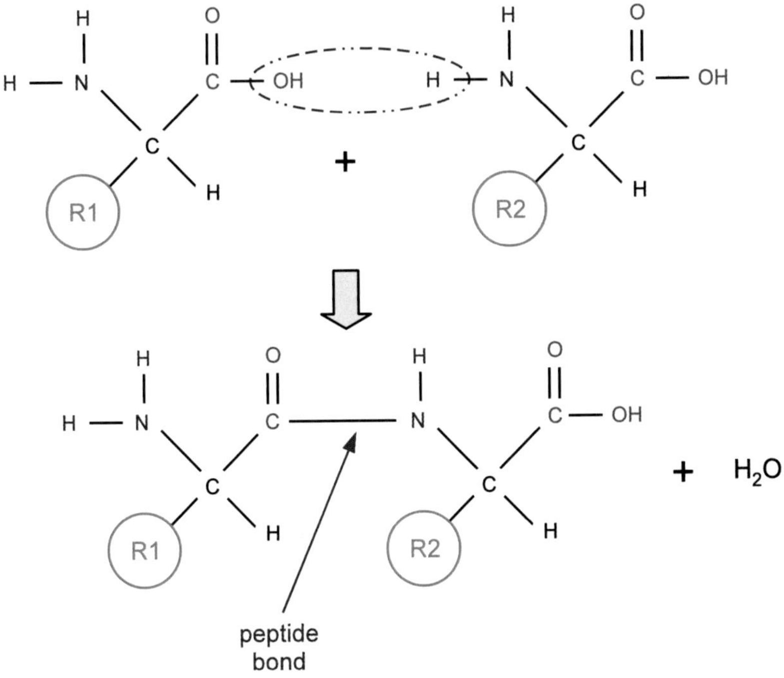

Figure 5.5 Peptide bond formation during the incorporation of amino acids into proteins.

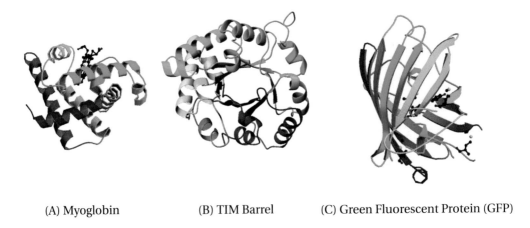

(A) Myoglobin   (B) TIM Barrel   (C) Green Fluorescent Protein (GFP)

Figure 5.6 Typical protein folding shapes. Image sources: (A) RCSB PDB 1MBN [910] (B) "8tim TIM barrel topview" by WillowW, Wikimedia Commons, http://commons.wikimedia.org/wiki/File: 8tim_TIM_barrel_topview.png (C) RCSB PDB 1EMA [647], www.rcsb.org

Table 5.1 The 20 common α-amino acids, with their full name, three-letter abbreviation, one-letter code, and side-chain formula. 'Ph' here stands for the phenyl group, the hexagonal ring shown in figure 5.4.

Name	Abbr.		Side-chain formula
Alanine	Ala	A	CH_3-
Arginine	Arg	R	$HN=C(NH_2)-NH-(CH_2)_3-$
Asparagine	Asn	N	$H_2N-CO-CH_2-$
Aspartic acid	Asp	D	$HOOC-CH_2-$
Cysteine	Cys	C	$HS-CH_2-$
Glutamine	Gln	Q	$H_2N-CO-(CH_2)_2-$
Glutamic acid	Glu	E	$HOOC-(CH_2)_2-$
Glycine	Gly	G	$H-$
Histidine	His	H	$NH-CH=N-CH=C-CH_2-$
Isoleucine	Ile	I	$CH_3-CH_2-CH(CH_3)-$
Leucine	Leu	L	$(CH_3)_2-CH-CH_2-$
Lysine	Lys	K	$H_2N-(CH_2)_4-$
Methionine	Met	M	$CH_3-S-(CH_2)_2-$
Phenylalanine	Phe	F	$Ph-CH_2-$
Proline	Pro	P	$NH-(CH_2)_3-$
Serine	Ser	S	$HO-CH_2-$
Threonine	Thr	T	$CH_3-CH(OH)-$
Tryptophan	Trp	W	$Ph-NH-CH=C-CH_2-$
Tyrosine	Tyr	Y	$HO-Ph-CH_2-$
Valine	Val	V	$(CH_3)_2-CH-$

serving as receptors for signaling molecules such as hormones, or antibodies that detect antigens in the immune system. A protein may bind to many different molecules (low specificity) or may bind only to a very few specific shapes (high specificity).

5.2.7 Nucleic Acids

A *nucleic acid* is a macromolecule composed of nucleotides. A *nucleotide* is a monomer composed of a base (or nucleobase), a sugar, and one to three phosphate groups. The nucleobases are classified into *purines* and *pyrimidines*. The purines are further subdivided into adenine (A) and guanine (G), and the pyrimidines into thymine (T) and cytosine (C), and uracil (U).

The most common nucleic acids are deoxyribonucleic acid (DNA) and ribonucleic acid (RNA). Both DNA and RNA are polymers composed of four types of monomers, identified by their type of base. In DNA, the sugar is a deoxyribose and the bases are adenine, guanine, thymine, and cytosine (AGTC). In RNA the sugar is a ribose and the bases are AGUC, i.e., uracil is used instead of thymine. Figure 5.7 shows the two sugars ribose and deoxyribose, and figure 5.8 shows the five nucleobases.

The DNA molecules found in cells are structured as two strands of nucleotides coiled together forming a double helix. The two strands are formed by base pairing: an adenine bonds only to a thymine, and a cytosine only to a guanine. The A-T and C-G bonds in DNA base pairing are hydrogen bonds, hence easy to break during the DNA duplication process. Figure 5.9 shows how

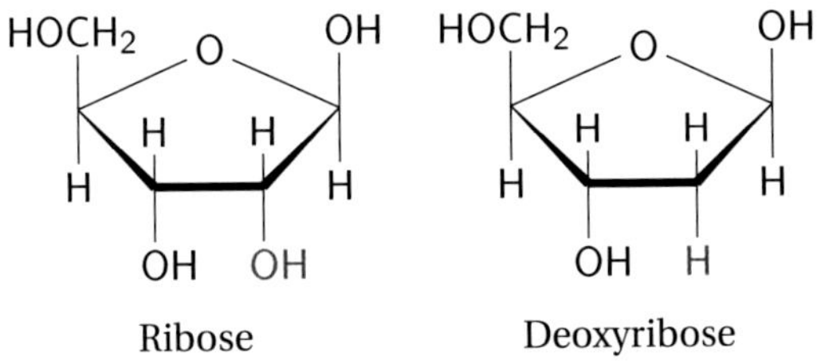

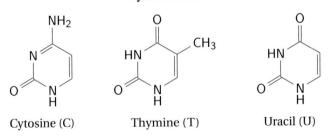

Figure 5.7 Nucleotide sugars: ribose, used in RNA; and deoxyribose, used in DNA. The difference between them is highlighted in red: deoxyribose has no oxygen at the second carbon of the ring, therefore its name. Adapted from [596]

Purines:

Adenine (A)   Guanine (G)

Pyrimidines:

Cytosine (C)   Thymine (T)   Uracil (U)

Figure 5.8 The five standard nucleobases: on top the purines adenine (A) and guanine (G); at the bottom the pyrimidines cytosine (C), thymine (T), and uracil (U).

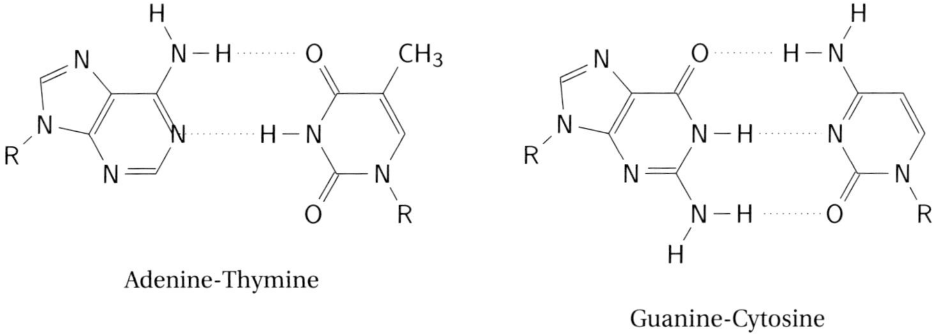

Adenine-Thymine

Guanine-Cytosine

Figure 5.9 Adenine-thymine and guanine-cytosine bond together via hydrogen bonds (dotted links). The placeholder '-R' indicates the attachment point for ribose or deoxyribose.

the AT and CG pairs bond together, and figure 5.10 displays the famous double-helix structure of the DNA.

Like proteins, DNA and RNA molecules may also fold into intricate 3D shapes: The DNA double helix forms coils of coils, folding itself into a compact structure. RNA molecules are usually single-stranded and may fold into various shapes, with segments resembling loops and hairpins. Figure 5.11 illustrates a few typical RNA shapes. The ladder segments are formed mainly by complementary base-pairing within the RNA strand: 'A' pairs with 'U', and 'C' pairs with 'G' (although other unconventional pairings sometimes occur). Loop regions remain where segments do not match by complementarity. Like proteins again, the shape of RNA molecules determine their function. Some of these RNA structures may have catalytic functions, such as cleaving or ligating other RNAs, and are therefore called "ribozymes" [844]. The folding of RNA is relatively well understood, and there are efficient algorithms able to predict RNA structure from sequence [389, 390]. However, the mapping from shape to function is still poorly understood. A hot topic of current research is to understand the functions of various types of newly discovered RNA molecules, as well as to design RNA molecules with desired functions, for instance, to serve as drugs in new therapies against cancer and other diseases.

Nucleic acids carry genetic information: DNA sequences store the genetic instructions used to create and maintain a living organism. These instructions are stored in DNA segments called genes. *Genes* are portions of DNA that code for specific proteins or play specific roles. During the process of gene expression, the information in a gene is transcribed to RNA sequences. The RNA sequence is then translated into proteins that will carry out various functions. During the reproduction process, the DNA information must be faithfully copied to the next generation. Rare errors in the copy process constitute mutations that are mostly lethal. Occasionally a mutation might lead to a protein shape that is better at performing some function, making the mutated organism more likely to survive and pass on the beneficial mutation to the offspring. In this way, living organisms are able to store and transmit information, and this information is subject to slow transformations that in the long run lead to the evolution of the species. In what follows, we will talk more about how the different building blocks of life may combine to form primitive cells, and how an evolution process may emerge, causing these cells to improve and increase in complexity, leading to multicellular organisms and to the rich ecology of species that we see today.

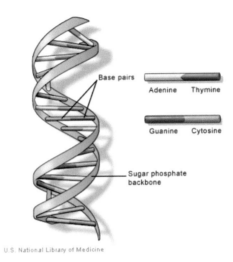

Figure 5.10 DNA double helix structure. Source: Genetics Home Reference, U.S. National Library of Medicine, http://ghr.nlm.nih.gov/handbook/basics/dna, August 25, 2014.

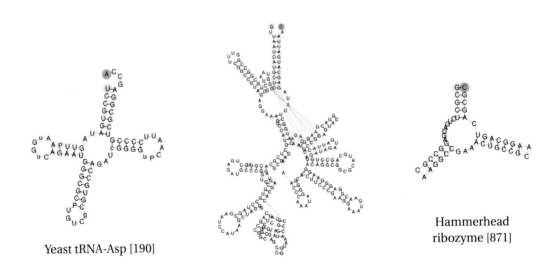

Yeast tRNA-Asp [190]

Bacterial ribonuclease P RNA [451]

Hammerhead ribozyme [871]

Figure 5.11 Some typical RNA shapes. Secondary structures generated using the Vienna RNA Package [349], http://www.tbi.univie.ac.at/RNA/, via the software RNA STRAND v2.0 - The RNA secondary STRucture and statistical ANalysis Database [38], http://www.rnasoft.ca/strand .

5.2.8 Composing Building Blocks and Biomolecules

In the midst of today's focus on genetic engineering, there is a tendency to overemphasize the role of DNA and proteins and to underestimate that of lipids and carbohydrates in the construction of life. In order to even out this imbalance and to stress the equally important role of all biomolecules, a unified view of the building blocks of life is proposed in [545]. The author points out that 68 molecular building blocks compose the four main classes of biomolecules. This set of 68 building blocks can be regarded as a sort of "alphabet" or "periodic table of elements" for biology. While the DNA of our genome encodes the instructions to manufacture proteins that direct the cellular processes, the full cellular operation relies on the four types of biomolecules, of which lipids and glycans (carbohydrates) are not directly encoded in DNA.

Several combinations of the four basic types of biomolecules also play important cellular roles, such as glycoproteins (proteins with added sugar chains), glycolipids (lipids with attached carbohydrates), and lipoproteins (proteins containing cholesterol). Glycoproteins and glycolipids are found on the surface of cell membranes, and play a fundamental role in cell communication and immune recognition. Lipoproteins act as lipid carriers in an aqueous environment, allowing energy to be transported between cells that store energy and those that need it.

5.3 The Organization of Modern Cells

Now that we have looked at the building blocks of life individually, we will briefly look at how they are combined into the intricately organized living cells that we see today.

Modern cells are divided into two main groups: prokaryotes and eukaryotes. Prokaryotic cells have no nucleus, and the genetic material floats in the cytoplasm. In eukaryotic cells, the genetic material is contained in the nucleus of the cell, and the cytoplasm is full of organelles carrying out various functions, such as mitochondria (for respiration in animals) and chloroplasts (for photosynthesis in plants). Bacteria are prokaryotes, whereas yeasts, amoebae and multicellular organisms are eukaryotes. Figure 5.12 shows schematic representations of a prokaryotic and an eukaryotic cell. In what follows, we will briefly review the basic inner workings of cells that are essential to keep them alive.

5.3.1 Cell Membranes

The structure of modern cell membranes is based on a lipid bilayer punctuated by proteins that implement molecular channels and receptors for cellular signals (figure 5.13). The channels selectively let nutrients enter the cell and let waste products leave the cell. Receptors on the surface of the membrane bind to ligand molecules, activating a chain of chemical reactions (signaling cascade) that transmits information to the interior of the cell. In this way, the membrane is selectively permeable to chemicals and sensitive to environmental information. These properties are useful for protecting the cell against external hazards and for cell-to-cell communication in colonies of cells or in multicellular organisms. As we will see in chapter 17, artificial molecular communication mechanisms have been designed based on natural membranes.

5.3.2 Metabolism

All organisms rely on harvesting, storing and using energy for their survival. The set of all chemical reactions that steer such energy flow is the metabolism. Two types of metabolic reactions

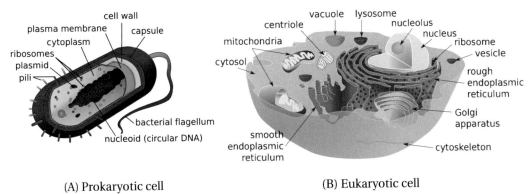

(A) Prokaryotic cell   (B) Eukaryotic cell

Figure 5.12 An example of a prokaryotic cell (A) and of an eukaryotic cells (B): prokaryotic cells are smaller (0.1-10 μm) and have no nucleus, while eukaryotic cells are much bigger (10-100 μm), and contain inner membrane-bound structures such as the nucleus and various organelles. Image Sources: (A) "Average prokaryote cell- en" by Mariana Ruiz Villarreal, LadyofHats, Wikimedia Commons http://commons.wikimedia.org/wiki/File:Average_prokaryote_cell-_en.svg (B) "Biological cell" by MesserWoland and Szczepan1990, Wikimedia Commons http://commons.wikimedia.org/wiki/File:Biological_cell.svg .

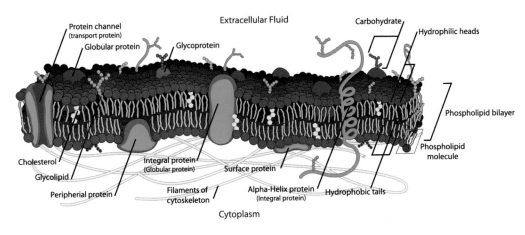

Figure 5.13 The structure of modern cell membranes: A lipid bilayer hosts several embedded structures, such as protein channels and carbohydrate signals. Image Source: "Cell membrane detailed diagram en" by LadyofHats Mariana Ruiz, Wikimedia Commons, http://commons.wikimedia.org/wiki/File:Cell_membrane_detailed_diagram_en.svg .

can be distinguished: catabolism and anabolism. *Anabolism* uses energy to drive various cellular processes, including the synthesis of various molecules needed by the cell (biosynthesis). *Catabolism* breaks down energy-rich food molecules (such as sugars and fats) in order to extract the energy they contain (this energy can then be used in anabolic processes). Anabolic reaction pathways are endergonic (absorb energy), whereas catabolic ones are exergonic (release energy). Living organisms therefore rely on a constant inflow of energy to drive the anabolic processes that are necessary for running the cells.

Enzymes play a fundamental role in the catalysis of most metabolic reactions: Without enzymes, the activation energy barrier for these reactions would be too high, and they would occur only at a rate much too slow to ensure survival.

Metabolism makes heavy use of coupled *redox* reactions. The flow of electrons from reductants to oxidants is at the heart of the energy transfer processes crucial to metabolisms: In catabolic pathways, cells obtain energy from the oxidation of energy-rich nutrients such as carbohydrates; conversely, anabolism relies on the oxidation of energy carriers such ATP to drive reducing processes that create macromolecules from small monomers.

Perhaps the most important set of metabolic reactions is the one responsible for photosynthesis. Photosynthesis is an anabolic process that captures solar energy to produce energy-rich organic compounds (such as sugars) from inorganic matter. In this way, plants and other photosynthetic organisms (called phototrophs) are able to produce sugars, fatty acids, amino acids, and nucleotides by extracting the necessary atoms from carbon dioxide, water, nitrogen and other inorganic compounds. Other organisms such as animals and fungi are left with no alternative but to prey on the organic compounds produced by phototrophs.

Photosynthesis occurs in the chloroplasts of plants, and involves several elaborate chemical reactions aided by a number of enzymes. The net reaction of a typical photosynthesis process can be simplified as follows:

$$6CO_2 + 6H_2O + \text{photons} \rightarrow C_6H_{12}O_6 + 6O_2 + \text{heat.} \qquad (5.6)$$

That is, a sugar (such as $C_6H_{12}O_6$) is produced out of carbon dioxide (CO_2) and water, under the action of light photons, with oxygen and heat being released in the process. This is a typical redox reaction, where CO_2 is reduced to produce sugar, while water is oxidized to produce O_2.

The photosynthesis process is divided into two phases: During the first phase, photons are used to activate energy carriers such as ATP. In the second phase, sugars are manufactured in order to store this energy for future use. The first phase can be summarized as:

$$2H_2O + 2\text{NADP}^+ + 3\text{ADP} + 3P_i + \text{photons} \rightarrow O_2 + 2\text{NADPH} + 2H^+ + 3\text{ATP.} \qquad (5.7)$$

In this phase, light photons hit chlorophyll (a photosensitive green pigment), causing an electron movement that ends up splitting the water molecules into hydrogen ions (H^+) and oxygen (O_2). The proton gradient created due to the H^+ flow enables the "unloaded" energy carrier molecules NADP and ADP to be "loaded" with electrons plus H^+ (in the case of NADP), and with P_i (in the case of ADP), adopting their energized forms NADPH and ATP (respectively). The energized NADPH and ATP can now drive the second part of the photosynthesis process.

The second phase is more complex, involving a cyclic reaction network called the *Calvin cycle*. It uses the energy in NADPH and ATP to perform *carbon fixation*, which is the reduction of carbon dioxide to intermediate organic substances that are later converted to sugars. This phase can be summarized as:

$$6CO_2 + 12\text{NADPH} + 18\text{ATP} \rightarrow C_6H_{12}O_6 + 12\text{NADP}^+ + 18\text{ADP} + 18P_i + 6H_2O. \qquad (5.8)$$

Here, the energized NADPH and ATP molecules fuel the production of sugar ($C_6H_{12}O_6$) from carbon dioxide, and become again unloaded in the process (reverting back to NADP and ADP). The phosphates and part of the water molecules used in the first phase are also recovered in the process. These molecules, together with the unloaded NADP and ADP can be reused in new photosynthesis reactions, as long as a fresh supply of water, carbon dioxide and light remains available.

The sugars and polysaccharides produced and stored in plants can be later used as food supply by the plants themselves or by animals eating these plants. The energy stored in these carbohydrates can be retrieved with the help of catabolic reactions. Catabolism occurs in two main stages: digestion and cellular respiration. During digestion, enzymes break complex biopolymers into their monomeric constituents: proteins are broken into amino acids, polysaccharides into sugars, and so on. Cellular respiration breaks the chemical bonds in nutrients and uses the released energy to load carrier molecules such as ATP. Cellular respiration can be aerobic, anaerobic, or hybrid. Aerobic respiration uses oxygen as oxidation agent, whereas anaerobic respiration uses other agents such as sulfate or nitrite. Hereby a rough summary of the reactions involved in aerobic respiration:

$$C_6H_{12}O_6 + 6O_2 + \text{unloaded carriers} \rightarrow 6CO_2 + 6H_2O + \text{loaded carriers.} \qquad (5.9)$$

This is again a redox reaction in which the sugar is oxidized to produce CO_2, whereas oxygen (O_2) is reduced to produce water; at the same time, carrier molecules are loaded with energy.

Two main stages can be distinguished in aerobic respiration: glycolysis and the Krebs cycle. Through a series of catalyzed reactions, glycolysis partially degrades glucose into intermediates called pyruvates, and loads the released energy onto carrier molecules. The pyruvates are then transported to mitochondria (in eukaryotes), where they are converted to intermediates able to enter the *Krebs cycle* (or *citric acid cycle*). This famous cycle breaks down the pyruvate derivates into CO_2 and water, loading more ATPs in the process. The Krebs cycle is very complex, and it is often cited to illustrate the complexity of metabolic pathways in cells. During this cycle, a number of precursor metabolites are also produced, which are later used in anabolic processes for the synthesis of various organic molecules in the cell. In simple terms, the Krebs cycle "burns" carbon compounds and uses the released energy to fuel itself and other metabolic processes further downstream. The catabolism of proteins and fats follow conceptually similar routes. Finally, catabolism also generates waste products that must be eliminated.

5.3.3   Gene Expression and Regulation

The information necessary for the cell to operate is encoded in its genetic material or *genotype*. The collection of observable characteristics or traits of an organism is called its *phenotype*.

In order to construct the genotype from the phenotype, the genotype must first be decoded and converted into molecules that are able to orchestrate the various operations of the cell. These molecules are mainly proteins (although RNAs also play a role). Some proteins act as construction "bricks" for the organism, others act as enzymes in metabolic reactions that produce other substances needed, such as lipids and sugars. An important class of proteins, the so-called *transcription factors* (TFs), act as genetic regulatory elements: they bind to given genes, activating or deactivating their expression, which causes other proteins to be produced or suppressed. RNAs also help to orchestrate various activities such as producing proteins and catalyzing specific reactions.

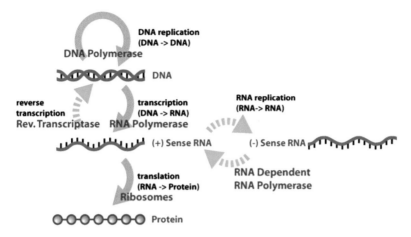

Figure 5.14 Central dogma of molecular biology. Image source: "Extended Central Dogma with Enzymes" by Daniel Horspool, Wikimedia Commons, `http://commons.wikimedia.org/wiki/File:` `Extended_Central_Dogma_with_Enzymes.jpg` .

The Central Dogma

Gene expression is the process of decoding the information in the gene to produce RNA and proteins. The assumption that information flows from nucleic acids (DNA or RNA) to proteins, and never in the reverse direction has been termed the *central dogma of molecular biology.* It assumes that DNA is *transcribed* to RNA, and RNA is *translated* into proteins, as shown in figure 5.14. DNA can replicate, and sometimes a reverse transcription from RNA to DNA occurs; however, proteins never revert back to DNA/RNA. In the light of newer experimental results, the central dogma has become disputed and is now considered a simplification of the actual processes happening.

The *transcription* from DNA to RNA is carried out by the RNA polymerase enzyme (RNAP). This enzyme recognizes and binds to a specific region of the DNA called a *promoter,* which marks the start of a gene. It then opens the double helix in order to initiate transcription downstream from the promoter. It then starts reading one of the single strands that acts as the template for the transcription. For each DNA nucleotide in the template sequence, RNAP attaches the complementary RNA nucleotide to the newly formed RNA sequence, thus elongating it, and then moves on to the next nucleotide. This process continues until a *terminator* DNA region is found, indicating the end of the translation. The new RNA strand is then released. The transcription is therefore one-to-one and reversible: information is not lost in the process, so the corresponding (complementary) DNA sequence can be recovered by reading the RNA in a similar way, a process achieved by reverse transcriptase enzymes used by some viruses (such as retroviruses, among them the HIV virus).

In prokaryotes, the RNA can be directly translated into a protein in a nearby ribosome. In eukaryotes, however, noncoding sequences or *introns* might be present in the RNA sequence. A process called RNA splicing then removes these introns before the RNA is ready to leave the nucleus and travel to a ribosome for translation. There are also sometimes more complex processes involved that are generally termed "RNA editing." The RNA that is ready for translation is called *messenger RNA (mRNA).*

Table 5.2 The Genetic Code: the rows to the left indicate the first base from the codon; the top columns indicate the second base; the third base is indicated at the right side of the table. Each cell displays the full codon together with the corresponding amino acid that it encodes (see Table 5.1 for the list of amino acids). "Stop" codons indicate the end of the translation.

Second nucleotide

5'	U	C	A	G	3'
U	UUU ⎱ Phe UUC ⎰ UUA ⎱ Leu UUG ⎰	UCU ⎱ UCC ⎬ Ser UCA UCG ⎰	UAU ⎱ Tyr UAC ⎰ UAA ⎱ Stop UAG ⎰	UGU ⎱ Cys UGC ⎰ UGA Stop UGG Trp	U C A G
C	CUU ⎱ CUC ⎬ Leu CUA CUG ⎰	CCU ⎱ CCC ⎬ Pro CCA CCG ⎰	CAU ⎱ His CAC ⎰ CAA ⎱ Gln CAG ⎰	CGU ⎱ CGC ⎬ Arg CGA CGG ⎰	U C A G
A	AUU ⎱ AUC ⎬ Ile AUA ⎰ AUG Met	ACU ⎱ ACC ⎬ Thr ACA ACG ⎰	AAU ⎱ Asn AAC ⎰ AAA ⎱ Lys AAG ⎰	AGU ⎱ Ser AGC ⎰ AGA ⎱ Arg AGG ⎰	U C A G
G	GUU ⎱ GUC ⎬ Val GUA GUG ⎰	GCU ⎱ GCC ⎬ Ala GCA GCG ⎰	GAU ⎱ Asp GAC ⎰ GAA ⎱ Glu GAG ⎰	GGU ⎱ GGC ⎬ Gly GGA GGG ⎰	U C A G

First nucleotide (left side label) — Third nucleotide (right side label)

The Genetic Code

The *translation* from RNA to proteins occurs in the *ribosome*, a sophisticated "molecular ma-chine" that produces proteins from mRNAs. There, special RNAs known as *transfer RNAs* (*tRNA*) read each sequence of three nucleotides (a *codon*) in the mRNA and map the codon to an amino acid, according to the genetic code. The genetic code, shown in table 5.2, is a translation table from all possible codons to all possible amino acids (plus a special "stop" codon that indicates the end of the translation process).

The tRNA molecule performs its task thanks to its special shape: one end is complementary to a specific codon, and the other end binds to the corresponding amino acid. There is one type of tRNA for each possible codon. The translation process within the ribosome is shown in figure 5.15.

The genetic code is universal across organisms (with very few exceptions), meaning that the same types of tRNAs are found in different organisms. The genetic code is also redundant: sev-eral codons may map to the same amino acid, a property denoted as *degeneracy*. Degeneracy has the consequence that information is lost during the translation process: For most amino acids, it is no longer possible to recover exactly which codon gave rise to it. Therefore, an exact "re-verse translation" from protein to RNA is no longer possible once the protein is made. This is the motivation behind formulating the central dogma, and it contributes to explain why phenotypic changes to an organism are not be propagated to its offspring.

Not all genetic material gets transcribed and translated. In fact, most does not end up in proteins. The majority of DNA in genomes was initially regarded as "junk DNA," but today it is believed to actually play important functional roles in organisms, although such roles remain

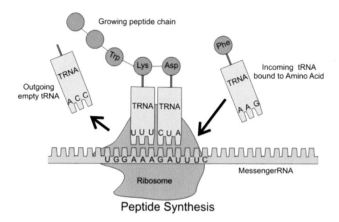

Peptide Synthesis

Figure 5.15 A messenger RNA being translated to a protein within a ribosome. Image source: "Peptide syn" by Boumphreyfr, Wikimedia Commons, `http://commons.wikimedia.org/wiki/File:Peptide_syn.png`

under investigation. The roles of the various RNA types are also only beginning to be explored and understood.

Genetic Regulatory Networks (GRNs)

Most organisms carry a large genome, with the potential of expressing a huge number of different gene products. However, only a few of these possible products are active at any given time. The organism adapts to its environment by expressing only the necessary genes. For instance, the expression of a gene involved in metabolism may be enhanced if there is abundance of food or water, while it may be suppressed under starvation or drought.

Gene expression is regulated by the interaction between genes, transcription factors, and signaling molecules coming from the environment. Such interactions form vast and complex *genetic regulatory networks* (GRNs) that still elude analysis. A lot of research today is dedicated to understanding and controlling the genetic regulatory processes in order to steer the response of organisms to various situations, which could lead to better methods to control or cure certain diseases, among other applications.

There are two types of transcription factors: *activators* and *repressors*. A repressor binds to the promoter site in the place of the RNAP and remains there for a while, preventing RNAP from binding there. As a result, transcription is inhibited during this period. An activator binds to a DNA region near a promoter that is otherwise poorly recognized by the RNAP, in such a way that it facilitates the binding of the RNAP to this site; in this way, it enhances the rate of expression of the corresponding gene.

The Lac Operon

In bacteria, a single promoter may trigger the transcription of a group of related genes. Such group is called an *operon*. Among these, the lactose (*Lac*) operon from the bacterium *E. coli* is a well-known model system for gene regulation. *E. coli* is able to digest both glucose and lactose; however, it "prefers" glucose. The Lac operon contains three genes that code for proteins that participate in the metabolism of lactose. Since this is not the first nutrient choice for the bacterium, this operon is switched off by default. The expression of the Lac operon is controlled by two TFs: the Lac repressor (LacI) and the activator CAP. The operon can be expressed only when two conditions are met: The repressor must be absent, and the activator CAP must bind to its activator binding site.

A schematic representation of the Lac operon is shown in figure 5.16. CAP can function as an activator TF only in association with another protein called cAMP, forming the CAP-cAMP complex. When glucose is abundant, the concentration of cAMP is low, which prevents the activator complex to form. The operon remains off (figure 5.16 (a)). The absence of glucose causes the concentration of cAMP to increase, thus enabling the formation of CAP-cAMP, which is able to bind to the Lac operon's activator binding site (figure 5.16 (b)). However, this is not sufficient to activate the operon, since the Lac repressor is always expressed, binding to the operon's repressor site, hence ensuring that the operon remains silent. In order to avoid starvation, the absence of glucose triggers other cellular processes that facilitate the uptake of alternative nutrients such as lactose. When lactose enters the cell, a lactose metabolite binds to the repressor, deactivating it (figure 5.16 (c)). The repressor then leaves its binding site, allowing the activator to operate. However, remember that the activator is only operational when glucose is absent. These two situations combined (absence of glucose with corresponding increase in the amounts of functional activator, and presence of lactose, resulting in the deactivation of the repressor) finally leads to the expression of the Lac operon, such that lactose can be digested (figure 5.16 (d)). Therefore, the operon can be expressed only when glucose is absent and lactose is present.

The Lac operon behaves like a logic gate "lactose AND NOT glucose," implemented with genes and proteins. Such logic naturally embedded in GRNs has inspired many researchers to look for ways to program cells using genetic circuits, as if they were dealing with electronic circuits. This topic will be further discussed in chapter 19 in the context of bacterial computing (section 19.4.1).

GRNs are of particular interest in the artificial chemistry domain for a number of reasons. First of all, the interactions among genes and their response to environmental signals can be modeled as an AC where molecules are DNA strands and proteins, and reactions include the binding/unbinding of a protein (transcription factor) to/from a gene, the construction of mRNAs and proteins using a machine-tape metaphor, together with the interaction between proteins and chemicals from the environment. Second, evolving GRNs are good examples of constructive ACs (see chapter 15), since mutated DNA sequences can give rise to different reaction networks implementing novel functions, with a potentially unlimited innovation potential. Finally, natural and artificial GRNs can be programmed or evolved to solve particular problems, falling in the chemical computing domain. Throughout this book, AC-based GRNs will appear in chapter 11, methods for modeling biological GRNs will be reviewed in chapter 18, and the programmability of GRNs will be covered in chapter 17.

5.4 Multicellular Organisms

In multicellular organisms, specialized cell types form various tissues that compose the different organs. At the beginning of the development, the single cell present in the fertilized egg divides into a number of cells forming a morula of identical cells. At some point in cell division, however, daughter cells differentiate into specialized tissue cells. Additionally, cell growth, division and death are controlled together with cell differentiation to give rise to the full organism shape, in

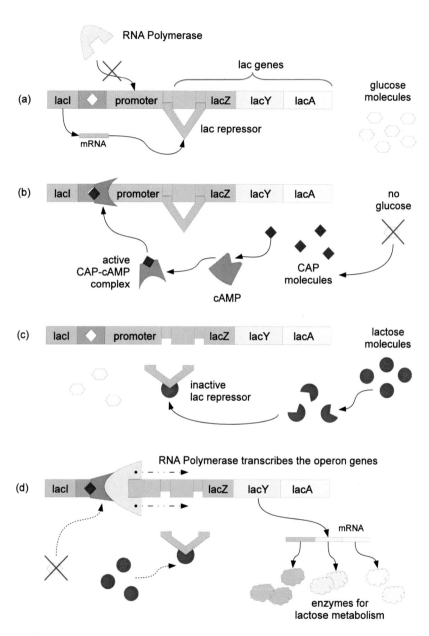

Figure 5.16 The Lac operon from *E. coli* encodes for proteins that help to metabolize lactose, a fallback nutrient for the bacterium. The operon is activated only when glucose is absent and lactose is present (bottom). (a) only glucose is present; (b) no glucose and no lactose; (c) both lactose and glucose are present; (d) lactose is present and glucose is absent.

a process called morphogenesis. The control of cellular development relies on communication signals that travel among cells.

Recent advances in the understanding of these three processes — cell differentiation, signaling, and morphogenesis — have spawned a surge of interest in these topics, not only within biology and medicine but also within the AC community. Such topics are gaining increasing prominence in the AC literature, where they are explored for essentially two purposes: The first and most obvious one is to model the development of real biological organisms; the second one is to obtain distributed algorithms inspired by morphogenesis, in which target structures can form in a self-organizing way and without central control. These two research directions will be further explored in chapters 18 and 17. The brief introduction provided here is intended to help understanding the research issues that will be covered in those subsequent chapters.

5.4.1   Cell Differentiation

Every multicellular organism starts with a single egg cell (zygote) that transmits its DNA to the daughter cells upon division. With few exceptions (such as germ cells for sexual reproduction, and some cells from the immune system), all cells in a multicellular organism have identical DNA. The process of cell differentiation is triggered by differences in expressed genes, in response to various input signals coming from the extracellular environment.

Once differentiated, a kind of "cell memory" is established: When a differentiated cell divides, it transmits its gene expression pattern to its daughter cell, which automatically adopts the same cell type. This mechanism guarantees that a liver cell will only give rise to another liver cell, and never to a cell belonging to a different tissue. Most often, however, the differentiated cell no longer divides, relying on the existence of undifferentiated *stem cells* for the production of further cells that can later differentiate into cells belonging to the target tissue.

Three types of stem cells can be distinguished [517]; at early embryonic stage, *totipotent stem cell* are present. These cells are able to generate all the embryonic tissues plus extraembryonic placental tissue necessary for embryo development. At a later stage of embryonic development, *pluripotent stem cells* can be observed. They can differentiate into all kinds of embryonic tissues, but no longer into placental tissue. In the adult organism, normally only *multipotent stem cells* are found, which can produce some but not all types of cells. One example of multipotent stem cells are *hematopoietic stem cells* found in the bone marrow, which are responsible for replenishing the organism with the various types of blood cells needed. Another example are intestinal stem cells, which continuously replenish the small intestine with its three types of epithelial cells. Figure 5.17 shows examples of adult stem cells and their differentiated offspring.

Although differentiated cells usually do not revert back to the stem cell stage, scientists have succeeded in forcing them to revert in the test tube upon injection of the right transcription factors. The stem cells artificially produced in this way are called induced pluripotent stem cells (iPS) [842], and they have promising therapeutic uses.

5.4.2   Cell Signaling and Transport

Several mechanisms for cell to cell communication can be found in nature [27, 607]. They can be based on passive or active transport of molecules. Passive mechanisms work downhill in the energy landscape, whereas active mechanisms require an external input of energy (in the form of ATP, for instance). Passive transport mechanisms involve molecules that diffuse in the

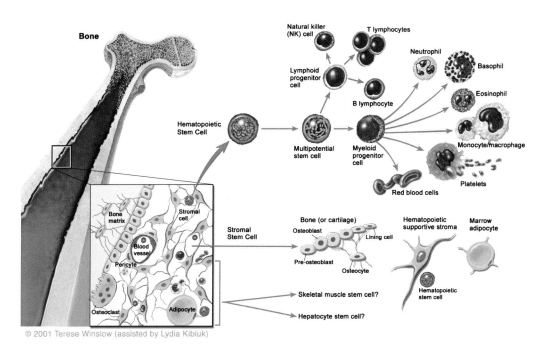

© 2001 Terese Winslow (assisted by Lydia Kibiuk)

Figure 5.17 Examples of adult stem cells: hematopoietic stem cells (HSC) differentiate into blood cells, including cells from the immune system; stromal stem cells give rise to cartilage, bone and adipocyte cells. © 2001 Terese Winslow (assisted by Lydia Kibiuk), reprinted with permission.

medium, either freely or through channels such as gap junctions and ion channels. Ion channels are protein-made pores in the cell membrane that allow ions to quickly enter or leave the cell following electrochemical gradients. They occur in neuronal synapses, cardiac, skeletal, epithelial, and a variety of other cell types. Gap junctions are protein channels embedded in the cell membrane that connect the cytoplasms of two cells together, allowing the passage of small molecules and ions from one cell to the other. Active transport mechanisms include molecular pumps that move substances against their electrochemical gradients, molecular motors such as myosin (involved in muscle contraction), and transport by bacteria (via the exchange of chemicals and genetic material). Figure 5.18 depicts an ion channel, a gap junction between two cells, and a myosin motor "walking" on top of a protein filament.

Signal transduction is the process by which extracellular signaling molecules trigger a cell response by binding to a receptor molecule on the surface of the cell, and initiating a cascade of chemical reactions that carry the signal to the interior of the cell, ultimately leading to a response to the stimulus.

Within cells, signaling cascades transmit information from the outside world to the interior of the cell, without actually carrying a messenger substance, but only by virtue of the transmission of information in the kinetic of enzymatic reactions that occur in cascades. Cell signaling networks involve such cascades plus networks of hormones that travel through the organism, delivering specific chemical signals to specific parts of the body.

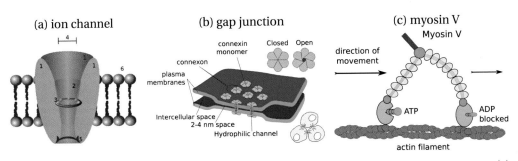

Figure 5.18 Examples of molecular communication channels (a),(b) and active molecular transport (c). (a) Ion channel; (b) gap junction; (c) myosin V molecular motor "walking" on top of a protein filament, powered by ATP (see [833] for details). Image sources: (a) "Ion channel" by Pawel Tokarz, Wikimedia Commons, `https://commons.wikimedia.org/wiki/File:Ion_channel.png` (b) "Gap cell junction-en" by Mariana Ruiz LadyofHats, Wikimedia Commons, `https://commons.wikimedia.org/wiki/File:Gap_cell_junction-en.svg` (c) drawn by the authors, based on [833].

5.4.3 Morphogenesis

The process of morphogenesis, responsible for guiding embryo development, is the result of a concerted effort to decide on the possible fates of a cell: a cell may grow, divide, differentiate, fuse with other cells, or die in a programmed way. Upon differentiation, the shape of a cell changes according to its function, for instance, some cells become very elongated (such as muscle and nerve cells) whereas others become tightly attached to their neighboring cells (such as epithelial cells), forming barriers that separate different regions of the body. Programmed cell death (apoptosis) also plays an important role in morphogenesis, by sculpting shapes such as the fingers of a hand. Figure 5.19 shows some examples of vertebrate embryo development in vertebrates.

Patterning genes are responsible for specifying the shape of an organism. Some of these genes express transcription factors, while others express proteins that have a role in cell adhesion or cell signaling.

Whereas the exchange of chemical signals and the resulting changes in gene expression patterns are crucial to guide the morphogenetic process, mechanical aspects play an equally essential role: cell motility, shape change (condensation, folding, swelling, or elongation of a cell), adhesion and mechanical forces between cells have an impact on the final shape of the organism.

5.4.4 The Immune System

The immune system [475] is responsible for defending the organism against pathogens. The immune system of vertebrates is extremely complex and much remains to be learned about it. However, not all organisms can count on such a fully developed defense mechanism. Plants, for instance, rely on intercellular signaling molecules and intracellular defensive compounds to fight invaders.

In this section we offer a very brief overview of the vertebrate immune system, just enough to follow some related work in Artificial Chemistries that will appear later in this book. We present the immune system from the perspective of [188, 214], who emphasize aspects of the immune

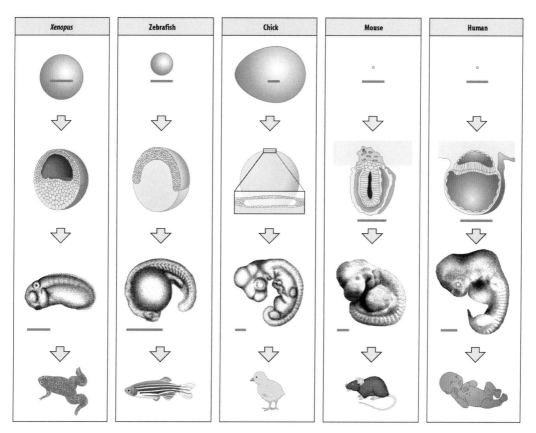

Figure 5.19 Examples of morphogenesis, illustrated through the developmental stages of various vertebrate embryos. Reproduced from figure 3.2 of [930], by permission of Oxford University Press, http://www.oup.com.

system that have been used as inspiration for artificial immune systems (AIS), a class of computer algorithms that are very related to ACs, as we will see in chapter 11.

The vertebrate immune system consists of a complex set of cells and molecules that interact in various parts of the organism in order to defend it against pathogens. Its major agents are leukocytes or white blood cells, produced in the bone marrow from hematopoietic stem cells. There are several types of leukocytes, each involved in a specific aspect of the immune response. These various cells communicate with each other through chemical signals called cytokines. Some of the main components of the immune system are listed in table 5.3.

Figure 5.20 shows a simplified view of the main components of the immune system and how they interact with each other. The activities of the immune system can be divided into two categories:

- The *innate immune system* offers a first line of defense against invaders that penetrate the body, for instance following tissue damage. It mounts a prompt reaction to the invading microorganisms, in the form of an inflammatory response, which increases the blood flow to the affected area, and recruits white blood cells to react to the damage. Among the

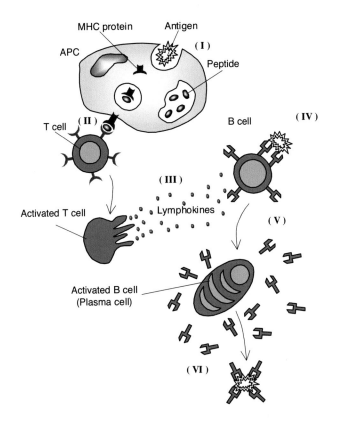

Figure 5.20 Simplified overview of the immune system, from [215]. Pathogens display antigens on their surface (I). Macrophages engulf and digest pathogens, breaking the antigens down into small peptides. These peptides bind to MHC proteins that move to the cell surface and display the pathogen's peptides to T-cells (II). Upon binding to the MHC/peptide complex, T-cells become activated and secrete lymphokines (III) that recruit B-cells to the infection site (IV). Upon binding to an antigen, B-cells become activated (V), and secrete large amounts of free antibodies that bind to the pathogen (VI), surrounding it and signaling it for destruction. Reproduced from [215] with the authors' permission.

Table 5.3 A glossary of basic immune system elements.

Name	Type	Short description and main role
antigen (Ag)	molecule	recognized by the immune system, potentially foreign
antigen-presenting cell (APC)	phagocyte	presents antigens on its surface, in the form of MHC/peptide complexes
antibody (Ab)	protein	released from the surface of B-cells to bind to antigen
B-cell	lymphocyte	exhibits antibodies on its surface, recognizes free antigens with high specificity
cytokine	signaling molecule	induces inflammatory response and fever
leukocyte	white blood cell	any of the cells from the immune system, including macrophages and lymphocytes
lymphocyte	leukocyte	mediates adaptive immune response
lymphokine, or interleukine (IL)	cytokine	secreted by lymphocytes
macrophage	APC	engulfs pathogen and presents its antigens for recognition by T-cells
major histo-compatibility complex (MHC)	protein complex	present on the APC surface, binds to antigen forming MHC/peptide complex
MHC/peptide complex	protein-ligand complex	found on the surface of APCs, recognized by T-cells
plasma cell	effector B-cell	secretes large amounts of free antibodies
phagocyte	leukocyte	engulfs and digests harmful material such as pathogens and dead cells
T-cell	lymphocyte	recognizes antigens on the surface of APCs in the form of MHC/peptide complexes

recruited cells there are phagocytic cells such as macrophages that engulf and digest microorganisms such as bacteria. These phagocytes are able to recognize and bind to common molecular patterns found on the surface of these microorganisms, but they do so in a nonspecific way, reacting in a similar way to any kind of microorganism encountered. Macrophages do not limit themselves to destroying the captured foreign material; they also display fragments of the digested microorganisms on their own cell surface, in the form of a MHC/peptide complex, such that they can be recognized by T-cells and trigger an adaptive immune response.

- The *adaptive immune system* mounts a targeted immune response against a specific infectious agent. There are essentially two ways to trigger an adaptive immune response: via T-cells that bind to antigen-presenting macrophages, allowing the body to eliminate infected and defective cells (such as cancer cells); and via B-cells that produce antibodies able to recognize free-floating antigens in a very specific way, directly attacking the pathogens and their byproducts. The cells of the adaptive immune system are able to collectively form an immune repertoire of previous infectious agents, constituting an immune memory of past attacks: when faced with a previously encountered pathogen, the immune response to it is much faster and more efficient than at the first encounter, preventing reinfection.

The innate and adaptive immune systems are closely intertwined and mutually regulate each other, working in a concerted way to restore the body to a healthy state. This happens through a complex network of interactions forming positive and negative feedback loops. Some of the aspects of this network are reviewed here.

Clonal Selection, Negative Selection, and Idiotypic Networks

The adaptive immune response is mediated mainly by two kinds of lymphocytes: T-cells and B-cells. Both types of cells are produced in the bone marrow, but while B-cells remain in the bone marrow until maturation, the T-cells migrate to the thymus for their maturation. During the maturation of both cell types, they undergo a *negative selection* process in which the lymphocytes are presented with self-antigens from the own body. Those lymphocytes that strongly bind to the self-antigens are eliminated. This maturation phase enforces a distinction between self and nonself that is crucial to make the immune system tolerant to the organism's own substances. The mature lymphocytes, called naive B- and T-cells, are sent out to the bloodstream, where they will be exposed to the actual antigens.

When a naive T-cell recognizes and binds a MHC/peptide complex on a macrophage, it becomes activated. Active T-cells divide, proliferate, and differentiate into effector and memory T-cells. Some of these effector cells secrete cytokines that further recruit more leukocytes to the infection site, creating a positive feedback loop. Other effectors have a cytotoxic activity, killing infected cells. The memory T-cells join the immune repertoire of the organism, providing future immunity against the same or similar pathogens.

Activated T-cells also secrete lymphokines, a class of cytokines that recruit B-cells to the site of infection. Each B-cell carries several copies of identical antibodies on its surface, targeting a specific antigen. A wide variety of B-cells are present in an organism, each with a specific target. The specificity of the recognition is given by the shape of the antibody's binding sites, that is complementary to the shape of the antigen that it recognizes, like a lock to a key. This lock-and-key principle of molecular recognition and binding is pervasive in biology, and it is the basis of functioning of many enzymes and cell signaling molecules. The part of the antigen that is recognized by the antibody is called an *epitope*, and the corresponding docking region on the antibody is called a *paratope*.

When a naive B-cell succeeds in binding to an antigen, it becomes activated. Activated B-cells undergo a second maturation phase called *affinity maturation*. During this phase, they proliferate via cycles of cell division, somatic hypermutation, and clonal selection. *Somatic hypermutation* is a mechanism that promotes an accelerated mutation rate in the genes encoding antibody paratopes. Recombination is accelerated, too. Among the mutants, those that bind to the antigen with the highest affinity are selected, and those that fail to bind die out. This process is called *clonal selection*, and it results in the amplification of the population of B-cells that are specific for a given antigen, and in an increasing refinement of their specificity. The selected B-cells further differentiate into effector B-cells (also called plasma cells) and memory B-cells.Plasma cells secrete big amounts of free antibodies that swarm to the damaged area and surround the infectious agents, signaling them for destruction. Memory B-cells join the immune repertoire of the organism. In contrast to memory T-cells which have a low specificity, memory B-cells are highly specific to their antigens.

T-cells regulate each other and other cells, and together with the other cells of the immune system they form complex networks of interactions called *idiotypic networks*. Such networks are involved in adaptation, learning, immune memory, and the maintenance of a homeostatic balance inside the organism.

5.5 Summary

In this chapter we have offered a broad overview of important aspects of the chemistry of life that are commonly encountered in artificial chemistries. While the discussion could not fully cover all biology and biochemistry, we hope that this chapter can guide newcomers to the field and give them a taste of the complex but fascinating world of biology. The background information provided in this chapter should help to understand the subsequent chapters of this book, starting with chapter 6, which examines the various theories about the origins of life.

Organisms are different from machines because they are closed to efficient causation.

ROBERT ROSEN, LIFE ITSELF, 1991

THE ESSENCE OF LIFE

The basic atoms of life (carbon, hydrogen, oxygen, and nitrogen) are everywhere. They are among the most common and cheapest chemical elements found on Earth. And yet, mixing them all together in a container does not result in anything near organic. The key to life is the way these atoms can form basic organic building blocks, and then how these building blocks can be combined into highly organized complex structures.

The organization of current life forms is very complex indeed. How does such complexity emerge? Would it be possible to reduce it to the bare essential? Would it be possible to put together a very simple cell out of the basic building blocks of life, one that is much simpler than today's organisms, but still functional and able to evolve into more complex forms? What is then the essence of life?

Now that we have looked at how the building blocks of life are combined into the complex organisms that we see today, we will look at how they could be combined (hypothetically) to build the simplest possible living cell. Initially, assume that building blocks such as fatty acids, amino acids, and nucleotides are available in abundance. The possible origin of these compounds, and how they could have led to life on Earth, will be discussed in section 6.2.

6.1 A Minimal Cell

Recall from the operational definition of life (chapter 5) that a minimal cell requires three basic subsystems: a compartment, an information carrier, and a metabolism. Different origin-of-life theories differ in their hypotheses about which of these subsystems may have emerged first, and how they may have been integrated to form a living cell. This topic will be covered in section 6.2. For now, we will look at how each individual subsystem may be put together using the building blocks of life introduced in section 5.2. We will then look at how the subsystems may be combined into a minimal cell, as an outcome of a natural or artificial process, without any necessary connection to what may have actually happened when life first appeared on the primitive Earth. A brief summary of the various theories about the actual origin of life on Earth will then follow.

6.1.1 Membrane Compartment

A compartment defines the boundaries of the cell, protecting its information and metabolic content from the external environment. A compartment must also perform the role of a communication channel, selectively importing nutrients into the cell, exchanging signaling molecules, and expelling waste and toxic substances. The structure of modern cell membranes offers an example of a sophisticated compartment: a number of specialized proteins are embedded in the membrane's phospholipidic bilayer, forming selective pores that actively facilitate the passage of certain molecules while blocking others. Are there simpler and yet feasible types of compartments?

One of the easiest ways to create simple compartments is by throwing fatty acids into water. As outlined in section 5.2, fatty acids are amphiphiles with a simpler molecular structure than phospholipids, but are also able to form micelles and vesicles similar to those shown in figure 5.3. Micelles are small spherules, while vesicles are larger bubbles. Such structures "self-assemble," that is, they form spontaneously; hence, they are stable yet dynamic, and easily rebuild when shaken apart. Free fatty acids may be incorporated in the structure, causing it to grow. As it grows, its shape may become irregular, and it may be divided by mechanical forces.

Vesicles have a hollow interior which is also filled with water, making them a suitable container to hold the other subsystems. Vesicles made of fatty acids have also been shown to be permeable to small molecules such as nucleotides, amino acids, and small food molecules [839]. At the same time, bigger molecules such as biopolymers tend to remain inside the vesicle.

Since such vesicles are able to self-maintain, grow, and divide, can we call them alive, even if empty or just full of water inside? The answer is no, because they are not able to carry genetic information in a reliable way, and to transmit this information to the next generation. There is no "program" controlling the maintenance of the cell. The growth and division processes are erratic, subject to external factors, and prone to disruptions that can easily destroy the compartments. However, this answer is not so straightforward as it may seem. It is also possible to build roughly spherical aggregates out of different types of lipids [761]. These aggregates can also grow and divide, and their composition ("compositional genome" or "composome") can be regarded as information that is transmitted during division, albeit less accurately and with a lower carrying capacity than what can be achieved with linear polymer sequences. The idea of such aggregate-based information carriers gave rise to the Lipid World hypothesis for the origin of life, further discussed in Section 6.2.4.

Since it is very easy and cheap to obtain compartments by self-assembly using amphiphiles, this technique is being used to explore alternative designs for artificial protocells [691], see chapter 19.

6.1.2 Genetic Information

The information subsystem is at the core of all forms of life. Organisms use this information to maintain themselves, and they must transmit it from one generation to the next. Offspring resemble their parents. The classical view of information subsystems is anchored upon life on Earth: DNA and RNA molecules carry the genetic information needed to construct the organism, and this information is encoded in their polymer sequence.

If we wished to "design" the simplest possible cell, what kind of information system would we use? Certainly not the complex machineries of today's cells. We could then ask ourselves whether we would use only DNAs, only RNAs, or perhaps even only proteins. Researchers have debated this topic extensively over the years, and have reached a rough consensus that proteins by themselves would not make a good information storage substrate, since they are unable to replicate in general, and their alphabet of about 20 digits is overly complicated. Similarly, DNAs are not good candidates either, because they form too stable structures that need several enzymes to be replicated. This leaves us with RNA molecules: RNA molecules form less rigid structures that can be copied by template replication. Furthermore, some RNAs have shown to have catalytic properties ("ribozymes") [458, 657, 844]. Hence, if we are looking for a simple way to carry information within a cell, RNA seems to be a good candidate to start with. The idea would be to construct a system made only of RNAs, where some RNAs would store information, others would help to replicate them, and others would help with the various other processes necessary to maintain the cell. This idea gave rise to the RNA World hypothesis for the origin of life [327].

However, storing and transmitting information in RNAs is not as simple as it might appear. The polymerization of RNA strands in water is unfavored; more problematically, side reactions and errors in the replication process hinder the accuracy of information transmission. In modern cells, the replication of genetic material counts on sophisticated error correction processes aided by specialized enzymes. In the absence of such enzymes, replication is unfaithful and tends to accumulate too many errors. This compromises the fidelity of the information transfer and generates offspring that is unable to survive, ultimately leading to the mass extinction of the species. In the 1970s, Eigen [249] showed that there is a threshold in the amount of replication errors, above which the species cannot survive. He also showed that this error threshold imposes a maximum sequence length beyond which the error threshold is crossed, leading to an "error catastrophe" (mass extinction). Therefore, without the aid of error-correcting enzymes, only small sequences could be replicated with sufficient accuracy. But longer sequences are needed to form enzymes, and if the replication process relies on these enzymes, then we are left with a chicken-and-egg problem known as the "Eigen paradox." Small molecules could survive by nonenzymatic replication, but the evolution of more sophisticated functions would require such primitive replicators to elongate while crossing the error threshold. How to achieve this in a hypothetical RNA World and in synthetic protocells are subjects of current research, further discussed in Sections 6.2.2 and 19.2.

6.1.3 Metabolism

As explained earlier, all life forms require some source of energy for their survival. A living system can be seen as an open dynamic system kept out of equilibrium by a continuous inflow of energy and outflow of waste. The metabolic processes in today's organisms take care of harvesting the necessary energy, storing and transforming it along several metabolic pathways such that it can be retrieved when needed and used for the various activities of the cell. Without a metabolism, organisms would have to rely entirely on the prompt delivery of all sorts of needed raw materials from the outside, and would quickly die as soon as these raw materials were depleted. Moreover, the complexification of life forms would be hindered by their inability to manufacture increasingly more sophisticated molecules.

Most of the metabolic reactions in today's organisms rely on specialized enzymes that accelerate their rates to acceptable levels. Moreover, it is not sufficient to have a serial reaction pathway, made of a chain of reactions occurring one behind the other, transforming nutrients to useful components for the cell. Such a serial construction would easily break down due to a missing intermediate and would be seriously limited by the rate of the slowest reaction. In order to construct robust metabolic networks, one needs redundant metabolic pathways (such that

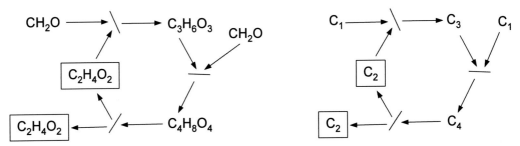

Figure 6.1 Schematic representation of the formose reaction, a possible base for a minimalistic metabolism. Left: organic compounds reacting in an autocatalytic cycle that takes two molecules of formaldehyde (CH_2O) and produces two molecules of glycolaldehyde ($C_2H_4O_2$), that can then be used to produce other organic compounds such as sugars and lipid components. Note that the component $C_4H_8O_4$ of the cycle is already a simple sugar (tetrose). Right: abstract representation of the cycle on the left, with C_i labels representing the number i of carbons in the actual molecule. Adapted from [305].

at least one of them remains functional when others are knocked out), cyclic reaction networks (able to regenerate and reuse components), and more especially, autocatalytic reaction networks (able to replicate their constituents).

What would a minimalistic metabolism look like? Researchers in biology and artificial life have considered this question for a long time. One possibility might be something like the formose reaction (figure 6.1). The formose reaction is actually not a single reaction but an autocatalytic cycle of chemical reactions that is able to produce sugars such as tetrose and ribose (a nucleotide component), as well as glycerol (a lipid component, see section 5.2.5) from formaldehyde, a common organic compound in the universe. It does all this without the need for biological enzymes or ATP input. For these reasons the formose reaction is a good candidate constituent of primitive metabolic processes in early life forms, before the appearance of specialized enzymes and energy carriers such as ATP.

The next question to ask is, how could metabolisms emerge and evolve? This question is a difficult one with no clear answer so far, and may be asked in various contexts: the first one is within origin of life hypothesis, and will be addressed in section 6.2.3 when looking at "metabolism first" theories for the origin of life. Another context is the synthesis of artificial metabolisms in the laboratory for various purposes, and will be discussed in chapter 19. Finally, it is interesting to look at abstract models of metabolic processes, expressed as artificial chemistries. These will be covered in section 6.3.1.

6.1.4 The Chemoton

In order to compose a living cell using the three subsystems described earlier, a useful exercise is to imagine an abstract chemistry able to give rise to entities with lifelike properties and to support them afterward. Someone had already thought about this before: Back in the 1950s, Tibor Gánti proposed the *Chemoton*, one of the first abstract models of a living cell [305, 306]. It was conceived around 1952 and published in Hungary in 1971, but became known in the Western world only much later. Since that time, the progress in molecular and cell biology has been enormous. Synthetic biology has emerged, and numerous alternative designs for synthetic protocells

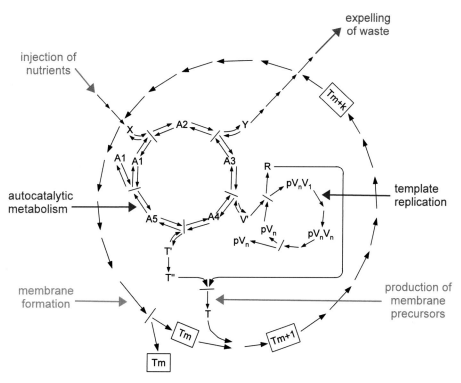

Figure 6.2 The Chemoton by Tibor Gánti [305, 306], an early abstract protocell model: a membrane consisting of molecules of type T encloses the cell's genetic and metabolic subsystems. The genetic material is made of single-stranded (pV_i) and double-stranded (pV_iV_j) polymers. Molecules A_i run the metabolic subsystem that autocatalytically produces template and membrane molecules from nutrients X, expelling waste Y in the process. Adapted from [305, 306].

have been proposed (see [692, 787] for an overview). Yet, the basic concepts behind the original Chemoton model remain valid and are very instructive to understand the principles behind the organization of life in a simple and abstract way, without resorting to all the complex details of real chemistry.

The Chemoton is shown in figure 6.2. One can distinguish a surrounding membrane made of m molecules of type T. Nutrients are taken up from outside in the form of X molecules, that enter an autocatalytic cycle representing the metabolism of the cell.

The metabolic subsystem uses the X molecules to run a cyclic reaction network that transforms molecule A_i into molecule A_{i+1} (modulo the size of the cycle, which is 5 in the model). Each turn of the metabolic cycle produces raw materials for the other subsystems: precursors of T molecules for the membrane, and V molecules for the genetic subsystem. It also generates waste molecules Y that are expelled out of the cell. The last step of the metabolic cycle splits one molecule of A_5 into two molecules of A_1, therefore the cycle is autocatalytic. This cycle is a conceptual simplification of the formose reaction (figure 6.1), already mentioned as a candidate component of a primitive metabolism.

The genetic material takes the form of pV_n molecules, where p indicates that the molecule is a polymer made of n monomers of type V. The genetic information is encoded in the pV_n molecules, and is copied by a process called *template replication*. Template replication is a simplified form of what occurs today in DNA replication. First of all, a single-stranded polymer pV_n serves as a template for the binding of free monomers to the monomers already in the chain, forming a double-stranded ladder pV_nV_n. The double strand then splits into two single strands that can again serve as templates, and the cycle repeats, duplicating the number of template molecules with each turn. It is therefore another example of an autocatalytic cycle.

In the Chemoton, the template replication cycle also generates R molecules that participate in membrane production. These R molecules react with the precursors T'' produced by the metabolic cycle to produce the actual T molecules to be incorporated in the membrane. In this way, the speed of membrane formation is directly coupled to the speed of replication of the genetic material, and to the metabolic rate with which the "A-wheel" turns.

Starting with a given amount of materials inside the cell, when the genetic material duplicates, the membrane surface doubles. At the same time, the overall amount of substances within the cell also doubles, so the volume of the cell doubles as a consequence, but this is insufficient to compensate for the surface doubling. It follows that the cell can no longer maintain its spheric shape due to osmotic pressure, and it ends up dividing into two daughter cells, each with roughly half of the material of the mother cell. Each daughter cell then goes back to the initial state, taking up X molecules from the environment to produce more membrane and template constituents, again reaching an osmotic tension that ends up splitting the cell in two, and so on. The chemoton is "alive" in the sense that it maintains its three subsystems, and is able to reproduce.

Gánti's Chemoton model offers a computational vision of the cell. The word *Chemoton* stems from "chemical automaton." Gánti refers to his Chemoton as a "program-directed, self-reproducing fluid automaton," in contrast to von Neumann's self-replicating machines [893]. The program in the Chemoton is the set of autocatalytic reactions that drive the survival and replication of the cell. It is fluid, as opposed to solid electric or mechanical devices. Its fluid nature makes it more flexible and easy to replicate, since new cells may easily move to other areas in fluid space, something that is rather difficult to achieve when trying to copy solid parts. Gánti's vision of the minimum cell is also that of a machine doing useful work: The cell is organized as a set of interconnected chemical cycles, each of which performs useful work by producing molecules that will be used in other parts of the cell.

The Chemoton was ahead of its time, and it is a very elegant model of a simple cell. It has been subject to extensive studies [204, 271] and has been recently shown to exhibit complex dynamics [604], being to some extent robust to noise [880, 949] and environmental changes such as food shortage [273], and able to differentiate into species [159]. However, the Chemoton is sometimes too simplistic and unrealistic. For instance, Gánti acknowledges that at least four different types of template monomers are needed to constitute a functional information system (akin to the four nucleobases in DNA and RNA). These different types are not explicitly modeled in the Chemoton, where all template monomers are just represented as V molecules. Actually, the genes in the Chemoton do not play such a crucial role as the genes in modern cells. The Chemoton relies on the synchronized growth and replication of its internal materials and membrane surface, and this synchronization is achieved via stoichiometric coupling between the different subsystems. This makes the Chemoton fragile in the face of changes in the internal concentration of its compounds.

The Chemoton was designed by a scientist using a thorough knowledge of chemical reaction networks and their mathematical modeling. In contrast, how would a minimum cell emerge spontaneously in some primordial ocean on Earth or elsewhere? Would it look like the Chemoton, or would it be based on an entirely different set of reactions? Would there be a variety of feasible primitive cells, all with potentially different internal architecture, and perhaps competing for resources? How could it be set to evolve to more complex life forms? The Chemoton does not provide direct answers to these questions. For this we must turn to two interrelated and complementary exploration fields. The first of these is an investigation into the possible origin of life on Earth, looking at our past in search for these answers. The second field is the exploration of alternative emergent protocell designs from the bottom up [691], either by computer simulations using artificial chemistries, or by attempting to obtain synthetic cells *in vitro*, thus seeking these answers by looking toward the future. These different research directions will be surveyed in the remainder of this chapter and in chapter 19.

6.1.5 Autopoiesis

Another early conceptual model of living systems was put together by Varela, Maturana, and Uribe in the 1970s [557, 881]. It is based on the notion of *autopoiesis*, which means "self-production." In contrast to others, Varela and coworkers emphasize the self-maintenance aspect of living systems, rather than their reproductive capabilities. A system can be alive if it is able to maintain itself autonomously, that is, if it is autopoietic. It could remain alive forever, without ever reproducing or dying. Neither is evolution by natural selection an essential property of minimal life forms.

According to [881] an autopoietic unit must satisfy six conditions. It must have *identifiable boundaries (1)* with inner constitutive elements or *components (2)*; the interactions and transformations of these components form a *mechanistic system (3)*; the unit's boundaries are formed through *preferential interactions (4)* between the elements forming these boundaries; all the components of the unit, including its *boundary components (5)* and *inner components (6)* are produced by interactions between the unit's own components.

A computer simulation of an autopoietic system satisfying these conditions was then designed in [881]. It consists of a two-dimensional grid where particles of three types (substrates, catalysts, and links) move and interact. Link particles are membrane constituents that can bond to each other. A catalyst molecule converts two substrate molecules into a link molecule. As they bond, links may form closed membranes that are nevertheless permeable to substrate molecules. Link molecules also decay spontaneously. Using this simple model, the spontaneous formation of an autopoietic unit was demonstrated in [881], as depicted in figure 6.3. In the same paper, the authors also show that the boundary can be regenerated after the decay of some of its constituent molecules, illustrating the self-maintaining features of the autopoietic system.

Following that pioneering work, Maturana and Varela later published a book on the topic [559], and subsequently attracted the attention of numerous researchers who tried to reproduce their results and improve upon their computer experiments [136, 156, 282, 421, 437, 560, 569, 637, 639–641, 729, 730]. These ideas were also brought into synthetic biology for the creation of minimal cells in the laboratory [531, 533–536, 801].

6.1.6 Robert Rosen's ideas

Robert Rosen was one of the pioneers of the now widespread opinion that biology must be approached from a complex systems perspective, and that reductionism is not sufficient to understand biological phenomena. Rosen argued that biological organisms can be distinguished

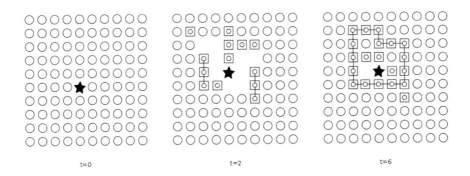

Figure 6.3 Computer simulation of autopoiesis: Initially ($t = 0$) only one catalyst molecule (*) is present, surrounded by substrate molecules (O). Link molecules [O] are formed in a catalytic reaction $*+2O \rightarrow *+[O]$. This can be seen at $t = 2$. Link molecules tend to bond to each other, eventually leading to the formation of a closed boundary resembling a cell membrane (visible at $t = 6$). Adapted from [881].

from simple physical systems by their "organization," a result of the complex interactions between the organism's several parts, which cannot be reduced to the parts alone: When we break the system apart in order to study it, we destroy its organization and therefore cannot see how it functions [725]. This basic property of organization is what in systems theory, and in models of complex systems today in particular, is referred to with the saying "the whole is more than the sum of the parts." From this perspective, Rosen can be considered as one of the founders of systems biology.

Rosen's concept of organization is linked to self-organization and thermodynamics: *"a system is organized if it autonomously tends to an organized state"* (p. 115 of [725]). We know from the second law of thermodynamics that a closed system tends to a state of thermodynamic equilibrium, associated with a state of maximum entropy or maximum disorder. Therefore, in order to be organized the system must be open and thus out of equilibrium. One can measure the degree of organization of a system by measuring its distance from equilibrium, or equivalently, by measuring the improbability of its state.

Rosen also claimed that an organization is independent from its material support. Therefore, a complex living system could be studied in an abstract way through the modeling of its organizational properties. He then set out to explain how such living organizations could work. For this purpose, he proposed an abstract mathematical model of a living organism called *metabolism-repair (M,R) system* [720, 723, 724]. An (M,R)-system is a formalism intended to capture the minimal functionality of a living organism, without details of its biochemical implementation. In an (M,R)-system, metabolism (M) and repair (R) subsystems interact to keep the organism alive (see figure 6.4). Rosen's metabolic subsystem is a mathematical abstraction of biological anabolic and catabolic functions. The repair subsystem is an abstraction for a genetic subsystem, containing the information necessary to construct the (M,R)-system, regenerating and replicating it when needed. (M,R)-systems contain no membrane abstraction, consistent with Rosen's goal of a very

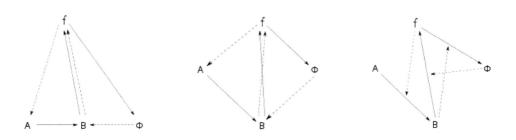

Figure 6.4 Rosen's (M,R) systems are closed to efficient causation. The notion of these systems has developed over the years. A, B, f, Φ are components (material and functional) of the system. A is transformed into B with the help of f; B is transformed into f with the help of Φ; f is transformed into Φ with the help of B. Left: Rosen's original unsymmetrical conceptualization [725]. Middle: Symmetrical depiction as proposed by Cottam et al [201]. Right: More conventional use of reaction and substrate notation as proposed in [343]. Line arrows symbolize material cause, broken line arrows efficient cause.

minimalistic formalism. Self-replication follows as a consequence of requiring the avoidance of an infinite regress of repair systems.

After an initial graph-theoretical formulation of (M,R)-systems in [720], a more powerful formulation based on category theory was proposed in [723, 724]. About a decade ago, a relation between (M,R)-systems and autopoietic system was discovered: Both theoretical frameworks abstract away from particular features of their parts/components, and emphasize the circular character of causation in these conceptual models of living systems [509]. The authors conclude that autopoietic systems are a subset of (M,R)-systems. Certain features of (M,R)-systems, one being the controversial noncomputability feature proved by Rosen for (M,R)-systems would therefore carry over to autopoietic systems and render them noncomputable as well [180, 725].

6.2 Origin of Life

For many centuries, people believed in the theory of spontaneous generation: Life (or at least microbes) could arise spontaneously from a mix of nutrients. Rudolph Virchow proposed in 1855 that life could only come from life, what he expressed as "omni cellula ex cellula." This idea was corroborated by Louis Pasteur, who in 1862 proved beyond any doubt that earlier experiments that seemed to have proven otherwise were flawed due to inadvertent contamination. But if this is so, how did the first life form appear? It was then postulated that if life once arose out of inanimate matter, this must have happened a very long time ago, under conditions that were totally different from the ones found nowadays.

There are various hypotheses about how life might have originated on Earth, based on the early chemical composition and atmospheric conditions of the prebiotic Earth. There are hypotheses focusing on each of the elementary subsystems as the one that could have emerged first: compartment, metabolism or information first. And there are hypotheses combining subsystems. Probably we will never know for sure how exactly life appeared on Earth,[1] but the various theories help us not only to shed light on the possible past history of our planet, but also to learn about the physicochemical processes responsible for producing life from inanimate matter. Such fundamental knowledge can then be applied to various areas of science and technology, and ultimately, it can be used to seek life in other planets.

In this section we give a brief overview of the existing theories on the origin of life, with special attention to those aspects of these theories that have been investigated with the help of artificial chemistries. For more comprehensive recent reviews on the topic, see [48, 140, 449, 534, 710, 763].

6.2.1   The Prebiotic World

About four billion years ago, the Earth was cooling down, and water vapor in the atmosphere condensed to form the oceans. It is believed that some basic organic building blocks of life formed "shortly" after that, around 3.8 billion years ago.

In 1952, through a series of famous experiments, Miller and Urey [587, 588] showed that various organic compounds could be formed under conditions similar to those assumed to have occurred in the early Earth. Based upon previous theoretical works by Oparin and Haldane [299, 644], they assumed that the primordial atmosphere of the Earth was composed of hydrogen (H_2), ammonia (NH_3), methane (CH_4), and water vapor (H_2O). In their experiments, these components were mixed in a flask, and an electric discharge was applied to them. As a result, amino acids such as glycine and alanine, and a number of other organic compounds were formed.

Subsequently, other scientists showed how to synthesize other building blocks of life, including adenine, guanine, and ribose. Today, plausible chemical pathways have been discovered for synthesizing most building blocks of life from inorganic compounds (see [184] for an overview). However, there are still gaps in this knowledge. For instance, sometimes the yield of a given compound is too low, or too many undesirable side-products are formed. Worse, the conditions for the formation of some building blocks sometimes differ from those of the formation of others, questioning how they could have been put together into early life forms. Even the composition of the prebiotic oceans and atmosphere is still subject to debate [950]. Some claim that hydrogen would have easily escaped to outerspace, and that methane and ammonia would have been quickly destroyed. Others sustain that the early atmosphere contained significant amounts of volcanic gases such as carbon dioxide (CO_2) and nitrogen (N_2), conditions that do not favor the formation of organic compounds. These difficulties make some believe that a large portion of the organic material needed to bootstrap life arrived on earth from outerspace via comets and meteorites (the panspermia theory). Others suggest that the basic building blocks started underwater, near hydrothermal vents (volcanic fissures that release hot water).

Whatever their origin, these organic compounds would have ended up in the oceans and ponds of the primitive Earth, forming a prebiotic "soup," that would later lead to the formation of the first cells. Once the basic building blocks were there, floating in the primordial soup, how did they combine to form the first primitive organisms? Several theories have been proposed to explain the emergence of life, each with their strong and weak points, and no overall consensus has been reached so far.

One of the most famous theories is the RNA World hypothesis, according to which primitive nucleotides polymerized into RNA molecules or some variant thereof. Some of these polymers

---

[1]Some ideas discuss that life originated elsewhere in the universe, e.g., on Mars, or even farther out.

would present catalytic properties (ribozymes). Eventually some ribozymes became able to copy other ribozymes, and get copied too, giving rise to the first replicators.

Another theory is the Lipid World hypothesis, according to which lipids, or even simply fatty acids, self-assembled into compact aggregates, or into hollow vesicles, that have the natural ability to growth and divide [760].

A third theory is the metabolism-first hypothesis. It holds that life was bootstrapped from a web of chemical reactions forming a primitive metabolism able to use energy to maintain itself. Such web of reactions would typically contain autocatalytic cycles such as the formose reaction, which could later be accelerated by primitive catalysts. The energy flow within the organism would be ensured by coupled redox reactions.

These theories are often combined to produce hybrid explanations for the origin of life: For instance, a primitive metabolism could have supplied building blocks for the first self-replicating information molecules. Szostak [839] proposes a protocell made of a primitive replicator enclosed in a membrane, without an explicit metabolic network. Shapiro [769] highlights a boundary as the first requirement in a metabolism-first scenario; the boundary is needed to allow the organism to increase its internal order while the external entropy increases.

6.2.2 The RNA World

Cells rely on genetic information contained in the DNA to carry out their vital functions. Today, this information is transcribed from DNA to RNA, and then translated from RNA to proteins. At some point in time, a first molecule or group of molecules able to make crude copies of themselves must have emerged in the primitive Earth. DNA is too stable and its replication requires a complex mechanism guided by enzymes. Therefore, various theories on the origin of life generally assume that more primitive molecules played the role of information carriers in early life forms. The *RNA World hypothesis* says that RNA was such a molecule: A primordial mechanism for RNA replication could have been based on *ribozymes*, RNA molecules able to act as enzymes, cleaving or ligating other RNA molecules. Such built-in catalytic ability is a crucial first step to explain how RNAs could have replicated by themselves, without the existence of protein-based enzymes that fulfill this role in modern cells.

According to the RNA World hypothesis [327,433], RNA molecules able to self-replicate originated in the prebiotic Earth and progressively grew in number. Random errors in the replication process introduced occasional mutations, and natural selection favored groups of sequences able to replicate more efficiently, giving rise to evolution. This theory puts information transfer through RNA sequences at the center of the process that gave birth to life. Moreover, it highlights the dual role of RNA as blueprint and construction worker for building life.

Figure 6.5 depicts the emergence of RNAs and ribozymes in a hypothetical RNA World. Once the first nucleotides became available, mineral surfaces such as montmorillonite clays [137,645] could have catalyzed the formation of short RNA oligomers with random nucleotide sequence (figure 6.5). Some of these oligomers could have folded into secondary structures that conferred them a catalytic ability, giving rise to the first ribozymes: some ribozymes would be able to ligate two RNA strands together (ligases), others to cleave a strand in two pieces, yet others to cleave portions of themselves (figure 6.5). The ligase activity is especially important in order to create longer polymers able to perform more complex catalytic functions [137]. Some ligase ribozymes would ligate segments that are base-pair complements of themselves, giving rise to replicase ribozymes [657], able to produce copies of themselves by template replication (figure 6.5). Different variants of such self-replicators would compete for resources, and selection would

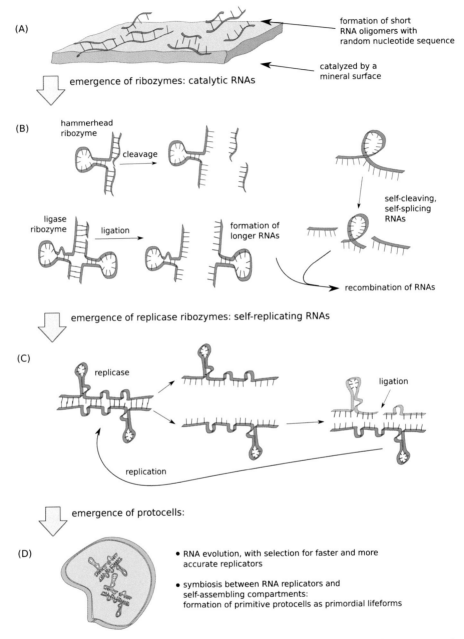

Figure 6.5 A schematic picture of the RNA World hypothesis: (A) formation of short RNA oligomers with random nucleotide sequence, possibly catalyzed by minerals such as clays. (B) Emergence of ribozymes (catalytic RNAs). (C) Emergence of self-replicating (replicase) ribozymes. (D) Primordial replicators enclosed in self-assembling vesicles become the first protocells.

favor faster and more accurate replicators, leading to the first evolutionary steps. More complex cross-catalytic replicators and autocatalytic networks of ribozymes could form [458, 878]. These primordial replicators could have become trapped inside lipid vesicles that would offer them protection, forming the first protocells (figure6.5). A more detailed overview of the RNA World hypothesis can be found in [137].

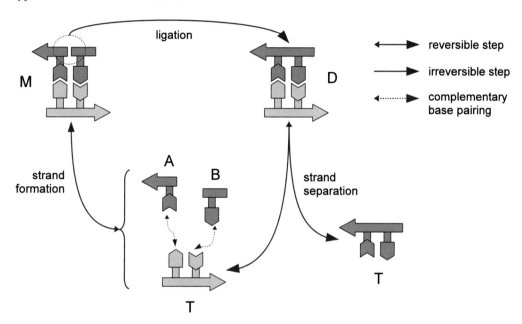

Figure 6.6 A model of the minimal self-replicator demonstrated by Von Kiedrowski [890]: The two *T* molecules are complementary templates (hexamers in the experiments). Each of them base-pairs with units *A* and *B* (trimers in the experiments), to form an intermediate molecule *M* which is almost a double strand: the still missing link between *A* and *B* is then catalyzed by *T* itself, forming the double strand *D*. *D* then splits into two single strands *T* and the cycle repeats. Adapted from [890].

A crucial point in this theory is to explain exactly how the first RNA replicators could have arisen. There are two possibilities: One is a self-replicating RNA molecule based on template replication; another is an autocatalytic network of ribozymes. Template replication is hard to achieve without the help of enzymes. Von Kiedrowski [890] reported the first nonenzymatic template replication of a nucleotide sequence in laboratory. In his experiment, a sequence of six nucleotides (the template) catalyzes the ligation of two trimers together, each trimer being paired with half of the template sequence by complementarity, as depicted in figure 6.6. This apparently simple experiment was actually a landmark in the field, as it demonstrated that template replication was possible without enzymes. Subsequently, several other theoretical studies and experiments on template replication were reported.

The experiments in [890] were based on deoxynucleotides. Concerning RNAs, a self-replicating ligase ribozyme was demonstrated in [657]. It was obtained experimentally using *in vitro* evolution, and is depicted in figure 6.7: RNA strand T is made of two segments A and B. A and B are designed such that they bind by complementary base-pairing to consecutive portions of T, forming the complex **A•B•T**. T acts as a ligase ribozyme to help forming a covalent bond between A and B, creating a new copy of itself. The two T strands then separate (for instance, by raising the temperature), and each of them is ready to ligate another A with another B, repeating the cycle. This ribozyme was further extended in [458] to provide cross-catalytic function in a system of two templates that catalyze each other's ligation from smaller segments.

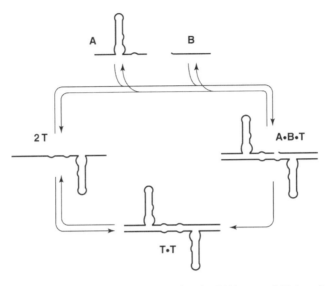

Figure 6.7 The self-replicating ligase ribozyme from [657]: RNA strand T is a ligase ribozyme that catalyzes its own formation from shorter RNA segments A and B. © 2002 National Academy of Sciences, U.S.A., reprinted with permission.

Once such a simple replicator appeared, how could it have evolved to store more information and to perform more complex functions? Actually, this problem remains largely unsolved; more information requires longer strands, but strand elongation is equivalent to a mutation, or error, in the copy process, so must be kept at a minimum. Fernando and others [275] showed that elongation comes along with a potential "elongation problem" that can lead to an "elongation catastrophe" (analogous to Eigen's "error threshold" leading to an "error catastrophe" [249]). During template replication, imperfect duplexes often form, leading to an elongated replica. The elongation problem is that, the longer the double strand is, the more stable it gets, so it becomes more difficult to split it into single strands able to enter a new replication cycle. The replication of such longer strands can be achieved with the help of enzymes, but these were not available then. Over time, the accumulation of such elongated replicas leads to an "elongation catastrophe," where information is not faithfully transmitted, and the longer strands are no longer able to separate, leading to a collapse of the replication process both due to too many errors and due to a halting copy speed. Recent work [888] emulates template replication in hardware using a group of robots, and points out that the evolvability of self-replicators based on template replication must lie in between the rigidity of fixed-length replicators and the excess freedom leading to the elongation catastrophe.

Instead of template replication, another possibility for RNA replication would be that of an autocatalytic network of reactions, catalyzed by ribozymes [878]. According to this view, one ribozyme would act on some neighbor ribozyme, which would act on another one, until an autocatalytic cycle would eventually form, leading to the replication of the molecules in the cycle. Since one ribozyme could act as a catalyst to several others, and in turn, receive catalysis from various others, a graph of catalytic interactions would form. Such graphs are similar to the autocatalytic sets of proteins by Kauffman [446] (see section 6.3.1), with RNAs replacing proteins. In short, the main objection against the autocatalytic set view is the information loss due to leakings and errors caused by side reactions and stochastic noise. Eigen introduced the hypercycle [249,252] (see also chapter 14) as a candidate structure to overcome the error threshold. Boerlijst later showed that hypercycles could resist the invasion of parasitic reactions when placed on a surface [125]. However, it was later discovered that hypercycles in space are also vulnerable to parasites [750]. Besides parasite invasion, other problems such as collapse due to member extinction and chaotic dynamics could make it difficult for hypercycles to survive in general. On the other hand, related dynamic phenomena could also help hypercycles to fight parasites and evolve, under some conditions. The dynamics of hypercycles is complex and still a subject of research [742, 743].

More generally, the question is still open regarding whether and how self-replication of single molecules, or collective replication via autocatalytic networks could sustain themselves sufficiently well to constitute a reliable information system for the first life forms.

As a further obstacle to an RNA World, even the feasibility of spontaneous RNA formation under prebiotic conditions is questioned [645]. Recently, the synthesis of RNA bases under prebiotic conditions was achieved [673], especially the pyrimidines (cytosine and uracil), which had been difficult to obtain so far. Such findings help to counteract some objections to the RNA World theory, but other issues still remain. Many say that some other form of primitive molecule is likely to have existed before RNAs [645]; however, evidence of such molecule has yet to be found.

Szostak [839] points to other types of nucleotides that are able to polymerize spontaneously, forming new templates for replication. New monomers attach to the single-stranded template, forming a double-strand polymer that can unzip into two single strands under higher temperature. Under lower temperature, these single strands could catch new monomers, forming double strands again, and so on, repeating the template replication cycle. According to Szostak, these alternative nucleotides could have been present in the prebiotic world and could have been at the origin of early replicators, which were later replaced by the ones we encounter today. Later, these RNA-like molecules could have become trapped within self-assembling lipid compartments, giving rise to primitive protocells without an explicit metabolism.

6.2.3 Metabolism-First Theories

The complications and difficulties around the RNA World theory left several researchers skeptical about it, and led to the formulation of alternative hypotheses for the origin of life. One of these hypothesis is that a set of chemicals reacting together could form a collectively autocatalytic entity able to metabolize energy and materials from the environment in order to replicate itself, leading to the first primitive life forms. Kauffman's autocatalytic sets of proteins [446] and the autocatalytic metabolisms by Bagley and Farmer [49] both offered enticing theoretical ideas for the origin of metabolisms, but both were based on abstract mathematical and computational modeling, and lacked a real chemistry support.

In 1988, Wächtershäuser [895] surprised the research community with a new theory for the origin of life, comprising a concrete chemical scenario for the emergence of autocatalytic metabolisms. He named his theory "*surface metabolism*," and it later became also known as the

"Iron-Sulfur World" [898]. According to his theory, the precursors of life were self-maintaining autocatalytic metabolisms running on the surface of pyrite (FeS_2). Pyrite is a mineral crystal that could form abundantly in deep sea hot springs, from redox reactions between ferrous ions (Fe^{2+}) and hydrogen sulfide (H_2S). Pyrite surfaces have a slight positive charge, which would attract negatively-charged molecules that would attach to this surface by ionic bonding. Ionic bonding is strong enough to keep the molecules on the surface with a small propensity for detaching, while weak enough to allow molecules to move slowly on the surface, enabling sufficiently diverse interaction possibilities. Autocatalytic cycles could then form as a result of these interactions, leading to the first autocatalytic metabolisms able to feed on inorganic nutrients from the environment, (such as CO, CO_2, HCN, and other gases that could emanate from volcanic exhalations), and to grow by colonizing nearby vacant mineral surfaces. Molecules able to bind more strongly to the pyrite surface would have a selective advantage. Occasionally, a different, novel molecule could form. Some of these novelties could lead to new autocatalytic cycles, which could be regarded as a form of mutation. A crude evolutionary process would then arise [895, 896], selecting for longer molecules able to bind to the surface more strongly and on several points. Such a mineral surface would facilitate the synthesis of larger organic molecules from smaller ones, and could then lead to the progressive formation of sugars, lipids, amino acids, peptides, and later of RNA, DNA, and proteins. Wächtershäuser also sketches possible chemical pathways that could explain the origin of lipid membranes, of RNA and DNA replication, of RNA and protein folding, of the genetic code and its translation mechanism, and of full cellular structures able to emancipate themselves from the pyrite surface and to conquer the vast oceans.

Since then, experimental evidence for Wächtershäuser's theories is being gathered [409, 897–899]. For instance, the synthesis of some amino acids and related compounds has been achieved from volcanic gases as carbon sources and iron-group metals as catalysts [899], at high temperatures in a volcanic hydrothermal setting that is considered a plausible prebiotic scenario. However, much remains to be done, from the synthesis of more complex compounds up to the demonstration of fully self-sustaining autocatalytic metabolisms and their evolution.

Other authors have also advocated the importance of being able to harvest energy first [216, 616, 769], prior to processing information as required by an RNA World. For instance, ATP and other nucleotides such as GTP, CTP, and UTP could have played an early role in energy transfer (a role that ATP still keeps today) before becoming building blocks for more complex information-carrying molecules such as RNA [216].

6.2.4   The Lipid World

One of the hypothesis for the origin of life is that of a "compartment first" world where early amphiphile molecules such as lipids or fatty acids would have spontaneously self-assembled into *liposomes*, which are simple vesicles akin to soap bubbles. The plain laws of physics would have caused these primitive compartments to grow and divide. Growth could happen naturally due to the incorporation of free amphiphiles into the structure. Division could be triggered by mechanical forces, such as marine currents, osmotic pressure, and surface tension caused by a modification of the volume to surface relation as the membrane grows. These "empty shell" membranes would have constituted the first self-replicators in a *"lipid world"* scenario for the origin of life, illustrated in figure 6.8.

The *Lipid World hypothesis* [534, 760, 762, 763] states that liposomes could have given rise to the first cell-like structures subject to a rudimentary form of evolution. It is very plausible indeed

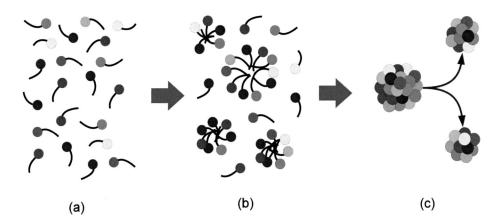

Figure 6.8 Lipid world scenario: (a) floating amphiphiles of various types (indicated by the different colors of their hydrophilic heads) self-assemble into lipidic aggregates (b), that then grow and divide (c)

that these aggregates could have formed in the primitive Earth. This is the main argument in favor of a lipid world: Liposomes form spontaneously in water, whereas the spontaneous emergence of RNA or similar biopolymers is very unlikely, let alone their maintenance in a hostile environment subject to numerous side reactions.

The first amphiphiles could have come from two sources. They could have formed in prebiotic synthesis reactions using energy from light; they could also have been delivered by the numerous meteorites and comets bombarding the primordial Earth.

Liposomes could be composed of several types of amphiphiles with different properties. Their multiset composition can be regarded as a kind of rudimentary genome, called *"compositional genome"* [761, 762]. A crude, still error-prone form of inheritance could then be provided by the rough preservation of the composition structure of the aggregate during division. Such inheritance is illustrated in figure 6.9.

Some of these liposomes could possess catalytic properties, enhancing the rate of reactions that produce precursors for the assembly, or facilitating the incorporation of additional molecules to the assembly, hereby accelerating its growth. Such catalytic liposomes have been named *"lipozymes,"* in analogy with RNA ribozymes. Since their catalytic properties could facilitate their own growth, such aggregates could become *autocatalytic lipozymes*, able to self-replicate collectively by a growth and division process. Lipozymes with more efficient autocatalytic properties would grow faster than the others, leading to selection for efficient ensemble autocatalysis. Combined to the rough propagation of compositional information during division, a primitive evolutionary process would arise, in which the units of evolution would be ensemble replicators.

Later, lipozymes could have trapped free-floating RNA molecules, significantly increasing their ability to carry and transmit information to the next generation. A protocell with information-compartment subsystems could then form. Note that in such a protocell structure, the compartment subsystem also plays a metabolic role, since it catalyzes the growth of the membrane.

The evolution of liposome compositions has been investigated computationally using the GARD model (graded autocatalysis replication domain) [764, 765]. Figure 6.11 depicts the basic

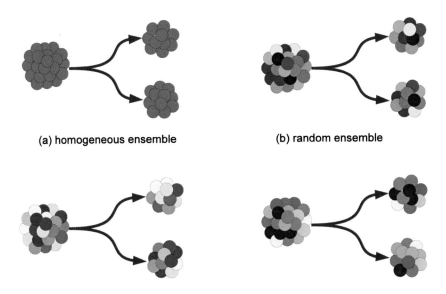

(a) homogeneous ensemble   (b) random ensemble

(c) two examples of inheritance of information during ensemble division

Figure 6.9 Various degrees of compositional inheritance in a Lipid World scenario: (a) Only homogeneous ensembles are present (formed by the same type of lipid), hence no information is carried. (b) When ensembles are made of too many different types of lipids, their fission results in random aggregates; in this case, no useful information is carried either. Things become more interesting in (c): when ensembles are made of a few types of lipids, in a way that they attract more lipids of their type to the ensemble, fission results in "daughter" ensembles that are similar to their parents (in the picture, a yellowish ensemble gives rise to yellowish offspring, and a blue ensemble to blue offspring); these two types of ensembles could grow at different rates, giving rise to competition and selection that could lead to a primordial form of evolution.

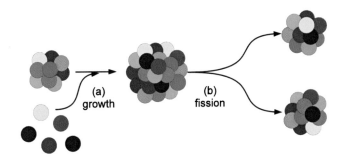

Figure 6.10 Growth and fission of liposomes: Growth (a) happens by incorporation of lipids into the aggregate; (b) fission simply splits the aggregate into two parts, that then start to grow again, repeating the cycle.

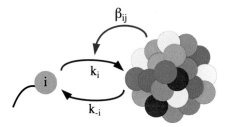

Figure 6.11 GARD mutual catalysis scenario: each lipid of type i joins the lipid aggregate at a basal rate k_i, and leaves it at rate k_{-i}. This join/leave process can be accelerated (catalyzed) by other lipids of type j already present in the aggregate. The amount of catalysis that a lipid of type j offers to a lipid of type i is β_{ji}.

GARD model: at any time, lipids join and leave the compositional assemblies spontaneously, but this happens at a slow rate. This rate can be accelerated with the help of other lipids that are already in the assembly; these lipids then act as catalysts for the incorporation and release of lipids from the aggregate. This model can be expressed by the following reaction:

$$L_i + CL_j \quad \overset{L_j}{\underset{}{\overset{\downarrow}{\underset{k_{-i}}{\overset{k_i}{\rightleftharpoons}}}}} \quad CL_j L_i \tag{6.1}$$

Reaction 6.1 expresses the GARD mutual catalysis model of figure 6.11 in chemical reaction format: lipid L_i enters aggregate CL_j (representing a lipid ensemble where at least one lipid molecule of type L_j is present) at rate k_i and leaves it at rate k_{-i}. The incorporation of lipids into the aggregate occurs in a noncovalent fashion. This reaction is facilitated by the presence of catalyst L_j, which accelerates the reaction by a factor β_{ji} in both directions. Since catalysis accelerates both forward and reverse reactions (here, respectively, join and leave the assembly), there is no change in the equilibrium state of the system with and without catalysts. Therefore the system must be kept out of equilibrium by an inflow and outflow of lipids. Moreover, the rates β_{ji} must be carefully chosen following the receptor affinity distribution model [490, 726], such that the probability of a lipid catalyzing a given reaction is maximal between similar but not identical lipids.

Simulations of the GARD model have been performed using Gillespie SSA [764, 765]. They have shown that when autocatalysis is favored, the faster autocatalyst ends up dominating, as expected in a system consisting of competing species that do not interact. These experiments also show that mutually catalytic networks emerge when catalytic activity is assigned at random, showing a behavior similar to that of random catalytic networks [792], albeit extended to the lipid aggregate domain.

Further computer simulations reveal the emergence of homeostatically stable ensemble replicators or *composomes* [761, 762], also referred to as quasicompartments, or quasistationary states (QSSs). As mutations accumulate, these composomes can get replaced by new ones, like ESSs being replaced by a fitter mutant (see also section 7.2.8). Unfortunately, however, the low accuracy of replication characteristic of such molecular assemblies compromises their evolvability [884]. This is one of the reasons why these compartments are primarily regarded as shelters for information-carrying molecules, rather than as standalone units of evolution.

In spite of the evolvability issues raised in [884], the Lipid World hypothesis and its simulated GARD model remain interesting ideas worth further exploration, since they open up the possibility of alternative ways to encode and transmit the information necessary for life, different from more traditional sequence-based approaches.

A number of extensions of the GARD model have been proposed in order to address its shortcomings. Polymer GARD [771] is an extension of the basic GARD model in which polymers can form by the combination of basic monomers, therefore supporting the dynamic creation of new molecular species. Another extension of the GARD model called EE-GARD (environment exchange polymer GARD) was proposed in [773] to study the coevolution of compositional protocells with their environment. A spatial version of GARD is presented in [772], where limitations in assembly size give rise to complex dynamics, including a spontaneous symmetry breaking phenomenon akin to the emergence of homochirality.

6.3   Artificial Chemistry Contributions to Origin of Life Research

One of the initial motivations for artificial chemistries was exactly to study the origin of life. Since it is difficult or even impossible to actually demonstrate the origin of life in the laboratory, abstract computer models are valuable tools to obtain insights into this complex problem within feasible time frames.

AC research themes related to the origin of life include the exploration of various types of autocatalytic sets, the investigation of how evolution might emerge in chemical reaction networks, and the formation of protocellular structures akin to chemotons. We shall now focus on autocatalytic sets and the formation of protocells with ACs, whereas ACs related to evolution are covered in chapter 7.

6.3.1   Autocatalytic Sets

Catalysis is important for life, otherwise chemical reactions would simply be too slow to be able to keep an organism alive. Moreover catalysis also plays a regulatory role in modern cells, acting as a mechanism to control which reactions should occur and when. Autocatalysis is even more interesting, because it can be regarded as a form of molecular self-replication. As we will see in chapter 7, the combination of replication with some variation, subject to a selection pressure can lead to evolution. Strictly speaking, autocatalytic reactions do not actually exist in chemistry; no molecule is able to replicate itself in a single step. Instead, autocatalysis occurs in cycles of individual reaction steps, such as the formose cycle (figure 6.1), or the template replication cycle shown in figure 6.6. An *autocatalytic cycle* contains a reaction that duplicates one of the members of the cycle, and as a result, the other members of the cycle are also duplicated. Examples of autocatalytic cycles in biochemistry include the Calvin cycle (part of photosynthesis) and the reductive citric acid cycle (a reverse form of the Krebs cycle that could also have played a role in prebiotic metabolism). Cycles can also be nested, forming *hypercycles* [249]: the output of one autocatalytic cycle feeds another cycle, and so on, until the last cycle feeds the first one, closing the hypercycle.

More generally, one can have a set of chemicals that together form a *collectively autocatalytic* reaction network, in which every member of the set is produced by at least one reaction catalyzed by another member of the set. Such a set is called an *autocatalytic set* [260, 446].

Autocatalytic Sets of Proteins

Motivated by the formation of amino acids in the Miller-Urey experiments, in the 1980s Kauff-man [446] investigated the hypothesis that life could have originated from an *autocatalytic set of proteins*. Proteins fold into 3D shapes that may confer them catalytic properties. One protein could then catalyze the formation of another protein, and so on, leading to a network of cat-alytic interactions between proteins. Kauffman then argued that for a sufficiently large amount of catalytic interactions, autocatalytic sets would inevitably form.

Proteins are formed in condensation reactions that join two amino acids together, releasing one molecule of water in the process:

$$A + B \overset{(E)}{\rightleftharpoons} C + H \qquad (6.2)$$

In this reaction, peptides A and B are concatenated to produce peptide C, and a molecule of water (H) is released. Molecule E is an optional catalyst for this reaction. In an aqueous envi-ronment, the reverse reaction (cleavage of the protein, consuming a water molecule) is favored; therefore, the chemical equilibrium tends toward small peptides. One solution to allow the for-mation of large peptides is to keep the system out of equilibrium by a steady inflow of amino acids or other small peptides, which can be accompanied by the removal of water molecules in order to favor the forward reaction. Another (complementary) solution to favor polymerization is the addition of energy, for instance, in the form of energized molecules.

In order for these reactions to proceed with sufficient speed, they must be catalyzed. Since proteins can have catalytic abilities, the catalysts can be the proteins themselves, therefore au-tocatalytic sets containing only proteins could be envisaged. Figure 6.12 shows an example of autocatalytic set as idealized by Kauffman. In his model, peptides are represented as strings from a two-letter alphabet $\{a, b\}$. The graph of figure 6.12 shows the strings as nodes and the re-actions between them as arcs, with the action of catalysts represented as a dotted arrow pointing to the reaction that it accelerates.

Starting from a "food set" of monomers and dimers (molecules a, b, aa and bb), progres-sively longer polymers can form. Moreover, autocatalytic cycles can form, such that the concen-trations of the reactants in the cycle gets amplified as the expense of other molecules that can quickly get broken down into smaller ones, until they end up feeding the cycles. One such cycle is highlighted in figure 6.12: molecule "abb" catalyzes the formation of "$baab$" which in turn catalyzes the formation of the long polymer "$aabaabbb$" on top of the figure. This long string then catalyzes the cleavage of "$abab$" into two molecules of "ab," each of them a catalyst for "abb," closing the cycle.

Using a graph-theoretic analysis, assuming that all proteins had a given probability P of cat-alyzing the formation of another protein in a condensation-cleavage reaction (equation 6.2), Kauffman calculated the minimum probability of catalysis above which autocatalytic sets would form with high probability. In a subsequent study [260] Farmer and colleagues estimated the probability of catalysis as a function of the size of the food set and the size of the polymer al-phabet: they showed that for a polymer alphabet of size B and a firing disk (food set) contain-ing about $N = B^L$ polymers of maximum length L, a critical probability of catalysis P_{crit} occurs around

$$P_{crit} \approx B^{-2L} \qquad (6.3)$$

For $P > P_{crit}$, the formation of autocatalytic sets is favored due to the dense connectivity of the reaction graph, which causes its fast expansion from the food set as new polymers are produced.

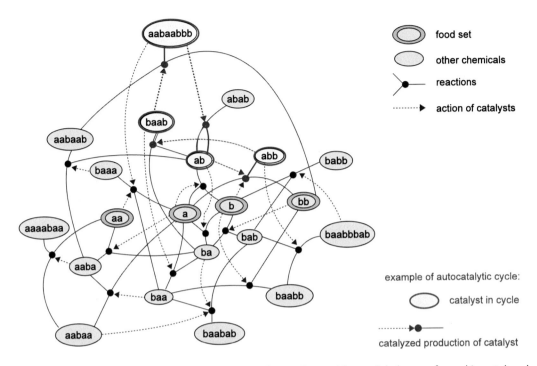

Figure 6.12 An example of an autocatalytic set: polymers from a binary alphabet are formed in catalyzed reactions, from monomers and dimers that constitute the food set. An example of autocatalytic cycle within this set is highlighted in red. Adapted from p. 54 of [260] or p. 323 of [447].

For $P < P_{crit}$, the reaction graph stops growing very soon, because nodes tend to end up isolated in disconnected islands. A large firing disk (due to a large B or a large L) greatly reduces P_{crit}, hence a small probability of catalysis is enough to form autocatalytic sets, which is consistent with a prebiotic scenario dominated by inefficient catalysts. Note that, according to this model, since a larger B is more favorable for the formation of autocatalytic sets, proteins (with $B \approx 20$) are more suitable to form such sets than RNA or DNA (with $B = 4$) for the same maximum length L of the food set polymers.

Autocatalytic Metabolisms

Kauffman [446] has pointed out that autocatalytic sets could also be used to study the emergence of metabolisms: Proteins could be used to catalyze not only the formation of other proteins but also the formation and breakdown of other organic compounds in metabolic reactions. The model of an autocatalytic metabolism would then be essentially the same as an autocatalytic set. More complex compounds are formed from simple building blocks provided in the food set, and autocatalytic sets would emerge with high probability when the connectivity of the reaction graph exceeds a threshold, determined by the probability of catalysis of each reaction. He then conjectured that primitive metabolisms could arise in conjunction with autocatalytic sets of proteins, such that the latter could be used to produce the catalysts needed to maintain both sets of reactions. Accordingly, Farmer's polymers in [260] are abstract strings that could represent proteins or other organic compounds.

The studies in [260, 446] are mostly based on graph-theoretical, static considerations, without taking dynamics nor chemical kinetic aspects into account, although an ODE formulation is presented in [260] together with preliminary results.

In 1992, Bagley and Farmer [49] extended the autocatalytic set model of [260, 446] to a dynamical system model of autocatalytic metabolisms, which still follows the same condensation-cleavage scheme of reaction 6.2), where only a subset of the reactions are catalyzed. To emphasize the importance of dynamics, the authors define an *autocatalytic metabolism* as an auto-catalytic set in which the species concentrations are significantly different from those expected without catalysis. This model highlights the role of catalysis in focusing the mass of the system into a few species that constitute the core of the autocatalytic metabolism. Such *catalytic focusing* only occurs when the system is kept out of equilibrium by a steady inflow of food set members. This is because catalysis does not change the chemical equilibrium of a reversible reaction: it accelerates both forward and backward reactions by an equal factor, thereby keeping the equilibrium unchanged.

Starting from a "firing disk" of small polymers, Bagley and Farmer [49] simulated the dynamics of their system under a continuous supply of small food molecules. Under various parameter settings, they observed the emergence of autocatalytic networks able to take up the food molecules provided and turn them into a reduced set of core molecules that formed a stable autocatalytic metabolism. In a subsequent paper [50] the authors studied how such a metabolism would react to mutation events and concluded that such metabolisms had the potential to evolve by jumping from one fixpoint to a different one.

Formalizing Autocatalytic Sets

Jain and Krishna [425–427] proposed a network model of autocatalytic sets in which only the catalytic interactions between species are depicted on a graph. This model is a simplification of Kauffman's graphs in which only the catalytic action (dotted arrows in figure6.12) is explicitly represented. The dynamics of the system is simulated in a way that species that fall below a given threshold concentration get replaced by newly created species. This model will be covered in more detail in chapter 15 in the context of novelty creation in ACs.

Hordijk and Steel [402] devised a formal model of autocatalytic sets based on the concept of a RAF set (reflectively autocatalytic and F-generated). According to the RAF framework, a set of reactions R is reflexively autocatalytic (RA) when every reaction in R is catalyzed by at least one molecule involved in any of the reactions within R. The set R is F-generated (F) if every reactant in R can be constructed from a small set of molecules (the "food set" F) by successively applying reactions from R. A set R is then a RAF when it is both RA and F. In other words, a RAF set is a self-sustaining set that relies on externally supplied food molecules; thus, it is an autocatalytic set satisfying the catalytic closure property.

The RAF formalism was used to devise a polynomial-time algorithm (polynomial in the number of reactions) to detect autocatalytic sets within a given chemical reaction network [402]. The algorithm is then applied to study the phase transitions from parameter settings where random reaction networks remain highly probable, to those favoring the emergence of autocatalytic sets.

The RAF formalism was extended in [599] to take inhibitory effects into account. Using this extended model, the authors show that the problem of finding autocatalytic sets becomes NP-hard when inhibitory catalytic reactions may occur. Hence, the algorithm in [402] cannot be at the same time extended to support inhibition and kept polynomial.

In a review paper [400] about the RAF framework, the authors highlight the important role of their framework in adding formal support to Kauffman's original claim that autocatalytic sets arise almost inevitably in sufficiently complex reaction networks. The RAF framework quantifies the amount of catalysis needed to observe the emergence of autocatalytic sets: according to computational and analytical results [402, 599], only a linear growth in catalytic activity (with system size) is required for the emergence of autocatalytic sets, in contrast with Kauffman's original claim that an exponential growth was needed. However, the authors also acknowledge the limitations of the RAF approach: It does not take dynamics into account, does not address compartmentalized chemistries, and does not directly look at heredity and natural selection.

Side Reactions and Stochastic Effects

In practice, a cyclic reaction network is never perfect. Numerous side reactions take place, draining resources from the main cycle. Autocatalysis can help compensating for losses and errors due to side reactions [565]. In the absence of enzymes, reactions that are not autocatalytic have trouble to "survive" against a background of numerous types of reactions and reactants, leading to a dead "tar" of chemicals or to a combinatorial explosion of reactions that end up in the same dead tar after exhausting all the available resources. In contrast, autocatalysis provides not only a means to "stand out" against the background of various reactions, but also to replicate substances, therefore it can be seen as a first step toward reproduction and Darwinian evolution.

Unfortunately, however, autocatalytic reaction networks are also plagued by side reactions. Moreover, they seem to have limited capacity to evolve [836]. A vast amount of research literature is available on attempts to find solutions to make primitive autocatalytic metabolisms actually "jump to life."

Kauffman's original argument was based on the probability of catalysis in a static graph of molecular interactions. Therefore, it is largely criticized for overestimating these probabilities under realistic situations where dynamic and stochastic effects cannot be neglected. Bagley and Farmer [49] showed the emergence of autocatalytic sets in a dynamic setup; however, the emergence of such sets under stochastic fluctuations has only recently been extensively studied [277, 278]. As expected, these more recent studies [277, 278] confirm that autocatalytic sets emerge more rarely in stochastic settings than in deterministic ones, and that the sets formed are often unstable and easily disappear due to stochastic fluctuations. The main objections against the autocatalytic set theory thus still remain the leakings and errors caused by side reactions. In summary, the spontaneous emergence and maintenance of autocatalytic sets is not so straightforward to observe in practice, and this topic is still subject of active research.

6.3.2   Emergence of Protocells with ACs

The above discussion about autocatalytic sets teaches us that life is very unlikely to emerge from "naked" chemical reactions in a prebiotic "soup." Given the spontaneous formation of amphiphile vesicles in water, a more plausible scenario for the origin of life could be the emergence of primitive protocellular structures that could harbor the chemical reactions that would then support life. Much attention from the AC community has been devoted to investigate this hypothesis.

Earlier we have already looked at the GARD model as an example of artificial chemistry that aims at studying the origins of life using amphiphilic molecules, in what is termed the Lipid World hypothesis. Related ACs that simulate the self-assembly of amphiphiles into micelles and vesicles include [246, 270].

Autopoietic Protocells by Ono and Ikegami

Following from the computational models of autopoiesis, Ono and Ikegami devised an artificial chemistry able to spontaneously form two-dimensional protocells in space [637, 639, 641]. In their model, particles move, rotate and interact with each other in a hexagonal grid. Three types of particles are considered: hydrophilic, hydrophobic, and neutral. Hydrophilic and hydrophobic particles repel each other, whereas a neutral particle may establish weak interactions with the other two types. Hydrophobic particles may be isotropic or anisotropic. Isotropic hydrophobic particles repel hydrophilic particles with equal strength in all directions. Anisotropic particles, on the other hand, have a stronger repulsion in one direction. Figure 6.13 illustrates the difference between isotropic and anisotropic repulsion fields.

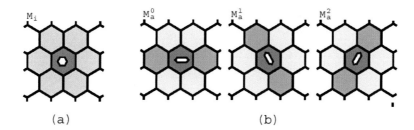

Figure 6.13 Repulsion field surrounding two types of hydrophobic particles in the two-dimensional protocell model by Ono and Ikegami [641]: (a) isotropic particles M_i form a uniform repulsion field around them; (b) anisotropic particles M_a have stronger repulsion in a given direction (a darker hexagon represents higher repulsion). From [641], © Springer 2001, reprinted with permission.

Five chemical species, A, M, X, Y, and W, react with each other, forming a minimal metabolic system. Particles of species A are autocatalysts that replicate themselves by consuming food molecules X. A particles may also produce molecules of M, which represent membrane constituents. W is a water molecule. All particles (except W) decay into the "waste" species Y. An external energy source recycles Y molecules into new food particles X. A and W are hydrophilic, M is hydrophobic, whereas X and Y are neutral.

After some preliminary exploration [639], in [641] the authors show the formation of membrane-like clusters of particles, with initially irregular shapes. Some of these membranes eventually form closed cells that are nevertheless permeable to some small food particles that cross the membrane by osmosis. Some of these protocells may even grow and divide. The presence of anisotropic particles is crucial for protocell formation in this model; when only isotropic particles are present, clusters of particles form but end up shrinking and disappearing due to their inability to sustain an internal metabolism. The advantage of anisotropic over isotropic particles is especially visible under low resource supply conditions, as shown in figure 6.14, which depicts the main results of [641]. Note, however, that the metabolic system inside these

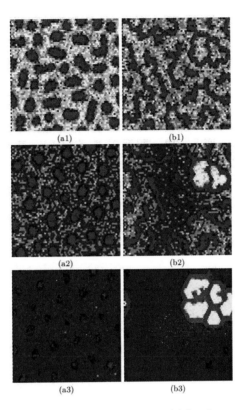

Figure 6.14 Emergence of protocells in the autopoietic model by Ono and Ikegami [641], under low supply of X food molecules. Left (a1, a2, a3): when all M particles are isotropic, clusters resembling cells form, but are unable to sustain themselves and end up dying out. Right (b1, b2, b3): when M particles are anisotropic, irregular membrane filaments form initially, and some of them form closed protocells able to grow, divide and sustain their internal metabolism. Time flows from top to bottom. Blue: Water (W) particles; red: membrane (M) particles; yellow: autocatalysts (A). From [641], © Springer 2001, reprinted with permission.

cells is extremely simple; there is no explicit genetic material and no evolution mechanism, and therefore this system is expected to have a limited ability to create more complex organisms.

A subsequent study [637] goes further in depth into the conditions for protocell emergence in this model. The model has also been extended to three dimensions [538], revealing a variety of self-maintaining cellular structures including the formation of parallel membranes with tubular structures connecting them, and intertwined filaments that organize themselves into globular structures.

6.4 Summary

Although considerable progress has been made in understanding the chemical processes that could potentially lead to life, the actual origin of life on Earth remains surrounded by mystery. The experimental results so far have relied on extreme simplifications and only succeed in demonstrating fragments of the puzzle. A full demonstration of a scenario for the origin of life under plausible chemical conditions which is compatible with a prebiotic Earth still remains to be shown. We can be sure about one point, though: Many exciting new discoveries still lie ahead of us.

Even at this stage, where there are still more open questions than firm answers, the investigations around the origin of life open up several doors to other connected research domains, including related exploratory efforts such as astrobiology [313, 667], and the construction of synthetic life in the laboratory. More about the latter will appear in chapter 19. Meanwhile, in the next chapter we shall turn our attention to how life evolves once it has emerged.

Evolution is cleverer than you are.

L. ORGEL'S SECOND RULE

EVOLUTION

After having mentioned evolution several times in past chapters, in this chapter we can finally delve into the actual mechanism of Darwinian evolution, or evolution by natural selection: what evolution is, why and how evolution works, its strengths and limitations, and of course, its relation to ACs.

Why should a book about artificial chemistries dedicate a full chapter to evolution? One answer is because ACs can facilitate the study and understanding of evolution, since evolutionary processes can be modeled with artificial chemistries. Moreover, ACs can help to shed new light on the origin of evolution. Indeed, one of the main motivations for artificial chemistries is to understand the dynamical processes preceding evolution and leading to the origin of life, and later enabling evolution to keep producing increasingly sophisticated organisms. By abstracting away from biochemistry details, ACs enable the experimenter to focus on the core dynamics of molecular interactions responsible for the evolutionary phenomena of interest.

At first glance, it seems hard to believe that pretty featureless and primitive protocells could have evolved toward complex living organisms and unfolded into the extraordinarily diverse biosphere that we see today. Indeed, since Charles Darwin formulated the theory of evolution by natural selection in the nineteenth century, the controversy on this topic has not ceased. In spite of some remaining skepticism, the scientific progress in molecular biology and genetics has accumulated overwhelming support for the thesis that evolution is indeed the main driving force behind the highly organized and elaborate functional structures of living beings. Although other related driving forces such as self-organization and self-assembly also play a key role in shaping organisms, ultimately these related forces are also steered by evolution.

Beyond biology, processes with evolutionary characteristics are encountered in various other domains, such as the evolution of human languages, technological advances, economy, culture, and society.

Since the natural evolution of any organism beyond viruses and bacteria occurs at timescales ranging from decades to millions of years, the mathematical and computational modeling of evolutionary processes is of paramount importance in understanding and assessing the mechanisms of evolution from a quantitative perspective.

Outside of natural evolution, abstract models of evolution can also be used as optimization tools to search for solutions to difficult problems. In the wet lab, *in vitro* artificial evolution can shed light on natural evolution and can be used to engineer optimized molecules for various biochemical usages.

In this chapter we introduce the basic concepts of evolutionary theory from a mathematical and computational perspective, in relation to ACs. There are, however, limitations at such an elementary level. With the current state of the art, no computer model of evolution has been able to reproduce the apparent everlasting potential for complexity and innovation attributed to natural evolution. The search for open-ended evolutionary growth of complexity in artificial evolutionary systems is a major research topic in artificial life and artificial chemistries, and it will be covered in chapter 8.

7.1   Evolution: Taming Combinatorics to Improve Life

Monomers such as amino acids and nucleotides can combine in infinitely many ways to form macromolecules such as DNA, RNA, and proteins. Of all these possible monomer combinations, only a few lead to viable lifeforms.

Since there are infinitely many possible sequences, each of them folding into a particular shape with a specific function, the space of possible sequences, shapes and functions to explore is huge. So at least in theory, there is plenty of room for new molecular structures to support new life forms. However, only a few of those possible sequences may lead to useful functions. Moreover, different sequences may fold into a similar shapes with equivalent function, a property known as *neutrality*.

Since there is so much choice to look into, a search mechanism for feasible polymers leading to viable life forms must be a very efficient one: It must quickly focus on the promising candidate solutions while discarding unsuitable ones; and it must keep a continuous balance between a flow of novelties to explore and a stable population of viable organisms. In nature, the mechanism for this search is evolution. Natural selection causes fitter organisms to survive at the expense of more poorly adapted cousins; these fitter organisms then produce offspring resembling themselves; during reproduction, modifications in the genetic sequences (through occasional mutations or recombination of genes) may give rise to new variants that are again subject to natural selection. The result of this process is a fast, beamed search into the vast combinatorial space of possible solutions leading to viable life forms. In this sense, evolution can be seen as an *optimization* process in which the *fitness* of the organisms in the population is maximized [285,395]. Accordingly, such a beamed search process is the basic inspiration for evolutionary algorithms [248,590], which have been extensively used in computer science to solve difficult combinatorial problems in various domains. Some of these algorithms are formulated in a chemical form, showing that artificial chemistries can be used for (evolutionary) optimization, as we will see in Chapter 17.

It is important to realize that although variation in the genetic material may result from random phenomena such as mutations, the evolutionary process as a whole — characterized by selection, heredity and variation working together in synchrony — is far from random. Instead, it is a very focused process whose overall behavior and qualitative outcome can be predicted to a certain extent. Indeed, the observation of artificial evolution against the background of random drift is an important tool to measure the actual impact of evolutionary adaptation [142,407].

Evolution is essential to life. Without the possibility of evolution, unavoidable errors in the genetic material would accumulate until organisms, species, and ultimately the entire biosphere could become extinct. All the existing error correction mechanisms for the genome have limitations and would end up breaking down in the face of accumulated errors. Moreover, without evolution, changes in the environment that could be fatal to certain species would end up exterminating them, without any chance for other species to emerge and occupy the ecological niches left vacant.

Even though evolution is essential to life, evolution cannot explain life itself. Evolution requires heritable genetic variation, which only became possible after sufficiently complex replicators were formed. Other order-construction mechanisms must have led to these primordial replicators. These preevolutionary mechanisms are studied under the various origin of life theories. It has been shown that preevolutionary dynamics is weaker than full evolutionary dynamics, and is easily displaced once replicators capable of Darwinian evolution emerge in the population [625]. The actual mechanisms for such transition from primordial chemical dynamics to Darwinian evolution are subject of much interest in the AC community. Some of the research in this area will also be surveyed in this chapter.

7.2 Evolutionary Dynamics from an AC Perspective

The best way to show that evolution really works and how it actually works is through the mathematical and computational modeling of evolutionary processes. The mathematical modeling of evolutionary processes is the domain of evolutionary dynamics. Research in theoretical biology, epidemiology, evolutionary ecology and other domains make use of essential concepts from evolutionary dynamics to understand and predict the evolution of pathogens, the spreading of infections, the effects of animal migrations, and even to look into the possible origins of life.

Evolutionary dynamics has been defined in [755] as a composition of three processes: population dynamics, sequence dynamics, and the genotype-phenotype mapping. *Population dynamics* describes the changes in numbers of individuals and is usually expressed in a *concentration space*, where individuals are analogous to molecules floating in a reactor. *Sequence dynamics* (or *population support dynamics*) looks at how genomes get modified throughout the generations, moving within a *sequence space*. The *genotype-phenotype mapping* defines how genetic sequences are translated to the phenotype or shape of an individual; hence, it defines a map from sequence space to *shape space*. From this perspective we can clearly see how evolution may be mapped to ACs: molecules have an internal structure (sequence) that unfolds into some shape that has some function (recall, for instance, the matrix chemistry of chapter 3).

In this section we provide a brief overview of evolutionary dynamics at textbook level [621], but put it into the perspective of chemical kinetics and artificial chemistries. We start with the basic population dynamics, whereas more sophisticated sequence and mapping issues will be introduced later in this chapter, and will be covered more extensively in chapter 8.

Evolutionary dynamics arises in a population of individuals as a result of a concerted action of three basic operators: *replication, variation,* and *selection*. Individuals have a genome that is passed to the offspring during reproduction. A distinction between reproduction and replication is sometimes emphasized [779]: replication is associated with the notion of identical copy (without errors), whereas reproduction is replication with a potential slight variation (such as a small mutation or a recombination of genes from parents when producing an offspring). The amount of variation must be small enough such that offspring still resemble their parents. Hence reproduction is replication with inheritance. Selection is usually driven by resource limitation: for instance, a competition for scarce food may lead to the death of weaker organisms at the expense of stronger ones, creating a selective pressure toward aggressive or defensive behavior, toward efficient food foraging skills, or a combination of traits that would allow some organisms to win over the competition. We will see how populations may evolve as a result of the combined action of reproduction and selection.

7.2.1 Replication and Death

In the real world, resources are finite; therefore, replication is often accompanied by later death. Replication can be expressed as a simple autocatalytic reaction, and death as a simple decay reaction:

$$\text{Replication}: \quad X \xrightarrow{b} 2X \tag{7.1}$$

$$\text{Death}: \quad X \xrightarrow{d} . \tag{7.2}$$

Here X is a species, and can represent an actual molecular species (such as a self-replicating organic molecule), a cell, an entire organism, or any other entity subject to autocatalytic production. Using the law of mass action, the differential equation for the net production of X can be easily written as:

$$\dot{x} = (b - d)x \tag{7.3}$$

where x is the concentration, or population, of species X. The solution for this simple differential equation is easy:

$$x(t) = x_0 \, e^{(b-d)t} \tag{7.4}$$

where x_0 is the initial population size.

Three situations may happen depending on the parameters b (birth rate) and d (death rate):

- if $b > d$, the population of X grows exponentially to infinity.

- if $b < d$, the population decays exponentially until it goes extinct.

- if $b = d$, the population size remains constant at its initial amount, but this situation is unstable: a slight inequality leads either to infinite growth or to extinction.

As simple as it may be, this solution holds one of the keys for evolutionary dynamics to work: exponential growth, which is a precondition for Darwinian evolution, as we will see later.

7.2.2 Selection and Fitness

Selection operates whenever different types of individuals (*species*) replicate at different speeds. The *rate of reproduction* of a given individual is directly interpreted as its *fitness*. For two species X and Y:

$$X \xrightarrow{a} 2X \tag{7.5}$$

$$Y \xrightarrow{b} 2Y \tag{7.6}$$

with corresponding differential equations and their solutions:

$$\dot{x} = ax \quad \Rightarrow \quad x(t) = x_0\, e^{at} \tag{7.7}$$

$$\dot{y} = by \quad \Rightarrow \quad y(t) = y_0\, e^{bt} \tag{7.8}$$

the strength of selection can be measured by looking at the ratio of X over Y in time:

$$\rho(t) = \frac{x(t)}{y(t)} = \frac{x_0}{y_0}\, e^{(a-b)t} \tag{7.9}$$

- if $a > b$, then ρ tends to infinity, thus X outcompetes Y, i.e., selection favors X over Y.

- if $a < b$, then ρ tends to zero; thus, selection favors Y over X.

- if $a = b$, then $\rho = \frac{x_0}{y_0}$ and there is no selection: Both species coexist in amounts proportional to their initial concentrations.

In this example, a and b are the *fitness* of X and Y, respectively, which are the rate coefficients of their growth reactions (7.5) and (7.6). These coefficients indicate how fast one individual from that species can reproduce. The notion of fitness is therefore associated with reproduction efficiency. The outcome of selection in this example is the *survival of the fittest* species: in the long run, the population of the less fit of the two will become vanishingly small compared to the fitter (winner) species, so for all practical effects the loser species will become extinct, even in the presence of infinitely abundant resources, and in the absence of death.

This deterministic selection result is oversimplified, and does not hold for the important case when stochastic fluctuations occur. However, such simple initial formulation is useful to highlight the fundamental nature of selection: selection does not require finite resources (note that no death rate was used in the example); selection does not assume a specific growth mechanism, nor a specific death mechanism. It simply happens by the very nature of differences in production rates. However, the outcome of selection does depend on the shapes of the growth and death curves for the competing individuals: in the previous example, both X and Y grow exponentially; behaviors other than exponential might lead to different outcomes as we will see later.

7.2.3 Fitness Landscapes

How does an organism achieve a sufficiently high fitness in order to be able to survive and reproduce in a competing environment? Although the survival of an organism ultimately relies on chance events, much of its ability to struggle for survival is encoded within its genes. At the unicellular level, genes code for proteins that perform most cellular functions ensuring adaptability and reproductive fidelity. At the multicellular level, genes encode the information necessary to grow a full organism out of an embryo, in a developmental process. Hence the genotype of an individual unfolds into its phenotype, and that in turn determines its fitness. The way the genotype maps to the phenotype is referred to as the *genotype-phenotype map*. Since the phenotype can then be associated with a certain fitness value (at least in a simplified, abstract way), a direct map from genotype to fitness can be derived from the genotype-phenotype and then phenotype-fitness maps.

Genomes are sequences of length L over an alphabet of size B. In nature, $B = 4$, corresponding to the four bases A, C, T, G. Binary alphabets ($B = 2$) are very common in artificial chemistries due to their simplicity and direct correspondence to computer bit strings. The set of all possible sequences of length L can be seen as a L-dimensional space, where each dimension can take any of B possible values. Any of the possible B^L genotypes can now be specified as a point in this sequence space. Assuming that the fitness of each genotype can be measured somehow (through the resulting phenotype and its environment), the fitness value can be described as a point in a *fitness landscape*, a function that maps sequences to their respective fitness values.

For a short genome of length $L = 2$ and sufficiently large B, the fitness landscape can be visualized as a 2-dimensional surface in 3D space, similar to a natural mountain landscape, with peaks and valleys. If there is only one peak, then the landscape is *unimodal*. A landscape with several peaks is referred to as *multimodal*. The landscape is *rugged* when there are numerous peaks of irregular shapes and sizes. A *neutral* fitness landscape is characterized by plateaus with the same level of fitness.

Evolution then operates by walking on the fitness landscape. A harmful mutation sends individuals to valleys, and a neutral one moves then horizontally on a plateau, whereas a beneficial one causes individuals to climb a mountain, hopefully reaching a peak, which can be either a local optimum or the global optimum. The global optimum is located at the highest peak, that is, it corresponds to the genotype value for which the fitness is maximal.

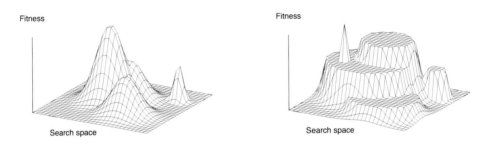

Figure 7.1 Examples of fitness landscapes. Left: multimodal. Right: neutral. Courtesy of Sébastien Verel [886], reprinted with the author's permission.

Figure 7.1 shows two examples of fitness landscapes: a multimodal one with several local optima peaks, and a neutral one with several plateaus and a narrow highest peak. Of course, real fitness landscapes in natural and artificial evolution are highly multidimensional, living in a space of dimension $L + 1$, and as such cannot be so easily visualized. However, the fitness landscape concept is a useful and widely used metaphor to imagine how evolution works.

7.2.4   Resource Limitations

Systems where resources are limited are, of course, more realistic and are of practical interest. A straightforward way to enforce resource limits is to impose a mass conservation constraint for each individual chemical reaction. In this way, unlimited growth is not possible, since it must rely on a limited supply of nutrients. However, this solution often requires a larger number of species to be modeled, in order to ensure mass balance.

Another solution is to enforce resource limits by the nature of the interactions among species. For instance, consider the following reversible reaction:

$$X \underset{d}{\overset{r}{\rightleftharpoons}} 2X \qquad (7.10)$$

This is a simple variation of the replication and death reactions (7.1)—(7.2), but now, instead of a unimolecular "death by decay," we have a "death by fight": two individuals interact and only one of them survives. Let us set the coefficient of the reverse reaction to $d = r/K$, where r is the rate coefficient of the forward (replication) reaction, and K is the carrying capacity of the system. Using the law of mass action, this system of reactions leads to the *logistic growth* equation:

$$\dot{x} = rx(1 - \frac{x}{K}) \qquad (7.11)$$

This equation has a well-known solution which is a sigmoid function. Figure 7.2(left) shows a numerical integration of eq. 7.11, where it indeed exhibits a sigmoid shape with a plateau at the value K. Therefore the population cannot grow beyond the carrying capacity K. Note however that in a more realistic, stochastic simulation of the same reaction (7.10), the population fluctuates around K, often exceeding it. Therefore, for small populations subject to stochastic fluctuations, reaction 7.10 cannot guarantee a strict population bound.

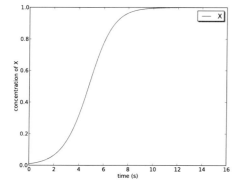 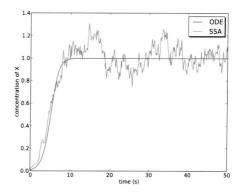

Figure 7.2 "Replicate and fight" model for resource limitation: the reversible chemical reaction (7.10) can be seen as a chemical implementation of the well-known logistic equation. Left: the numerical integration of the logistic equation (7.11) shows that the population cannot exceed the carrying capacity K, here $K = 1$. Right: a comparison between numerical ODE integration and the stochastic simulation (using Gillespie SSA) of reaction (7.10) reveals that the population fluctuates around K, often exceeding it.

Another often encountered solution to model resource limitations, especially in ACs, is to impose a maximum carrying capacity (usually normalized to one) for the whole reaction system, and to implement an outflow ϕ (also called a dilution flow) that compensates any overproduction beyond this maximum capacity by throwing away excess molecules, as illustrated in figure 7.3. This method has the advantage that it also works well in stochastic simulations, where it is usually trivial to implement: it suffices to make sure that the total amount of molecules does not exceed a given maximum. We will show in section 7.2.5 how this idea of an outflow can be used to model selection under resource limitations.

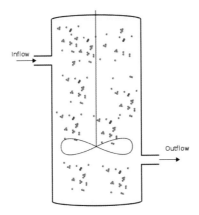

Figure 7.3 Well-stirred flow reactor with dilution — inflow and outflow of material.

Growth can also be regulated via predator-prey interactions among species, without explicit resource constraints, as will be shown in section 7.2.8.

7.2.5 Selection under Resource Limitations

The concept of outflow ϕ provides a useful way to model selection under resource limitations, as an extension of a simple birth-death process. For a system with n species $X_i(t)$, each with concentration $x_i(t)$ and with fitness $f_i \geq 0$, $i = 1, ..., n$, the *selection equation* can be written as:

$$\dot{x}_i = x_i(f_i - \phi) \tag{7.12}$$

Note that this is a simple extension of the elementary birth/death equation (7.3), where f_i is the birth rate and ϕ is the death rate.

The outflow ϕ must ensure that the total population size remains constant:

$$\sum_{i=1}^{n} x_i = 1 \quad \text{and} \tag{7.13}$$

$$\sum_{i=1}^{n} \dot{x}_i = 0 \tag{7.14}$$

For this to happen, ϕ must be:

$$\phi(\vec{x}) = \sum_{i=1}^{n} f_i x_i \tag{7.15}$$

Thus, ϕ must be equal to the *average fitness* of the population, in order to ensure that the maximum capacity will not be exceeded.

It is well known that the selection dynamics equation (7.12) admits a single globally stable equilibrium. This means that starting from any valid concentration vector $x = (x_1, x_2, ...x_n)$ satisfying eq. 7.13, the population will converge to an equilibrium where only one species has survived: the one with largest fitness f_k, where $f_k > f_i \forall i \neq k$. All the others have become extinct. This is called competitive exclusion, or survival of the fittest.

Note that eq. (7.12) implies that, without resource limitations (no outflow, thus $\phi = 0$), the population would grow exponentially. Other growth laws different from exponential lead to different survival outcomes. This can be seen by generalizing eq. (7.12) to:

$$\dot{x}_i = f_i x_i^c - \phi x_i \qquad (7.16)$$

Without resource limitations ($\phi = 0$), the growth is subexponential for $c < 1$, leading to the "survival of everybody," i.e., the various X_i coexist in the system. Conversely, for $c > 1$ the growth is superexponential (hyperbolic), leading to a "survival of the common" [838] (or survival of the first [621]) situation: for the special case $c = 2$ (bimolecular collision between two X_i molecules), whoever had the highest product $p_0 = f_i x_i$ at $t = 0$ wins, i.e., the outcome depends on the initial conditions $x_i(t = 0)$; quite scaringly, the population reaches infinite numbers in a finite time $t = 1/p_0$. Such dangerous population explosion can be prevented with an outflow ϕ that limits the population size to a maximum. Yet even in the presence of an outflow, the qualitative outcome of the competition does not change: Survival of everybody occurs for $c < 1$ and survival of the first for $c > 1$ (see [621, 838] for details). None of these outcomes (for $c \neq 1$) is favorable for evolution.

It has been shown that the case of $c = 0.5$ (parabolic growth) applies to the template replication of small oligonucleotides (akin to short DNA or RNA sequences) [838, 891]; therefore, this could have constituted an obstacle against the evolution of the first replicators leading to primordial life forms. On the other hand, it has also been shown that if such primordial genetic material is enclosed in a cell able to replicate, an exponential growth rate can be preserved at the cell level [837, 838]. Therefore it has been argued that Darwinian evolution could only have occurred during the early life on Earth once the DNA became enclosed in a protective membrane compartment such as a lipid bilayer, forming a protocell that as a whole could replicate at exponential rate.

A well-mixed reactor is obviously not a good representation of reality where individuals are spread in space. The dynamics of replicators in space [125, 892] departs significantly from the idealized notions depicted. However, the idealized well-stirred case is still useful from a theoretical perspective, as it lays the foundations for the more complicated and realistic cases.

7.2.6 Selection under Stochastic Effects

The simplest way to consider stochastic effects in evolutionary dynamics is through a *Moran process*, which is a simple stochastic birth-death process: a population of size N contains individuals of two species A and B that reproduce at the same average rate, that is, they have the same fitness. At each time step, one individual is chosen for reproduction, while another one dies to make room for the offspring. There is no mutation. A mathematical analysis of the Moran process tells us that even if we start with equal amounts of A's and B's, ultimately only one species will survive [621]. This result is in stark contrast with the deterministic competition case stemming from equation 7.9, where two species with the same fitness could coexist. This phenomenon is called *neutral drift*; even in the absence of selection, stochastic fluctuations may drive species to extinction.

In the presence of selection, a random drift also occurs which may wipe out unlucky species even if they are fitter than the others. Note that fitness must now be defined in probabilistic terms: it is the expected (average) reproductive rate. Again in contrast with the deterministic case, a higher fitness offers no survival guarantee in a stochastic setting. Instead, a *fixation probability* can be calculated to quantify the chances of a new mutant to replace a population dominated by individuals of a given species.

7.2.7 Variation

Variation can be introduced by two mechanisms [236]: *external* or *passive*, such as random mutations induced by radiations; *internal* or *active*, resulting from programmed molecular interactions such as genetic recombination in sexual organisms.

When introducing random mutations in the growth-selection process described in Section 7.2.2, instead of a single surviving species, it has been shown that a family of similar species survives. This family is called *quasispecies* [249]. The selection equation modified to include the possibility of mutations is called the *quasispecies equation*:

$$\dot{x}_i = \sum_{j=0}^{n} x_j f_j q_{ji} - \phi x_i \quad , \quad i = 1, ..., n \tag{7.17}$$

Each individual now carries a genome which might be copied with errors (mutations), in which case an individual of type X_j may produce a mutated offspring of type X_i. The probability that such a mutation happens, converting a genome j into a genome i, is given by q_{ji}.

Equation 7.17 can be expressed in chemical form as:

$$X_j \xrightarrow{f_j q_{ji}} X_j + X_i \tag{7.18}$$

$$X_i \xrightarrow{\phi} \quad . \tag{7.19}$$

Reaction 7.18 expresses the production of mutant X_i whereas reaction 7.19 expresses the elimination of a molecule X_i by the dilution flow ϕ (still described by eq. 7.15) that compensates any overproduction.

Now the growth of X_i is not contributed by X_i directly, but by all those species X_j mutating into X_i (which might include X_i itself if $q_{ii} > 0$). If there are no errors ($q_{ii} = 1$ and $q_{ji} = 0 \ \forall j \neq i$), the quasispecies equation (7.17) reduces itself to the selection equation (7.12).

Note that the equilibrium quasispecies does not necessarily maximizes the average fitness ϕ, since it is composed of a family of mutants with different fitness values.

Reproduction is replication with variation. The *error threshold* [249] provides an upper bound on the amount of mutation under which the system is still able to sustain inheritable variation, thus the evolutionary potential, without collapsing under a wave of deleterious mutations (*error catastrophe*). For a genome sequence of length L, the error threshold μ is given by: $\mu < c/L$, where c is a constant (typically $c = 1$ for biological genomes), i.e. evolutionary adaptation can only be sustained when the mutation probability per base is less than μ, or one mutation every L nucleotides on average. This result is valid for the deterministic (ODE) case, whereas for the stochastic case in finite populations the error threshold is smaller [627], that is, the accuracy of replication must be larger (the tolerance to errors decreases) due to stochastic fluctuations that may wipe out species in spite of a fitness advantage.

Figure 7.4 shows examples of stochastic simulations of the quasispecies reactions 7.18—7.19, using Gillespie SSA, for a population of $N = 100$ molecules. Here molecules are binary strings of length $L = 10$ bits, and their fitness is the proportion of bits set to one. Results are shown for three mutation rates: below ($m = 0.2$), above ($m = 2.0$) and at the theoretical (deterministic) error threshold ($m = 1.0$). The actual mutation probability per bit is $p = m/L$. We know from [627] that the actual stochastic error threshold for this case must lie below the deterministic value, since smaller population sizes require higher replication accuracy. Nevertheless, let us look at how evolution manages to make progress under these three different conditions. Starting from a

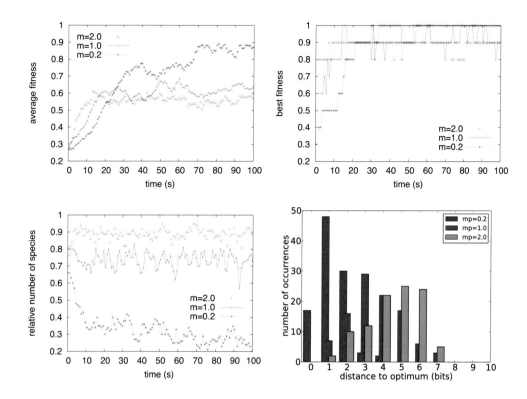

Figure 7.4 Stochastic simulation of the quasispecies model under low ($m = 0.2$), medium ($m = 1.0$) and high ($m = 2.0$) mutation rates, for a population of $N = 100$ binary strings of length $L = 10$ bits. Top: average and best fitness among all individuals in the population. Bottom left: number of species relative to the total population. Bottom right: histogram of species at the end of the simulation, grouped by their distance to the optimum (number of bits that differ from the optimum sequence '1111111111').

random mix of molecules (that is nevertheless biased toward low fitness), under a low mutation rate the system converges to a much smaller set of molecules that are near the optimum. For higher mutation rates, initially the system seems to make more progress toward the optimum, and the diversity of species remains high, but the average fitness of the population quickly stagnates at low values. For $m = 0.2$, the optimum is found and survives to the end of the simulation, whereas for the higher mutation rates the optimum is occasionally found but repeatedly lost, and does not survive to the end of the simulation.

7.2.8 Interacting Species

The selection equation (7.12) and the quasispecies equation (7.17) consider individuals in isolation: An individual may reproduce at a given rate and with a given fidelity. They do not take into account possible interactions among objects, such as molecular collisions, recombination of genomes, competitive interactions, ecological interactions such as predator-prey, parasite-host, and so on. These interactions cannot be neglected, and can sometimes even boost the evolution process.

Two famous equations that take interactions into account are the *replicator equation* from evolutionary game theory, and the *Lotka-Volterra equation* from ecology. In spite of coming from two different fields, it is known that these two equations are equivalent [621, 792].

The Replicator Equation

The *replicator equation* has the same shape as eq. (7.12), but the objects X_i are now *strategies* of a game, the variables x_i are their frequencies in the population, and f_i is the expected *payoff* of strategy i when interacting with the other strategies:

$$f_i = f_i(\vec{x}) = \sum_{j=1}^{n} a_{ij} x_j \quad , \quad i = 1, ..., n \tag{7.20}$$

where a_{ij} is the payoff for strategy i when interacting with strategy j.

The fitness values f_i are no longer constant (in contrast to what had been assumed in eq. (7.12)): they now vary as a function of the amount of objects $\vec{x} = x_1, ..., x_n$ in the population.

The replication equation can be obtained by substituting eq. (7.20) in eq. (7.12):

$$\dot{x}_i = x_i (\sum_{j=1}^{n} a_{ij} x_j - \phi) \quad , \quad i = 1, ..., n \tag{7.21}$$

Looking at this equation from a chemical perspective, we can see that it is equivalent to a description of bimolecular reactions between objects X_i and X_j subject to the law of mass action:

$$X_i + X_j \xrightarrow{a_{ij}} 2X_i + X_j \quad , \quad \text{for} \quad a_{ij} > 0 \tag{7.22}$$

$$X_i + X_j \xrightarrow{-a_{ij}} X_j \quad , \quad \text{for} \quad a_{ij} < 0 \tag{7.23}$$

Equation 7.21 thus sums up the effects of all the reactions in which X_i participates. For $a_{ij} > 0$ (cooperative interaction), object X_j acts as a catalyst for the replication of X_i, without being consumed in the process. For $a_{ij} < 0$ (defective interaction), X_j destroys X_i. The collision is elastic for $a_{ij} = 0$. Note that it is perfectly legal to have $a_{ij} \neq a_{ji}$. For instance, when $a_{ij} > 0$ and $a_{ji} < 0$, molecule X_i benefits from the interaction while molecule X_j gets harmed by it. Such different situations reflect different games in evolutionary game theory.

As before, the system is subject to a dilution flow ϕ, still described by eq. (7.15), which flushes out any overproduction and keeps the population size constant. However, the fact that the growth is now based on interactions among objects is reflected in the dilution as well. This becomes apparent by substituting (7.20) in eq. (7.15), obtaining:

$$\phi = \sum_{i=1}^{n} \sum_{j=1}^{n} a_{ij} x_i x_j \tag{7.24}$$

One concept from evolutionary game theory is especially important in this context: An evolutionarily stable strategy (ESS) [563, 564, 621] cannot be invaded by new mutants arriving in small quantities. Formally, for an infinitely large population of a strategy X_i (the wild type) challenged by an infinitesimally small quantity of a potential invader X_j (the mutant), strategy X_i is an ESS

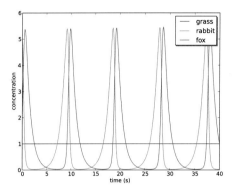 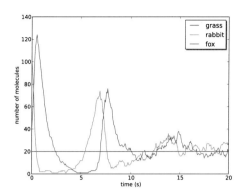

Figure 7.5 Typical oscillations in predator and prey populations that interact according to the Lotka-Volterra model (reactions 7.25-7.27). Left: numerical ODE integration for $k_a = k_b = k_c = 1$, with initial concentrations $x_0 = 5$, $y_0 = 2$, $g_0 = 1$. Right: stochastic simulation (using Gillespie SSA) with the same parameters (in a small volume $V = 20/N_A$, resulting in microscopic rate constants $c_a = k_a/(N_A V) = 0.05$, $c_b = k_b/(N_A V) = 0.05$, $c_c = k_c = 1$ according to [329, 926] and initial molecule counts $n_x = x_0 N_A V = 100$, $n_y = y_0 N_A V = 40$, and $n_g = g_0 N_A V = 20$).

if $a_{ii} > a_{ji}$, or $a_{ii} = a_{ji}$ and $a_{ji} > a_{jj}$ for all $j \neq i$; that is, its payoff when interacting with itself (a_{ii}) must be greater than all the other payoffs, or, if there are other challenging strategies X_j with matching payoff ($a_{ji} = a_{ii}$), then the payoff of each challenging strategy when interacting with itself must be smaller than the payoff that the wild type X_i gets when interacting with X_j. If this condition holds, then any new mutants will be quickly wiped out. For the more realistic case of a finite population, the stochastic fluctuations will actually play in favor of the ESS, since most mutants will not stand a chance against these fluctuations. It can be shown that when the new mutant has a lower fitness than the established population, and its fixation probability is less than the one under random drift, then the mutant will die out for sure [621].

The Lotka-Volterra Equation

The *Lotka-Volterra equation* describes the interactions among species in an ecology, and it is typically illustrated with a simple predator-prey example:

$$X + G \xrightarrow{k_a} 2X + G \tag{7.25}$$

$$Y + X \xrightarrow{k_b} 2Y \tag{7.26}$$

$$Y \xrightarrow{k_c} . \tag{7.27}$$

These reactions can be interpreted as follows: eq. (7.25): rabbit X eats some grass G and reproduces. For simplicity, the amount of grass is assumed to remain constant. eq. (7.26): fox Y eats rabbit X and reproduces; eq. (7.27): fox Y dies. This simple model is well known to produce oscillations in predator and prey populations under some conditions, as shown in figure 7.5.

The differential equations for these reactions can be easily derived using the law of mass action:

$$\dot{x} = k_a x - k_b y x \tag{7.28}$$

$$\dot{y} = k_b y x - k_c y \tag{7.29}$$

By setting equations 7.28 and 7.29 to zero, it is easy to find out that this system has an equilibrium point at $x = c/b$ and $y = a/b$. A stability analysis reveals oscillations around the equilibrium. Figure 7.5 shows an example of such oscillatory behavior, comparing the idealized ODE case with a more realistic stochastic simulation. Whereas the ODE scenario shows sustained oscillations, in small populations contingency coming from stochastic fluctuations causes erratic oscillations and frequent extinctions.

The generic form of the Lotka-Volterra equation for n interacting species is given by:

$$\dot{x}_i = x_i (r_i + \sum_{j=1}^{n} b_{ij} x_j) \quad , \quad i = 1, ..., n \tag{7.30}$$

where:

- r_i is the spontaneous growth (if $r_i > 0$) or death (if $r_i < 0$) rate of species i (independently of interactions with other species, i.e., the rate of growth (resp. death) of i caused by unimolecular reactions)

- b_{ij} is the strength of interaction between species i and j:

  - $b_{ij} = 0$: no interaction
  - $b_{ij} > 0$: positive (cooperative) interaction: j makes i grow
  - $b_{ij} < 0$: negative (defective) interaction: j "eats" i

Note that it is possible to have $b_{ij} \neq b_{ji}$, leading to the same types of interactions as in the replicator equation:

- both b_{ij} and b_{ji} are positive: the interaction between X_i and X_j is mutualistic (mutually beneficial).

- both b_{ij} and b_{ji} are negative: the interaction is competitive (mutually destructive)

- $b_{ij} > 0$ and $b_{ji} < 0$: the interaction is antagonistic (individual X_i benefits from it, whereas X_j is harmed by it); for $b_{ij} = 1$ and $b_{ji} = -1$ we have the predator-prey case, as in reaction 7.26, where X_i is the predator and X_j is the prey.

It has been shown that, after some transformation of parameters, the generalized form of the Lotka-Volterra equation (7.30) with n species is equivalent to a replicator equation (7.21) with $n + 1$ strategies [621, 792].

Typically, the replicator equation is used in the context of evolutionary games to find ESSs and to predict if a certain mutant will be able to invade a population. The Lotka-Volterra equation is typically used to study foodwebs in ecosystems.

The Lotka-Volterra model has been extended in many directions. It has been studied in a wide variety of settings, including well-mixed and spatial scenarios, as well as deterministic and stochastic settings [37, 238, 593]. Generalized predator-prey systems involving multiple species have been proposed, including cyclic interactions (the predator of one species is the prey for another, and so forth, forming a food chain in an ecosystem), as well as mutations of one or more species, leading to adaptations in individual behavior [5, 292, 712, 834].

Albeit simple, Lotka-Volterra models display a rich set of behaviors, from the already mentioned oscillations to explosion of prey populations, extinction of both species, waves of predators chasing preys, clustering of species, and so on. When including mutations, predator and prey species are allowed to coevolve; for instance, predators may evolve an improved ability to track and capture prey, whereas prey may evolve more efficient escape strategies. In section 8.2 we will look at coevolution in a bit more detail, and in section 8.2.3 some ACs modeling ecosystems will be presented.

7.2.9   Random Catalytic Reaction Networks

The quasispecies equation includes only self-reproduction with mutation, without interactions among individuals. On the other hand, neither the replicator equation nor the Lotka-Volterra equation take variation into account (e.g., mutation, recombination). Therefore, none of these equations contains all the evolution ingredients.

A slightly more generic model was presented by Stadler and colleagues [792], who formalized a generalization of the replicator equation for systems in which two entities interact to produce a third, potentially different one.

In [792] the so-called *random catalytic reaction networks* are defined as systems containing reactions of form:

$$X_i + X_j \xrightarrow{\alpha_{ij}^k} X_i + X_j + X_k \tag{7.31}$$

In this reaction, molecules X_i and X_j interact to catalyze the production of another potentially different molecule X_k. They are not consumed in the process, but they do not reproduce directly either, in general. However, later in the process, a given molecule X_k might produce X_i or X_j back again, resulting in an *autocatalytic cycle*, or more generally a graph, i.e., an *autocatalytic network*.

Note that molecule X_k could be regarded as a mutation of X_i or X_j. In contrast to the quasispecies model however, here the production of a different molecule type X_k is a deterministic result of the chemical rules governing the reaction between molecules X_i and X_j, as opposed to an error due to external noise.

This system is described by the *catalytic network equation*:

$$\dot{x}_k = \sum_{i=1}^{n} \sum_{j=1}^{n} \alpha_{ij}^k x_i x_j - x_k \phi(t) \quad , \quad k = 1,...,n \tag{7.32}$$

The nonspecific dilution flux $\phi(t)$ is still described by eq. (7.24), which is the equation describing dilution for interactive models. In this case, the coefficients a_{ij} from eq. (7.24) are given by:

$$a_{ij} = \sum_{k=1}^{n} \alpha_{ij}^k \tag{7.33}$$

so a_{ij} gives the total rate coefficient for all reactions involving X_i and X_j, independently of what X_k they produce. Direct replication occurs when $k = i$ or $k = j$.

The authors [792] show that the catalytic network equation (7.32) contains the replicator equation (7.21) and its equivalent form, the Lotka-Volterra equation (7.30), as special cases. Note however that, whereas equation 7.32 is generic, reaction 7.31 only captures the case $\alpha_{ij}^k > 0$. The collision is elastic when $\alpha_{ij}^k = 0$. The case $\alpha_{ij}^k < 0$ can be implemented by the reverse reaction:

$$X_i + X_j + X_k \xrightarrow{-\alpha_{ij}^k} X_i + X_j \tag{7.34}$$

but was not the focus of the experiments by [792]. In those experiments, the parameters α_{ij}^k were randomly generated, with probabilities adjusted to shape the chemical reaction network as desired, from sparsely connected (most α_{ij}^k coefficients set to zero) to densely connected (most α_{ij}^k set to positive values), with or without direct self-replication (direct replication can be easily disabled by setting $\alpha_{ij}^i = \alpha_{ij}^j = 0 \ \forall i, j$).

Starting from a random initial mix of molecules, and simulating the reactions (7.31) using the ODE expressed by eq. (7.32), the authors [792] showed that even when direct self-replication is disabled, such random mix converges to a stable fixpoint in which only a subset of species has survived due to their mutually catalytic interactions. Later it became clear that this kind of fixpoint is an instance of an *organization*, in the sense of chemical organization theory (see Part IV).

7.3 Artificial Chemistries for Evolution

Since ACs were motivated by Artificial Life studies, most ACs have some relation with evolution, either by assuming an intrinsic evolutionary dynamics, or by looking at how such dynamics could emerge out of (bio) chemical reactions. The range of ACs to mention here would be too broad, perhaps even encompassing most existing ACs. Hence we will restrict our discussion to a few examples that illustrate the potential of ACs to help understanding the origin of evolutionary processes and their dynamics. We focus on two basic aspects: the emergence of evolution, and algorithms for simulating evolutionary dynamics. Other AC topics related to evolution appear in chapter 8.

7.3.1 Emergence of Evolution in Artificial Chemistries

The investigation of how evolution could originate out of simple chemical reactions is perhaps one of the most exciting and central themes in AC. ACs that focus on the origin of evolution include [49, 63, 233, 281, 419, 566, 651, 790].

Note however that most of these chemistries already assume some underlying elements of evolutionary dynamics following the principles discussed in earlier in this chapter. Some of them [651] follow the general scheme of the unimolecular quasispecies reaction (7.18), where the individuals in the population do not interact directly, and reproduce by a process akin to cell division in unicellular organisms. Others [63, 233, 281] follow the general scheme of the bimolecular catalytic reaction (7.31), in which two catalysts interact to produce a third one, without being consumed in the process. This second reaction scheme can be regarded as the operation of a machine (first molecule) on a tape (second molecule), producing a third molecule as the result of the operation. In both cases (reactions 7.18 or 7.31) the reaction is usually accompanied by a nonselective dilution flow, that is, for every new molecule created, a random molecule is destroyed in order to keep the total population size constant.

An example of an AC that does not assume an underlying evolutionary dynamics is the auto-catalytic metabolism by [49], discussed in chapter 6. This chemistry is based on simple conden-sation and cleavage reactions, sometimes aided by catalysts.

Here we will analyze each of these three cases (unimolecular, bimolecular, and metabolic replicators), trying to distinguish the built-in evolutionary elements from the emergent ones.

Unimolecular Replicators in the Amoeba Platform

In the Amoeba platform by Pargellis [650–652], computer programs inhabit processor slots on a two-dimensional lattice. Each program seeks to replicate itself by allocating memory in a neigh-boring slot and copying its own code to the child location. Programs are not always successful at replicating themselves, and children are mutated with some probability. If all slots are occupied with other programs, a random slot is chosen to hold the child program. The overall process fol-lows the scheme of reactions 7.18 — 7.19 from the quasispecies model, with the enhancement that the reacting molecules are located on a two-dimensional space. Therefore Amoeba can be regarded as an example of "unicellular chemistry" in which molecules are computer programs, and unimolecular reactions create new molecules, whereas an existing molecule is removed to make space for the newly created one.

Amoeba belongs to the family of assembler automata (see chapter 10), systems in which assembly-language programs compete for CPU and memory resources in a parallel computer. This family includes well-known ALife platforms such as Avida [19, 567], Tierra [701], and Core-World [696, 697].

Starting from randomly generated sequences of instructions, self-replicating programs emerge spontaneously in Amoeba [651], a result that turns out to be much harder to achieve in other assembler automata. At first, these initial replicators are extremely inefficient, but they get more proficient with the passing generations, and tend to cluster together in colonies. These descendants of the first successful replicators evolve rapidly toward a very fit subgroup akin to a quasispecies. Such population is generally unable to diversify sufficiently to form a complex and self-sustaining ecosystem.

Bimolecular Replicators

The matrix chemistry from chapter 3 [63] follows exactly the same reaction scheme as reaction 7.31; therefore,s it is an example of random catalytic reaction networks. Indeed, the emergence of organizations is also observed in the matrix chemistry. An important difference with respect to [792] is that in the case of the matrix chemistry, the internal molecule structures are explicitly represented, and the dynamics is played by the folding of operators and their action on operands.

Other chemistries based on reaction 7.31 include [233, 281]. As pointed out in [792], when direct self-replicators are disabled, such systems tend to settle to collectively autocatalytic fix-points, where species replicate indirectly via cooperative cycles of chemical reactions. Indeed, the formation of autocatalytic networks is reported in these chemistries. While this result may appear straightforward, it is not exactly so, considering the fact that these ACs are stochastic, as opposed to the ODE model of [792]. Moreover, in contrast to [792], where the system was closed and the molecules had no explicit internal structure, the ACs investigating the origin of evolution generally aim at the constructive aspects of the system. In such open and constructive systems, new molecule types can always be created and might have a qualitative impact on the overall dynamics of the system. The ACs will then look at what kind of new behaviors emerge as a result of the interactions between new molecular structures as these new types arise.

Throughout this book, many other ACs that also make use of the same or similar reaction schemes will appear, and many of them are also able to produce organizations. However, the use of reaction 7.31 alone is not a sufficient recipe for success; the chemistry must also be able to generate molecular structures that react in a way that favors the emergence of such organizations, as opposed to an inert system or a system that degenerates to a random unorganized soup. There is currently no firm recipe for the discovery of such "right chemistries," and the researcher is often left with not much more than intuition as a guide.

Other chemistries follow the machine-tape concept but depart from reaction (7.31) with respect to the products of the reaction. For instance, in [566] the machine may modify the tape; in [419] the machine also changes state while altering the content of the tape; whereas in [790] a variety of products may be formed, depending on the operators (in that case, combinators) present. In spite of these differences, a wide variety of emergent phenomena have also been reported for these chemistries, showing that the scheme of reaction (7.31) is only one among many possible ones leading to the emergence of phenomena leading to evolutionary dynamics. For instance, the emergence of organizations is also reported in [790]. In [419], according to the amount of random mutation allowed, simple autocatalytic loops, parasites, hypercycles, and more complex autocatalytic networks emerge. The initial work [566] was later expanded to a lattice system running on parallel hardware [136, 843], where self-replicating strings, parasites, hypercycles, and even Chemoton-like cellular structures have been reported to emerge.

From a chemical point of view, the machine-tape metaphor expressed by reaction (7.31 along with its variants [419, 566, 790] actually describes a quite complex process, where the operation of a machine on a tape can be compared with the activity of an enzyme on a biopolymer. In the case of reaction (7.31), two catalysts cooperate to build a new molecule, using raw materials that are not explicitly modeled. The new molecule produced can be seen as an "offspring" of the two catalyst "parents." Certainly this cannot be implemented as a single reaction in real chemistry. Reaction (7.31) abstracts away from this complexity, but by doing so, it also prevents us from finding out how such reproductive-like scheme could arise out of simpler, more realistic chemical processes. Other chemistries are needed to investigate the emergence of evolution at such a fine granularity, and one example is perhaps the metabolic chemistry discussed in the following pages.

Emergence of Evolution in Primitive Metabolisms

The chemistry by Bagley and Farmer [49] is an idealized model of a metabolism where molecules are binary strings that react by condensation and cleavage, sometimes aided by catalysts. Small "food" molecules are supplied to the system at a constant rate, from which more complex molecules can be built. Autocatalytic networks (called "autocatalytic metabolisms" in [49]) are shown to emerge in this system under a range of parameters. Furthermore [50], when subject to occasional random mutations, these metabolisms jump from one fixpoint to another, showing first signs of evolution.

In spite of these encouraging results, further studies showed that the spontaneous emergence and maintenance of autocatalytic networks is difficult to observe, especially in stochastic settings. Moreover, autocatalytic sets are prone to leakings and errors due to side reactions (see chapter 6 for a more comprehensive discussion).

Although many insights into the emergence of evolution can be derived from these experiments, a clear overview of the general conditions for such emergence remains an open problem.

7.3.2 Algorithms for Simulating Evolutionary Dynamics

As discussed before, the computational modeling of evolutionary dynamics must go beyond numerical ODE integration in order to capture the stochastic nature of the evolutionary processes and its innovation potential. Furthermore, evolution is a process that can be described very naturally by an artificial chemistry.

One of the advantages of describing the system as an AC is that, given the chemical reactions rules together with their rates, both deterministic ODEs and their corresponding stochastic simulations can be easily derived, as we have learned from chapters 2 and 4.

For the simplest cases where all the reactions are bimolecular and happen with the same probability, the simple algorithm for simulating individual molecular collisions shown in section 2.3.3 might be sufficient.

For more generic cases, one of the algorithms covered in chapter 4 can be chosen. However, these simulation algorithms will need some adaptations to deal with evolutionary dynamics. This is due to the stiffness often observed in evolutionary dynamics. Such stiffness is caused by at least two factors: the sporadic nature of mutations, and the different time scales for species interaction and evolutionary change. The former refers to the fact that some mutations might have no effect, others might be quickly eliminated, yet others might cause waves of change that sweep through the population and then settle into long periods of low activity. The latter refers to the separation of time scales when both short-term interactions between individuals and their long-term genetic evolution are modeled.

In order to deal with this problem, a hybrid algorithm based on SSA and τ-leaping is introduced in [957], where periods of low activity are modeled with τ-leaping and the system switches to SSA when a high activity (due to a spreading mutation) is detected. The algorithm is used for simulating the evolutionary dynamics of cancer development.

The next mutation method [548] is another recent algorithm for simulating evolutionary dynamics. Inspired by the next reaction method (NRM) and taking into account that mutations are rare events, it aims at reducing computation effort by jumping from one mutation to the next.

In the evolving ecology model of [290], most mutations are deleterious, so the probability of a viable mutation is very small. In this context, SSA is used to jump to the next successful mutation, in order to avoid wasting computation time on simulating unsuccessful ones.

7.4 Summary and Open Issues

Darwinian evolution is a process that drives a population of individuals toward higher reproductive ability or fitness. It does so via a combination of replication, selection, and variation. Evolution works within certain ranges of replication rate, selection strength, and variation amount. Within these ranges, evolution tends to converge toward a mix of species (quasispecies) with genotypes close to the fittest individual.

Evolution and artificial chemistry are closely related. An evolutionary process can be described by an AC, and multiple ACs have been dedicated to the study of fundamental aspects such as the emergence of evolutionary processes out of a mix of reacting chemicals. In spite of the enormous diversity of these ACs, most of them follow reaction schemes compatible or easily leading to evolutionary dynamics, such as the quasispecies (mutation-only) scheme, or the random catalytic network scheme. In other words, they facilitate the emergence of evolution in order to focus on specific phenomena of interest within the evolutionary dynamics, such as the emergence of new ways to replicate the genetic information, or to produce genetic variations. Therefore, the problem of the emergence of evolution entirely from scratch remains mostly open, though we believe ACs are the place to look for a solution.

Even when evolutionary dynamics is embedded into the fabrics of the system, the emergence and maintenance of evolution is by no means guaranteed in a constructive AC: molecules with unpredictable effects may arise at any moment, leading to catastrophic events that dominate the dynamics and can even cause the system to collapse [224, 579]. The search for molecule representations and reaction rules leading to more robust and evolvable systems is an active research topic, and it will be further discussed in chapter 8.

The present chapter is only an introduction to the problem of evolution. The next chapter 8 will present a few additional insights into more advanced topics related to evolution, and the rest of the book will provide numerous examples of ACs that address evolutionary questions.

COMPLEXITY AND OPEN-ENDED EVOLUTION

Despite the obvious power of natural evolution, so far artificial evolution has not yet been capable of displaying the full richness of natural evolution. Typically, artificial evolution tends to stagnate at some point, while natural evolution seems to always find a way to move forward. Critics of evolutionary theory often use such weakness as an argument in their quest to prove it false. However, to most experts in the field, this only shows that our understanding of evolutionary processes is still far from complete. Indeed, evolution is a complex stochastic process driven by feedback and influenced by several parameters. The search for open-ended evolutionary growth of complexity in artificial systems similar to that shown in nature is still one of the major research topics in artificial life, and it touches artificial chemistries as well, as we shall see in this chapter. Many ACs have been motivated by a desire to understand evolution, from its origins to the production of increasingly sophisticated forms of life. The AC contributions to this area range from looking at the chemical processes leading to complex organisms to the dynamics of species interacting in ecosystems.

We start the chapter with a discussion about the interplay between evolution and self-organization (section 8.1). After that we look at how species coevolve in ecologies (section 8.2) and examine the delicate balance between robustness and evolvability (section 8.3). Section 8.4 discusses the still controversial topic of complexity growth in the biosphere, and section 8.5 reviews current efforts to obtain open-ended evolution on the computer.

8.1 Evolution: Steering Self-Organization and Promoting Innovation

A self-organizing system (chapter 14) constructs order from the inside. For this to happen, an inflow of energy is usually needed to drive the system into an organized state, and as a result the system expels some high-entropy waste to the outside world. The system is open and exports entropy as it generates internal order. In this way, it is able to obey the second law of thermodynamics. All living organisms are complex self-organizing systems relying on a continuous energy input and exporting entropy to their surroundings.

Such self-organization could allow the system to remain in a steady state and never change in time, so energy would be only used to maintain it in an autopoietic way. However, there is more to an evolving system than just the flow of energy and entropy. Openness to matter and openness to information are two other necessary ingredients. In particular, in living systems, matter and information are important. What happens in living systems is that the information to maintain their autopoietic organization, which is encoded in genes, may suffer changes over the generations through different kinds of transmission errors such as mutation, recombination, and horizontal gene transfer. Occasionally, a new more successful offspring might arise, replacing older variants, and the struggle for survival ends up selecting fitter organisms. In this way, evolution steers the population of self-organizing living systems from one autopoietic organization to another, resulting in organisms that are potentially improved and more adapted to their environment.

Evolution has been described as a movement in the space of organizations [555] that can be formalized in the context of chemical organization theory (see part IV). The study of evolution from an organization theory perspective allows us to formalize the transition from one autopoietic state to the next. However, this approach has been criticized because organizations are closed by definition, whereas living organisms are open systems. There is general agreement, though, that our tools for the mathematical modeling of evolutionary processes still need further refinement.

While evolution in the Darwinian sense is essential to life, it cannot explain the existence of life itself. Evolution requires heritable genetic variation, which became possible only after sufficiently complex replicators were formed. Other order-construction mechanisms must have led to these primordial replicators. These preevolutionary mechanisms are studied under the various origin of life theories (see chapter 6). It can be shown that preevolutionary dynamics is weaker than full evolutionary dynamics, and is easily displaced once replicators capable of Darwinian evolution emerge in the population [625].

The increase in complexity and diversity of species in the biosphere is an emergent phenomenon whose underlying mechanisms are still poorly understood. There is a general feeling that natural selection alone cannot explain such increase, and that other self-organizing phenomena caused major evolutionary transitions to higher levels of biological organization, such as the emergence of complex multicellular organisms [447, 565, 755]. Moreover, multiple species coevolve in an ecosystem, leading to a competitive arms race that results in increasingly adapted species, interwoven in an increasingly complex biosphere. In any case, evolution remains the driving force that steers these combined self-organizing processes, acting at various levels of organization. Hence, evolution itself should be regarded as a self-organizing process that steers self-organizing systems (living beings). Furthermore, since it may create a potentially unlimited amount of different types of organisms, natural evolution qualifies as an instance of a constructive dynamical system (chapter 15).

As a self-organizing, emergent, and constructive process, evolution suffers from the same current limitations in formal mathematical analysis that affect other self-organizing systems, emergent phenomena, and constructive dynamical systems, not to mention complex systems in general. In spite of this difficulty, much progress has been made in the study of evolutionary processes through the theory of evolutionary dynamics, as highlighted in chapter 7. However much remains to be understood, in particular, concerning the complex interactions between species and the emergence of totally different organizations of the system, such as the engulfing of microorganisms that later became chloroplast and mitochondria, or the association of independent unicellular organisms to form multicellular organisms with interdependent cells. The remainder of this chapter reviews such new frontiers in evolutionary science and the AC efforts in contributing to this burgeoning area.

8.2 Coevolutionary Dynamics in Ecologies

The simple models of evolution covered in chapter 7 did not consider the combination of inter-species interactions with intraspecies evolutionary changes. Yet, in nature several species interact and coevolve in an ecosystem. Such interactions can have a profound effect on the outcome of the evolutionary process.

Evolutionary ecology is the field that studies the evolution of species within ecosystems of interacting species. In a coevolutionary model, the fitness of a species is influenced not only by its own ability to survive in a given environment, but also by its interactions with other species. Interactions can be competitive or destructive, antagonistic (such as predator-prey and parasite-host), or cooperative (such as symbiosis and mutualism). The network of interacting species together with their environment form an ecosystem, which is in constant change as the populations of each species fluctuate, and new species appear while others go extinct.

One important consideration to take into account when modeling ecosystems is the environment where the various species interact. In this context, the notions of space and resources are particularly relevant. The coevolutionary dynamics will be influenced by how different species move and spread in space, as well as by the strategies that they evolve to escape predators or to more efficiently locate prey. Food webs may form involving multiple species feeding on each other. Ecological niches may emerge, inside of which a delicate balance between interdependent species is established.

In this section we review some computational approaches to coevolution with ACs, as well as some closely related approaches. We start with Kauffman's well known NKC model, an early abstract model of an ecosystem. Then we briefly discuss the tangled nature model, which is very similar to an AC. This facilitates the transition to a discussion of some ACs that model ecologies.

Although complex, the coevolution models that will be discussed are still extremely simplified when compared with natural ecosystems. One of the obstacles against modeling evolving ecosystems is the difference in time scale between the short-term species interactions and the long-term evolutionary changes. Therefore, many models are restricted to either the interactive aspect (considering species as static) or the evolutionary aspect (ignoring intraspecies variations). As computation power increases, more realistic combined models become possible [179, 290]. Even then, the existing models are still far too simplified compared to reality, leaving plenty of room for more research and improvement. For a review of the topic, the reader might look at [178] and for some more recent results we refer to [290].

8.2.1 The NKC Model

The NKC model [450], proposed by Kauffman and Johnsen in 1991, was an early model of coevolution that enabled a systematic investigation of the combined effects of long-term coevolution together with short-term population dynamics. The NKC model is an extension of the NK family of fitness landscapes [447], a well-known abstract model of epistatic gene interactions, which is able to generate rugged fitness landscapes in a controlled way. In the NKC model, a number of species interact in an ecosystem. For simplification, all the individuals of the same species are assumed to be identical, that is, they carry the same genome with N genes or traits. Each gene makes a fitness contribution to the overall fitness of the organism. The fitness contribution of each gene is influenced by K other genes in the genome of individuals of the same species, and by C genes from other species. Additionally, each gene may also be coupled with W features of the external world. At each time step, each species tries a random mutation, and the mutant replaces the current one only if it is fitter (simulating the fixation of a fitter mutant, equivalent to a coarse-grain jump from one ESS to the next in a single step).

In a spatially extended version of the NKC model where species interact with their neighbors on a square grid representing the ecosystem, the authors show that *avalanches* (cascades of changes) may propagate through the ecosystem, causing the system to move from one equilibrium to another. They also show that for a critical value of K, the distribution of avalanche sizes is a power law, meaning that most avalanches are very tiny, whereas a few of them are very large, indicating a self-organized critical state (see chapter 14). Furthermore, when extended to include the abundance of each species together with its growth rate, the authors showed that mutualism can emerge for an optimum K value, and that the system often exhibits long periods of stasis separated by short bursts of change after a fitter species initiates an avalanche. Such behavior is an indication of *punctuated equilibria*.

Punctuated equilibria are phenomena observed in the fossil record, in which most species remain unchanged during long periods of time, with sudden changes occurring very rarely, leading to cascades of speciation and extinction events that move the ecosystem to another a very different composition. This view contrasts with the traditional interpretation of the original theory of evolution by Darwin, which seemed to imply gradual changes throughout evolutionary history.

Numerous further models of coevolution followed [51, 179, 713], most of them also reporting some form of self-organized criticality and punctuated equilibria. Among these, we highlight the tangled nature model [179, 360], for its similarity with an artificial chemistry.

8.2.2   The Tangled Nature Model

In the tangled nature model [179], individuals are binary strings of length L that may interact in a collaborative, competitive, antagonistic, or neutral way. The fitness of an individual is a function of the strength of the interactions of the individual with others in the ecosystem, and the abundances of these interacting individuals. Individuals may reproduce sexually or asexually, and their growth is limited by the total carrying capacity of the environment. In the case of sexual reproduction, two individuals may mate if the Hamming distance between their genomes is less than a threshold, hence simulating a reproductive barrier between different species. Random mutations and death of individuals may also occur with some probability.

In a purely asexual reproduction scenario (with a mutation probability under an error threshold [360]), starting from a population of identical individuals, the system quickly diversifies into multiple genotypes, and later it reaches a stable configuration that the authors named quasi-ESS or q-ESS. In contrast to an ESS however, the q-ESS is not perfectly stable because stochastic fluctuations due to mutations can suddenly destabilize the system, triggering a period of quick reconfiguration, until another q-ESS is found. Again, this can be compared to punctuated equilibria. Overall, the periods of fast changes are shown to be much shorter than the q-ESS periods. Moreover, while the average duration of sudden transitions tend to stabilize with time, the average duration of the q-ESS epochs increases with time, leading to a more stable ecology. This trend is maintained in the sexual reproduction case, where *speciation* can also be observed, with the emergence of clusters of genotypes separated by a Hamming distances larger than the mating threshold.

8.2.3 Ecological Modeling with Artificial Chemistries

Coevolutionary phenomena occur not only at the level of ecosystems, but also at the molecular level, with autocatalytic reactions or sets of reactions playing a reproductive role. Artificial chemistries have many similarities with ecosystems, even more so if they contain the notion of space. Different types of molecules could be considered akin to species in an ecosystem, and their reactions with each other form a network akin to a food web in ecology. Further, introducing a notion of space will allow many different species to survive and thrive in different environments, thus generating a level of diversity not possible without the niches offered by spatial restrictions to reactivity.

Natural coevolution has inspired numerous artificial ecosystems in Artificial Life, some of which are ACs or can be interpreted as ACs, such as [134, 198, 315, 745, 795]. Natural coevolution has also inspired a range of coevolutionary optimization algorithms [213, 337, 648, 672] that have been used to solve a variety of real-world computation problems with a number of advantages when compared to traditional evolutionary algorithms, such as diversity maintenance [337, 338] and cooperative problem solving [213, 648, 672]. Some of these algorithms can be considered artificial chemistries.

In this subsection, we will introduce a few examples of ecological models using ACs.

Some Early Models

An early computer model of an artificial ecosystem was proposed by Conrad and Pattee in 1969 [191, 195]. In their system, virtual organisms live in a one-dimensional world, where they must collect materials for their own maintenance, interact with other organisms, reproduce via conjugation (in analogy to certain bacteria), and eventually die (with mass-conserving recycling of the materials used). The organism's behavior is dictated by which state, out of a set of possible internal states, it adopts in response to some input. State transition rules are encoded in the organism's phenotype, which is in turn determined from its genotype. Recombination of genetic material occurs during reproduction, enabling the evolution of new behaviors. One of the frequent outcomes of their system was the evolution of a homogeneous population where the organisms were very efficient at using materials, the ecosystem was collectively stable, and no further evolution occurred.

Subsequent developments led to the EVOLVE series of ecosystems [134, 135, 714], in which virtual organisms are also subject to energy and matter conservation. They live in a 2D world where they respond to their environment, secrete chemicals, harvest energy in the form of light, and use this energy for various metabolic and genetic processes. Enzymes and other proteins are represented as strings with critical sections that determine their shape. Their function is then fetched from a table by a matching mechanism that weights the critical section more heavily than the rest of the string, mimicking the importance of the critical region in determining the protein's function from its lock-and-key binding properties. Starting with a population of only autotrophs, scavenger populations emerge, go extinct, and reappear, showing the emergence of an ecological niche aimed at decomposing the metabolic remains of autotrophs.

Generalized Lotka-Volterra in Ecolab

The Ecolab platform by Standish [795, 796] runs the generalized Lotka-Volterra equation (eq. 7.30), with mutations of the parameters r_i and b_{ij}. The mutation functions are carefully chosen such that offspring resemble their parents and populations remain bounded. The algorithm is analogous to a numeric ODE integration where population counts are kept as integers with the help of a probabilistic rounding rule [795]. In spite of its simplicity, Ecolab shows self-organized criticality with a power law distribution of species lifetimes [796], like many of the coevolutionary systems discussed in Section 8.2.

Digital Ecosystems of Assembly-Language Programs

Rich ecosystems have been obtained with the Avida platform, where virtual organisms are assembly-language programs that self-replicate in a two-dimensional world [17, 198]. Starting from a single self-replicating program, the system diversifies into an ecology of interacting programs, forming host-parasite relationships and a wealth of complex dynamics.

Avida followed from predecessors such as Tierra [701] and CoreWorld [696, 697]. More about these and other chemistries from the assembly automata class can be found in chapter 10. Avida can be considered so far as the most successful digital ecosystem based on assembly-language programs that is able to show such a large variety of evolutionary phenomena, and at the same time solve concrete problems in domains such as software engineering [337, 567] and computer networks [468].

Evoloops: An Ecosystem of Self-Replicating Loops

In the Evoloop platform [745], an ecology of self-replicating loops evolves in a cellular automaton (CA). Starting from a single ancestor loop able to reproduce, mutate and dissolve (die), a population of descendant loops eventually occupies the whole CA space. As the various loops compete for space, smaller loops that have a reproductive advantage emerge and dominate. In simulations with sufficiently large spaces, a diversity of loop types emerges, and the various loop species tend to cluster into islands. Thanks to their ability to dissolve, these loops have some degree of robustness to underlying hardware errors, which is not always the case in CA. A comprehensive description of the evoloop system can be found in [744]. A through study of its resistance to hostile pathogens and mass killings is carried out in [739]. Further investigations on its evolutionary dynamics and genetic makeup are reported in [738]. Cellular automata and self-replicating loops are further discussed in chapter 10.

ARMS: Modeling Ecosystems with Membrane Computing

Ecologies have also been modeled with ARMS (Abstract Rewriting System on Multisets) [828], an artificial chemistry based on membrane computing. In membrane computing or P systems [678, 681], molecules are placed in a hierarchical structure of nested compartments enclosed by membranes, analogous to the structure of modern cells (see sec. 5.3). The ARMS chemistry is a variant of a P System specifically designed to investigate complex behaviors in Artificial Life. In [828, 829] ARMS is used to study the dynamics of an ecosystem consisting of plants, herbivores, and carnivores. A set of chemical reactions within ARMS defines the interactions between species. Simulation results reveal that the ecology is more robust when plants produce a herbivore-induced volatile substance that attracts carnivores. Both ARMS and membrane computing will be further discussed in chapter 9.

Urdar: Bacteria Cross-Feeding Ecologies

The ALife platform Urdar [315] is a more recent artificial chemistry where where artificial organisms akin to bacteria "metabolize" binary strings, converting them to higher-entropy forms. It is intended to investigate the evolution of a cross-feeding ecology where one bacterial strain feeds upon the metabolic waste of another strain, and so on, forming a network of interdependence similar to a foodweb. In Urdar, individuals contain cellular automata rules that operate on the binary strings, in such a way that only those individuals which contain rules that increase the entropy of the "metabolized" strings are able to replicate and hence to survive in the long run. The authors show that coexistence is a common outcome of the evolutionary dynamics in Urdar, where rich ecologies with high biodiversity emerge. The investigation of pairwise interactions between species from an evolutionary game perspective reveals a curious formation of rock-paper-scissor (intransitive) dominance relationships: say that species A can invade species B, and B can invade C; it occurs often that A cannot invade C, but rather the opposite: C can invade A. Several such cases were found in the system. Such intransitive relations are known to promote biodiversity. In a subsequent study [314] the authors show that diversity increases when the inflow of new energy-rich strings into the system is reduced, leading to energy-efficient organisms. A mathematical analysis to determine the conditions for coexistence and mean fitness improvement was carried out in [537].

Building Virtual Ecosystems from Artificial Chemistry

In [240] Dorin and Korb propose a more general approach to ecosystem modeling based on artificial chemistries. The authors point out that usually, the building blocks of organisms forming part of the environment escape detailed modeling efforts in virtual ecosystems, despite their extreme importance in real ecosystems. In their modeling approach, photosynthetic autotrophs, chemosynthetic autotrophs, and decomposers are part of the ecology, and their ability to store and release chemical energy is taken into account. They define an artificial chemistry consisting of virtual atoms from the set A, B, C, O that float in a 2D space. Atoms may bond covalently, and energy is required to make or break bonds, the result of which may either be to capture or to release energy. There are four types of catalysts in the system: chlorophyll-like catalysts help to produce sugar from sunlight (making A-B bonds, breaking A-O and B-O bonds); another enzyme breaks down sugar (by breaking its A-B bonds) and releases the chemical energy stored in these bonds. Separate enzymes decompose "organic" bonds (by breaking C-C bonds) other than sugar and "inorganic" bonds. A reaction that requires more energy than is available in its neighborhood cannot proceed. On the other hand, energy released from a chemical bond must be used in that time step or it is released in non-recoverable form. Finally, the only way energy can be stored is in complex molecules. The authors believe that such modeling does more justice to ecosystem mechanism than other related ALife-type modeling of ecosystems.

8.3 Robustness and Evolvability

In our discussion of evolutionary dynamics, the variation operators are mostly assumed as leading to viable individuals, or lethal mutations are simply not modeled since the corresponding individuals die immediately on the timescale of the simulation. Often, the the effects of variation is abstracted away by using simple parameters such as the mutation rate or the interaction strength. Although this helps us to look at an evolutionary process from a macroscopic perspective, it still does not explain how the underlying mechanisms leading to such variation work, such that heritable genetic variation becomes possible over the course of many generations. It is not only a matter of an error threshold (quantitative) but also a matter of a viability threshold (qualitative): for instance, if most mutations are lethal, although infrequent and small, or if offspring are frequently very different from their parents in terms of phenotype (which determines fitness), then the evolutionary process will be very slow and erratic.

Here we summarize the delicate interplay between robustness and evolvability in natural and artificial evolution.

8.3.1   Robustness and Evolvability in Nature

Evolvability is the ability to evolve. In the context of natural evolution, evolvability is defined by Kirschner and Gerhart as "the capacity to generate heritable, selectable phenotypic variation" [461]. Wagner [901] further complements this definition with the notion of innovation: an evolvable system must also be able to "acquire novel functions through genetic change, functions that help the organism survive and reproduce." He further argues that "the ability to innovate is the more profound usage of evolvability."

Natural evolution is considered to be "naturally evolvable," by virtue of the coevolution of biological variation mechanisms together with the organisms themselves over millions of years. Indeed, a multitude of examples of the evolvability of natural organisms can be observed [461, 462, 670].

But what makes a lineage of individuals evolvable, given that many random mutations have lethal effects? Intuitively, the underlying variation mechanisms must offer some protection against damaging mutations while letting harmless ones proceed, that is, the process must have some robustness against deleterious mutations. Variation, on the other hand, is needed for evolution, but this variation can turn out to be good or bad. Such dichotomy is expressed in the paradox of robustness versus evolvability [902].

Robustness in biological systems is defined by Wagner as the ability to "function, survive, or reproduce when faced with mutations, environmental change, and internal noise" [901]. While robustness may shield organisms from deleterious mutations, it may also hinder evolvability by preventing changes in order to keep the system stable. There is thus a tension between robustness (needed for stability and survival in the current environment) and evolvability (needed for adaptation to new conditions).

The balance between robustness and evolvability has been extensively studied by Wagner [900–902]. In [901] the role of neutrality in keeping this balance is analyzed: "the more robust a system is, the more mutations in it are neutral, that is, without phenotypic effect". He then argues that "such neutral change — and thus robustness — can be a key to future evolutionary innovation" because "a once neutral mutation may cause phenotypic effects in a changed environment." Thus one may conclude that robustness hinders evolvability in the short term, but may promote it in the long term, when environmental conditions change, such that a once neutral mutation might become a beneficial one. This is how "neutrality can become key to innovation." Such a double role of robustness is further quantified in his subsequent work [902], where he proposes a solution to the apparent conflict between robustness and evolvability by quantifying these two properties at the genotypic and phenotypic level: he points out that whereas genotypic robustness indeed hinders genotypic evolvability, phenotypic robustness actually promotes phenotypic evolvability. A review of the interplay between robustness and evolvability can be found in [547].

In [461] several examples of mechanisms leading to biological evolvability are presented. Such mechanisms decrease the interdependence between the various components of the system, such that the impact of changes remains localized, increasing the chances of nonlethal variation that could then accumulate, allowing evolution to proceed. Among the mechanisms cited, we highlight compartmentation, redundancy, and weak linkage. An example of compartmentation is the expression of a group of genes according to the cell type, such that changes within such a given group do not affect other cell types. An example of redundancy is gene duplication and divergence: A duplicated gene initially provides a redundant function, but with time this gene may mutate into something different, leading to a different function that might prove useful later. Weak linkage refers to the properties of some proteins to bind to several types of substrates with a low selectivity, such that several signals can activate the same protein, leading to some tolerance to the change in some of these signals. Whereas [461] covers mostly eukaryotes, in [670] the evolvability of both eukaryotes and prokaryotes is extensively discussed and compared. A notable difference between the evolvability strategies of these two taxa of organisms is highlighted in [670]: Whereas unicellular prokaryotes are capable of quick and diverse metabolic adaptations, multicellular eukaryotes focus on morphological changes, with core metabolic processes that remain mostly unchanged. A comprehensive discussion of evolvability cases and sources of novelty in evolution can be found in [462] and [903].

8.3.2 Robustness and Evolvability in Artificial Evolution

Nature has ended up producing both robust and evolvable organisms, simply because only those species possessing both characteristics could survive in a competitive and changing environment. Is it possible to achieve the same in artificial evolution? As a matter of fact, current research aims at learning how to design artificial evolutionary systems able to produce similarly robust and evolvable solutions. In artificial evolution (be it in ACs, evolutionary algorithms, synthetic biology, or any other form of man-made system intended to display evolutionary properties) we often wish to accelerate the evolutionary process such that it can be accomplished within a reasonable time. For this purpose, we wish to boost the production of beneficial mutations while decreasing the probability of lethal mutations.

In the context of artificial evolution, evolvability is defined by Altenberg as "the ability of a population to produce variants fitter than any yet existing" [28]. The focus here is on producing improvement rather than simply surviving. As such, it is important to be able to quantify and measure evolvability [407, 408, 708, 902] in order to detect its increase, as well as to understand and promote the *evolution of evolvability* [28, 95, 241, 423, 905]: how evolvability itself might evolve and hence improve with time.

Inspired by biology, several techniques for improving the evolvability of artificial systems have been envisaged [407, 461, 900]. These techniques seek to shield the system from deleterious mutations while promoting beneficial ones. For this purpose, the system must be designed such that the genotype representation is evolvable. Recall from section 8.3.1 above that an evolvable representation produces viable offspring with high probability under mutation, and moreover the offspring tend to be fitter than their parents. The representation problem is defined in [905] as the critical way in which the mapping from genetic variation onto phenotypic variation impacts evolvability.

One design principle for evolvable representations is to minimize the interdependence between components. One way to achieve this is through a modular design. Altenberg [29] points out that newly created genes should exhibit low pleiotropy (interdependence between genes) such that they are more likely to produce viable individuals. In a modular design, pleiotropic effects are mostly constrained within groups of genes related to the same function, with few pleiotropic interactions between genes coding for different functions. Therefore, it is argued in [905] that modularity "is expected to improve evolvability by limiting the interference between the adaptation of different functions." Modularity and its evolution will be further addressed in section 8.4.4.

Another technique to improve evolvability is to exploit redundancy in the representation, either in the form of gene duplication and divergence, or in the form of alternative genotypes mapping to the same phenotype, such that mutations among these redundant genotypes still fall within the same function.

A survey of techniques inspired by biology to improve evolvability in artificial evolution can be found in [407]. ACs aimed at evolution are among the artificial systems that can benefit from such techniques. However, the representation problem is far from fully solved, and only general guidelines are currently available. Indeed, evolvability is also a major concern for these ACs, and still an active research topic. Next we will focus on the current developments in the evolvability of ACs.

8.3.3   Evolvability of ACs

In the case of ACs, the instruction set used to construct new molecules (programs) plays an important role in the evolvability of the overall system: in order to lead to evolvable systems, the chemical programming language used must produce viable and fertile individuals with high probability. This is an area where there is currently no firm recipe for success, although some guidelines have been offered [628, 820, 822].

What features would the instruction set of a programming language have to possess, in order for programs to be evolvable? Design guidelines for evolvable computer languages that allow for universal computation are presented by Ofria and colleagues [628], building upon the success of the Avida platform. Ray observed that it is difficult to evolve programs using an instruction set with "argumented" instructions, mainly for the reason that instructions and their arguments might be mutated independently, leading in the overwhelming majority of cases to lethal or at least detrimental program variations. Ofria and colleagues [628] point out that 99.7 percent of mutations in effective code lead to such effects, with most of the rest leading to neutral variations. This is not exactly a recipe for empowering evolution.

The authors claim, however, that it is actually possible to design instructions sets that evolve better than others in the Avida platform. Key steps are to remove arguments from instructions and to establish memory protection, something Ray's Tierra already achieved. In a more recent study, Avida adds a local grid of location-bound instructions, which cannot be accessed directly by other programs, thus providing a spatial mechanism for diversity protection. In a more recent contribution [139], even broader collection of instruction sets is examined in terms of providing evolvability. The authors subjected each of the designed architectures/instruction sets to a number of different environments (computational tasks to be solved) and tested their evolutionary potential. Figure 8.1 shows how these variants performed in terms of evolutionary potential.

The evolvability of string rewriting ACs has been considered in [820, 821], and the experimental optimization of such string systems is reported in [822]. The evolvability of collections of self-replicating molecules in an AC is presented in [411], and further extended to self-reproducing cells in a two-dimensional AC [412].

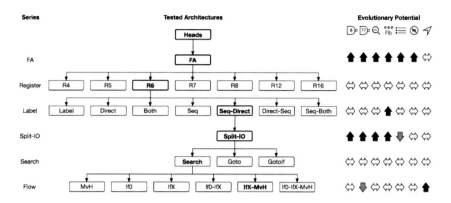

Figure 8.1 Instruction set variants are depicted in boxes, with variants in different series on different lines. Boxes in bold indicate best performing variant. At the far right are the results for each of the individual computational tasks reported. For details, we refer to [139]. (From [139]), with permission.

Overall, there remains much to be done in regard to the examination of the evolutionary potential of artificial chemistries.

8.4 Complexity Growth

We have seen how evolution can steer a population of individuals toward fitter variants through the combined effects of replication, variation, and selection. However, can evolution alone explain all the rich complexity and variety of the organisms observed in the biosphere? This is one of the biggest questions in evolutionary biology today, for which there is no firm answer so far.

The path toward the answer lies at a deep investigation of how variation mechanisms operate at the molecular level, in order to figure out under which conditions genotypic variation can lead to viable phenotypic traits with varying degrees of novelty, ranging from minor changes to truly novel phenotypic traits with a clear selective advantage, such as eyes or wings; and more, when would something so novel arise that it could trigger a *major evolutionary transition* [565], such as the creation of the genetic code, the emergence of multicellular organisms, cell differentiation with division of roles, with consequent invention of organs and tissues, vascular systems, nervous systems, immune systems, etc. Some of the possible mechanisms behind such important transitions will be discussed in this section.

Whatever the answer is, it is generally agreed that the level of complexity usually achieved in artificial evolution falls far behind that of the biosphere. As such, the search for open-ended evolution, or the unbounded growth of complexity, is a kind of "Holy Grail" in artificial evolution. In 2000, Bedau identified open-ended evolution as one of the great challenges in Artificial Life [93]. We will return to this quest in Section 8.5.

8.4.1 Complexity Growth in the Biosphere

Although it seems obvious that the biosphere has become more complex over time, from primitive cells to humans, the existence of a trend toward complexification (that is, an "arrow of complexity" [90]) is not yet entirely agreed upon [90, 161, 571, 670]. Neither are its underlying

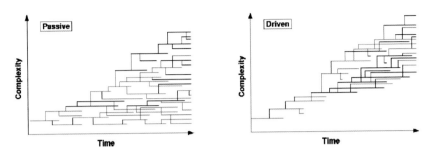

Figure 8.2 Passive versus driven trend of complexity growth in evolution. Left: Maximum complexity grows over time, while average complexity might stay the close to the lower boundary, especially if weighted with biomass. No selection pressure toward higher complexity. Right: Both maximum and average complexity grow due to selection pressure that weeds out low complexity entities. (From [583], ©2008 MIT Press, with permission.)

mechanisms entirely known. The controversy is complicated by the fact that there is not even a consensus on what complexity is, let alone on a way to measure it. Several quantitative metrics for complexity have been proposed [591], but they all meet criticisms. For the purpose of our discussion, the qualitative definition by Bedau [90] looks suitable enough: the complexity of an organism increases with its number of different functional properties.

The otherwise obvious trend toward complexity growth in the biosphere has been criticized for various reasons [571, 670], such as the tendency toward reduction of complexity observed in evolution, because smaller genomes can be replicated faster, illustrated by the frequent emergence of parasites; and the lack of empirical support, due to significant gaps in the fossil record. On the other hand, theoretical support in favor of the arrow of complexity is based upon two arguments [161]: *(1)* There must be a *passive trend* toward increase in complexity, since reduction can only proceed down to what fits in a certain minimum genome size; such passive trend is consistent with the fact that relatively simple organisms such as bacteria coexist still today with complex ones such as mammals. *(2)* The emergence of major innovations such as multicellularity and modular design cannot be explained by a passive trend alone; thus, there must be an *active trend* too, driving evolution in a biased way toward more complex organisms. The possible underlying molecular mechanisms for such active trends are extensively discussed by Wagner [903].

In an interesting contribution to this debate, Miconi has proposed to clarify some of the terms of this discussion [583]. He proposes to consider the *arrow of complexity* as the tendency of the maximal complexity to increase over time (as opposed to an overall, or average increase). He further clarifies that a passive trend toward increase in complexity versus a driven trend toward complexity (if one exists) can be discerned on the basis of measuring whether low complexity disappears over time or can coexist. The difference can be illustrated by figure 8.2, where we can see a gradual elimination of low-complexity species due to selection in the driven example.

On the basis of the fossil record and biomass measurements of extant life, one can conclude that the trend toward higher complexity in the biosphere globally can only be passive. However, there might be pockets of complexity increases in certain areas of the tree of life, driven by the evolutionary forces of variation and selection, as no doubt the growth of brain size in early hominids testifies. Miconi performs a series of thought experiments with a design system in order to arrive at three simple requirement that need to be fulfilled to allow a sustained growth of complexity, i.e., the existence of the arrow of complexity in a system.

1. Arbitrarily complex adaptive designs must exist in the design space explorable by the design generator;

2. The search heuristic must not impose a hard upper bound on the functional complexity of designs;

3. At any time, there must be more possible jumps toward successful designs of higher and equal complexity than toward designs that would make the starting points of these jumps unsuccessful.

Especially in the case of Darwinian evolution as the search heuristic, the last requirement is key to seeing a sustained growth in maximum complexity, according to Miconi.

In summary, it is generally believed that the arrow of complexity trend exists, although it is merely a general trend or tendency, never a certainty, since complexity may fluctuate in time and in species space due to the stochastic nature of the evolutionary process. Moreover it is believed that natural evolution is *open ended*, with no foreseeable limit in the potential amount of innovation that may be produced. Arguably, such open-endedness is facilitated by the interplay between two phenomena: the interlocking of species in coevolutionary webs, and the emergence of new levels of complexity on top of existing ones, in a hierarchical way. We have already explored the first phenomenon in section 8.2. The second phenomenon, hierarchization, will be discussed here. We start by explaining levels of evolution, and then look at the current theories about how nested structures can emerge and how evolution can operate on them.

8.4.2 Conditions and Levels of Evolution

If a system produces new levels of complexity in a hierarchical way, such as the emergence of multicellular organisms out of colonies of interacting cells, then evolution must also operate in a hierarchical way, since it now must act on entire organisms as opposed to merely at the level of individual cells. Understanding how evolution can operate on groups and also on whole hierarchies is key to understanding the evolution of new levels of complexity. The concepts of *unit of evolution, group selection*, and *multilevel selection* are useful in this context.

A *unit of evolution* is defined as an entity satisfying three conditions, quoting from [838]:

1. *Multiplication*: entities should give rise to more entities of the same kind.

2. *Heredity*: A type entities produce A type entities; B type entities produce B type entities

3. *Variability*: Heredity is not exact: occasionally A type objects give rise to A' type objects.

Objects satisfying conditions 1 and 2 only are *units of selection*, i.e., they may be subject to natural selection but do not evolve toward more sophisticated ones. Examples include objects belonging to the replicator or Lotka-Volterra category, described by equations (7.21) and (7.30).

A population of units of evolution undergoes evolution by natural selection when it contains objects of different types which present a hereditary difference in their rates of reproduction: selection operates due to such difference, and heredity ensures that the variability that is created is transmitted intact most of the times to the next generation.

A fourth condition for evolution by natural selection is given by [76]:

4. *Finite resources* should cause competition among the units of evolution.

This condition is clear from our previous discussion on resource limitations (section 7.2.4).

Moreover, a fifth condition must be fulfilled if the system is to exhibit an unlimited evolutionary potential [838]:

5. Replicators must display *potentially unlimited heredity*. This means that the possible number of different types of objects must be much larger than the number of objects actually present in the system at any given time.

Note that this fifth condition is common to a constructive dynamical system, which is consistent with what we had already pointed out before: an evolving population with unlimited evolutionary potential is an instance of a constructive dynamical system.

The five conditions above summarize what we had already discussed in sections 7.2 and 8.3. With these in mind, we are now ready to move to the next step: how evolution might steer organisms toward an increasing level of complexity. For this we have to consider new hierarchical levels where evolution can be applied.

Sometimes the unit of selection is not an individual object, but a group of objects that cooperate. In this case, we talk about *group selection*. When selection acts at the individual level, and also at the group level (potentially involving multiple levels in a nested way), we have *multilevel selection*. The existence of group selection per se is highly disputed: How could evolution operate only at one specific level and cease to operate at the other levels? On the other hand, there is a wide consensus on the existence of multilevel selection. A simple example is a minimum cell (see section 6.1). It has been shown that the evolution of the genetic material inside the cell is facilitated by its enclosure in a cell container: The cell then becomes the unit of evolution, and selection acts at the level of the cell, for instance, by selecting those cells that replicate faster. However, the consequence of such selection also impacts the genetic level (for instance, leading to the proliferation of genes that cause the whole cell to replicate more efficiently). Hence, both genetic and cell levels are subject to evolution: by acting directly at the cell level, evolution ends up acting indirectly at the genetic level too, leading to multilevel selection.

The stochastic corrector model by Szathmáry and Demeter [838] demonstrates multilevel selection in a simple way, and is illustrated in figure 8.3. It works as follows: Imagine a population of cells, where each cell contains two types of self-replicating molecules, slow and fast replicators. The cell as a whole can only survive if the two types of molecules are present. In addition, the cell replicates more quickly when the two types of molecules are present at the same amount. Cells may enclose only small amounts of molecules; therefore, stochastic effects play an important role in the model. Cells grow and divide; upon division, roughly half of the molecules go to each daughter compartment at random. In a world without cells, faster replicators would quickly outnumber the others. Within cells, however, both types are "forced" to coexist, or else they die with the cell. This simple model shows how the outcome of multilevel selection can differ from that of "flat" selection: by making use of multilevel selection, the stochastic corrector model is able to preserve the information inside compartments (information here is defined in terms of the proportion of molecular types present, in a way similar to the compositional genome discussed in section 6.2.4).

The stochastic error corrector model has been shown to provide a solution to Eigen's paradox (see sections 7.2.7 and 6.1.2), enabling the transmission of genetic information across generations of cells. Moreover, it has been shown to provide protection against the spread of parasitic

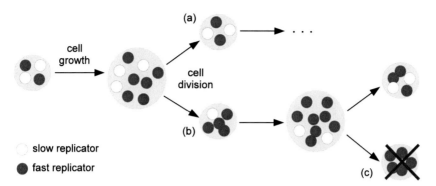

Figure 8.3 The stochastic corrector model [838]: cells rely on two types of self-replicating molecules for their survival, one of which replicates faster than the other. However each cell as a whole replicates more efficiently when both types are present at equal amounts. This leads to a coexistence of replications inside the cell, in spite of their intrinsically different replication rates. In the example above, the cell (a) will continue growing normally, whereas (b) will grow more slowly due to an uneven ratio of inner components, until cells start dying (c) due to the lack of an essential component.

mutants. Because of these interesting properties, many artificial chemistries have made use of the stochastic corrector model to promote the evolution of replicators with desirable characteristics. Some of these chemistries will appear throughout this book.

8.4.3   Complexity Encapsulation and Emergent Hierarchies

In their book *The Major Transitions in Evolution* [565], Maynard Smith and Szathmáry try to explain how and why complexity can increase during evolution. The authors identify the origin of the first self-replicating molecules, the first cells, the emergence of the genetic code, of chromosomes, of sexual reproduction, of multicellular organisms, as well as the formation of animal societies and the emergence of human language as major transitions.

   According to some researchers the significant qualitative changes represented by these major transitions require other self-organizing phenomena operating together with evolution in order to make such important transitions possible [447, 755]. Many of these transitions create new hierarchical levels of biological organization such as cells after self-replicating polymers, or multicellular organisms after unicellular ones. These levels can be referred to as *dynamical hierarchies* [689]. Self-assembly plays a key role in dynamical hierarchy formation, as stressed in [689]. Self-assembly is the spontaneous formation of an ordered structure, such as vesicles formed from fatty acids thrown in water (section 5.2.5), without any extra input of matter or energy. The combination of vesicle self-assembly with self-replicating template molecules encapsulated within such vesicles could explain the origin of the first primitive cells capable of evolution (see also section 6.1.1).

   Once the transition occurs, evolution then operates at the newly created higher level of hierarchy in order to optimize the population of individuals at that level. This is possible when the individuals at the new level form a new unit of evolution, which in turn is possible if the newly composed objects satisfy the conditions listed earlier. Crucial to the establishment of such transitions is therefore a concerted operation of genetic variation and selection at different levels of biological organization. In this way, evolution steers the process of self-organization toward higher levels of complexity. Note that the self-organizing forces contributing to complexity increase (such as self-assembly) could be considered as a kind of variation operator, however one that not only changes the genotype a little bit, creating a new variant of the same species, but one that potentially creates something radically new. Watson investigates how to obtain such operators for artificial evolution, in an approach called compositional evolution [911]. One of us has also contributed to a solution of this problem by proposing a new cooperation operator for evolutionary computing, that orchestrates the collaboration of a number of individuals on a new level [932].

It would be outside the scope of this book to discuss the transitions to higher levels of biological organization in detail, especially when books such as [462, 565, 903] already do an excellent job in that. We recall that the emergence of self-replicating structures, primitive metabolisms and cells has already been briefly summarized in chapter 6. Thus we will discuss in more detail only one example that can be considered close to artificial chemistries and constructive dynamics: the emergence of modularity.

8.4.4 Modularity

Modularity is defined by Wagner and Altenberg [905] as a genotype-phenotype map in which most interdependencies (pleiotropic effects) fall within genes coding for the same function, whereas genes coding for different functions have few interdependencies among themselves. Modular design can improve robustness and evolvability by confining the impact of a genetic change mostly within the affected module.

Modular design plays a key role in maintaining nested levels of hierarchy in biological organization, as well as in producing innovative biological designs [755]. The emergence of new phenotypic traits is investigated in [630] using the Avida platform. The authors confirm through simulations that indeed complex traits appear mostly via combinations and extensions of existing traits, thus significant innovations emerging from scratch are extremely unlikely. This requires such traits to be encoded in a modular way within the genome, such that genome sections corresponding to these modules can be recombined to give rise to more complex functions.

The conserved core processes in multicellular organisms exemplify such hierarchical scaling: core processes such as DNA and RNA processing, the genetic code, and the basic metabolic functions remain mostly unchanged and are largely conserved across species. The genes encoding such core processes can be regarded as building blocks or modules that can be reused in many ways to produce higher-level functions, either via genetic recombination, gene duplication, or via changes in gene expression. The molecular mechanisms behind the encapsulation of these core processes in a way that protects them against lethal mutations, and the regulation of their use in a modular way result in a new level of hierarchy for evolution to operate: The variation operators now act at the level of modules, recombining them in different ways, and no longer at the level of nucleotides.

Evolution tends to produce nonmodular structures because these are often more efficient than modular ones; direct connections between elements might help to increase fitness. Therefore, under ordinary circumstances there is no selective pressure to maintain a modular solution, so that modular solutions tend to quickly disappear once discovered. A number of conditions may facilitate the emergence of modularity, though. Noisy environments subject to unpredictable changes and extinction events seem to promote the spontaneous emergence of modularity during evolution [442, 513, 755]. Such studies provide evidence of the role of modules as an evolutionary response to noise. One reason for this is that a noisy or rapidly changing environment requires a quick switch from one configuration to another in order to respond to changes. Modular design is one way to solve this problem by switching from one module to another, for example, via a signaling system that causes a change in gene expression patterns.

Vinicius [887] claims that the tendency of certain evolutionary systems to evolve modular solutions that can be later encapsulated and might lead to the development of new codes transmitting information between the levels on which evolution takes place, is actually key to the complexity growth and open-endedness of natural evolutionary systems. In his view, an aggregation model of complexity growth should be surpassed by a model that emphasizes information transfer between levels using codes.

- The RNA world is the source of the genetic code and organic modularity

- Protein modularity is the origin of the developmental code via transcription factors

- Neural cells and brain modularity are the cause for behavioral modularity

- Specific modular signaling behaviors (e.g., vocalizations) are the origin of human language which itself is a modular system.

In a nutshell, the idea of Vinicius is that subsequent levels of genotype-phenotype relations developed in the course of evolution. Each time specialized phenotypes turned into information carriers, i.e., became genotypic elements of a higher level phenotype. The new phenotypes themselves were adding additional levels and resulted in a more powerful expressions of evolution.

8.5 Toward Open-Ended Artificial Evolution

The quest for open-ended evolution, or the unbounded growth of complexity is one of the main open issues in artificial evolution [93]. A system exhibiting open-ended evolution would never stop evolving and innovating. So far artificial evolutionary systems, especially computational ones, still fall short of achieving this. "Open-ended evolution is characterized by a continued ability to invent new properties — so far only the evolution of life on Earth (data partly from the fossil record) and human technology (data from patents) have been shown to generate adaptive novelty in an open-ended manner" [693].

Ideally, one would set up the environment, initial conditions, and rules of the game as in an artificial chemistry, and let the system evolve to produce individuals that are increasingly better adapted to their environment. As the environment changes, new generations of individuals would again become adapted to the new environment. The system would never stop evolving; it would always be able to produce novel individuals carrying novel adaptations. In practice however, a stagnation is often observed in artificial evolution. Sooner or later, the system reaches a plateau where nothing interesting seems to happen.

Bedau and colleagues [90, 91, 94, 97] formulated a series of metrics for quantifying evolutionary dynamics, in the hope of identifying open-ended evolution. *Evolutionary activity* is the rate at which useful genetic innovations are produced and adopted in a population. These metrics are then used to classify systems into four classes of activity [97]:

1. Unadaptive evolution: too many mutations occur, individuals have no time to adapt between them.

2. Adapted, uncreative evolution: the system is already highly adapted, almost no mutations occur, so it remains unchanged.

3. Creative, adaptive evolution: the mutation rate is right to allow interesting phenomena to happen such as the emergence of new species.

4. Unboundedly creative evolution: this generates open-ended novelty, such as the biosphere.

Maley [540] devises a series of toy models of evolution intended to assess the prospects for artificial open-ended evolution. Two of these models show unbounded evolutionary activity according to Bedau's metric [97]: Urmodel3 simulates a simplified process of evolutionary adaptation of a parasite species to a (static) host. Urmodel4 simulates the coevolution of parasite and host species. However, both models are clearly and deliberately oversimplified and, as such, do not qualify as open-ended. Maley therefore concludes that the metric [97] is insufficient to detect open-endedness. Channon arrives at a similar conclusion using the Geb platform [170, 171] and proposes a new, more stringent metric for measuring evolutionary activity [171].

Standish [797] applies Bedau's evolutionary statistics [97] to the Ecolab platform (introduced in section 8.2.3) and points out that connectivity (the number of interacting species) and diversity are inversely related in Ecolab. Therefore, in order to increase diversity, connectivity must be decreased. In subsequent work [798], he further elaborates on the connectivity-diversity relation and derives two ways in which it is possible to achieve unbounded evolutionary activity in such generalized predator-prey systems. The first is to reduce connectivity by biasing the ecosystem toward the creation of specialists (species that feed upon a restricted set of food sources) as opposed to generalists (species that are able to feed upon many different food sources); the second is when the migration rate (in a spatially extended version of Ecolab) is set to an optimum value called the "resonance" value.

Drawing from the fact that complex systems can be expressed as networks of interacting components, a complexity measure for networks is proposed in [799], based on the information content of the system. The metric is applied in [800] to compare the network complexity of three platforms: Tierra, EcoLab, and Webworld. None of them showed substantially more network complexity than the control networks, which confirms related observations from evolutionary activity statistics [97].

Several other ALife platforms have aimed at open-ended evolution, such as Tierra, Avida [628], Framsticks [470], Geb [170, 171], division blocks in Breve [788], Polyworld [933, 934], and Cosmos [845, 847], among others. Evolutionary algorithms aimed at achieving and exploiting open-ended evolution have been proposed [502]. Artificial chemistry platforms have also attempted to reach open-ended evolution, including Ecolab (discussed above), Squirm3 [529], Swarm Chemistry [747], the artificial economy by Straatman [816], and approaches combining genetic regulatory networks with morphogenesis [431]. Open-ended evolution in ACs has also been pursued by exploring niche construction [471], and the use of energy to drive diversity [405].

These attempts have reached varying degrees of success according to certain metrics. In general, however, it is agreed that none of the artificial evolution results so far has been able to really achieve open-ended evolution to its full extent, such that the richness of artificial life forms created could compare at least to primitive forms of the natural biosphere. Although these artificial systems can sometimes display apparent complexity increase, according to proposed quantitative metrics, qualitatively they still have not shown impressive innovations "crawling" out of the "digital test tube." For a recent survey and reflection on the progress in this area, see [410].

8.6 Summary

As we have seen, complexity growth is tightly connected to open-ended evolution. Without it, open-ended evolution could end up meandering in a mostly empty configuration space, always discovering novelty, yet never something truly qualitatively different. On the other hand, equally important driving forces for open-ended evolution, i.e., permanently changing environments and coevolutionary pressures between different species in a spatially organized ecology are necessary as well to drive evolution to new levels.

Complexity growth requires the production of nested dynamical hierarchies, which in turn rely on self-organization mechanisms other than variation and selection, such as self-assembly or aggregation, and the development of new codes through modularization. Unfortunately, to this day, artificial systems have not been able to bring about open-ended evolution and the unbounded growth of complexity to any satisfactory level. Thus, a major open issue for artificial evolution remains to be solved in the future.

Part III

Approaches to Artificial Chemistries

Rewriting is the theory of stepwise, or discrete transformations of objects, as opposed to continuous transformations of objects. Since computations in computer science are typically stepwise transformations of objects, rewriting is therefore a foundational theory of computing.

TERESE, TERM REWRITING SYSTEMS, 2003

CHAPTER

9

REWRITING SYSTEMS

So far, we have only discussed a very special numerical example of an artificial chemistry in chapter 3. It is now time to widen the horizon and look at what kind of systems have been used in the literature to capture the essentials of ACs. The next three chapters are dedicated to that purpose.

We will start with production or rewriting systems, a very general class of formal systems in mathematics, computer science and logic, in which rules for replacing formal (sub)structures with others are repeatedly applied. We shall see that these systems provide a broad basis for artificial chemistries and have been used in the literature repeatedly to elucidate aspects of ACs. Chapter 10 will then discuss abstract behavioral models of machines. There is a long history of this area in computer science under the heading of automata theory. Again, we are interested to see whether and how these models can be used to describe artificial chemistries. We will also discuss artificial chemistries close to the "bottom" of a computer, i.e., close to the actual processor or machine that provides the computational engine for all sorts of calculations. This will enable us to see how close indeed notions of AC can become to computation. Finally, chapter 11 will exemplify a few artificial chemistries conceived by looking at biological systems. Some of them could be classified as string rewriting systems, others as graph-based systems; others again have an element of mobility in them. The general characteristic that distinguishes them from the discussion in earlier chapter is their complexity: They are much more complex because they are aimed at implementing specific functionalities.

But first things first. Rewriting systems come in many forms depending on what kind of objects they rewrite. Examples are term rewriting, string rewriting, equation rewriting, and graph rewriting systems, among others. They have been at the foundation of programming languages in computer science and of automated deduction in mathematics. String rewriting systems, first considered by Axel Thue in the early part of the twentieth century [131, 228], have a close relationship with Chomsky's formal grammars [736]. Many rewriting systems have the goal of transforming particular expressions into a normalized form, deemed to be stable in some sense. Prior to reaching this stable form from a starting expression, the rewriting process goes through a series of transient expressions. For example, mathematical equations can be formally rewritten until they reach a form in which it is possible to read off solutions.

We can see, therefore, that rewriting allows a sort of dynamics of symbolic systems. For example, de Morgan's law of Boolean algebra can be expressed in this way[1]:

$$\neg(s \wedge t) \rightarrow \neg s \vee \neg t \tag{9.1}$$

and interpreted as the rewriting of the left side into the right (getting rid of the parentheses). There is not much difference formally between this activity of rewriting logical symbols s and t and the "rewriting" happening in a chemical reaction

$$s_1 + s_2 \rightarrow s_3 + s_4, \tag{9.2}$$

where $s_1, ..., s_4$ are symbols for atoms or molecules, and the equation describes the reaction of s_1 with s_2 which produces s_3 and s_4. In notation at least, the same type of formalism seems to be able to capture both of these instances. We can see here that our earlier statement of widening the perspective of mathematics to include dynamical symbol manipulation can be generalized to contain logic as well. Again, we want to move from a static picture of equality or identity of expressions to one where an act of replacement takes place, an operation that requires time.

Rewriting systems provide the basis for the λ-calculus [80], which has been used early to implement artificial chemistries. A simpler form of rewriting is behind other systems discussed later in this chapter.

9.1 Lambda Calculus

The λ-calculus is a formal system for the definition of functions and the (possibly recursive) application of functions to functions and therefore provides a way to define an implicit structure-function mapping. It has been used by Fontana to define artificial chemistries in the early 1990s [281].

Before we can understand how this method is used for ACs, we have to understand the λ-calculus itself. The objects of the λ-calculus are so-called λ-expressions. A λ-expression is a word over an alphabet $A = \{\lambda, ., (,)\} \cup V$, where $V = \{x_1, x_2, ...\}$ is an infinite set of variable names and A provides some special symbols important in evaluating expressions, the parentheses used for ordering the evaluation and λ and . used for abstract operations.

Suppose we define the following objects: $x \in V, s_1 \in S, s_2 \in S$. The set of all λ expressions S can be obtained in the following three simple ways:

$$x \in S \quad \text{a variable is a } \lambda \text{ expression;}$$
$$\lambda x.s_2 \in S \quad \text{a } \lambda \text{ abstraction is a } \lambda \text{ expression; and}$$
$$s_2 s_1 \in S \quad \text{an application is a } \lambda \text{ expression.}$$

We have already encountered x, symbolizing a variable. What does the λ abstraction $\lambda x.s_2$ mean? It can be interpreted as binding variable x in s_2. Bound variable, as opposed to free variables, have a value. That value in the λ abstraction is determined by what follows it. A variable x is bounded if it appears in in a form like $...(\lambda x. ... x ...) ...$. The λ calculus does work without naming functions, and its functions have only one argument. This is, however, not a restriction,

---

[1]where $\neg$ is the negation ("NOT") operator, $\wedge$ is the logic AND and $\vee$ is the logic OR operator.

since functions with more arguments can be rewritten as equivalent functions with one argument via what is called "currying."

The expression $s_2 s_1$ can be interpreted as the application of s_2 on s_1. This can be formalized by the β-rule of reduction (our first rewriting rule):

$$(\lambda x.s_2)s_1 = s_2[x \longleftarrow s_1], \tag{9.3}$$

where $s_2[x \longleftarrow s_1]$ denotes the term that is generated by replacing every unbounded occurrence of x in s_2 by s_1.

There are three forms of reduction of λ-expressions, α, β, and η. An α-conversion changes bound variables, which is possible because variable names are only conventions. Thus, $\lambda x.x$ is the same expression as $\lambda y.y$. A β-reduction we have already encountered in the previous example. It captures the idea of function application, and is the process of substitution of all free occurrences of a variable with an expression. An η-conversion can be applied to convert between $\lambda x.f$ and f if it is sure that both expressions give the same results for all arguments.

The final topic we must understand is the notion of a normal form (symbolized here by NF). A normal form in any type of rewriting system is the form of the expression that cannot be rewritten any more. Thus it consists of terminal symbols only. The pure untyped λ calculus (as defined before) does not have the property to be strongly normalizing, i.e., to have normal forms for all expressions. Thus, there is the possibility that the rewriting never ends. However, so-called typed λ calculi, refinements of the general case,[2] have this feature. They are the basis for programming languages in which it is guaranteed that a program terminates. Going back to the last section, it is this termination we meant by saying that a stable form of an expression can be reached, that is a form of the expression that cannot be rewritten any further.

We are now in a position to see how an artificial chemistry can be defined using the λ calculus: Molecules will be the normal forms of λ-expressions. This will achieve that the objects we consider are of a stable nature. The simplest way to define the equivalent of a reaction is a scheme that allows two molecules s_1 and s_2 to collide by applying s_1 to s_2 and then normalizing the resulting expression:

$$s_1 + s_2 \longrightarrow s_1 + s_2 + NF((s_1)s_2). \tag{9.4}$$

The procedure NF reduces its argument term to the normal form.[3] Thus, we can write

$$NF((s_1)s_2) = s_3, \tag{9.5}$$

considering the normal form as a new object. This then allows us to rewrite equation 9.4 as

$$s_1 + s_2 \longrightarrow s_1 + s_2 + s_3, \tag{9.6}$$

which we recognize as equation 3.10.

Now that we have seen what the objects are and how reactions can be implemented, the dynamics of the reactor vessel must be determined. Fontana and Buss [281,282] have performed a substantial number of experiments based on what Fontana originally called the λ-chemistry or AlChemy. They have used explicit molecules and simulated each reaction one by one in typical

---

[2]Some argue that it is vice versa, i.e. that the untyped λ calculus is a special case.

[3]Given that the pure λ calculus cannot guarantee normalization, the reduction process is bound by a maximum of available time and memory. If these resources are exceeded before termination, the term is considered to be not normalizable, the reaction is considered elastic and no reaction product results.

experiments with reactor sizes of $M = 1000$ to $M = 3000$, using the algorithm of section 2.5.1. In these experiments the reactors were initialized by randomly produced λ-expressions as the starting molecules.

With this simple reaction scheme and algorithm it was found that the diversity of the initial population reduced quickly, often leading to only one surviving self-replicating species of molecules / λ-expressions. The organization structure of the surviving ensembles is called L0-organization (for level-0). In general, an L0-organization consists of a few closely coupled replicators that form a hypercycle. However, these types of organizations are easily perturbed, according to [282], with a collapse of the cycle ultimately leading to one replicator. In part IV of the book we will look more closely at how an organization can be defined generally and what to expect under certain circumstances.

In order to arrive at more complex reaction networks additional filter conditions had to be introduced, similar to what we have done with the matrix chemistry in chapter 3. For instance, reactions that would replicate one of the participating λ-expressions were prohibited by redefining them as elastic. Thus, replication was not allowed. This is, of course, much more radical than what we have seen previously in the matrix chemistry, where only one species was prohibited from emerging in reactions. But the result of these additional filters was a reaction network composed of a huge, possibly infinite number of molecules, called L1-organizations (for level-1).

Because the reactor size was small compared to the size of a L1-organization, an L1-organization was only present by some if its species in the reactor and continuously regenerated other species of the organization (thus effectively not producing anything new if considered over the long term). These organizations were quite stable, in that a perturbation through injection of randomly generated species inserted into the reactor would not be incorporated into the organization. Only in rare cases did an incorporation happen and the resulting new organization was typically extended in layers.

Further, L2-organizations (for level-2) were defined which consist of two (or more) coexisting L1-organizations. Stable coexistence is characterized by the two conditions that, (i), the organizations interact; and (ii), this interaction is moderated by some intermediate species called glue. The spontaneous emergence of a L2-organization from a randomly initialized population is extremely rare. However, it can be synthetically generated by merging two independently emerging L1-organizations.

A key observation of [281] was that species dominating the reactor after a transient phase have a similar syntactical structure. In a sense, the system quickly exhausts its potential for innovation. This, unfortunately is typical for this type of ACs and has to do with the closure of a system. We shall see more about closure in part IV.

9.2   Gamma

The Gamma language was introduced in 1990 by Banâtre and Métayer as a new way of approaching parallel computation using chemistry as a paradigm [56, 58]. Gamma stands for *g*eneral *a*bstract *m*ultiset *ma*nipulation, and it has since given rise to the so-called γ-calculus [55], a generalization thereof. The central concern of Gamma is to eliminate any unnecessary sequentiality of computer programs often erroneously thought of as the key concept of the von Neumann machine, the register-based hardware in use today in computers' CPUs. In order to relax requirements on the sequentiality of programs, Gamma works with multisets. More to the point, it is the transformation of a multiset that is of interest.

Gamma has a particular set of transformations which constitute program statements. They are made up of (reaction action, reaction condition) pairs. The reaction action is a replacement operation that takes elements of the multiset and replaces them with others. This operation is, however, restricted to cases where these elements fulfill the reaction condition. To give a simple but very typical example: the task of finding the maximum of a set of numbers would read in Gamma very simply

$$max: x, y \rightarrow y \Leftarrow x \leq y, \tag{9.7}$$

which translates into: Let's assume all numbers constitute a multiset. Take any two elements of the multiset, x, y. If the condition on the right of the equation is fulfilled, i.e., if $x \leq y$, then remove x from the multiset. This happens until the cardinality of the multiset is reduced to 1, and only one number is in the multiset, the largest of them all. It is easy to see that, whatever the order we pick the numbers, at the end only one number will "survive," and it will be the maximum of all values. Moreover, pairs of numbers could also be compared in parallel "tournaments," until only the winner number remains. In contrast, the traditional way to perform this task in a first-year computer science class is to go through each number in sequence and check whether it is larger than the most recently discovered largest number. This simple example illustrates the spirit of Gamma to take out sequentiality from programs wherever possible, by allowing operations to happen in whatever order they happen to occur. That means, of course, that they can happen in parallel, which is interesting from an AC perspective.

Another very typical problem to demonstrate the features of this language is number sorting. Suppose you have a sequence of numbers which we represent as a set of pairs of numbers $(index, value)$. The Gamma program for sorting this sequence is simply

$$sort: (i, x), (j, y) \rightarrow (i, y), (j, x) \Leftarrow (i \geq j) and (x \geq y). \tag{9.8}$$

In other words, reactions happen only if the two pairs are not in order. The algorithm terminates with a fully sorted set, which starts from index $i = 1$. Again, this is an extremely simple way of doing sorting, without any sequentiality, except for the local comparison between every two pairs that participate in the reaction.

We have earlier discussed so-called filter conditions (see chapter 3 and the previous section 9.1). These filter conditions in Gamma are called reaction conditions, written on the right side of the statement. The reaction or replacement operation in Gamma is what we discussed previously as rewriting. In the context of Gamma, it is a rewriting of the multiset. There is one peculiarity with Gamma, however, that needs to be mentioned: The inventors imagine a computation being terminated (and a result ready) if no reactions happen any more. Thus, the result is only available after all reactions have ceased. Note, however, that since the reaction rules remain active, at any time it is possible to reinject new input values into the system, and the system will automatically recompute the operation, obtaining new results that can be collected again, once the chemical system becomes inert. This is an interesting property that is usually not present in conventional programming languages, and can be exploited to design systems that keep themselves constantly up to date upon arrival of new information [57].

A final example of applying Gamma concludes our section: The calculation of prime numbers already mentioned earlier as an example of an AC. Suppose we have a multiset of numbers. A very simple implementation of calculating the prime numbers in this multiset in Gamma would be

$$prime: x, y \rightarrow y/x \Leftarrow y \bmod x = 0. \tag{9.9}$$

Take two numbers x, y and divide y by x if an integer results. Replace y by the result.

A number of extensions to Gamma have been published in the years since its introduction. The γ-calculus for example, is a higher-order extension of the original Gamma language where each of the programs is itself considered a molecule [55]. So, similar to the λ-calculus, it is recursive. A program is a solution of molecules, a variable x can represent any molecule, and a γ-abstraction is $\gamma(P)\lfloor C\rfloor.M$ with P being a pattern determining the type of the expected molecule, C being the reaction condition, and M being the result of the reaction.

9.3   The Chemical Abstract Machine

Another extension of Gamma is the chemical abstract machine (CHAM) by Berry and Boudol [113, 422], which is again inspired by the chemical information processing metaphor to model concurrent computation. In the case of a CHAM, the intention is to implement known models of concurrent computation called process calculi or process algebras [589].

A CHAM consists of molecules $s_1, s_2, \ldots$ which are terms of an algebra. These molecules can be grouped into populations represented by multisets $M, M', \ldots$. Transformation rules determine relations $M \to M'$ that allow to rewrite populations. There are two types of transformation rules, general rules playing the role of laws applicable to all CHAMs, and specific rules defining a particular CHAM. The specific rules are the type of elementary rewriting rules we already know

$$s_1, s_2, \ldots, s_k \longrightarrow s_1', s_2', \ldots s_l', \tag{9.10}$$

while the general laws define how these rewriting rules can be applied to a multiset.

A population itself can be considered a molecule if it is appropriately separated from the rest of the multiset by what is called a membrane. Technically, if a membrane operator is applied to a multiset, it becomes such a newly defined molecule. Another operator called airlock and symbolized by $\lhd$, allows molecules to pass membranes, e.g. $s_1 \lhd M$.

As indicated by the choice of populations as multisets, they are well stirred, so that we can speak of a solution of molecules. It is imagined that random motion of molecules will allow them to collide with other molecules within their respective solution. Subpopulations are independent of each other, so that reactions may take place in parallel.

Four general laws govern how a CHAM can behave. These general laws define how rewriting rules can be applied to a multiset. They determine what molecules in a (sub)population that matches the left side may be replaced by the molecules on the right side ((1) reaction and (2) chemical law). As well, they state that a subpopulation may develop independently ((3) membrane law), and that molecules may enter or leave a subpopulation ((4) airlock law).

In order to define a special CHAM an arbitrary set of special rules of the form (9.10) are to be added to the general laws (1) — (4). To ease handling of the special rules the authors distinguish three classes of special rules, namely, heating rules, cooling rules, and reaction rules. A heating rule decomposes a molecule into its components and a cooling rule allows the joining of two molecules. Thus, heating rules are those rules that lead to a decay of molecules

$$a \to b \cdots c, \tag{9.11}$$

whereas cooling rules lead to the buildup of molecules

$$a \cdots c \to b. \tag{9.12}$$

Heating and cooling is considered to be reversible. Reaction rules are irreversible and are usually applied to heated molecules. They change the information content of a population [113].

At each time step a CHAM may perform an arbitrary number of transformations in parallel provided that no molecule is used more than once to match the condition side of a reaction law. A CHAM is nondeterministic if more than one transformation rule can be applied to the population at a time. In this case the CHAM selects one of them randomly.

9.4 Chemical Rewriting System on Multisets

The chemical abstract machine has been developed to model concurrent computing. Suzuki and Tanaka demonstrated that it can be utilized to model chemical systems [830, 831]. They defined an ordered abstract rewriting system on multisets called chemical ARMS, for *a*bstract *r*ewriting *s*ystem on *m*ulti*s*ets.

An ARMS in their definition consists of a pair $\Gamma = (A, R)$ where A is a set of objects (symbols, for instance) and R is a set of rewriting rules. As in the Chemical Abstract Machine, molecules are abstract symbols and reactions take place by applying rewriting rules. The reaction vessel is represented as a multiset of symbols. Now, however, there is also a set of input strings and a rule order that specifies in what order the rules are processed. This will allow particular sequences of rewritings to take place and therefore be conducive to the emergence of defined temporal patterns. The kinetic constants of reactions are modeled by different frequencies of rule application. The authors discriminate between three different types of rewriting rules; sequential, free, and parallel rewriting rules or transitions. If a multiset is subjected to transformations, those orderings emerge from the interplay between the starting condition of the multiset and the rewriting rules.

We shall mostly follow the exposition of [830] to discuss an example. Consider an ARMS defined as

$$\Gamma = (A, R) \tag{9.13}$$

$$A = \{a, b, c, d, e, f, h\} \tag{9.14}$$

$$R = \{a, a, a \rightarrow c : r_1, b \rightarrow d : r_2, c \rightarrow e : r_3, d \rightarrow f, f : r_4, a \rightarrow a, b, b, a : r_5, f \rightarrow h : r_6\}. \tag{9.15}$$

Suppose now that there is a rule order determined, such that rewriting rules are executed according to index, i.e., r_i is executed before r_j if $i < j$. The starting state of our system, i.e., the content of the multiset at the beginning, is

$$M_0 = \{a, b, a, a\}. \tag{9.16}$$

We now write the temporal sequence of the multiset transformations as follows:

$$\{a, b, a, a\} \rightarrow \{c, d\} \rightarrow \{e, f, f\} \rightarrow \{e, h, h\}. \tag{9.17}$$

After the last rewriting step, no rule exists that can be applied to the multiset, and nothing can happen any more. The system has "halted."

Consider now just the opposite rule order, i.e., r_i is executed before r_j if $i > j$. The rewriting transformations lead to an ever expanding multiset:

$$\{a, b, a, a\} \rightarrow \{a, b, b, a, d, a, b, b, a, a, b, b, a\} \rightarrow \cdots \tag{9.18}$$

The close connection between initial conditions of a multiset and the order of application of rewriting rules has been examined in [830] by generating rewriting rules randomly and deriving

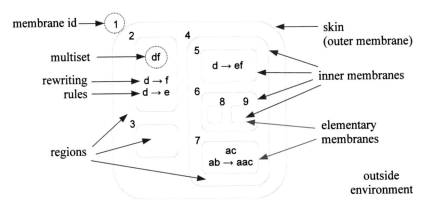

Figure 9.1 An example of a P System (adapted from [681]).

criteria for the emergence of cycles. The authors identify an order parameter they call λ_e in analogy to Langton's λ-parameter. It calculates the ratio between the number of heating rules to the number of cooling rules. They find that for small and for large λ_e values the dynamics of ARMS is simple, i.e., the rewriting system terminates and no cycles appear. For intermediate λ_e values, however, cycles emerge.

As can be imagined, an ARMS can be used easily to model chemical oscillators like the Brusselator. For reader interested in that oscillator, chapter 19 discusses the Brusselator model in more detail.

9.5 P systems

P systems or membrane systems [678, 681] are another class of rewriting systems inspired by chemistry. As in Gamma and CHAM, reaction rules determine the way to rewrite reactant symbols into products, and represent intermediate steps in the computation. P systems have another source of inspiration, besides chemistry: the biological organization of a cell. All living cells are enclosed within membranes that define the cell boundaries. The membrane protects the cell against external hazards, and selectively controls the flow of material in and out of the cell. In eukaryotic cells, a further membrane encapsulates the genetic material in the nucleus. Eukaryotic cells also contain multiple organelles such as mitochondria and chloroplasts, which can be seen as smaller subcells within the main cell.

In analogy with an eukaryotic cell, a P System is structured as a hierarchy of nested membranes, which delimit regions containing multisets of objects. Figure 9.1 shows an example: the outer membrane (skin) delimits the cell from the outside environment. It has id 1 and encloses membranes 2, 3, and 4. Membranes 3, 5, 7, 8, and 9 are elementary membranes, since they enclose no other membranes inside. All membranes delimit internal regions where molecules and reaction rules can be placed. Reactions are expressed as multiset rewriting rules.[4] Some of these molecules and reactions are depicted inside membranes 2, 5, and 7. For example, membrane 7 encloses a multiset composed of one instance of molecule a and one of c, together with a rewrit-

---

[4]Note that in the P systems literature, the reaction rules are often termed *evolution rules,* a term that we refrain from using in this book in order to avoid confusion with evolution in the Darwinian sense.

Figure 9.2 Dissolving a membrane in a P system with the help of the δ operator.

ing rule $ab \rightarrow aac$, which is equivalent to a chemical reaction $a + b \rightarrow 2a + c$, that is, it takes one molecule of a and one of b and produces two molecules of a and one of c.

Membranes define compartments that localize the computation and direct the flow of material. This provides a structured way of encapsulating complexity in order to express large and highly organized systems, something that would be difficult to obtain with well-mixed chemistries. This expressive power of P systems has attracted attention to them for use in Artificial Life, synthetic biology, and the modeling of biological systems (see also chapter 18). The ARMS system described in section 9.4 is an example of a P system specialized for Artificial Life experiments, which has also been used to model biological systems [832].

In a P system, objects are represented as symbols from a given alphabet Σ, e.g. a, $a \in \Sigma$, $\Sigma = \{a, b, c\}$. Multisets are represented as strings $s \in \Sigma^*$, e.g. $s = aabbbc = a^2 b^3 c$ represents the multiset $\mathcal{M} = \{a, a, b, b, b, c\}$. A membrane is denoted by $[_i \ldots]_i$, inside which other membranes may be nested. Using this notation, the membrane structure of Figure 9.1 can be represented as:

$$[_1 \quad [_2 \quad]_2 \quad [_3 \quad]_3 \quad [_4 \quad [_5 \quad]_5 \quad [_6 \quad]_6 \quad [_7 \quad]_7 \quad]_4 \quad]_1. \tag{9.19}$$

Besides modeling chemical reactions, reaction rules in P systems may also express the transport of objects from one region to another, and operations on membranes, such as dividing or dissolving membranes. Moving objects from one region to another is specified as follows: $abc \rightarrow a_{here} b_{out} d_{in_j}$. In this example, object a remains in the same place;[5] object b is expelled to the outer membrane; and object c is converted to d, and at the same time injected into a specific inner membrane j (if j is not specified, a random inner membranes is chosen).

The special symbol δ denotes the dissolution of a membrane. When a rule containing δ is fired within a given region, it causes the membrane enclosing this region to be removed: its objects now belong to the outer region. For example: when fired, the rule $a \rightarrow b\delta$ in figure 9.2 converts molecule a into b and at the same time causes membrane 2 to dissolve and release its content (in this case, b) to the outer membrane, which in this case is the skin membrane 1. Note that the skin membrane can never be dissolved.

A membrane division can be roughly expressed as: $[a] \rightarrow [b][c]$ where a denotes the content of the original membrane and $\{b, c\}$ denote the multisets that go to each daughter membrane (see [680, 683] for more details). The ARMS system makes use of membrane growth and division to simulate artificial cells.

Formally, a P system is defined as [680, 681]:

$$\Pi = (\Sigma, \mu, w_1, \ldots, w_m, (R_1, \rho_1), \ldots, (R_m, \rho_m), i_o), \tag{9.20}$$

where:
  Σ is the alphabet of objects;

---

[5]The indication *here* is often omitted as this is the default location.

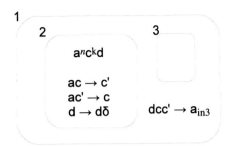

Figure 9.3 An example of a computation with a P System: deciding whether n is a multiple of k (adapted from [150, 680]).

μ is the hierarchical membrane structure containing m membranes indexed $1, 2, ... m$;
w_i are the initial multisets present in regions i of μ;
R_i are the sets of rewriting rules associated to regions i of μ;
ρ_i are priority relations over rules;
i_o is the id of the output membrane where the computation results can be collected.

The default algorithm for the execution of a P system is to fire all rules in a maximally parallel way at each time step, and to proceed like this until no further rules can be fired. Sometimes, however, priorities between rules have to be defined in order to enforce a desired order in the choice of rules. This is expressed by the priority relations ρ_i, as will become clear in the example below.

A simple example of computation with P systems

Consider a P system that must decide, given two positive integer numbers n and k, whether n is a multiple of k. The following P system, also depicted in figure 9.3, solves this problem [150, 680]:

$$\Pi = (\{a, c, c', d\}, \mu, \emptyset, a^n c^k d, \emptyset, (R_1, \emptyset), (R_2, \rho_2), (\emptyset, \emptyset), 3), \qquad (9.21)$$

where:

$\mu = [_1 \, [_2 \,]_2 \, [_3 \,]_3 \,]_1$
$R_1 = \{r_4 : dcc' \to a_{in_3}\}$
$R_2 = \{r_1 : ac \to c', r_2 : ac' \to c, r_3 : d \to d\delta\}$
$\rho_2 = \{r_1 > r_3, r_2 > r_3\}$

In this example, membrane 2 holds the input molecules and a set R_2 of three rewriting rules, one of which (r_3) dissolves the membrane. After dissolution, membrane 1 collects the intermediate results and injects them into membrane 3 (using rule r_4), from where the output of the computation can be collected. The priority relation ρ_2 establishes that, within membrane 2, rules r_1 and r_2 have priority over rule r_3.

This solution works as follows: Initially, n molecules of a and k molecules of c are present in membrane 2, plus one molecule of d. Let us first assume that $n \geq k$. This causes rule r_1 to fire k times, until the c molecules are depleted. This is the first computation step, and it replaces all molecules of c with c'. If molecules of a are still available, the presence of c' causes rule r_2 to fire in the second computation step, reverting all molecules of c' back to c. Each of these steps consumes k molecules of a. The computation proceeds in this way until all a molecules are consumed. If n is multiple of k, only c or only c' molecules will be left in membrane 2. Otherwise (and also when $n < k$), a mix of c and c' will be present. In any case, rule r_3 fires after the stock of a is depleted, when neither r_1 nor r_2 can be fired. Rule r_3 causes membrane 2 to dissolve and release its content to membrane 1. If n is not a multiple of k, rule r_4 then applies, since at least one c and one c' will be present, causing one molecule of a to be injected into membrane 3, which until now was empty. If n is a multiple of k, rule r_4 does not fire, and membrane 3 remains empty. Afterward the system becomes inert, and the result can be read from membrane 3: n is multiple of k if a is absent from membrane 3. A more sophisticated version producing explicit "yes" and "no" answers can be found in [680].

Obviously, this example looks way too complex a solution for a simple problem. The actual strength of P systems does not lie in implementing traditional algorithms, but in performing computation where parallelism is useful, or, more recently, in modeling biological systems. Some of these applications will appear throughout this book.

Many variants of P systems are computationally universal, i.e., equal in power to Turing machines. Moreover they can express highly parallel computations. In this context, trading space for time is a commonly exploited trade-off: in P systems allowing exponentially increasing numbers of objects, it becomes possible, at least in theory, to solve exponential problems in polynomial time, at the cost of storage resources for the large amount of objects created. Theoretically, such exponential growth can be achieved by replicating strings (P systems with worm objects [164]), or by creating new membranes, for instance via membrane division (P systems with active membranes [679, 683]). In practice however, physically limited resources will always constrain the growth of compartments or objects, reducing the actual amount of parallelism that can be achieved.

9.6 MGS

MGS[6] [317, 320] is a computation model based on a chemical metaphor that focuses on spatial considerations. MGS integrates elements from cellular automata, CHAM, P systems, and Lindenmayer systems into a unified rule system that operates on generalized collections of objects.

MGS is based on the notion of topological collections, which are generalizations of data structures with spatial arrangements, based on the mathematical theory of algebraic topology. In practice, topological collections include monoidal collections (sets, multisets, and sequences), maps or dictionaries (sets of key-value pairs), and labeled graphs able to express spatial arrangements such as regular grids. Computations proceed by transformations of objects in collections. These transformations can be regarded as rewriting rules that operate on a collection, selecting objects that satisfy some conditions, and rewriting them to a resulting collection of objects.

MGS has been designed to model dynamical systems (DS) with a dynamical structure $(DS)^2$. In a normal dynamical system, the behavior of the system is described by a fixed and well-known amount of state variables, such as the predator-prey system described by essentially two variables: the amount of predator and the amount of prey at a given moment (see also chapter 7). In a $(DS)^2$, in contrast, the number of state variables may grow as the system changes in time, such as an embryo that develops from a single cell. Note that the notion of $(DS)^2$ is close to the notion of constructive dynamical systems, with the difference that in $(DS)^2$ the emphasis is placed on

---

[6]The acronym MGS stands for *"(encore) un modèle général de simulation (de système dynamique)*, "yet another general model for the simulation of dynamical systems".

the topology of the space where the DS variables evolve (such as the shape of the embryo and the proximity relations between the cells) whereas constructive dynamical systems focus on content of the rules and the objects that cause the complex structure to unfold (such as the regulatory network of genes and proteins that cause the embryo to develop into its final shape).

The MGS software package provides a rule-based programming language and graphical simulator to model dynamical systems in multiple domains such as mathematics, physics, biology, and music. We show a couple of examples that give a flavor of what MGS programming looks like.

Example: Sort

This simple transformation in MGS is able to sort a sequence of elements (such as numbers) [317]:

```
trans sort = { x, y / x > y => y, x } .
```

The keyword 'trans' defines a transformation. The transformation 'sort' acts on a sequence: 'x, y' means a sequence of two elements, x followed by y. The rule reads: take any subsequence of two neighbors 'x, y', compare them; if the first element is greater than the second, swap them. By successively applying this rule until it is no longer possible, the sequence ends up sorted, producing the result of the algorithm. In this solution, any two adjacent numbers can be swapped at any position in the sequence, and transformations could in principle also run in parallel.

Notice the difference with respect to both traditional programming and the Gamma example: the traditional way to implement a sort would be to compare the numbers in a predetermined order (typically sequential); whereas in Gamma the sort solution needs to operate on pairs of numbers (*index, value*), since a multiset has no intrinsic sequence. The MGS solution needs neither of these artifacts, and is able to apply rules on a sequence structure without an imposed order.

Example: DNA cleavage by a restriction enzyme

This example shows the combination of string processing with multiset processing in MGS, through the modeling of the restriction enzyme EcoRI that cleaves a DNA strand at a specific position. The cleaved strands are then inserted in a test tube. EcoRI cleaves the strand at the first nucleotide of the recognition sequence 'GAATTC':

```
collection DNA = seq;;
collection TUBE = bag;;
EcoRI = X+, ("G","A","A","T","T","C"), Y+
        => (X,"G") :: ("A","A","T","T","C",Y) :: ():TUBE ;
```

First of all, the collections "DNA" and "TUBE" are created, representing the input DNA sequence and the output test tube, respectively. The patterns X+ and Y+ match the rest of the strand to the left and to the right of the recognition sequence, respectively. The actual matching segments are named X and Y, respectively. When applied to a DNA sequence, the EcoRI rule will split the sequence at the first nucleotide of the recognition point. For example, a sequence 'CCCGAATTCAA' will result in two segments 'CCCG' and 'AATTCAA.' It will then insert the result into the test tube "TUBE" using the operator ' : : ' which adds an element to a monoidal collection. See [317, 321] for the complete program.

These two simple examples are meant to capture essential aspects of MGS that show its kinship to ACs. They are far from illustrating the full power of MGS. However a detailed explanation of the theory and features of MGS is outside the scope of this book. For more information, a thorough description of the MGS language can be found in [319] and an overview of more recent applications can be found in [322].

9.7 Other Formal Calculi Inspired by a Chemical Metaphor

Besides the various calculi covered in this chapter, a number of other formal calculi inspired by a chemical metaphor have been proposed, such as Brane calculi [157], the Kappa calculus (κ-calculus [208]), the Port Graph calculus [36], and BioAmbients [704]. These approaches stem from the world of process algebras, formalisms used to express concurrent systems, such as the π-calculus and Mobile Ambients. There is an increasing trend to apply these formalisms to model biological systems [158, 279, 674].

As an example, in the κ-calculus [208] objects are proteins and rewriting rules express the binding of proteins to each other when forming complexes, and their dissociation from these complexes. Proteins may have several binding sites and form large complexes, which are represented as graphs. Rewriting operations in this calculus hence correspond to graph rewriting rules. The κ-calculus has been used to model gene regulation, signal transduction, and protein-protein interaction networks in general.

Another example are Brane calculi [157], which actually represent a family of process calculi based on a nested membrane abstraction similar to P systems. In contrast to P systems, however, Brane calculi emphasize operations on membranes and performed by membranes, as opposed to computation inside the compartments. Typical actions in Brane calculi include endocytosis (engulfing another membrane), exocytosis (expelling a vesicle), budding (splitting a membrane), and mating (fusion of membranes). Brane calculi have been applied to model viral infections, molecular pumps, and ion channels. More recently Fellermann et al. [269] have applied them to model the transport of molecular cargo by synthetic containers in the context of wet biochemical computation.

9.8 L-Systems and Other Rewriting Systems

In a more general Artificial Life context, transformations based on rewriting can be used to describe growth and developmental processes. For example, besides transforming complex expressions into normal form, another approach uses the method of rewriting to generate complex objects starting from simple objects. This has provided the formal basis to studies of biological development through what were later called L-systems [512]. The difference between L-systems simulating growth and development and formal grammars lies in the order of rewriting. While formal grammars consider rewriting in a sequential fashion, L-systems consider parallel (simultaneous) rewriting, much closer to the asynchrony of the growth of real cells. But central to all these models is the recursive nature of rewriting systems [376]. Lindenmayer and, later, Prusinkiewiecz and Lindenmayer [677] have pioneered the use of turtle commands of computer languages like LOGO to visualize plant patterning and growth.

Other rewriting systems worth mentioning are the reflexive artificial chemistry by Salzberg [737] which deconstructs finite state machines and their input data into a graph that rewrites a graph, and in principle itself; and the combinator chemistry by Speroni [790] which makes use of combinators as a replacement for λ-terms. Combinators are strings from a finite alphabet or combinators themselves, on which an algebra can be defined [383].

9.9   Summary

In the meantime, a large number of applications of rewriting systems have appeared. For example, such classical Artificial Life systems as *Tierra* have been reformulated as rewriting systems [818]. Reference [318] describes applications of rewriting systems to the modeling of biological systems, and virtual cities have been modeled with the help of L-systems [443], too, just to name three examples.

In this chapter we have seen that rewriting systems provide a fruitful area for formulating artificial chemistries. The operations employed by rewriting systems are very close to chemistries on symbols, and it is for this reason that we can expect more rewriting AC variants to appear over the coming years. The next chapter will deal with automata and machines, active entities well known to computer scientists.

CHAPTER 10

AUTOMATA AND MACHINES

I n this chapter we shall look at another class of systems that can be used for the realization of artificial chemistries. Historically, these systems have been among the earliest used for implementing ACs, because, among other things, they provide a level of analogy to real chemistry that is remarkable.

Like a real machine, its abstraction is a manufacturing device. It takes input, I, and produces output O, presumably consuming energy (or the equivalent in an abstract space). Thus, we can write the action of a machine or automaton M as

$$M + I \rightarrow M' + I + O. \tag{10.1}$$

Without loss of generality, we assume here that M is undergoing a change to M' by the operation, although this change might be abstracted away or not relevant for its future operations. In any event, the input, being information, is not changed by the operation, but output O is generated. A better expression in the context of abstract machines might be to say that O is calculated by the action of M.

We can immediately see the connection to the activity of molecules that attach to other molecules and perform operations on them. For instance, if we take the action of a molecule of polymerase which attaches to DNA and reads from it information to produce mRNA, we have a simple real system that might be formalized by an abstract machine or automaton. More specifically,

$$Poly + DNA \rightarrow Poly + DNA + mRNA. \tag{10.2}$$

In this case, we have summarized the macromolecules of DNA and mRNA without giving more details about their content. We have also neglected that the material for synthesizing mRNA must come from somewhere. Often, in this sort of consideration, the raw materials are suppressed. In reality, of course, we would need a source of material on the left hand side of equation 10.2.

In the following, we shall discuss a number of different abstract models of machines and automata, that provide powerful tools for the implementation of artificial chemistries, finite state machines, Turing machines, von Neumann machines, and cellular automata. The rest of the chapter is then dedicated to the discussion of artificial chemistries realized with these automata and machines. Let us start with the simplest one, finite state automata.

10.1   Finite State Automata

The theory of abstract automata defines finite state automata as behavioral models of machines that possess a finite number of states, a set of transitions between those states, and machine actions (such as "read input" or "write output"). Based on these notions, an entire field of computer science (automata theory) has been formulated which addresses their mathematical features and the use of these entities.

More formally, a finite state automaton, sometimes also called finite state machine, is a quintuple

$$FSM = (\Sigma, S, \delta, \Gamma, \omega), \tag{10.3}$$

where Σ is a set of input symbols, S is a set of states, including state s_0, the start state of the machine and state f, the final state of the machine, δ is the transition rule or transition function

$$\delta : S \times \Sigma \to S \tag{10.4}$$

that takes input and state of the machine to produce a new state, and Γ is the set of output symbols produced by a further function, the output function ω,

$$\omega : S \times \Sigma \to \Gamma. \tag{10.5}$$

As mentioned before, the concept of an abstract automaton or machine can be used in artificial chemistry since input is transformed into output by the help of S. The state transition and output functions S and ω respectively, can be defined in tabular form, since, by definition, the set of possible states and inputs/outputs is finite. If we consider a multiset of inputs as the starting condition of a "reaction vessel" powered by the finite state machine, and define an algorithm for how input is selected from the multiset, then we can interpret the action of the FSM as an artificial chemistry producing a multiset of outputs from the initial conditions. Thus, we consider inputs and outputs also as molecules.

FSMs therefore can be considered as a very basic example of molecular machines. Notably, the fact that an FSM acts locally can be seen as inspired by physical machines [737]: An FSM reads a particular input and transforms it based in its integrated memory of previous behavior (represented by its state) into an output. Again, the machine doesn't disappear in the process, so its actions are similar to what an enzyme does that transforms educts into products without itself being consumed by the process.

Just as in the molecular world, where molecular machines are itself built from molecules, the abstraction in the form of a finite state machine can be described by symbols. Notably, many finite state machines operate on binary data. It is natural to represent molecules (inputs and outputs) as collections of bits organized into binary strings. The state transition function, on the other hand, of the finite state machine can be represented by a lookup table which itself could be represented as a collection of binary strings.

This opens the door to changing the actions of the automaton itself by manipulating the description of its transition function. This feature lies at the core of the similarity between molecular systems and abstract machines.

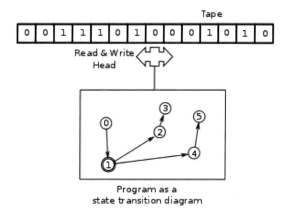

Figure 10.1 Turing machine with Tape and Machine (Read & Write Head, program) sliding along. Program box contains states and transitions between states, 0 starting state, 3,5 terminal states. Machine is currently in state 1.

Before we move on to a more complex form of machine, we should mention that the FSMs we have discussed here are of the deterministic variety. Thus, any time a particular input is given to the machine, the same output is produced by virtue of the transition function. However, FSMs can also come in a nondeterministic variety, in which case the output (and state transition) function is not unique, but determined by a probability function

$$\delta : S \times \Sigma \to P(S). \tag{10.6}$$

While this change requires a more complex description, the principles of action of FSMs remain the same.

10.2 Turing Machines

A Turing machine is one of the first concepts for a computational machine and was proposed by Alan Turing in 1936 [867]. An illustration of the action of the Turing machine on the tape is given by Figure 10.1. In his own words, such a machine would consist of: "an infinite memory capacity obtained in the form of an infinite tape marked out into squares, on each of which a symbol could be printed. At any moment there is one symbol in the machine; it is called the scanned symbol. The machine can alter the scanned symbol and its behavior is in part determined by that symbol, but the symbols on the tape elsewhere do not affect the behavior of the machine. However, the tape can be moved back and forth through the machine, this being one of the elementary operations of the machine" [869].

Looked at in more detail, a Turing machine consists of ...

- a tape, which is divided into cells with each cell holding a symbol from a finite symbol set. The tape is extended in both directions as much as is needed to perform the operations.

- a head, which is able to read from a cell of the tape or write to a cell.

- an action table of instructions, which allow the machine to use the currently read symbol and its own state to "behave," i.e., to output a new symbol, move the tape, and/or assume itself a new state.

- a state register, which stores the state of the Turing machine.

Formally, this can be summarized [397] by saying that a one-tape Turing machine is a 7-tuple M,

$$M = (Q, \Gamma, b, \Sigma, \delta, q_0, F), \tag{10.7}$$

where Q is a finite set of states, Γ is a finite set of tape symbols, $b \in \Gamma$ is the blank symbol, $\Sigma \subseteq \Gamma \setminus b$ is the set of input symbols, $q_0 \in Q$ is the initial state of the Turing machine, $F \subseteq Q$ is the set of final or accepting states, and

$$\delta : Q \setminus F \times \Gamma \longrightarrow Q \times \Gamma \times \{L, R\} \tag{10.8}$$

is the transition function which determines what state the machine will assume after the transition, what symbol it will choose for the tape, and whether it will move left or right. The subset of states F is excluded from transitions because those are "accepting" states, meaning the Turing machine halts computation.

How can the Turing machine formalism be used for an artificial chemistry? We shall discuss an example of its use in a later section, but for the moment, let's just look at the principle. Both, the Turing machine and the tape could be considered as molecules. In such a setting, a collision of molecules would be achieved by the Turing machine's head moving to a particular cell of the tape and reading its symbol content. A reaction could then be envisioned by the Turing machine making a transition according to its transition function, and the tape being written to with a possibly different symbol. An elastic collision would happen if no transition of state and no change in tape content would ensue, and the machine would simply move left or right to encounter a different cell on the tape.

In the example given, each interaction of the Turing machine with the tape would be equivalent to a chemical reaction. There is, however, an entirely different possible interpretation of the action of a Turing machine. It would consider a Turing machine including its state transition function (the program) as a molecule, the entire tape with its content of filled and blank cells a molecule, and the reaction between those molecules happening through the interaction of the Turing machine and the tape in its entire sequentiality. In other words, the reaction would last for whatever number of computation steps is necessary to produce an accepting state on the Turing machine, i.e., the Turing machine to halt, given the tape. Thus, the collision between molecules would be the encounter of a particular tape with a particular Turing machine, which would only detach from each other after the halting state has been reached.

Because Turing machines are at the foundation of computer science, they provide a rich substrate to think about artificial chemistries, a substrate that has not been fully explored yet in its usability for ACs.

10.3   Von Neumann Machines

While the Turing machine has provided the theoretical basis for computing machines, the practical implementation of computing in electric components was left to another concept, the von Neumann machine.

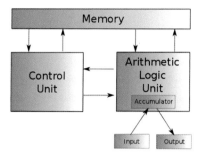

Figure 10.2 Architecture of a von Neumann machine with a central processing unit (control and arithmetic/logic units), memory, and input/output.

The von Neumann machine, or stored program digital computer, is the mainstream computer architecture in use today. It uses a central processing unit to provide the computational engine, together with a single separate storage unit, the memory. A schematic drawing can be seen in figure 10.2.

One of the key points of the von Neumann architecture is that programs and data can be stored in the same memory, i.e., there is no material difference between information being processed (data) and information being used to process the data (program). This is an extremely important point, since it allows us to manipulate programs the same way as data. In other words, it is thinkable that a von Neumann machine modifies its own program. This power of the von Neumann concept, implemented through memory and CPU, is at the core of its usability in many different contexts.

The von Neumann machine has some similarity with the Turing machine, though, in that the program stored in memory is executed through the CPU in a sequential order, directed by a device called the program counter. This device always points to the next instruction to be executed, and is updated, once an instruction is complete, to point to the next instruction. Thus, in the same way as the tape of a Turing machine moves left of right to a particular cell on the tape, the program counter of a von Neumann machine points to another memory location where the next instruction resides. Technically, memory is realized by registers in the von Neumann machine, and the arithmetic logic unit (ALU) is realized by some digital logic circuits.

Again, we can ask how a von Neumann machine can be used to implement an artificial chemistry. The most direct way of addressing this question is to say that we can interpret the access of an ALU to a data register (using its program) is the collision between two molecules, and the subsequent calculation is a reaction. This would be analogous to the Turing machine acting on a cell of its tape. In the case of a von Neumann machine, however, the interaction is considerably more complex, since there are many ways an ALU can interact with data registers. For example, there are 0-operand, 1-operand, 2-operand, and 3-operand instructions. Let's just take the example of a 3-operand instruction.

$$R_3 \leftarrow R_1 + R_2 \tag{10.9}$$

is an addition instruction that uses the content of register R_1 and adds it to content of register R_2, with the result of that operation stored in register R_3. What this means is that the content of

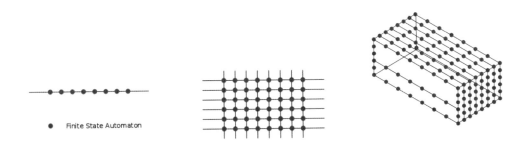

Figure 10.3 Cellular automata in different dimensions. Left : 1-dimensional CA. Middle: 2-dimensional CA. Right: 3-dimensional CA. Dots represent finite state automata (usually identical), residing as cells on a grid. Grids could also obey other geometries, like triangular or hexagonal. Automata need not be necessarily homogeneous.

registers R_1, R_2 is left undisturbed, and a third location R_3 is written into, by way of performing an addition operation on the data. We can say, the ALU performs a reaction between "molecules" (the data residing in R_1, R_2), and produces a new molecule that is subsequently stored in R_3. Register addresses could be interpreted as spatial coordinates, the ALU operation as the type of chemical reaction that is performed, and the content of registers as being the types of molecules.

Another way of looking at this process is to say that whatever is in a particular register is a type of molecule. Thus, we have the address of the register as its type, and the content of the register as a molecule count. In this case, not all instruction operations of the von Neumann machine should be permitted, but just a subset that would be allowed to manipulate the number of molecules in a consistent way.

Finally, one could look at a particular program in its entirety, and its encounter with a set of data as the collision of molecules. In this way, a program executes all its steps (as part of a chemical reaction) and leaves results in the form of data somewhere in memory. This would constitute just one reaction, with many reactions happening through the repeated action of programs on bodies of data.

10.4   Cellular Automata

A Cellular automaton (CA) is composed of a large interconnected collection of component finite state automata. These possess, besides states and transitions, an additional feature, a location within a neighborhood, often constituted as a grid. The neighboring "cells" or "nodes" which are themselves finite state automata with a different location, receive the state of the particular automaton as input, and also, reciprocally, provide their own states as inputs that inform neighboring automata which state transition to realize. Although the word "cell" here gives rise to the term "cellular automaton," note that it is *not* being used in a specifically biological or technical sense; it rather just means a discrete, spatially demarcated component of a larger, static, locally interconnected, array. Figure 10.3 shows sample arrangements for different dimensions.

CA are typically defined to be structurally homogeneous, i.e., all cells contain the same finite state automaton and all share the same local neighborhood pattern. Figure 10.4 shows different

1D radial neighborhood   2D von Neumann neighborhood   2D Moore neighborhood

Figure 10.4 Cellular Automata neighborhoods in one and two dimensions.

neighborhoods for different dimensions of the cellular automaton. In one dimension, it is really only a radial neighborhood, with coordinates of neighboring cells being addressed relative to the center cell $N = \{-r, -(r-1), \cdot, -1, 0, 1, \cdot, r-1, r\}$, with neighborhood size $2r + 1$. In two dimensions, more complex neighborhoods can be defined. The best known are the von Neumann neighborhood and the Moore neighborhood (see Fig. 10.4). In these cases, the neighborhood can be defined as

$$N_{vN} = \{\{0, -1\}, \{-1, 0\}, \{0, 0\}, \{+1, 0\}, \{0, +1\}\} \tag{10.10}$$

$$N_M = \{\{-1, -1\}, \{0, -1\}, \{+1, -1\}, \{-1, 0\}, \{0, 0\}, \{+1, 0\}, \{-1, +1\}, \{0, +1\}, \{+1, +1\}\}, \tag{10.11}$$

where the central cell has coordinates $\{0, 0\}$ and neighboring cells are distinguished relative to the central cell by at least one coordinate different from 0. Von Neumann neighborhoods have a total of 5 cells, and Moore neighborhoods count 9 cells, respectively.

Of course, cell *states* vary dynamically in time and across the cells of the automaton; and it is such spatiotemporal patterns of states that are used to represent so-called *embedded* machines or automata. Such a model can thus be used to model spatially extended systems and study their dynamical behavior. Cellular automata (CA) have also been used as a modeling tool to capture behavior otherwise modeled by partial differential equations (PDEs); and indeed can be regarded as a discrete time, discrete space, analogue of PDEs.

For example, a one-dimensional partial differential equation of the form

$$\frac{\partial A}{\partial t} = \kappa \frac{\partial^2 A}{\partial x^2} \tag{10.12}$$

can be reformulated as a one-dimensional cellular automaton with a_x^t symbolizing the state of the automaton at cell index x:

$$a_x^t = a_x^{t-1} + \kappa \frac{\Delta t}{(\Delta x)^2} (a_{x-1}^{t-1} - 2a_x^{t-1} + a_{x+1}^{t-1}). \tag{10.13}$$

This can be updated once for each cell independent of the other, provided an initial condition is fed into the system.

Cellular automata play a particularly useful role in computation, in that they possess the twin features of being applicable for synthesis of system behavior, and for analysis of system behavior. They have been thoroughly examined in one-dimensional and two-dimensional realizations [924,925], and have been considered in the fundamental contexts of computability [923] and parallel computing [382,853]. The discrete nature of states, space and time available in cellular automata has led to a number of simulation models. For example, the patterning of seashells has been modeled using discrete embodiments of nonlinear reaction-diffusion systems [573]. Scientists in urban planning have started to use the tool of cellular automata to model urban spread [917].

The computer pioneer John von Neumann, following a suggestion of the mathematician Stanislaw Ulam, originally introduced CAs as a basis to formulate his system of self-reproducing and evolvable machines. In fact, von Neumann's contributions stand at the cradle of the field of Artificial Life by showing, for the first time, a mathematical proof that an abstract "machinelike" entity can be conceptualized which can construct itself, with that copy being able to achieve the same feat in turn, while also having the possibility of undergoing heritable mutations which can support incremental growth in complexity. This was a break-through showing that above a certain threshold complexity of an entity this entity is constructionally powerful enough not only to reproduce itself (and thus maintain that level of complexity) but to seed a process of indefinite complexity growth.

While von Neumann achieved this result in principle, a number of scientists have since succeeded in pushing down the complexity threshold. Much simpler CAs with smaller numbers of states per cell have been shown to support self-reproducing configurations, many of which are 2D loops [495, 779]. In fact, the smallest currently known example in this form is a cellular automaton with 8 states per cell where the self-reproducing configuration is of size 5 cells [705]. However, in many of these cases, the core von Neumann requirement to embody a "general constructive automaton" in the self-reproducing configuration has been abandoned: so the potential for evolutionary growth of complexity is correspondingly less general.

Cellular automata also have been used in the design of the Game of Life [309, 310] which draws inspiration from real life in that replication and transformation of moving entities, again in a 2D CA, were observed. We shall look at this more closely in one of the following sections.

10.5   Examples of Artificial Chemistries Based on Turing Machines

In this and the following sections we shall discuss a few examples of artificial chemistries built using automata and machines. It can be freely stated, however, that this area is conceptually very diverse. There are extremely many possibilities how to combine automata and machines into systems that act like ACs. Most authors make use of a number of different aspects which allow them to produce the effects they desired. Due to this engineering-style construction, there are many elements in these artificial chemistries that seem unnecessary or even misleading when looked at from a purist's point of view. Other authors have proposed models not originally intended to be artificial chemistries, but which can be classified as such upon analysis. Here again, naturally no effort was made in most cases to stay as minimal as possible with respect to AC functionalities.

By discussing these examples, we hope to help readers develop a sense for the more or less arbitrary choices of authors of these models. This should allow them to design their own artificial chemistries, perhaps by stripping away functions of these models, and by inventing new, less complex models that nevertheless convey the essential points of an AC. We argue here for a minimalist approach, because experience in science has shown time and time again that minimalist approaches allow us to better isolate and study interesting phenomena and to derive lessons from them that can be subsequently applied in other contexts.

With that said, the first examples are models that look at the implementation of ACs as Turing machines in interaction with their tapes.

Figure 10.5 Turing machine — Tape interaction in Laing's model Tape contains binary numbers "0" and "1," while Turing machine is made of instructions w(1), w(0), TT, CT, L, R, NOP, built from an interpretation of another binary string. Reprinted from [486] Journal of Theoretical Biology, 54(1), R. Laing, Some alternative reproductive strategies in artificial molecular machines, pages 63–84, © 1975, with permission from Elsevier.

10.5.1   Artificial Molecular Machines

Richard Laing proposed artificial molecular machines as a model for life, interestingly modeled after the Turing machine able to compute universally. He argued for the development of a general theory for living systems [484] based on *artificial organisms*. Laing suggested a series of artificial organisms [484–488] that were intended to study general properties of life, foreshadowing the theme of Artificial Life of "life as it could be." Laing's artificial organisms consist of different compartments containing molecules from an artificial chemistry.

The molecules were strings of symbols, usually binary strings. Molecules could appear in two forms, binary data strings (passive) or machines (active). In machine form the molecules were interpreted as a sequence of instructions with the sequence possessing a three-dimensional shape. In analogy to the Wang implementation of a Turing machine, which foresaw jumps between instructions, however, the machine form of a molecule would contains branching points and loops [486].

For a reaction, two molecules encountered and attached to each other such that they touched at just one position of both. One of the molecules was considered as the active machine able to manipulate the passive data molecule. For running the machine the instruction at the binding position was executed. Depending on the content of machine at the binding site, the binding position was moved to the left or right on the data molecule, or the data entry was read or a new entry was written into that position. On the machine molecule, the binding position was moved in order to point to the next instruction.

A picture of how Laing imagined the interaction of Turing machine and tape is shown in figure 10.5. Laing has not considered an explicit dynamics, probably because a dynamical simulation of his artificial chemistries has been beyond the computational means available back then. However, he proved that his artificial organisms are able to perform universal computation [485] and he also demonstrated that self-reproduction of machines can happen based on the capability of self-inspection and self-description [486–488]. Thus, steps toward life's self-replication and self-reproduction tendency were sketched.

Table 10.1 Typogenetics: Mapping from base-pairs to amino acids in Hofstadter [391] (left) and Varetto [882] (right). The letters s, l, r carried as indices by the operations specify the contribution of that amino acid to the binding preference of the typoenzyme (its binding inclination). Table adapted from [785].

		Second Base			
		A	C	G	T
First Base	A		cut_s	del_s	swi_r
	C	mvr_s	mvl_s	cop_r	off_l
	G	ina_s	inc_r	ing_r	int_l
	T	rpy_r	rpu_l	lpy_l	lpu_l

		Second Base			
		A	C	G	T
First Base	A		cut_s	del_s	swi_r
	C	mvr_s	mvl_s	cop_r	off_l
	G	ina_s	inc_r	ing_r	int_l
	T	rpy_r	rpu_l	lpy_l	lpu_l

10.5.2 Typogenetics

Typogenetics is the name of another system closely resembling the machine tape-interaction of Turing machines. In his seminal book *Gödel, Escher, Bach: An Eternal Golden Braid* [391] Douglas Hofstadter introduced this system as a way to exemplify the formal characteristics of life. Here, the two entities necessary to start the system dynamics are strands of formal nucleotides A, C, T, G which can form both single-stranded or double-stranded character sequences. These strands can in turn be manipulated by operators called typographic enzymes, or typoenzymes for short, which would attach to strands and perform operations like copying, cutting, etc. A translation code between strands and typoenzymes allows for self-reference and potential self-reproduction in the system.

Typogenetics was examined in more detail in [482, 597, 598, 882, 883]. Morris looks at formal logical and philosophical implications of the concept while Varetto investigates a modified version of the original system in simulation. He finds emergent strands that are able to self-replicate, as well as autocatalytic cycles he called "tanglecycles."

Molecules come in two forms, single-stranded and double-stranded. Single-stranded sequences consist of the four letters A, C, T, G, which are called bases. These may act as complementary pairs in A-T and G-C pairings for the formation of double-stranded sequences. In line with molecular biology there is a 3' and a 5' end of a single strand, with double strands consisting of strands that have their 3' (5') ends at opposite sites of the string. One further term is useful in connection with typogenetics: A "unit" is the location of a particular letter on a strand. This is important when the other sort of agents comes into play in typogenetics: typoenzymes. Typoenzymes consist of "amino acids" and perform operation on strands at particular units where they bind. The sequence of amino acids forming a typoenzyme are coded in bases and translated by a "genetic" code (see table 10.1). They can perform operations on a strand, like cutting, switching, moving, and eliminating. In addition, each typoenzyme has a "tertiary" structure, which determines to which unit it can bind.

Reactions occur through the binding of typoenzymes to units on the strand and the subsequent operations performed by the enzyme. A typical example of such a complex reaction is given by figure 10.6.

A strand codes for several enzymes that are separated by the "duplet" AA. To translate a character sequence into a sequence of operations a pair of two characters codes for one operation. For example AC codes for "cut" which cuts a strand at the current binding position of the enzyme (see figure 10.6). The initial binding position of an enzyme depends in a nontrivial way on its coding sequence. In figure 10.6 the first enzyme binds initially to G. If the strand contains more than one character of this kind, one will be randomly chosen.

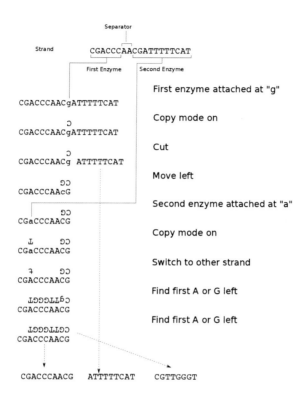

Figure 10.6 Typogenetics: Typical reaction processing a string using typoenzymes. Two enzymes coded by the strand CGA....AT operate on a copy of the same strand. The result are three reaction products. The current binding position of an enzyme is depicted by a lower case character. Example from [597].

Double-stranded molecules can be formed, too. There are two operations that create double-stranded molecules: "copy mode on" and "copy mode off." If the copy mode is active, movement of the enzyme creates an inverse copy of characters it is touching. The enzyme may also switch from one strand to its parallel strand by a "switch" operation (figure 10.6).

The overall system consists of strands which are interpreted one by one. After interpretation and before moving on to the next strand, all operations coded in the string are performed (on the string) to generate the complex reaction products following from the interpretation of one string. Thus, the system really only contains "self-reactions." In variants, also, a lethal factor was introduced.

Varetto [882] finds short self-replicators that start to grow in numbers based on an exponential law. In general, typogenetic is very complex because the size of molecules is variable, a reaction may produce a variable number of products and the reaction mechanism can be applied nondeterministically.

10.5.3   Polymers as Turing Machines

McCaskill investigated the self-organization of the genetic apparatus by a model of a reactive polymer chemistry implemented in silico [566]. He later realized the AC in a specially designed parallel computer [136, 247, 843].

Linear macromolecular heteropolymers provided a simple class of molecules with combinatorial complexity. The polymers consisted of finite sequences of length l of bases of just two monomers with a bounded automaton. Each molecular string could be processed as a linear molecule but also coded for an automaton.

Binding was realized using a collision table based on pattern matching of sequences. The general reaction scheme of two strings s_1, s_2 can be written as:

$$s_1 + s_2 \longrightarrow s_1 + s_3, \text{ if } p_1 \subseteq s_1 \text{ matches } p_2 \subseteq s_2. \tag{10.14}$$

Here, p_1 and p_2 are substrings (patterns) of the polymers, with p_1 the pattern of the processor and p_2 acting as the pattern of data. To encode the degree of matching between two sequences, patterns consisting of $\{0, 1, \#\}^*$ (# is a "don't care" symbol) were used. The number of "#"'s in the pattern determined the specificity of the recognition.

The reactor consisted of a two-dimensional domain where polymers were placed at random locations. The neighborhood of a polymer was defined by the distance between polymer locations in the reactor.

These ACs were later implemented and simulated with reconfigurable FPGA hardware [136, 843]. This implementation realized the "pattern processing chemistry," which was potentially universal: it constituted a probabilistic and constructive variant of a Turing machine operating on a polymer tape. A special error rate controlled the precision of elementary steps of the Turing machine. This error rate can be regarded as a "mutation."

It was observed that the system produced strings with the ability to self-replicate. In coevolution with parasites, an evolutionary arms race started among these species and the self-replicating strings diversified to an extent that the parasites could not coadapt and went extinct. Further details can be extracted from the references given earlier.

10.5.4   Machine-Tape Interaction

As a last example, we discuss the work of Ikegami and Hashimoto [419], who developed an abstract artificial chemistry based on binary strings where the interaction is defined by a Turing machine with a finite circular tape.

There are two molecular types, tapes and machines. Tapes are circular binary strings (7 bits long). Machines are binary strings composed of a head (4 bits) a tail (4 bits) and a state transition lookup table (8 bits). Tapes and machines form two separate populations in the reactor.

Reactions take place between a tape and a machine according to the following reaction scheme:

$$s_M + s_T \longrightarrow s_M + s_T + s_M' + s_T'. \tag{10.15}$$

A machine s_M is allowed to react with a tape s_T if its head matches a substring of the tape s_T and its tail matches a different substring of the tape s_T. The machine operates only between these two substrings (called reading frame) which results in a tape s_T'. The tape s_T' is folded (translated [419]) into a machine s_M' by reading 16 bits in a circular way starting at the head binding position. Thus, a tape s_T' can be translated into different machines if there are machines that bind to different positions of the tape s_T.

The molecules are represented implicitly by an integer frequency vector. During the dynamics simulation the population size of machines and tapes is kept constant, like in [233,284]. Updates are performed by replacing a fraction of c (in [419] $c = 60\%$) molecules of the reactor by reaction products. The composition of the reaction products is calculated using rate equations with real-valued variables and parameters. In order to do this, the integer frequency vector is converted into a real-valued concentration vector and the result of calculations using the rate equation are rounded to return to integer values.

Ikegami and Hashimoto showed that under the influence of low noise simple autocatalytic loops are formed in such systems. When the noise level is increased, the reaction network is destabilized by parasites, but after a relative long transient phase (about 1,000 generations) a very stable, dense reaction network is formed, called *core network*, [419]. A core network maintains its relative high diversity even if the noise is deactivated subsequently. The active mutation rate is high.[1]

At low noise levels, reaction cycles are formed that are similar to the hypercyclic organizations proposed by Eigen and Schuster [252]. Ikegami and Hashimoto called them Eigen/Schuster type networks, which have the following form:

$$s_M + s_T \longrightarrow s_M + s_T + s'_M + s_T$$
$$s'_M + s'_T \longrightarrow s'_M + s'_T + s_M + s'_T. \qquad (10.16)$$

A reaction network like eq. (10.16) does not perform any active mutation.

At higher noise levels a different type of reaction cycle emerged, called *double autocatalytic network*. Cycles of this type maintain a high active mutation and do usually not perform exact replication. The following network is an example for a double loop type network:

$$s_M + s_T \longrightarrow s_M + s_T + s'_M + s'_T$$
$$s'_M + s'_T \longrightarrow s'_M + s'_T + s_M + s_T \qquad (10.17)$$

The core networks emerging under the influence of external noise are very stable so that there is no further development once they have appeared. In order to promote evolution reaction systems can be encapsulated into cells (compartments) where substances are allowed to diffuse. Ikegami and Hashimoto showed that this can result in complex, coevolutive behavior [420].

10.6 Artificial Chemistries Based on von Neumann Machines

Von Neumann machines are the most widespread computer architecture in use today. From an efficiency point of view, adapting artificial chemistries to these architectures is wise, since it will facilitate the simulation or larger systems and therefore the observation of more complex phenomena. We shall start our discussion with a system that is at the transition between Turing machines and von Neumann machines and move on to cleaner von Neumann machine implementations of ACs.

10.6.1 Automata Reaction

The finite state machine developed in [233], is a command driven register machine combined with a Turing machine, inspired by Typogenetics [391]. The reactions of this artificial chemistry

---

[1]Active mutation rate refers to the tendency of the system to change itself.

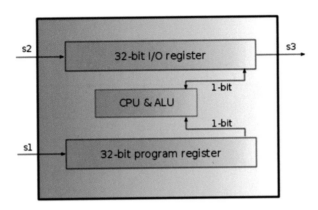

Figure 10.7 A schematic diagram of the automata reaction. The CPU executes the program specified by s_1. 4-bits are interpreted as one command. The CPU may read the program and I/O register but can only write to the I/O register. With an appropriate program the contents of the program register can be copied to the I/O register.

have been designed with a particular view on "respecting the medium" [701], referring here to the von Neumann architecture, As a result, it is fast and efficient and allows large population sizes of the order of 10^6 molecules to be simulated for longer simulation times of the order of 10.000 generations.

The molecules are binary strings $S = \{0, 1\}^{32}$ of length 32 bits. This allows easy and efficient explicit simulations, as that is the word length in many processors. Reactions are string interactions of the form

$$s_1 + s_2 \longrightarrow s_1 + s_2 + s_3, \tag{10.18}$$

with educts $s_1, s_2 \in S$. A reaction is performed in two steps: (1) s_1 is mapped into a finite state automaton A_{s_1} by interpreting s_1 as a group of 4-bit machine codes; (2) the automaton A_{s_1} is then applied to s_2 to generate new output s_3. s_1, s_2 are not changed in the course of the reaction. Each sequence could act as automaton and/or be processed by an automaton with equal probability. However, there is no separate population of automata like in the system [419].

Figure 10.7 shows the structure of the automaton-machine. There are two 32-bit registers, the *I/O register* and the *operator register*. At the outset operator/program string s_1 is written into the operator register and the operand s_2 into the I/O register. The program is generated from s_1 by simply mapping successive 4-bit segments into instructions. The resulting program is executed sequentially, starting with the first instruction. The automaton halts after the last instruction has been executed. During program execution the automaton can modify bits in the I/O register, but only read bits from the operator register. There are logic and arithmetic operations which take as input two bits, one from the I/O and one from the operator register, respectively. The result is stored in the I/O register.

Several experiments have been performed with this AC using explicit single molecule collisions and explicit populations with and without a topological structure. Typical phenomena that have been observed were: dominance of self-replicators, the formation of catalytic networks, complexity increase if exact replication is forbidden by defining them as elastic collisions and syntactic similarity of molecules forming in emergent organizations. These observations support the findings of Fontana et al. [281], especially the phenomenon of syntactic and semantic closure. In large systems of this AC, evolutionary phenomena have been observed, notable because no explicit variation and no explicit fitness function were implemented. Variation and "natural" selection is driven by the molecules themselves. The phenomenon has been termed *self-evolution*. Moreover, further phenomena have been observed like the spontaneous emergence of recombination and the spontaneous transition from lazy to eager replicators.[2]

10.6.2 Assembler Automata

An *Assembler automaton* is a parallel computational machine consisting of a core memory and parallel processing units. It can be regarded as a parallel von Neumann machine. In this section we will describe the basic functionality of assembler automata and discuss how they fit into artificial chemistries.

The first implemented assembler automaton of competing computer programs is "Core War." Core War has been designed by Dewdney [229] as a competition game in which computer programs written by programmers struggle for computer resources and may fight each other. The system can be imagined as a (virtual) multiprocessor, shared-memory system. The idea is that all processes operate on the same linear and cyclic memory called *core* and try to spread there as much as possible. Core War programs are written in the language "Redcode" [440] made up of a small number of machine language instructions with one or two arguments. Redcode is run by a program called MARS (Memory Array Redcode Simulator). Since it is so foundational, we reproduce the basic instruction set in table 10.2.[3]

For a typical match of two programs playing against each other, the programs are stored at random locations in core memory. They are then started in parallel by assigning each program to a CPU with its own program counter (PC) and registers. The following example illustrates the use of the Redcode language for a program writing a zero at every fifth memory location which should overwrite and destroy code of its opponent.

cell no.	instruction	arg1	arg2	comment
0	DAT		-1	memory cell used as a variable, which is initialized by -1
1	ADD	#5	-1	adds 5 to the contents of the previous memory cell
2	MOV	#0	-2	writes 0 to the memory cell addressed indirectly by the memory cell 0
3	JMP	-2		jump two cells backward

This is one of the strategies to assure one's survival: Make sure that the opponent cannot survive. On the other hand, the simplest self-replicating program consists of only one instruction:

---

[2]A *lazy replicator* s_1 copies every molecule it collides with and can mathematically viewed as an identity function: $s_1 + s_2 \longrightarrow s_2$. An *eager replicator* s_1 creates a copy of itself: $s_1 + s_2 \longrightarrow s_1$.

[3]Redcode has gone through some revisions since its inception, using modifiers and special addressing modes.

Table 10.2 Instructions of the Redcode language. Arguments to an instruction are A and B. Further details refer to addressing modes and to instruction modifiers. Italic instructions added in later version of Redcode. (For details, see [432, 440]).

Mnemonic	Syntax	Semantics
DAT	initialize location to value B	kills the process
MOV	move from A to B	copies data from one address to another
ADD	add A to B, result in B	adds one number to another
SUB	subtract A from B, result in B	subtracts one number from another
MUL	multiply A by B, result in B	multiplies one number with another
DIV	divide B by A, result in B if A is not zero, else DAT	divides one number by another
MOD	divide B by A, remainder in B if A is not zero, else DAT	divides one number by another and give the remainder
JMP	execute at A	continues execution from another address
JMZ	execute at A if B is zero	tests a number and jumps to an address if the number is zero
JMN	execute at A if B is not zero	tests a number and jumps to an address if the number is not zero
DJN	decrement B, if B is not zero execute at A	decrement a number by one and jump unless the result is zero
SPL	split process, new task at A	starts a second process at another address
CMP	skip if A is equal to B	compares two instructions and skips the next instruction if they are equal
SEQ	skip if A is equal to B	compares two instructions and skips the next instruction if they are equal
SNE	skip if A is not equal to B	compares two instructions and skips the next instruction if they are not equal
SLT	skip if A is less than B	compares two instructions and skips the next instruction if the first is smaller than the second
NOP	no operation	does nothing
LDP	*load from p-space*	*loads a number from private storage space*
STP	*save to p-space*	*saves a number to private storage space*

MOV 0 1. This instruction copies the contents of the relative address 0 (the MOV 0 1 instruction) to the next memory cell (relative address 1). Thus, the program moves through memory leaving a trace of MOV 1 0 instructions behind. There is also an instruction to split execution which creates a new process illustrated by the following program which creates an avalanche of self-replicating MOV 0 1 programs:

cell no.	instruction	arg1	arg2	comment
0	SPL	2		split execution; creates a new process (CPU) starting at cell 2
1	JMP	-1		jump one cell backward
2	MOV	0	1	copy contents of current cell to the next cell

10.6.3 Coreworld

Rasmussen et al. have used the idea of Core Wars to build a number of different artificial chemistries called *Coreworlds* [696, 697]. These core worlds are made up of memory locations with periodic boundary conditions. Each address in the memory is occupied by an instruction, taken from the set of Core War instructions. The authors also introduce execution pointers that point in different locations to the instructions next to be executed. In addition *computational resources* are explicitly introduced in order to encourage cooperation between instructions: Each address holds an amount of computational resources, a number that is consumed by program execution but refilled through a constant influx of resources. Further, a maximum *operation radius* restricts relative addressing among instructions and promotes locality.

In order to avoid deterministic interactions, the authors introduced *random fluctuations* in two different ways:

1. New processes are started at random with randomly initialized execution pointers.

2. When a MOV instruction is executed the copied data may be mutated.

One of the problems of Coreworld is to identify an individual organism that might be defined as a collection of instructions working together. The smaller an individual is the harder it is to distinguish it from the environment. This situation is aggravated where different instruction pointers (which could be used to identify individuals) are intermingled and point to the same instructions. The Tierra system discussed in the next section has overcome this problem by introducing memory management instructions.

The Coreworld chemistry is different from other ACs because there are really no explicitly defined reactions between individual molecules (we use individual, organism, and molecule here as synonyms). When it is difficult to even identify molecules, this is all the more difficult for interactions between molecules. Implicitly, the execution of code belonging to another molecule/individual can be seen as a reaction.

Again, following our regular scheme of studying molecules, reactions and algorithm for implementing the dynamics, we turn to the dynamics. The system works as a multitasking CPU, using a queue of execution pointers. A maximum length of this queue allows only a limited number of execution pointers to survive. The system can be executed in a parallel manner where all execution pointers are updated before results are put back into memory (Venus I), or sequentially, when each pointer at the top of the execution queue is executed and modified data are put back into memory (Venus II).

Through various experiments, Rasmussen et al. observed that qualitative dynamic behavior of such systems crucially depends on the parameter setting. For low resource influx (called desert condition), only simple cyclic structures dominated the population once converged. For high resource influx (called jungle condition), diversity grows richer and more complex, and even cooperating structures emerge.

Systems running on Redcode have shown difficulty in becoming really evolvable, that is, evolving self-replicating entities that were seeded into the systems from the beginning, without decomposing their complexity first. More recent work in this direction has taken hints from the Tierra system and fed them back into the Redcode competitions [789, 894].

10.6.4 Tierra

An exciting and decisive step was taken by Tom Ray in the 1990s by substantial changes to the language underlying a system of self-replicating machine programs. *Tierra* was designed by Ray to model the origin of diversity of life (e.g., the Cambrian explosion) not its origin [701]. Tierra is similar to Core War with the following important modifications and extensions:

- A small instruction set without numeric operands:

  The instruction set of Tierra consists of 32 instructions requiring only 5 bits to code per cell. As a result, memory cells in Tierra are much smaller than in Core War where in addition to the opcode of the instruction the arguments have to be stored.

- Addressing by templates:

  Memory locations are addressed by templates of special no-op instructions, NOP0 and NOP1. Consider for example the following program sequence consisting of four instructions: JMP NOP0 NOP1 NOP0. The target for the jump instruction is the closest of the inverse patterns formed by the three no-op instructions NOP1 NOP0 NOP1.

- Memory allocation and protection:

  Programs can allocate memory that will then be protected by a global memory manager. Other programs are not allowed to write into protected memory space but may read or execute its instructions only. This method makes it much easier to define the confines of an organism in terms of the memory cells belonging to it.

- Artificial fitness pressure:

  Tierra also has an explicit fitness pressure by penalizing "illegal" or unwanted operations — instructions like a JMP with no target pattern or division operation with a zero as divisor. Penalties are accumulated and programs executing many unwanted operations are more likely to be terminated by a selection process called the "reaper" invoked when the allocated memory exceeds a predefined threshold. This selection mechanism create at the same time pressure toward more efficiently reproducing organisms.

Table 10.3 shows the instruction set used in Tierra. In typical experiments with Tierra, the core memory is initialized with a handwritten program, the *ancestor*, which is able to self-replicate. Figure 10.8 shows the ancestor code used by Ray in [701]. It consists of a self-inspection segment first, then a reproduction loop that is repeated multiple times. Inside the reproduction loop is a call to a subroutine that does the actual copying of instructions. Under the

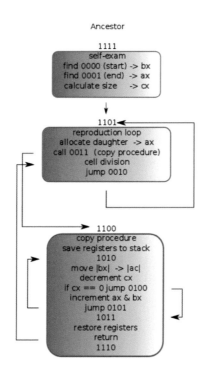

Figure 10.8 Tierra Ancestor.

influence of mutations and random fluctuations the population diversifies. Ray has observed extended processes of evolution in Tierra, with a surprising diversity of phenomena including optimization of program length, parasites, protection against parasitism, hyperparasites (see figure 10.9), and cooperative organizations. But different species are often similar to the ancestor and are likely generated by a few point mutations and other experiments have shown that in Tierra self-replicating entities like the ancestor do not appear spontaneously.

Tierra has led to a wide variety of related work. To begin with, Ray has proposed an extension to Tierra into a multiprocessor networked environment ultimately to be distributed across the internet, under suitable "sandbox" execution controls [702]. Adami and Brown have proposed the *Avida* system which introduced a spatial 2-D spatial world in which programs would colonize nodes, and also be evaluated for their ability to complete extrinsically defined computing tasks [19]. We are going to discuss this in more detail in the section on cellular automata. Then there is *Nanopond*,[4] a highly simplified implementation of a 2-D spatial program evolution system, drawing on elements of both Tierra and Avida, but implemented in less than 1,000 lines of C source code. It finally merits mention the system developed by Pargellis [650–652], which specif-

---

[4]http://adam.ierymenko.name/nanopond.shtml

Table 10.3 Instruction set of the Tierra system. Only register and stack-based manipulations. Search is understood to be done only to a maximum of 1024 instructions. If unsuccessful, next instruction is executed.

Mnemonic	Syntax	Semantics
or1	cx XOR 1	flip low order bit of register cx
shl	cx « = 1	shift left register cx once
zero	cx = 0	set register cx to 0
if_cz	if cx == 0 execute next instruction	jump over next instruction if cx is not 0
sub_ab	cx = ax - bx	subtract bx from ax
sub_ac	ax = ax - cx	subtract cx from ax
inc_a	ax = ax + 1	increment register ax by one
inc_b	bx = bx + 1	increment register bx by one
inc_c	cx = cx + 1	increment register cx by one
dec_c	cx = cx - 1	decrement register cx by one
push_ax		push ax on stack
push_bx		push bx on stack
push_cx		push cx on stack
push_dx		push dx on stack
pop_ax		pop top of stack to ax
pop_bx		pop top of stack to bx
pop_cx		pop top of stack to cx
pop_dx		pop top of stack to dx
jmp	move ip to complement of template after jmp	search both backward and forward and jump
jmp_b	move ip backwards to complement of template after jmp_b	search backward and jump
call		call a procedure
ret		return from a procedure
mov_cd	dx = cx	copy content of register cx to register dx
mov_ab	bx = ax	copy content of register ax to register bx
mov_iab		move instruction at address in bx to address in ax
adr	ax = address; cx = length of template	store address of nearest complement of template in ax, length in cx
adrb	ax = address; cx = length of template	search backward for complement of template and store
adrf	ax = address; cx = length of template	search forward for complement of template and store
mal		allocate memory for daughter cell
divide		cell division
nop_0	no operation	used in templates
nop_1	no operation	used in templates

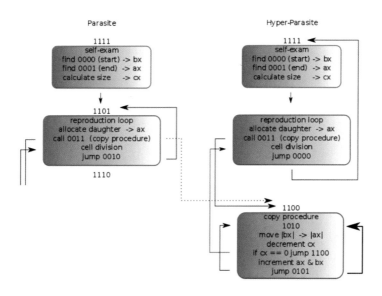

Figure 10.9 Tierra parasite and hyperparasite examples.

ically demonstrated the possibility of the spontaneous emergence of a self-replicating program in Coreworld-like systems.

Interpreting Tierra, as well as in Coreworlds and Core War as an artificial chemistry might have looked somewhat unconventional when it was first done. But the AC perspective opens up new avenues to think about these systems and has brought new variants that are quite compelling in their ability to realize evolutionary systems close to the onset of Darwinian evolution.

10.7 Artificial Chemistries Based on Cellular Automata

The notion of space brought into the picture by cellular automata is important for evolution as it provides more room for diversity and for protection of innovation as long as it is not yet able to compete with the best organisms of the world. A natural step, therefore, was it to extend the Tierra system to a spatial world of two dimensions, and allow interaction of individuals only when they are neighbors on the cellular grid.

10.7.1 Avida

This is what Adami and Brown did through developing *Avida* [19], a Tierra-like system put on a two-dimensional grid akin to a cellular automaton. In this environment, the maximum population size is now determined by the dimensions of the grid.

A molecule is an individual assembler program sitting in one of the cells of the cellular automaton. The program language is a reduced variant of Tierra's instruction set. Initial molecules are usually hand-written self-replicating programs.

In the basic Avida system, only unimolecular first-order reactions occur, which are of replicator type. The reaction scheme can be written as

$$s \xrightarrow{E} s' + s \qquad (10.19)$$

and thus are more similar to chemical interactions than in other assembler automata. In eq. 10.19, s' becomes identical to s if no mutation occurs. The program s' is created by execution of s, allocation of new memory in the immediate neighborhood and possibly explicit mutation of its output. E can be seen as a certain energy that is consumed during the replication and is modeled as CPU-time.

The additional principles of Avida in comparison to Tierra can be summarized by

- Cellular organization:

  The population consists of a 2-D lattice where each lattice site holds one program. If interactions between individual programs happen, they are restricted to the neighborhood.

- Parallel execution:

  Programs are executed in parallel and independent of each other through appropriate time-slicing.

- Use of templates as instruction modifiers:

  No-op interactions are of three types and can be used to modify instructions based on their templates.

- Environment and external fitness function:

  The environment in the Avida world is described by a set of resources and a set of reactions that can be executed in interaction with these resources. These resources might also have flows (inflows and outflows) attached to them.

  The resources can be used to differentially award better performing individuals (as judged by the computations they perform on input/output register transformations).

- Heads and input/output registers

  The virtual architecture of Avida contains four heads, besides a CPU, registers, stacks, memory and the already mentioned I/O registers. The heads are of the type read, write, flow control, and instruction. Heads can be considered pointers to memory locations that make absolute addressing unnecessary and thus assist in the growth of programs.

  Read and write heads are used in the self-reproduction of individuals, whereas the flow control head is used specifically in loops and for jumps.

Table 10.4 shows the instructions used in the Avida system. Figure 10.10 shows an ancestor able to replicate itself in Avida, and figure 10.11 is a snapshot of a typical run of Avida.

The system is seeded with an ancestor and evolves by adapting the population via mutation to an external fitness-landscape that defines the amount of CPU time (i.e., the energy) given to an individual. It is important that now, besides the simple ability to replicate based on efficiency,

Table 10.4 Instruction set of the AVIDA system. The actual syntax is critically dependent on instruction modifiers [507].

Mnemonic	Semantics
shift-r	Shift all the bits on a register one position to the right
shift-l	Shift all the bits on a register one position to the left
if-n-eau	test if two registers contain equal values
if-less	test if one register contains a value lesser than another
if-label	Test if a specified pattern of nops has recently been copied
add	Calculate the sum of the value of two registers
sub	Calculate the difference between the values in two registers
nand	Perform a bitwise NAND on the values in two registers
inc	increment a register
dec	decrement a register
push	copy the value of a register on the top of a stack
pop	pop value from the top of a stack and place into a register
swap-stk	Toggle the active stack
swap	Swap the contents of two specified registers
IO	Output the value in a register and replace with a new input
jmp-head	Move a head by a fixed amount stored in a register
get-head	Write the position of a specified head into a register
mov-head	Move a head to point to the same position as the flow-head
set-flow	Move the flow-head to a specified position in memory
h-search	Find a pattern of nop-instruction in the genome
h-copy	Make a copy of a single instruction in memory specified by heads
h-alloc	allocate memory for an offspring
h-divide	divide off an offspring specified by heads
nop-A	no operation, template, modifies other instructions
nop-B	no operation, template, modifies other instructions
nop_C	no operation, template, modifies other instructions

other criteria enter, for instance, certain formal calculations posed as problems to the system. Competition and selection occur because programs replicate different speed depending on their code and energy consumption related to their performance at the task.

Over time, the *Avida* group has investigated a number of interesting topics, like evolvability, the growth of complexity, the emergence of gene expression, and similarities to population dynamics in *E. coli* bacteria [16, 21, 628, 629]. It is probably adequate to call Avida one of the most influential systems of "digital biology" out there. The journal *Artificial Life* dedicated an entire issue to a discussion of its experimental results in 2004 [22]. We discuss a selected number of aspects of Avida in chapter 7.

```
# Setup
h-alloc      # Allocate extra space at the end of the genome to copy the offspring into.
h-search     # Locate an A:B template (at the end of the organism) and place the Flow-Head after it.
nop-C        #
nop-A        #
mov-head     # Place the Write-Head at the Flow-Head (which is at the beginning of the offspring-to-be).
nop-C        # [Extra nop-C command can be placed here without harming the organism.]


# Copy Loop
h-search     # No template, so place the Flow-Head on the next line (to mark the beginning of the copy loop).
h-copy       # Copy a single instruction from the read head to the write head (and advance both heads).
if-label     # Execute the line following this template only if we have just copied an A:B template.
nop-C        #
nop-A        #
h-divide     # Divide off offspring (note if-statement above!).
mov-head     # Otherwise, move the IP back to the Flow-Head at the beginning of the copy loop.
nop-A        # End label.
nop-B        # End label.
```

Figure 10.10 Avida Ancestor.

Figure 10.11 Avida world, screenshot from run. Courtesy BEACON Center for the Study of Evolution in Action.

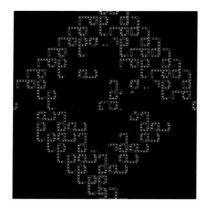

Figure 10.12 Example of a set of self-replicating loops in a finite space by Sayama [744]. Reproduced with permission by author.

10.7.2 Self-Replication in Cellular Automata from an AC Viewpoint

In this section we are going to discuss the action of a regular cellular automaton as a model chemistry. We shall simply interpret the state of a cell in the cellular automaton as a molecule (with the quiet state symbolizing the absence of a molecule). Movement of molecules and reactions are implemented by state transitions. In particular, the transition of a molecule state to the quiet state in one cell, and the transition to that molecule state in a neighboring cell can be interpreted as a movement of that molecule. An interaction between two molecules in neighboring cells that leads to a change in state can be interpreted as a reaction. There are only very special transitions that can be interpreted in this way; thus, only very particular CAs can be seen in this light.

Self-Replicating Loops

As we have already mentioned, cellular automata were invented to study self-replication in machines. Von Neumann was interested in finding a machine that was able to construct any other machine (a universal constructor). He defined a very complex CA in order to demonstrate that self-replication was possible. The von Neumann CA was a two-dimensional cellular automaton with 29 states per cell. According to Kemeny, von Neumann's core design for a genetically self-reproducing automaton would have occupied a configuration of about 200,000 cells (a size dominated by the "genome" component, which would stretch for a linear distance of about 150,000 cells) [453].

In subsequent work by others, simpler and more elegant cellular automata were developed able to display self-replicating patterns. One example is that of Langton's self-reproducing loop [495] who simplified the approach by dropping the universality requirement for the constructing machine. Thus, not any can Langton's self-replicating machine produce an arbitrary other machine, but only itself.

The self-replicating patterns in a cellular automaton like this can be interpreted as (possibly complex) molecular organizations. Due to the synchronous update rule of canonic cellular automata, once a loop has been generated as a pattern in the CA it remains present in the future.

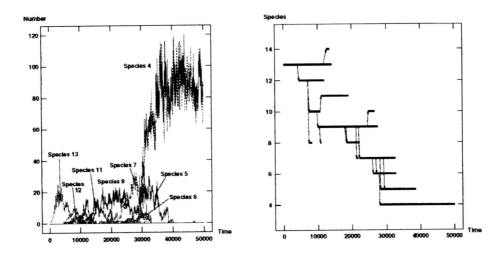

Figure 10.13 Changing concentrations of species of loops [744]. ©MIT Press 2000, with permission.

This can be interpreted as the self-maintenance of such an organization, again embodied in the transition rules of the CA.

With self-replication, however, we can easily see that another level of interpretation opens up for artificial chemistries. This time, the AC is seen not directly at the level of individual cells, but rather at the level of emergent patterns of the underlying CA transition rule system (in connection with starting conditions). In this context, Sayama [744] introduced an interesting twist to Langton's self-reproducing loop: structural dissolution. Loops can dissolve after replication (figure 10.12) and thus will have to permanently keep reproducing in order to hold their numbers at a certain concentration level. In a sense, they are now themselves "molecules" on a higher level. Sayama was able to show that in a bounded (finite) space competition between self-replicating loops occurs, which leads to evolutionary effects. At our higher level of interpretation, interaction among loops can be interpreted as chemical reactions. Figure 10.13 shows the changing concentration of these loops in a bounded space. One must realize, however, that the definition of a molecule or a reaction now depends on an emergent phenomenon in the underlying CA and could be dependent on an observer or on an interpretation.

Embedded Particles in Cellular Automata as Molecules

Even without self-replication, more abstract approaches can be taken in suitable CA systems. Thus, other emergent behavior of the CA patterns could be used to define an AC. For instance, a particle might consist of a number of cells of the cellular automaton. Reactions among those particles could be more complex to observe and to describe, since they would involve the interaction of entire fields of cells of the cellular automaton, and might not be attributed to the application of single rules alone.

For example, a moving boundary between two homogeneous domains in a CA could be interpreted as a molecule [398]. Figure 10.14 shows an example of a one-dimensional CA which produces boundaries. The figure shows the trace of the state-transition where two "particles"

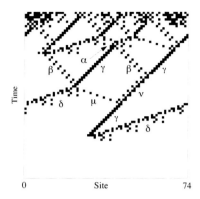

Figure 10.14 Filtered version of a space-time diagram of a one-dimensional, binary-state, $r = 3$ CA. Homogeneous domains are mapped to white, domain borders to black and domain borders can be interpreted as particles or molecules. From [399], ©Springer 1998, reprinted with permission.

collide. The major difference to the previous approach is that particles become visible as space-time structures and that they are defined as boundaries and not as a connected set of cells with specific states.

10.7.3 Bondable Cellular Automata

Another approach to using cellular automata in artificial chemistries is to use individual CAs as dynamical systems that show some behavioral similarity to atoms and molecules. It is well known that cellular automata can be studied as a class of dynamical systems [226]; thus, they could form a behavioral analogue to single atoms. While atoms are considered indivisible in chemistry, they still show some behaviors hinting at an internal dynamics, like the ability to absorb and emit energy or electrons, the ability to transfer energy from collisions into excited states, etc, that could be usefully captured by dynamical systems models. In other words, just as one would want to build higher order structure from atoms (organizations, complex molecules), one could aim at using atoms, composed of dynamical systems components, and observe their reactions. This approach is called a subsymbolic artificial chemistry, and it was first discussed in the context of an RBN system in [264].

The bondable cellular automata (BCA) model is an artificial chemistry that uses separate one-dimensional binary cellular automata for its base atomic elements. These atoms bond to form molecules and these molecules in turn can collide to form larger molecules or, alternatively, decay. The reaction rules are implicit, in that they depend on an observable of the dynamical state of the atoms and molecules participating in the collision. In [373] we considered mean polarity of the one-dimensional cellular automaton as this quantity. Atoms differ by type (which corresponds to the CA rule that is being used to run each CA), and reactions can then be observed under the right polarity conditions resulting from the dynamics of those rules.

Figure 10.15 shows the configuration of two bonded atoms. An atom is formed by a 1-dim binary CA with closed boundary conditions. If bonded, each cell in an atom is influenced by the neighborhood cells of the other (bonded) atom. This way, interaction at the level of the dynamics occurs and might lead to an increase in bonding or a decrease.

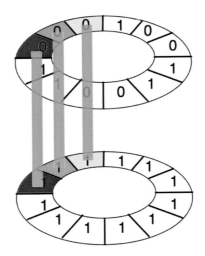

Figure 10.15 Two atoms, shown here as one-dimensional cellular automata with periodic (closed) boundary conditions interact. Green bars indicate bonding strength based on largest contiguous sequence of complementary binary numbers. Once bonded, the neighborhood in each CA is influenced by the other CA, so as to potentially change the overall dynamics. Figure from [373], reproduced with permission from the author.

One-dimensional cellular automata can be numbered from $0 \ldots 255$ where a number is assigned to each CA based on their transition rule (Wolfram's indexing system). This characterizes each atom type, thus there will be therefore 256 atom types in a system of binary one-dimensional CAs with nearest neighbor interactions. With the criterion of mean polarity upon settling of the dynamical system, these 256 types can now be ordered into the analogue of a periodic table which shows their dynamical behavior (see figure 10.16).

10.8   Summary

To summarize, the general experience of the investigation to date of artificial chemistries in automata and machines is that there is a broad variety of these systems. These systems can be viewed as formally similar to pure artificial "replicator chemistries," exhibiting similar phenomena of collective autocatalytic closure and self-maintenance [568]. Thus, both Turing and von Neumann style genetic self-reproduction and replication by simple copying or self-inspection, naturally lead directly into the problems of biological robustness, self-maintenance, and hierarchical organization discussed in chapter 7 and part IV of the book. The integration of self-reproducing or self-replicating programs with self-maintenance and individuation (autopoiesis), and the demonstration of sustained evolutionary growth of complexity in a purely virtual world remain two extremely important problems in the field of Artificial Life where artificial chemistries could contribute valuable insights.

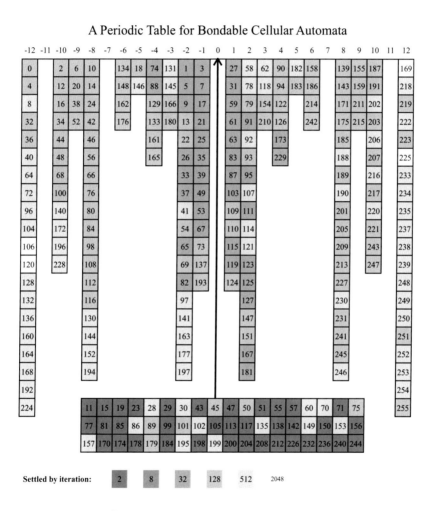

A Periodic Table for Bondable Cellular Automata

Figure 10.16 "Periodic table" of one-dimensional binary cellular automata of width 12 bits. Atom type number according to Wolfram's integer indexing system. The scale -12 to +12 shows the final value at which mean polarity settles. The shading of each box indicates how long that rule took for the value of its mean polarity to settle (i.e., the transition time of the dynamical system made up of that CA). [372], with permission from the author.

11

Bio-inspired Artificial Chemistries

In this chapter we are looking at examples of artificial chemistries that are inspired by biological systems. Some chemistries in this category have already appeared in chapters 9 or 10, such as P systems (inspired by the organization of eukaryotic cells), Typogenetics (inspired by how enzymes operate on DNA strands), and Coreworld, Tierra, and Avida (inspired by biological evolution). However, various other bio-inspired ACs exist, many of which do not fit purely into the rewriting or automata categories. This chapter is devoted to a closer look at some of these chemistries.

Bio-inspired ACs will generally result in very complex systems that show a multitude of phenomena rendering their analysis sometimes difficult. A great variety of mechanisms is recruited to arrive at the sought after effects. The list of systems we discuss is rich, and we are unable to do justice to this burgeoning field. Section 11.1 will focus on strings and their rewriting to produce effects that can be found on the cellular level of living systems, in particular self-replication, cell division and evolution. Section 11.2 will discuss chemistries inspired by the lock-and-key mechanism of biomolecular interactions. Section 11.3 will look at systems that make explicit use of the network paradigm, here in the context of two rather different levels: fundamental chemical reactions and cell growth/metabolism. In section 11.4 we look more closely at spatial aspects. We start with an example that illustrates cell division as a complex process orchestrated by (artificial) chemical reactions. This will bring us finally to appreciate the existence of mobility as a key ingredient in bio-inspired ACs, with two examples where the physics of the movements of molecules plays a crucial role in determining the behavior of the system.

11.1 String-Based Artificial Chemistries

A quintessential string-based artificial chemistry, Typogenetics was already discussed in the context of chapter 10. However, there are other, quite complex string-based ACs. One example is inspired by Holland's "broadcast language" proposed in his seminal book [395] and is based on classifier systems [396] that he considered somewhat later. It is called *molecular classifier system* and is discussed next.

11.1.1 Molecular Classifier Systems

The molecular classifier system (MCS) was developed in a series of contributions by McMullin and his collaborators [221, 222, 224, 225, 452, 570]. It is based on inspirations from cell signaling molecules that form complex interaction networks to distribute signals throughout a biological cell. More specifically, cells "talk to each other" by sending out signals in the form of more or less complex molecules that can be received by other cells and deciphered in a predefined way so as to inform the receiving cell of something going on in its neighborhood.

Molecular classifier systems belong to the class of string-rewriting systems we have discussed earlier. They are *reflexive* in the sense that a string can act as both an operator and an operand, a condition we have found with most artificial chemistries.

In the MCS variant called MCS.bl (MCS Broadcast Language), the strings in the system are *broadcast units* from Holland's Broadcast Language [395]. They consist of a condition and an action part, which are defined akin to the original broadcast language proposed by Holland. Strings encounter each other (analogous to chemical reactions in a reaction vessel) with one string possibly acting as an enzyme and the other as a substrate. The enzyme string tests whether its condition is fulfilled, by matching the condition string to the substrate string they encountered. If that is the case, the action inscribed in the action part of the string is "broadcast."

Thus, we have the usual situation of

$$S_1 + S_2 \rightarrow S_3 + S_1 + S_2, \tag{11.1}$$

with strings S_1, S_2 acting enzymatically and remaining present in the reaction vessel after the reaction. They produce a third string S_3 in the course of their interaction.

Each string called a broadcast device has at least one condition part and at least one action part, each consisting of a succession of symbols. The main informational symbols are "0" and "1." The condition and action part of the broadcast unit, i.e., the part of the broadcast device that is active is delimited by special symbols, "*" (for beginning of a broadcast unit) and ":" (as a separator for condition/action part). Figure 11.1 shows, on the left side, a typical "molecule" of this language. A broadcast device can contain more than one broadcast unit, if a proper sequence of "*" and ":" parts are found. Furthermore, there are some more symbols $\{\diamond, \triangledown, \triangle, '\}$ that are used to direct the behavior of molecules: $\diamond$ and $\triangle$ act as single character wildcards, while $\triangledown$ acts as a multiple character wild card. Wildcards are used in this way in the condition part of a string, but $\triangle$ and $\triangledown$ could also appear on both the condition AND action part of the string. In such a case, they symbolize an additional action, the transposition of what has been matched. Thus, $\triangle$ will transpose one character, while $\triangledown$ will transpose an arbitrary number of characters. The quote symbol $'$ is used to symbolize that the next symbol is not being interpreted.

Strings encounter each other with one acting as an enzyme, the other as a substrate. Table 11.1 shows the different types of operations that can be realized with this artificial chemistry.

During reactions, the operations prescribed in strings are not 100 percent faithfully executed. Rather, some degree of mutation is allowed at each symbol location. In other words, molecular processing errors are introduced into the reactions, based on a probability p_s that a symbol is changed in some way. Clearly, as this probability is given per symbol the total probability of a change in a string accumulates over its length, and increases the longer a string is. Technically,

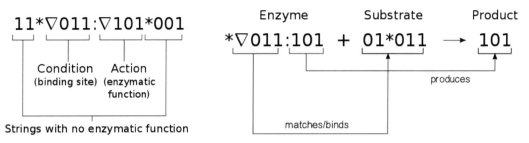

Figure 11.1 Left: A typical MCS string-molecule and its parts. Right: A typical reaction possible when two MCS string-molecules encounter each other. Adapted from [225].

three different mutations are introduced in the MCS.bl system: (1) symbol flipping, (2) symbol insertion, and (3) symbol deletion, each applied with equal probability.

A further interesting addition to the model that is important to understand before we can look at actual empirical outcomes: This is the fact that McMullin et al. apply multilevel selection, insofar as selection on two levels is applied (which is the minimum for "multi"-level). The idea of the authors is that there exist two levels of selection, a molecular level which determines the strength of reproduction of each of the strings (molecules), and a cellular level. This second level is implemented by simply considering multiple reactors (cells) that run as independent reaction systems, except that cells are allowed to compete with each other. Figure 11.2 depicts the process: Molecules reproduce independently in each of the cells, but some cells are more successful at growing their molecular content. At some threshold, cells with a large number of molecules spontaneously undergo division. From that point onward more cells would have to be in the entire system. However, since the total number of cells is kept constant, one of the remaining cells is eliminated. Thus, selection pressure on the higher level is for faster growth, based on molecular reactions.

There is one final twist to this model, which refers to higher-level selection. The previous discussion has shown that a cell that is able to grow its molecular content faster will prevail in the competition. However, this is a somewhat simple criterion which easily could be replace by other criteria. Recall that cell division is the cause of a selection event, thus if the criterion for division is modified, this will have an effect on when and what is selected. The authors' experiments showed that selection pressure can focus on particular molecular species. So, a cell division can only happen if a particular molecular species is found in a predefined number or concentration.

Table 11.1 Different operations implementable by MCS.bl.

Enzyme	Substrate	Product	Operation
$*\triangledown 1 : \triangledown 0$	$1:0$	$\emptyset$	No reaction
$*\triangledown 1 :' *\triangledown$	$0:1$	$*0:1$	Activation
$*' * 0 \triangledown : 0 \triangledown$	$*0:1$	$0:1$	Inhibition
$*\triangledown : \triangledown$	$*00:11$	$*00:11$	Universal replication
$*\triangledown 0 : \triangledown 0$	$*\triangledown 0 : \triangledown 0$	$*\triangledown 0 : \triangledown 0$	Self-replication
$*\triangledown 1 : \triangledown 10$	$*0:1$	$*0:10$	Concatenation
$*\triangledown 1 : \triangledown$	$*0:1$	$*0:$	Cleavage

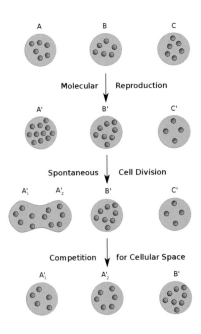

Figure 11.2 Multilevel selection mechanism implemented in MCS.bl: Molecular reproduction leads to cells spontaneously dividing, which sets off competition for cellular space.

This subtly shifts selection pressure towards cells that become more efficient in producing that particular molecular species.

Figure 11.3 shows a typical run with the system, using a sophisticated setup of 31 cells each run concurrently on a different CPU with wall-clock time used as racing condition. Each cell had a maximum carrying capacity of one million molecules. One molecule (rather than the full number) was selected as cell division threshold. As soon as 200 molecules of the target type (s_1) were present, the cell divided. We can see evolution at work in that, at different moments, a particular cell type c_1, c_2, c_3, c_4 (characterized by a population of molecules) becomes dominant. Each type consists of a number of string types that manage to self-maintain themselves. c_1 consists of strings s_1, s_2, s_3, s_4, whereas c_2 consists of s_1, s_2, s_5, s_6; for details see table 11.2. One cell type is depicted in figure 11.4 as a bipartite graph showing the reaction network with each reaction separately specified as a rectangle.

Table 11.2 Molecular types contained in dominant cell species of a typical MCS.bl run.

c_0	c_1	c_2	c_3
$s_1 = * \triangledown 0 : \triangledown 1$	s_1	s_1	s_1
$s_2 = * \triangledown 0 : \triangledown 0$	s_2	s_2	s_2
$s_3 = * \triangledown 1 : \triangledown 0$	$s_5 = * \triangledown \diamond : \triangledown 0$	$s_7 = * \triangledown \diamond : \triangledown \triangle 0$	$s_9 = * \triangledown \triangle : \triangledown 0$
$s_4 = * \triangledown 1 : \triangledown 1$	$s_6 = * \triangledown : \triangledown 1$	$s_8 = * \triangledown \diamond : \triangledown \triangle 1$	$s_{10} = * \triangledown \triangle : \triangledown 1$

Given that selection is for s_1 it is clear that at the very minimum, this string has to be present in all dominant cell types. In addition, the production efficiency varies over different cell types, which translates in them being able to dominate the cell population in subsequent periods. Details of this interesting model can be found in [221].

11.1.2 The Stringmol System

Another interesting AC system taking cues from biology is the Stringmol system developed at the University of York [379–381]. Bacteria were used as an inspiration here, together with the notion

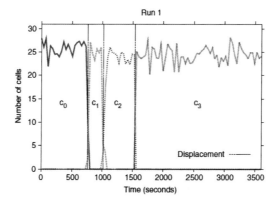

Figure 11.3 Typical run of the MCS.bl system, with dominant cell species shown in each of the cells. At certain moments, the dominant species is displaced due to mutation events that lead to a strengthening of the fitness of another species. From [221], ©World Scientific 2011, with permission.

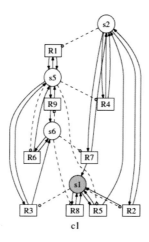

Figure 11.4 Cell species, c_1 with its reaction network depicted as a bipartite graph. Reactions shown as rectangles, with incoming and outgoing arrows symbolizing educts. Product indicated with dashed lines. From [221] , ©World Scientific 2011, with permission.

of the RNA world — at least for first empirical studies. Again, strings act as either substrate or enzymatic molecule, with the "program" of behavior being encoded in the string interaction akin to a processor-data pair.

The Stringmol system takes strings of symbols from a set Σ that consists of 26 template codes made up from the capital letters of the Latin alphabet, $A, \ldots, Z$ and of 7 function codes $\$, >, \hat{}, ?, =, \%,$. The function code make strings to execute certain behaviors, controlled by patterns expressed in the template code. Reactions between strings here correspond to the possible production of new strings, or the disassociation/decay of strings.

The system uses a cell container assuming some sort of topology, which is, however, not essential to the overall activity of the system. Two molecules have the potential to bind to each other, if they are sufficiently close to each other in this topology. This part of the system is a dispensable complication, however. Once two strings are determined to have the potential to bind (this happens with a certain probability), string matching determines what actually happens.

A sophisticated complementary string matching machinery is implemented in this system to determine whether and at which relative positions strings start reacting. This is based on the notion of complementarity between string symbols, which has been more or less arbitrarily to a distance of 13 in the template symbol code. Figure 11.5 shows the symbols and their complements, with function codes chosen to be self-complementary.

$$\begin{array}{c|c} \Sigma & \texttt{ABC\$DEF>GHIJ\textasciicircum KLM=NOP?QRS\}TUV\%WXYZ} \\ C(\Sigma) & \texttt{NOP\$QRS>TUVW\textasciicircum XYZ=ABC?DEF\}GHI\%JKLM} \end{array}$$

Figure 11.5 Table of complementarity of symbols from the set Σ: $C(\Sigma)$ is the respective complementary fit, with a graded neighborhood similarity, i.e., symbol G is complementary (exact match) to T, but also has a complement match to > and U, less match to S and V, etc. Function codes possess an exact complementary match to themselves. From [381], with permission from author.

With this complementarity measure of matching in place, a traditional Smith-Waterman algorithm [783] for string matching, as it is applied in bioinformatics can be employed to find the best matching (in the sense of complementarity) segments of two strings. Figure11.6 shows an example.

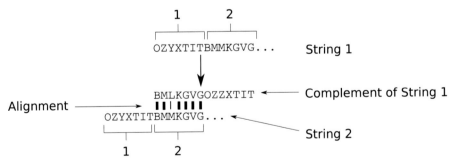

Figure 11.6 Example of complementary matching to determine where stringmol molecules bind to each other, which in turn determines the sequence of instructions begin executed. Adapted from [381].

From this point on, one of the strings (rather arbitrarily defined) becomes the enzyme string, whereas the other string acts as a substrate, providing template material. What makes the former an enzyme is its active role in guiding the reaction. This is realized by using execution control flow instead of three-dimensional shape as determining the function of the enzyme. However, everything, is encoded in the string of the enzyme itself. Technically, each string molecule M consists of a sequence s, a pointer vector p, an activity vector a and a status flag. s contains the information in the form of a sequence of symbols from the alphabet Σ. The pointer vector holds four pointers: a flow pointer, an instruction pointer, a write pointer, and a read pointer. These are used to inform the control flow of the execution model. The activity vector again holds four entries and determines which of the two string pointers is active in a molecular reaction (remember, two molecules have to participate in a reaction). Initially, the activity vector always contains numbers corresponding to the pointers of the enzyme string. But over time, in the execution of the program, this could change and the substrate pointer could be used, which is represented by a change in the activity vector to now point to the other pointer. Finally, the status flag holds either unbound, active or passive flags. The first of these shows the molecule is not in a state where it could react. Active means enzyme, and passive means substrate molecule.

Each of the function codes implements a particular operation, which is explained in some more detail in table 11.3. As can be seen, this is a complex set of operations allowing to implement, for example, replicating strings. This is precisely what Hickinbotham and colleagues have done. In the first instance they hand-designed a self-replicating molecule they call a "replicase."

Table 11.3 Different operations implementable by stringmol functions. $*X$ symbolizes a pointer.

Code	Name	Description
$	search	shift $*F$ to a matching template
>	move	shift pointers to the flow pointer
⌢	toggle	switch pointer to molecular bind partner
?	if	conditional single or double increment of instructor pointer
=	copy	insert at $*W$ a copy of the symbol at $*R$
%	cleave	split a sequence and generate a new molecule
}	end	terminate execution and break the bond between the molecules

The authors emphasize that there are plenty of possibilities to realize replicase, and the example given next is just one of many. Figure 11.7 shows a sample molecule in reaction with another slightly varied molecule (one mutation at the position "m") and able to produce a copy.

Having designed their own replicase, the authors perform evolutionary experiments where some of the string sequences are allowed to mutate away [380]. This is implemented through allowing errors to happen during copy (=) operations. With a probability of 10^{-5}, a symbol is replaced during the copy operation. More rarely still are errors which result in the deletion or insertion of a symbol.

Hickinbotham et al. are, however, not concerned with multilevel selection, so only one "reactor" or cell is used. The authors observe a number of phenomena in their experiments:

- Extinction events

  Molecular changes through mutation lead to a changed population that in the end is not able to maintain a molecular species. Runs come to an end. This happens through a sort of parasitic molecular species that relies on others to replicate, yet competes with them for space.

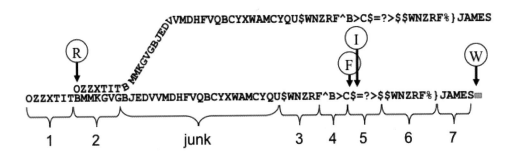

Figure 11.7 Example of replicase molecule, hand-designed (lower string). String binds to upper string, substrate, at the code segment aligned. Circles indicate initial position of pointers. Numbers indicate different sections on replicase used for different phases of the replication. Notable is section 5, which implements the copy loop. From [379], ©Springer 2011, with permission.

- Characteristic sweeps

  In sweeps, the dominant species is replaced with another species that has some fitness advantage and is able to outcompete the earlier dominant species.

- Drift

  In drift, the dominant species mutates away in various directions through neutral mutations.

- Formation of subpopulations

  Subpopulations are formed when some species are able to support themselves, though in smaller numbers, through mutual support or by acting as nonlethal parasites on the dominant species.

- Slow sweeps

  Slow sweeps tend to take much longer to develop to a dominant species. They are not as frequent as characteristic sweeps, but tend to appear regularly.

- Rapid sweep sequences

  Occasionally, mutants cause an entire cascade of species to become dominant one after the other.

- Formation of hypercycles

  Hypercycles can emerge when a subpopulation can couple itself to a dominant species. If that happens, and the dominant species looses its ability to self-replicate, both start to depend on each other. The class of hypercycles consists of some qualitatively different subclasses.

Figure 11.8 shows a run of the simulation where a number of these phenomena, which are quite similar to what appears in other well-mixed ACs, appear.

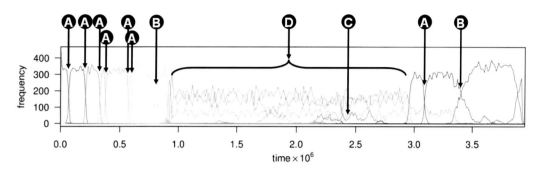

Figure 11.8 Example of stringmol system run. This particular run shows (A) characteristic sweeps, (B) slow sweeps, (C) subpopulations, and (D) multispecies hypercycles developing. From [380], reproduced with permission from author.

11.1.3 The SAC System

In a series of works Suzuki and Ono have also examined a string-based artificial chemistry, their SAC system [642, 823, 826]. Here again as in the molecular classifier systems, strings are only part of a system, with a higher level structure of cells also playing a key role in the competition for survival and reproduction. Within cells that do not have any spatial structure, strings ("genes") instruct the building of phenotypic molecules ("proteins") that can in turn operate on strings. Strings consist of rewriting operations of a simple string rewriting language as shown in table 11.4.

Table 11.4 Symbol set of the SAC rewriting language.

Symbol	Name	Function
.	Separator	Separation of the string
0, 1, 2, 3	Numbers	Non-functional pattern symbols
'	Deletion	Deletion of next symbol
"	Creation	Creation of next symbol
!, ?, %	Wild card	Wild card character that matches an arbitrary symbol
*	Wild card	Wild card character that matches an arbitrary (sub) string
/	Permutation	Permutation of two substrings delimited by by "///"
\	Suppression	Suppression of the function of the next symbol
&	Concatenation	Logical AND operation between rewriting rules
$	Concatenation	Logical AND-NOT operation between rewriting rules
M, E	Membrane symbols	Ingredients for membrane strings
L, R	Spindle signals	Active transport of the string starting with L/R to left and right

To see how this works, here is an example string P_0:

$$P_0 \equiv \backslash 0 \ !\$0'0''1 * 0'1''2 \Rightarrow [\backslash 0][\backslash !]\$[00 * 01 \rightarrow 01 * 02]. \qquad (11.2)$$

The string is first decoded by separating the rewriting operations into three substrings. Suppose now, this rewriting string reacts with a pure non functional string 002301. First, a search for characters 0 and ! is executed. While a 0 is found in the first symbol, no symbol ! is found. So the logical AND between these two rules gives a FALSE which allows the third rule to kick in.

$$P_0 : 002301 \rightarrow 012302.$$ (11.3)

This rule rewrites a string beginning with symbols 00 and ending in symbols 01, with arbitrary length by replacing the second and last symbol by 1 and 2 respectively.

Using these rewriting rules, Suzuki and Ono design a replication system that constitutes an example of a universal constructor. It consists of 12 strings that work together for replication. Since the operations are catalytic, a system of this type will produce more and more of itself. The authors then conduct experiments with variants of this system, studying the coordination and efficiency of replication. In one of these variants, each gene is acting independent of the others, and based on random collisions of strings. In a second model, the put all genes onto a chromosome which therefore allows a greater coordination of replication events. The final model the study is equipped with the equivalent of a cellular spindle that forces the replicated strings into different parts of the cell, such that a subsequent cell division does not inadvertently remove essential strings from one of the daughters (by inhomogeneous distribution of strings).

The experiments start with each functional set of strings put into a cell. A cell starts to grow by producing more strings, until a certain size is reached, upon which it divides into two daughter cells. Alternatively, a membrane string with a certain minimal number of membrane symbols "M" is required for cell division to occur. Cells have an age counter, which is incremented every time step and reset only when cell division occurs. After a maximum of cells has appeared, the number of cells is not allowed to grow any more. Every time a new cell is added, an old cell is removed.

Figure 11.9 shows the dynamics of exemplary runs with the different model systems. Due to the independent replication of genes in model (i), the initial growth spurt of replicators comes to a halt, resulting in a stagnating cell population. Models (ii) and (iii) show continued replication until the end of simulation times. The authors state that it is only in models with a coordinated replication, tied to cell division that sustainable cells can continue to reproduce. More details can be found in [823].

11.2 Lock-and-Key Artificial Chemistries

Many artificial chemistries explore the lock-and-key principle of recognition and binding that occurs between an enzyme and its substrate, or between a receptor and its ligand. The stronger the degree of lock-and-key complementarity between the two molecules, the tighter they tend to bind to each other. From a chemical perspective, such bindings are typically noncovalent and transient (reversible): the molecules stay bound just long enough to transmit a signal or to perform some operation, such as the transcription of a gene. They then dissociate and are free to bind elsewhere and perform other operations.

This notion of lock-and-key complementary matching between molecules as a prerequisite for their binding is extensively exploited in artificial chemistries. Indeed, all the string-based ACs illustrated in the previous section (MCS.bl, Stringmol, and SAC) rely on some sort of pattern matching for molecular recognition and binding:

- In MCS.bl, broadcast units specify patterns as strings of symbols. These patterns describe the condition that must be satisfied for a given enzyme string to bind to a substrate string, and the string rewriting action that will be undertaken once the binding occurs.

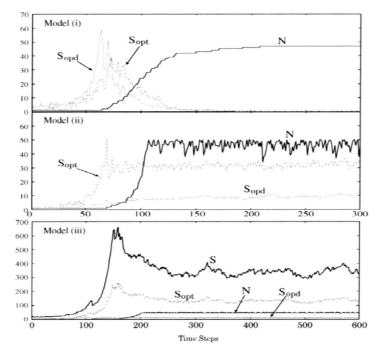

Figure 11.9 Cell number (N), average cell size (S), average number of operators/operands in a cell (S_{opt}/S_{opd}) over time. Model (i): Independently reacting genes; model (ii): Genes organized into a chromosome; model (iii): distribution of multiple chromosomes with a spindle controlled process. From [823], © Springer 2003, with permission.

- In Stringmol, a string alignment algorithm is used to determine the degree of complementarity between an enzyme string and a substrate string. The action to be performed by the enzyme on the substrate is decided based on the best matching segment between both strings.

- In the SAC system, a symbol rewriting language includes wild cards that match arbitrary symbols and substrings, a method that is conceptually similar to the one used in MCS.bl. Instead of enzymes and substrates, SAC uses the analogy of genes that encode proteins, which in turn may also operate on other genes and proteins.

The origins of the lock-and-key complementarity concept in ACs can be traced back to some ideas from the early 80s such as the Typogenetics system by Hofstadter [391], the enzymatic computer by Conrad [193], and the immune system model by Farmer and colleagues [262]. These systems were not originally proposed as ACs, but retrospectively they can be interpreted as AC examples (with Conrad's computer in the wet AC category). Typogenetics has already been introduced in chapter 10. There, the notion of complementarity was based on DNA base pairing.

In this section, we discuss the enzyme computer by Conrad and the immune system model by Farmer and colleagues, because their ideas are still influential in many contemporary pieces of research work. Conrad's enzymatic computing contributed to the foundations of the area of wet molecular computing (see chapter 19), and Farmer's immune system model is among the founders of the area of artificial immune systems (AIS) [214].

11.2.1   Pattern Recognition with Lock-and-Key Enzymatic Computing

The molecular computer proposed by Michael Conrad [193] was based on the lock-and-key interaction between an enzyme and its substrate. In nature, enzymes recognize their targets by their shapes and bind more strongly when the lock-key match is high. Upon binding, the conformation (shape) of the involved molecules may change, as the enzyme-substrate complex finds a more stable minimum-energy configuration. Conrad identified such lock-key binding as a form of pattern recognition, a task that to this day remains difficult for a traditional computer to solve. He viewed computation in general as a dynamical process, and enzymatic computing as a downhill dynamical process "falling" toward the solution as the system flows from a higher free energy state to a lower one [194]; so enzymatic devices solve a free energy minimization problem. Note that in 1968 Robert Rosen [722] had already drawn a parallel between finite automata and dynamical systems that fall toward basins of attraction. These ideas can be regarded as precursors of today's embodied computation paradigm [663, 664], in which some of the computation burden of a (robotic) system can be delegated to its physical parts and to its surrounding environment.

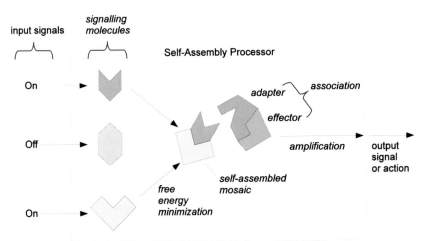

Figure 11.10 Schematic representation of Conrad's self-assembly processor. It is based on the lock-and-key principle of molecular recognition and binding: input signals trigger the release of molecules whose specific shapes cause them to be recognized by adaptor molecules, which in turn trigger effector molecules that lead to the amplification and further processing of the signal through a signaling cascade, until an output is produced. Adapted from [194].

Figure 11.10 shows an example of lock-and-key molecular device envisaged by Conrad. It builds upon the way information is transmitted to the interior of the cells via signaling cascades. Each input signal is associated with a molecule of a different shape. These different signaling molecules self-assemble into a macromolecular complex or mosaic that then binds to an adapter molecule, triggering a cascade of molecular bindings that processes the signal. Different combinations of signals define patterns that can be recognized due to a lock-and-key binding to the specific macromolecular complex that they form. Hence the device solves a pattern recognition problem using energy minimization of molecular conformations.

Conrad also pointed out that the search for the minimum energy state benefits from the natural parallelism found at the quantum level: Free energy minimization (as in protein folding, molecular lock-key docking, enzyme conformation change, etc.) is a quantum operation that exploits quantum superposition for parallel search of the minimum energy state. Since the correspondence between protein sequence and shape was (and still is) difficult to establish, such chemical computers would be difficult to program in the traditional way, so they would generally utilize learning (for instance, with an evolutionary algorithm) as the programmability method [194]. Conrad argued that such loss in direct programmability by humans should not actually be a big problem, since efficient parallel algorithms are also difficult to program in general. This discussion will be resumed in chapter 17 in the context of chemical computing principles.

Some of Conrad's ideas were later implemented in the wet lab, where logic gates based on lock-and-key conformation change were demonstrated [196, 954]. The resulting system was called enzymatic computing [954] or conformation-based computing [196]. More about wet artificial chemistries will appear in chapter 19.

11.2.2 A Bitstring Model of the Immune System

The vertebrate immune system is a highly adaptive and dynamic system, able to distinguish between self and nonself, and to mount a defensive response against invaders. It features a rapid mutation and selection process that refines the population of antibodies, such that they can more readily identify and eliminate antigens.

The ability of the immune system to recognize antigenic patterns and to keep a memory of previous infections has attracted the attention of computer scientists and has become a source of inspiration for many Artificial Intelligence algorithms, giving birth to the area of AIS [114, 211, 214, 262, 287]. AIS approaches encounter many applications in computer security, anomaly detection, machine learning, and combinatorial optimization, among others [368].

The binary string model of the immune system by Farmer and colleagues [261, 262] is considered as a pioneer work in the AIS field [214]. It will become clear from its description that it is an example of AC as well. In their model, antigens and antibodies are represented as binary strings that can bind to each other through lock-and-key interactions. Besides binding antigens, antibodies can also bind other antibodies, forming idiotypic networks that could potentially regulate the immune response and store an immune memory. The bindings between antibodies and antigens, and between antibodies in the idiotypic network, are modeled as chemical reactions with rate proportional to the strength of binding, which is itself proportional to the degree of complementarity between the bitstrings in the binding regions.

Each antibody is represented as a binary string containing two binding sites (p, e), one that can act as a paratope (p), and the other as an epitope (e). In the lock-and-key analogy, the epitope is the lock and the paratope is the key. When the paratope of an antibody j recognizes the epitope of an antigen i, the antibody j gets replicated whereas the antigen i is destroyed. The molecule i may also be another antibody of the idiotypic network.

Figure 11.11 shows an example of recognition and binding in this model: the paratope p_1 of antibody A_1 recognizes the epitope e_2 of antibody A_2 due to a good complementarity of the bitstrings p_1 and e_2. Upon binding, A_1 is stimulated, leading to the production of more copies of itself, whereas A_2 is destroyed.

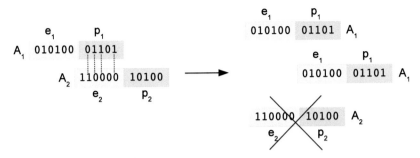

Figure 11.11 A recognition and binding reaction in the immune system model by Farmer and colleagues [261, 262]: the paratope p_1 of antibody A_1 recognizes the epitope e_2 of antibody A_2 and binds to it. As a result, A_1 is duplicated and A_2 is destroyed.

A paratope recognizes and binds to an epitope when the degree of complementarity between their respective strings is above a threshold s. Strings are allowed to match in any possible alignment, in order to model the fact that two molecules may react in alternative ways. The strength of binding between two antibodies i and j is calculated as the sum of the complementarily matching bits for all possible alignments between their respective binding sites e_i and p_j. A matrix of matching specificities m_{ij} between all possible pairs of binding sites is constructed for this purpose, where each element is calculated as follows:

$$m_{ij} = \sum_k G\left(\left(\sum_n e_i(n+k) \wedge p_j(n)\right) - s + 1\right) \tag{11.4}$$

where:

- s is a matching threshold: for an epitope e of length l_e and a paratope p of length l_p, s is the minimum strength necessary for p and e to bind to each other; hence, they bind when $s > \min(l_e, l_p)$;

- k determines the alignment position between e and p as the relative displacement between the two strings (k lies in the range $-l_p < k < l_e$);

- $e_i(n+k)$ is the $(n+k)$-th bit of epitope e_i;

- $p_j(n)$ is the n-th bit of paratope p_j;

- $\wedge$ is the boolean XOR operation that returns 1 if the two bits are complementary;

- G is the function: $G(x) = x$ for $x > 0$, $G(x) = 0$ otherwise; it makes sure that each alignment returns a positive value when it scores above the threshold s, and zero otherwise.

Upon recognition and binding, an antibody of type X_i may be stimulated (when acting as the key) or suppressed (when acting as the lock) by an antibody of type X_j. Moreover, an antibody X_i may also be stimulated by a foreign antigen Y_j, and may also die in the absence of interactions. These processes can be expressed in the form of chemical reactions as:

Using the law of mass action together with a few adjustment constants, it is easy to translate the above set of chemical reactions into the following set of differential equations from [261,262]:

$$\dot{x}_i = c\left(\sum_{j=1}^{N} m_{ji} x_i x_j - k_1 \sum_{j=1}^{N} m_{ij} x_i x_j + \sum_{j=1}^{M} m_{ji} x_i y_j\right) - k_2 x_i \tag{11.5}$$

$$\dot{y}_i = -k_3 \sum_{j=1}^{N} m_{ij} x_j y_i, \tag{11.6}$$

where:

- x_i is the concentration of antibody X_i;

- y_j is the concentration of antigen Y_j;

- N is the total number of antibodies in the system;

- M is the total number of foreign antigens present;

- m_{ij} is the strength of binding between epitope e_i of X_i and paratope p_j of X_j, as defined by equation 11.2.2;

- k_1 is the weight of the suppression effect with respect to the stimulation effect in antibody-antibody interactions;

- k_2 is the rate of spontaneous antibody decay;

- k_3 is the weight of the suppression effect with respect to the stimulation effect in antibody-antigen interactions ;

- and c is a constant that defines the importance of the the interactive part with respect to the spontaneous decay part.

Equation 11.5 contains four terms. The first term denotes the amount of stimulation that antibody X_i gets from recognizing the epitope of X_j. The second term quantifies the suppression of X_i when it gets recognized by the paratope of another antibody j. The third term represents the stimulation effect on X_i caused by external antigens Y_j. The fourth term denotes the spontaneous death of the antibodies. Equation 11.6 expresses the elimination of antigen Y_i by all the possible antibodies that may recognize its epitope and bind to it.

Such a system implements a clonal selection mechanism, leading to the selection and amplification of antibodies that are able to recognize and suppress others with high specificity. Those antibodies that fail to bind antigens are unable to replicate and die out by spontaneous decay. The system also causes the elimination of antibodies and antigens that get recognized and suppressed by other antibodies.

In addition to the selection process, antibodies and antigens may also undergo point mutations, crossover and inversion of their strings. This leads to a metadynamics in which new molecules are periodically introduced into the system. Hence, this chemistry can be considered as an instance of a constructive dynamical system. The procedure to simulate this metadynamics with an ODE is to check the concentrations periodically and to eliminate the molecules whose concentration fall below a threshold, making room for newly created molecules.

A constructive system based on binary strings and metadynamics was quite novel at the time, and although the authors did not report any concrete simulation results, their work spawned a surge of interest leading other authors to pursue several variants of their model, focusing on different aspects of interest such as learning or memory. Among these subsequent models we can find Forrest's approach to pattern recognition and learning using a bitstring immune system [287], a self-nonself discrimination algorithm to detect computer viruses [288], and various algorithms from the AIS literature (see [214, 368, 851] for an overview).

11.3 Networks

It can be argued that all chemical reactions are part of a huge network. This network would be formed by the substances that react with each other as vertices and the reactions between substances as the edges between vertices. Edges could be weighted by the reaction speed between different substances, and this network would be ever increasing due to the constant addition of new substances that chemists find / synthesize.

11.3.1 A Graph-Based Toy Chemistry

In a series of papers, Benkö et al. [102–105] have sketched out ideas for the formulation of a toy model of chemistry, based on graphs. In particular, they are interested in generic properties of these networks, like their connectivity and node degree distribution. The authors base their model on the overlap of outer orbitals of atoms/molecules, a notion that allows them to bring in energy as a consideration into the reaction graph, as well as the conservation of mass and atom types. Their motivation is to create an AC that has what they call a typical "look-and-feel" of real chemistry, in contrast to other AC models that have been proposed in a more abstract manner.

While details, especially those of the theoretical background of their model, need to be left to the original literature, figure 11.12 visualizes the concept. Starting from a structural formula of a chemical compound (figure 11.12 (b), lower right), a set of hybrid orbitals are generated using the well-established VSEPR rules of chemistry [333]. This representation, shown in figure 11.12 (a) is then abstracted into the orbital graph of figure 11.12 (b) that contains all the bonds between atoms, as well as information about energy content.

Equipped with this model, the authors define parameters for many of life's most relevant atomic types, like HCNOPS, and can proceed to produce realistic chemical reaction networks of molecules involving these atom types. The corresponding software implementation allows researchers to automatically produce differential equations for reaction networks and to study their reaction kinetics. The authors claim that this system is useful for AC researchers interested in more realistic settings of their chemistries.

11.3.2 Random Chemical Reaction Networks

One of the easiest ways to generate an artificial chemistry is by drawing random numbers for rate constants or by generating explicit reaction laws randomly. The random autocatalytic systems investigated in [792], the GARD system [764], or the metabolic networks in the multicellular organisms reported in [301] are examples of these nonconstructive approaches. It is also possible to generate constructive chemistries randomly by drawing parameters for the reaction laws "on the fly" when they are needed as, e.g., demonstrated by Bagley and Farmer [49].

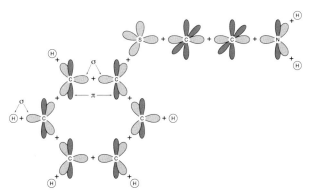

(a) Hybrid Orbital Model according to VSEPR rules. Atoms are labeled as H — hydrogen, C — carbon, N — nitrogen, O — oxygen, S - Sulfur. Corresponding orbitals are shown in tray. "+" indicates overlap of orbitals

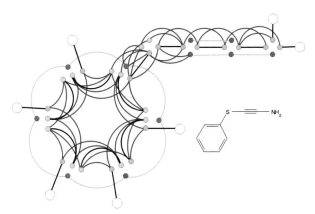

(b) Orbital Graph abstraction used for calculation of energies. Colored nodes indicate orbitals of different type; links indicate bonds with color code: Black — σ-bonds, Blue — π-bonds, Red — other (semi-direct, hyperconjugation, fictitious interaction). In small: Structural formula of compound.

Figure 11.12 Concept of ToyChem System. Figures from [105], reproduced with permission from authors.

In the following, we want to discuss in some detail one such system that is inspired by biology [117]. It investigates the properties of a large, conservative random chemical reaction networks, and in particular, their behavior as equivalent to thermodynamically realistic metabolic networks. Thus, a flow of external nutrients will be used to "feed" this network, with the results showing that in a very wide range of parameters, such networks will synthesize certain products which could be used as precursor molecules of a putative cell membrane.

The idea of Bigan et al. [117] is to start with an arbitrary set of chemicals, say A_1 to A_{10} which have random Gibbs free energy of their formation assigned to them. These chemicals are then assumed to be able to react with each other through the following mono- or bimolecular chemical reactions:

$$\text{Association} \quad A_i + A_j \quad \rightarrow A_k \tag{11.7}$$

$$\text{Dissociation} \quad A_i \quad \rightarrow A_j + A_k \tag{11.8}$$

$$\text{Transformation} \quad A_i \quad \rightarrow A_j. \tag{11.9}$$

Step by step, reactions are randomly generated between chemical types, with (again) random kinetic parameters assigned to these forward reactions while checking whether the network of reactions is still conservative. Backward reaction parameters are then assigned in such a way that the system is thermodynamically consistent. The authors consider both a mass-action kinetics and a so-called saturating kinetics, which is relevant in the context of living systems where molecular crowing is the rule rather than the exception [956].

At each step of the random network generation, a standard algorithm is run to check whether mass conservation still holds [756]. The condition for this is that the $N \times R$ stoichiometric matrix S (where N is the number of chemicals and R is the number of reactions), multiplied by a mass vector for all chemicals m_i, with $i = 1, \ldots N$ is still semi-positive definite: $S^T m = 0$. Once the number of reactions approaches N, the number of solutions decreases, until at a certain number above N only one mass vector is admissible any more for any given reaction network. This is when the authors stop the addition of new reactions.

The kinetic parameters are now assigned to each of these reactions based on a random parameters in forward direction and energy considerations in backward direction.

$$\text{Monomolecular forward constants} \quad k_r^f = k_{avg-mono} 10^{R(-s/2,s/2)} \tag{11.10}$$

$$\text{Bimolecular forward constants} \quad k_r^f = k_{avg-bi} 10^{R(-s/2,s/2)}, \tag{11.11}$$

where $R(x, y)$ is a function producing a random number between x and y. For backward constants and saturation kinetics the reader should consult the original paper.

An examination of the reaction system under equilibrium conditions yields that one of the chemicals ends up with the highest concentration, depending on system density (figure 11.13(a)). If one adds an external nutrient flow to the system, all chemicals except one (the dominant one) reach constant nonzero concentrations, whereas this one chemical grows in concentration unbounded (figure 11.13(b)). One could say that the nutrient flow is transformed into one particular chemical over time.

The authors then argue that most large conservative random networks with saturation kinetics function as "directed transformation machines," converting nutrient matter into one specific chemical. This chemical, if interpreted as a structural molecule (e.g., a membrane molecule) could thus be preferentially synthesized by such a network. The authors then consider the conditions under which such a system could sustain growth, even when this growth reduces the concentration of chemicals by enclosing a larger and larger volume.

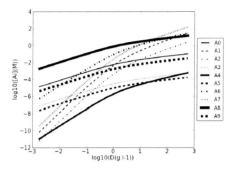
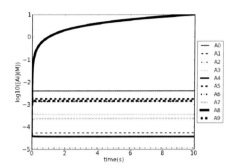

(a) Equilibrium concentrations as a function of density

(b) Nutrient A_5 added to the network flux, for saturating kinetics.

Figure 11.13 Random chemical network model after Bigan [117]. With permission from author.

It is interesting to see that such a simple toy model shows the ability to perform functions usually prescribed to very specially evolved networks.

11.3.3 Suzuki's Network AC

As we have seen in chapter 5, water plays a paramount role in the molecular interactions that enable life. The formation of hydrophobic and hydrophilic interactions is at the heart of the mechanisms of membrane self-assembly, DNA double helix formation, protein folding, and countless other essential cellular processes. However, it is computationally expensive to model each such interaction down to the atomic detail. The network artificial chemistry (NAC) proposed by Suzuki [824,827] models such interactions in an abstract way using a graph rewiring approach. In NAC, molecules occupy the nodes of a graph, linked by edges representing bonds. Nodes may be hydrophilic or hydrophobic. Covalent bonds and hydrogen bonds are represented by directed edges, and weak van der Waals interactions by undirected edges. Each molecule is a string that can be decoded into instructions for a von Neumann machine contained within nodes. Using this chemistry, Suzuki shows how node chains can fold into clusters that can be compared both to parallel computers and to enzymes with a catalytic site. He then designs a splitase and a replicase enzyme for his system and shows how they operate on other clusters: The splitase splits a cluster in two, and the replicase copies a node chain.

In [825] Suzuki extends the basic NAC approach with molecules that contain more sophisticated computer programs that are executed by the nodes where they reside. He shows that this new system is able to self-organize into a global hydrophilic cluster akin to a cell. Molecules akin to centrosomes (small intracellular structures that help to pull the chromosomes apart during cell division) send signals to cells. In response to these signals the cell divides into two daughter cells. Figure 11.14 shows an example of such a cell division process.

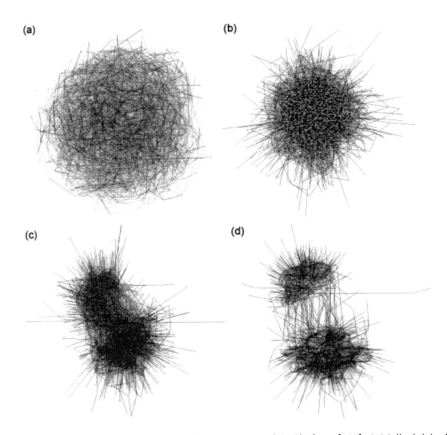

Figure 11.14 Cell division in the network artificial chemistry (NAC), from [825]. Initially (a) hydrophilic and hydrophobic nodes are randomly mixed in a network with randomly generated edges. After a few time steps, the hydrophilic nodes start to cluster at the center of the graph (b). Two centrosomes are placed in the system (c), and they start to pull the nodes apart, until they finally segregate into two separate clusters (d). Reprinted from [825], © 2008, with permission from Elsevier.

11.3.4   An AC Implementation of an Artificial Regulatory Network

As pointed out in chapter 5, genetic regulatory networks (GRNs) allow the same genome to express different behaviors in response to external conditions, and they are interesting from an AC perspective due to their constructive and computing potential.

In this section we shall examine the artificial regulatory network (ARN) approach proposed by one of us in [67, 68]. This AC is an abstract model of a GRN where genes and proteins are binary strings that interact by the complementarity of their bitstrings. The initial motivation behind ARNs was to endow artificial evolution with a developmental step, in order to render it more scalable and adaptive. Since then researchers have used ARNs for genetic programming [480, 521, 522] and for the modeling of biological systems [73, 478, 479, 504]. These applications of ARNs will be discussed in more detail in chapters 17 and 18, respectively.

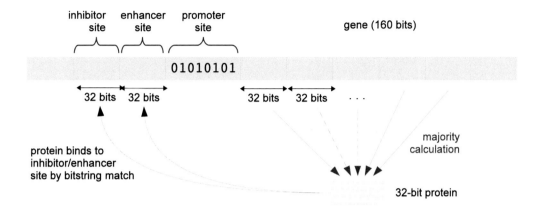

Figure 11.15 Structure of a gene in the artificial regulatory network (ARN): a promoter region with a well-defined bit pattern marks the start of a gene. The gene is divided into 32-bit portions, and encodes a protein via majority voting. Proteins act as transcription factors, binding to inhibitor or enhancer sites upstream from the promoter region of the gene. The binding strength is a function of the degree of complementarity between the bitstring of the protein and of the binding site, calculated using a XOR boolean function. Adapted from [479, 480].

The ARN Model

In the ARN model, a single bitstring contains the full genome, divided into 32-bit portions. The various proteins are represented as 32-bit numbers. A "promoter" sequence marks the beginning of a gene that codes for a protein. By scanning the genome for promoters, a number N of genes can be identified. Figure 11.15 shows the structure of a gene. Two regulatory sites are found upstream from the promoter: one binding site for activator TFs (enhancer site) and one for repressor TFs (inhibitor site). When a TF binds to the gene at the enhancer site, it enhances the rate of expression of this gene. When a TF binds to the inhibitor site, the expression rate of this gene is decreased.

There are no intermediate RNA molecules; thus, genes are translated directly into proteins, skipping the transcription step. Each gene i produces protein i, for $i = 1,...N$, following the translation process shown in figure 11.16: Each gene consists of 160 bits, divided into 5 segments of 32 bits. Each bit of of the protein is produced by majority voting among the corresponding 5 bits in the gene, one bit coming from each of the 5 segments: if the majority of the 5 bits at the k-th position (for $k = 1,..32$) is set to one, then the k-th bit of the protein is a one, else it is a zero.

Once expressed, proteins become TFs that bind to genes with a strength that increases with the degree of complementarity between the 32 bits of the protein and the binding site in the gene. The number of complementary bits is equivalent to the Hamming distance between the two bitstrings, and can be calculated with a XOR operation between protein and the binding site.

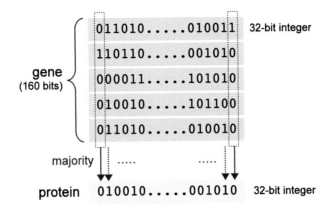

Figure 11.16 Direct translation from a 160-bit gene to a 32-bit protein in the ARN model. Each bit of the protein reflects the majority of a group of 5 bits from the gene.

The combined effect of activator and repressor TFs determine the rate of expression of the corresponding gene, which can be calculated using the following equations:

$$\frac{dc_i}{dt} = \frac{\delta(a_i - h_i)c_i}{\sum_j c_j} \tag{11.12}$$

$$a_i = \frac{1}{N}\sum_j c_j e^{\beta(d_j^a - d_{max})} \tag{11.13}$$

$$h_i = \frac{1}{N}\sum_j c_j e^{\beta(d_j^h - d_{max})}, \tag{11.14}$$

where:

- c_i is the concentration of protein i, for $i = 1,...N$; protein concentrations are normalized to one that is, at any time $\sum c_i = 1$, $\forall i$;

- a_i is the degree of enhancement of the production of protein i, caused by the combined effect of all TFs that may bind to the enhancer site of gene i;

- h_i is the degree of inhibition in the production of protein i, caused by all TFs binding to the inhibitory site of gene i;

- $0 \le d_j^a \le 32$ (resp. d_j^h) is the number of complementary bits between protein j ($j = 1,...N$) and the enhancer (resp. inhibitor) site of gene i;

- $d_{max} = 32$ is the maximum possible number of complementary bits between a TF and its binding site;

- β and δ are positive scaling factors.

Note that $d_j^a \le d_{max}$, $d_j^h \le d_{max}$, and $\beta > 0$ hence the quantities $\beta(d_j^a - d_{max})$ and $\beta(d_j^h - d_{max})$ are both negative for the proteins that have at least one complementary bit to the corresponding binding site, and zero otherwise. Hence the influence of a protein on a gene is a decreasing exponential on the number of matching (equal) bits: the stronger the complementarity between the protein and the binding site, the higher the values of the exponential parts of Equations 11.13 and 11.14, thus quantifying a more important contribution of this protein in the expression levels of the corresponding gene.

Equations 11.13 and 11.14 remind us of the Arrhenius equation already introduced in chapter 2, which expresses the rate of a reaction as a decreasing exponential function of the activation energy (E_a) necessary for the reaction to occur. Hence, reactions that have a high E_a barrier happen very slowly, whereas catalysts accelerate reactions by decreasing their E_a barrier. This is exactly what equation 11.13 does: The stronger a protein binds to the gene, the more it will enhance its expression rate. We can thus map $E_a = d_{max} - d_j^a$ and $\beta = RT$. Equation 11.14 works in a similar way, but now a reduction in E_a barrier is used to decrease the expression rate of the gene, due to the negative influence of h_i on the total rate of expression of gene i, as given by equation 11.12. We will make use of this correspondence in our simulations, but right now it suffices to notice that it provides a chemical basis for equations 11.12—11.14.

Exploring ARNs

The basic ARN model just described can be explored in three main directions: topology, dynamics, and evolution.

First of all, the static topology of a given ARN can be studied. Given an arbitrary genome, we can construct the graph corresponding to the GRN contained in this genome, as follows. We scan the genome for promoters, and identify the existing genes. After that, we assign a threshold of complementarity between protein j and binding site i, above which we consider that the influence of protein j on gene i is sufficiently high to deserve a link (j, i) connecting gene j to gene i. We can then analyze the connectivity properties of this graph, as done in [67, 478, 479].

A second experiment with the ARN model is to look at the dynamic properties of the underlying GRN, given its graph of interactions, already constructed in the first step above. We start with the identified genes and a very low concentration of the proteins, reflecting a basal rate of expression. We then calculate all the Hamming distances between all the proteins and binding sites, and use equations 11.13 and 11.14 to calculate the overall contribution of all the proteins on a given binding site. We repeat this for all the binding sites of all genes. We then apply a discretization of eq. 11.12 within a time step Δt in order to calculate the changes in concentrations of the various proteins, incrementing the corresponding concentration values accordingly. We repeat these steps successively, in what is clearly an instance of an ODE integration algorithm. In this process, we will be able to see if a subset of proteins end up dominating the system, if their concentrations converge to stable values or oscillate, and so on.

A third and more interesting experiment, which also integrates the other two, is to evolve a GRN to implement some target function. For instance, we can interpret one or more proteins as input values and a few others as output values, and establish a fitness function that evaluates how close the output values match a given output function, given the input values. We start with a population of random genomes (individuals) that can then be mutated or recombined to produce new genome variants. At each generation, the fitness of each individual is evaluated by expanding its graph, running it by ODE integration as explained above, and checking the resulting protein concentration dynamics against some target behavior. This has been done in [67, 479, 480], for instance. These experiments have highlighted the important role of gene duplication and divergence in the evolution of ARNs: In contrast with random ARNs where usually nothing interesting happens, in ARNs grown by gene duplication and divergence, small-world and scale-free topologies arise [478, 479], and the dynamics converges to attractors where a subset of the genes and transcription factors (TFs) play a crucial role [68, 479].

11.4   Spatial Structuring and Movement in Artificial Chemistries

More complex artificial chemistries are necessary to instruct the growth of an individual cell-like structure in detail. Tim Hutton's Squirm3 system discussed next exemplifies that work par excellence. The first step of this undertaking was to establish a system capable of self-replication. In a second step, evolutionary principles needed to be added, including competition between individual self-replicating entities. This would be difficult in a world of a well-stirred reactor, so localization through cell membranes, a typical example of spatial structuring with ACs, were studied by Hutton. Spatial structuring is a prerequisite of morphogenesis using artificial chemistries.

However, also the movement of cells and molecules must be considered. Two examples in this category are the Swarm Chemistry by Sayama (section 11.4.2), and the Flow artificial chemistry by Kreyssig and Dittrich (section 11.4.3). Models of morphogenesis using artificial chemistries will be discussed in chapter 18.

11.4.1   The Squirm3 Artificial Chemistry

The Squirm3 system proposed by Hutton [411] consists of atoms of different types, named after the alphabet: $T \in \{a, b, c, d, e, f\}$. Atoms are the indivisible basic units of his system, and they possess a state characterized by a number $S \in \{0, 1, 2, \ldots\}$ that might change in the course of reactions with other atoms. One could think of energy content as being an analogue of this state. Making and breaking bonds between atoms is what constitutes reactions in this system, we call these associations and dissociations. Some reactions also allow states to change in bonded molecules, these are called transformations.

There is a two-dimensional topology to the Squirm3 system of an either discrete of continuous form. In the discrete form, a Moore or von Neumann neighborhood on a grid is used, in the continuous form a radius is defined for atoms. If any other atom is closer that means a reaction can happen. Atoms can move in this 2D world, bumping into each other at random and, depending on reaction rules, either continue on with their travel, or engaging in bonding reactions. Once more than one atom are bonded, the entity is called a molecule, sticking together when moving.

Often, hand-designed reaction rules are applied to this system. An example is shown in table 11.5, which allows a molecule of a particular type to self-replicate.

If a particular molecule is used as a seed, and enough atoms in ground state $x0$ are available for reactions, a reaction system with these rules can self-replicate the molecule. Due to the rules, the seed molecule needs to start with atom $e8$ and end with $f1$, with any number of atoms in state $x1$ in between. Figure 11.17 demonstrates the sequence of reactions that leads to self-replication.

The system design is deliberate to allow for an arbitrary set of $a, \ldots d$ atoms in between the e, f atoms. The former can act as a genetic code that allows to produce another type of molecule, an enzyme coded in the sequence. The translation of a gene molecule into an enzyme specifies which reaction this enzyme can catalyze. Figure 11.18 shows the general setup for enzymatic reactions: A third atom $z(i)$ in state i must be present to catalyze a reaction between an atom $x(g)$ in state g and another atom $y(h)$ in state h. The bonding status between atoms is characterized

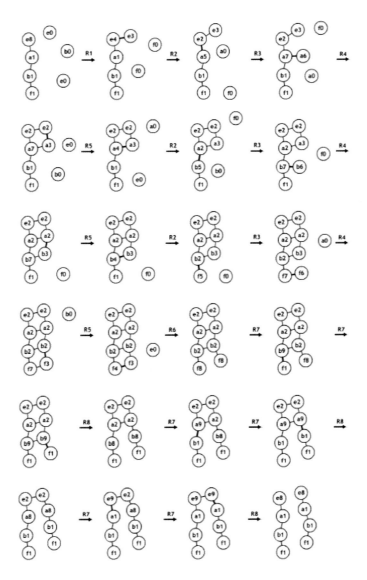

Figure 11.17 Self-replication of a seed molecule $e8 - a1 - b1 - f1$. Here, the reaction rule set is designed in such a way as to uniquely lead to the self-replication outcome. From [411], © MIT Press 2002, with permission.

Table 11.5 Reaction rules for a self-replicating entity in Squirm3.

Rule number	Reaction rule	Description
R_1	$e8 + e0 \rightarrow e4e3$	Specific association
R_2	$x4y1 \rightarrow x2y5$	General transformation
R_3	$x5 + x0 \rightarrow x7x6$	General homo-association
R_4	$x3 + y6 \rightarrow x2y3$	General hetero-association
R_5	$x7y3 \rightarrow x4y3$	General transformation
R_6	$f4f3 \rightarrow f8 + f8$	Specific dissociation
R_7	$x2y8 \rightarrow x9y1$	General transformation
R_8	$x9y9 \rightarrow x8 + y8$	General dissociation

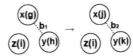

Figure 11.18 Enzymatic reaction between two atoms x, y (either bonded or not, symbolized by b_1, b_2) in states g, h, respectively. The entire reaction is specified in the number i, as given by formula 11.15. From [411], © MIT Press 2002, with permission.

by the variable b_1, b_2 which could assume values of "1" or "0," depending on whether they are bonded or not.

The atom $z(i)$ can only catalyze a reaction involving atoms whose state is characterized by the state number i. i is calculated as

$$i = 2(2(|T|(|T|(|S|(|S|(|S|g+h)+j)+k)+x)+y)+b_1)+b_2+S \qquad (11.15)$$

i in turn can be read off the sequence of genetic atoms which codes numbers in a base-4 code with a, b, c, d standing for numbers $0, 1, 2, 3$ and an offset to the number of states $|S|$. Thus, a genetic string like $ebdcaf$ would translate into the number $i = 1320_4 + |S|$, which will then allow one particular reaction to catalyze.

The entire system of enzymatic reactions in a membrane like envelope was formulated in [412] containing 41 reactions of different subtypes. Hutton also examined the potential of his Squirm3 system to evolution. To this end, he introduced a simple competition into his system by a flooding mechanism that would periodically erase molecules from areas of the world and replenish them with raw materials in the form of single atoms. Lucht gives a lucid account of the sequence of Squirm3 models and their features [529].

One key ingredient to a real morphogenetic system, however, is still missing in this example. As is well known from the development of multicellular organisms, motion is an important further step toward arriving at truly powerful self-organizing processes [242]. Motion, however, is not only found on the level of individual cells finding their place in a multicellular organism, but already on the level of molecules organizing a single-celled organism. So the next section will discuss bio-inspired artificial chemistries that possess — besides the notion of a neighborhood of molecules required for reaction — that feature that molecules can move and have velocity states.

Table 11.6 Types of kinetic parameters used in Sayama's swarm chemistries.

Symbol	Min	Max	Description	Units
R^i	0	300	Radius of local perception range	pixels
V_n^i	0	20	Normal speed	pixels / step
V_m^i	0	40	Maximum speed	pixels / step
c_1^i	0	1	Strength of cohesive force	$1 / \text{step}^2$
c_2^i	0	1	Strength of aligning force	$1 / \text{step}$
c_3^i	0	100	Strength of separating force	$\text{pixels}^2 / \text{step}^2$
c_4^i	0	0.5	Probability of random steering	$---$
c_5^i	0	1	Tendency of self-propulsion	$---$

11.4.2 Swarm Chemistry

Sayama's swarm chemistry [746, 748] is a different type of chemistry, with molecules moving about in a continuous 2D or 3D world. Thus, the reactor vessel has a topology, and reactions are more of a physical type assuming the ability for particles (molecules) to self-propelled movement and sensing of the kinetic parameters of other particles with a view to swarmlike alignment or to an exchange of parameters. Sayama studies these systems in regard to their capabilities of showing self-organized patterns of behavior (e.g., segregation, regular collective movement patterns). Different types of particles are characterized by different kinetic parameters, and reactions are really kinetic interactions between particles that are close to each other in the topology.

Reynold's Boid system [709] has inspired the swarm-based behavior of particles:

- If there are no other particles in the neighborhood (perception range) R_i of a particle: Stray randomly

- Else:

  - Steer to move toward center of nearby particles
  - Adjust velocity vector toward average velocity vector of nearby particles
  - Steer to avoid collision with nearby particles
  - Steer randomly

- Approximate normal velocity of own type

Figure 11.19 shows the kinetic interactions between particles that may lead to the formation of interesting patterns, depending on the individual features of particle types, which are listed in so-called recipes for the swarm under investigation. Table 11.6 shows the different kinetic parameters in Sayama's system, with minimal and maximal values, and figure 11.20 shows the coded particle behavior.

A swarm comprises a number of particles of different types, and can be compactly written down as a "recipe" like figure 11.21, which contains the number of particles and the specification of their parameters.

Sayama's experiments examined the behavior of homogeneous swarms, but mostly of heterogeneous swarms with parameter choices often evolved using interactive evolution [747]. Predominant phenomena are known from other systems [889]. They include oscillatory movements, rotation, segregation into different areas, and the like. Altogether these collective motion

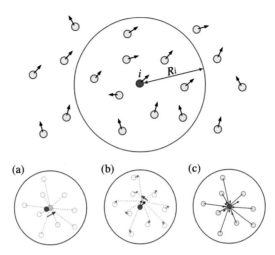

Figure 11.19 Kinetic interactions between particles in a swarm chemistry: Particle i (dark blue) perceives other particles within radius R_i of its current position and may (a) move towards the center of nearby particles (orange); (b) adjust its velocity toward the average velocity vector of nearby particles (big red arrow); (c) avoid collision. Courtesy of Hiroki Sayama, reproduced with permission from author.

```
 1: for all i ∈ particles do
 2:     N ← {j ≠ i that satisfies |x⃗_j − x⃗_i| < R^i}  // Finding other particles within its local perception range
 3:     if |N| = 0 then
 4:         a⃗ ← (r_{±.5}, r_{±.5})  // Straying
 5:     else
 6:         ⟨x⃗⟩ ← Σ_{j∈N} x⃗_j/|N|  // Calculating the average position of nearby particles
 7:         ⟨v⃗⟩ ← Σ_{j∈N} v⃗_j/|N|  // Calculating the average velocity of nearby particles
 8:         a⃗ ← c_1^i(⟨x⃗⟩ − x⃗_i) + c_2^i(⟨v⃗⟩ − v⃗_i) + c_3^i Σ_{j∈N}(x⃗_i − x⃗_j)/|x⃗_i − x⃗_j|^2  // Cohesion, alignment and separation
 9:         if r < c_4^i then
10:             a⃗ ← a⃗ + (r_{±5}, r_{±5})  // Randomness
11:         end if
12:     end if
13:     v⃗_i' ← v⃗_i + a⃗  // Acceleration
14:     v⃗_i' ← min(V_m^i/|v⃗_i'|, 1) · v⃗_i'  // Prohibiting overspeed
15:     v⃗_i' ← c_5^i(V_n^i/|v⃗_i'| · v⃗_i') + (1 − c_5^i)v⃗_i'  // Self-propulsion
16: end for
17: for all i ∈ particles do
18:     v⃗_i ← v⃗_i'  // Updating velocity
19:     x⃗_i ← x⃗_i + v⃗_i  // Updating location
20: end for
```

Figure 11.20 An algorithm for producing particle behavior in Sayama's swarm chemistry. Reproduced with permission from author.

$$
\begin{array}{ll}
97 \; \ast & (226.76, \; 3.11, \; 9.61, \; 0.15, \; 0.88, \; 43.35, 0.44, \; 1.0) \\
38 \; \ast & (57.47, \; 9.99, \; 35.18, 0.15, \; 0.37, \; 30.96, \; 0.05, \; 0.31) \\
56 \; \ast & (15.25, \; 13.58, \; 3.82, \; 0.3, \; 0.8, \; 39.51, \; 0.43, \; 0.65) \\
31 \; \ast & (113.21, \; 18.25, \; 38.21, \; 0.62, \; 0.46, \; 15.78, \; 0.49, \; 0.61)
\end{array}
$$

Figure 11.21 Parameter collection or "recipe" for all types of particles participating in a particular swarm of Sayama's swarm chemistry. Sequence of parameters are: Number of particles of a type, $R^i, V_n^i, V_m^i, c_1^i, c_2^i, c_3^i, c_4^i, c_5^i$.

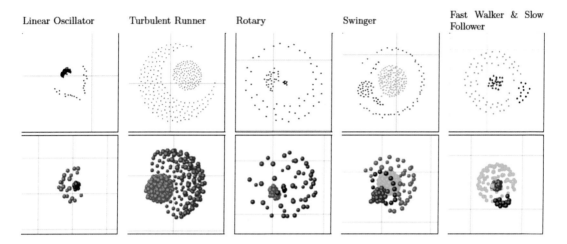

| Linear Oscillator | Turbulent Runner | Rotary | Swinger | Fast Walker & Slow Follower |

Figure 11.22 Different swarm chemistry behavior in 2D and 3D. With permission from the author.

phenomena allow the structuring and possibly morphogenetic patterning of particle types. Figure 11.22 shows a few examples of behavior of a 2D and a 3D swarm chemistry [748,749]. In [259] a particles swarm with two types of particles has been shown to perform cell division-like behavior.

11.4.3 Flow Artificial Chemistry

In the previous subsection we discussed molecules possessing self-propelling features. However, motion can also be realized through an external stirring mechanism subjecting passive molecules to move. Flow artificial chemistries aim at this variant of realizing local interactions together with movement of molecules. For example, transport processes could be produced by pumps, or by other mechanical means of producing particular movement of a set of molecules. In addition, there could be molecular processes whose raison d'être is to transport molecules along defined paths. From biology it is known that such processes exist and are driven by proteins like kinesin [385].

One simple idea is to provide vector fields of molecular flow (regardless of how such vector fields are generated) and place an artificial chemistry into such a vector field [474]. The authors provide an example of vector fields of motion, together with a simulation of the reactions that

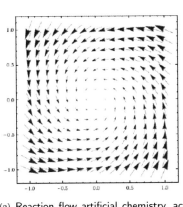

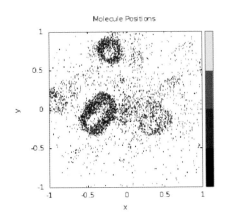

(a) Reaction flow artificial chemistry, according to [474]

(b) Reaction flow vector field produced by a swirl coded mathematically as $V_1(x, y) = \frac{1}{\sqrt{2}} \left((1 - \sqrt{2}) - y, x + (1 - \sqrt{2}) y \right)$.

Figure 11.23 Stochastic simulation of a chemistry with four molecular species, color coded, snapshot after 400 iterations. Clearly inhomogeneous distributions of molecules are visible. From [474], with permission from the authors.

take place in their artificial chemistry. A number of methodological challenges have to be overcome to be able to simulate such systems, using either partial differential equations or stochastic molecular simulation. Figure 11.23 shows an example: left, the flow field, right, the simulation of an artificial chemistry with four color-coded species of molecules.

11.5 Summary

The bio-inspired ACs discussed in this chapter are more complex artificial chemistries than the ones we have encountered earlier under their computational models. They have been designed to fulfill particular purposes connected to living systems and have been specialized and become elaborate. Nevertheless, we hope to have demonstrated that the same methods for ACs that we discussed earlier have been applicable here. There are many more bio-inspired AC models, but we were forced to make a selection. Readers interested to learn more about ACs inspired by biological systems might want to consult chapter 18 or the primary literature.

Part IV

Order Construction

CHAPTER

THE STRUCTURE OF ORGANIZATIONS

With Pietro Speroni di Fenizio

The term *organization* is widely used in the sciences and humanities, from the social sciences and economy to physics, chemistry, and computer science. In nearly all these areas, *organization* has a specific meaning, which is sufficient for qualitative statements on a system embedded in a specific context. Often further qualifications apply, such as in self-organization, an area of research that has become very active in the last decades [70]. However, once quantitative measurements are sought, an exact definition of organization is necessary.

This chapter discusses theoretical concepts of a framework that is useful to define "organization." We will use precise mathematical statements, intended to give a means for accurately describing organization in a large class of systems, independent of their constituting parts, be they molecules, stars of a galaxy, or living beings.

The concepts developed here are not entirely new. In the 1970s, Varela, Maturana, and Uribe [557, 881] investigated the basic organization of living systems, using the concept of autopoiesis, Greek for self-production. Their approach can be seen as one of the early conceptualizations of a theory of organizations of (artificial) living systems (see also [282, 437, 835]). Central to autopoiesis, as to later approaches to organization, is the notion of *self-maintenance*: A system consisting of components and their interactions must be able to produce those components and their interactions itself. It is only then that it is able to sustain itself in a dynamical equilibrium. We will later see how this can be used as the basis for a mathematical definition.

In the words of Maturana [557]: "A living system is properly characterized only as a network of processes of production of components that is continuously, and recursively, generated and realized as a concrete entity in the physical space, by the interactions of the same components that it produces as such a network."

Another root of the studies of organizations lies at the very center of the dichotomy between quantitative and qualitative studies and, in a sense, can be characterized as the search for a quantitative analysis of qualitative changes. Fontana and Buss have done substantial research in this direction by working on systems that produce elements and components which are different from those producing them [282, 284]. These systems have been termed "constructive" dynamical systems to set them apart from other dynamical systems, whose components are known and present in the systems from the start.

Dynamical systems are often modeled as systems of differential equations. These differential equations are supposed to capture the time development ("dynamics") of the system under consideration. One arrives at these differential equations by abstracting from the material that constitutes the observed system, and represents observable quantities by the variables governing a set of differential equations, one equation per observable. For instance, the content of a chemical reaction vessel could be represented by concentration variables, changing with the development of the chemical reactions over time.

Studies of dynamical systems then consist of examining how the components of a system interact with each other and produce the dynamical behaviors we observe for their components. A more basic question, however, is usually not addressed: Whether components are present in a system at all, and what happens when they disappear (interactions cease) or when they appear the first time (new interactions start to happen).

One might argue that an interest in the quantitative description of a system requires that components are observed to be present in the first place, which is a qualitative observation. Otherwise, one would certainly not have the idea to measure their time development. However, there is a broad class of systems which show qualitative changes in that components appear or disappear.

In nature, for example, we often face the unexpected. A system that produces new components, which subsequently start to interact with the existing system's components producing totally unexpected results. Differential equations, in their traditional form, are unfit to describe such systems. If we are to write a differential equation system that spans all the possible elements that might be generated by a system ever, it would likely be too large to be solvable. The number of equations is even potentially infinite. Further, given that a new component would normally arise in very small numbers initially, the tool of differential equations with its assumption of describing possibly infinitely small quantities is not appropriate to capture the essentials of the process. The smallest number at which a new component could arise is the discrete number of one exemplar.

So besides changing quantitatively, constructive dynamical systems change qualitatively, they evolve. *What* constitutes the system and *which* components are present will be of main interest in such a study, rather than "only" the question of *how much* of a particular component is being present in the system at a given time. It will therefore be necessary to study the qualitative move of a system from one set of components to another set. Since each set of components will have its own rules of interaction, the system itself will evolve.

If we take regular dynamical systems as an inspiration, one of the first questions that can be posited is, whether a system has an equilibrium, or fixed point. This is defined as a point in state space where there are no changes any more to the variables of the differential equations. In a similar vain, one might ask whether constructive dynamical systems have equivalent, let us call them *qualitative fixed points*, points in other words, that are invariant with respect to time. As we move from quantitative studies to more general qualitative studies we are interested in which particular sets of components do not change qualitatively with time.

We can now see the parallels between the work of Maturana and Varela and the work of Fontana and Buss. In both approaches a key element is stability in the face of an ever changing context. This stability is achieved by a continuous production of key components, self-maintenance, and a closure that will assure that the system does not disperse. While the latter authors consider logical or symbolic entities, the former consider material or physical entities. In nature we tend to see both of them, but physical closure, a boundary that separates an organism from its environment, is never complete. There always is some form of interaction, be it in the form of energy or molecules that cross the boundary. On the other hand, logical closure has in general been seen as an all-or-nothing quality.

In the next section we shall describe the general idea of a fixed point for a constructive dynamical system. From now on we will call such fixed points *organizations*. We will come to understand that the two properties of *closure* and *self-maintenance* are needed to cause the sort of stability in the dynamics of interactions that generate an organization and that both, closure and self-maintenance, are crucial notions to understand the structure and dynamics of organizations in general, independent of their particularities.

12.1 Basic Definitions

In this chapter we will only concentrate on the static structure of organizations and will neglect their dynamics. We shall tie these concepts to notions of chemistry, as these seem most appropriate to talk about organization. Let us now introduce the concept of an *algebraic chemistry* which we understand as an artificial chemistry without any dynamics. The molecules of chemistry have here been replaced by elements of a set, and we have to discern elements mentioned in this context from the notion of elements traditionally employed in chemistry.

Definition 12.1.1 *(Algebraic Chemistry)*
Let S be a set of elements and R a reaction rule of arity n such that $R : S^n \to S \cup \{\emptyset\}$. We call $\mathbb{S} = \langle S, R \rangle$ an algebraic chemistry.

What does this mean? The elements $s_1, s_2 \in S$ are different types of (abstract) molecules that are allowed to react according to reaction rule R. The result of an application of this reaction rule is either a new element of the set S, or the empty set, which symbolizes that no reaction will take place between the molecules. The arity of the reaction rules describes how many molecules participate in the reaction. In many cases, the arity will be $n = 2$, and we imagine, as in previous chapters, two molecules s_1 and s_2 colliding and reacting to another element, say s_3. Again, we speak of an "elastic" collision if the empty set is the result of the reaction.

Usually, an artificial chemistry is simulated by setting up a population of molecules taken from S. Recall that in the simplest case — the well-stirred tank reactor — the population is a multiset P with elements from S. This means that there are possibly multiple copies of elements of S in P. Thus, P consists of, for instance, one copy of s_1, two copies of s_2, three copies of s_3 and so forth: $P = \{s_1, s_2, s_2, s_3, s_3, s_3, ...\}$. Here we have named each molecule by the name of its type. We can also just number the molecules in the multiset, which would read $P = \{p_1, p_2, p_3 p_4, p_5, p_6, ...\}$. Technically, we can produce a multiset P from S by sampling, with replacement.

For a simulation step the population P is updated by using the reaction rule[1] R. Normally what a simulation step does to P is to extract some of its elements (usually two, say s_1 and s_2), apply one to the other, and add the result to the multiset. Some elements are then eliminated to keep the multiset from becoming too large (called *dilution flow*). The specific "reactor" algorithm

---

[1] Note that sometimes the term "reaction rule" refers to a rule that is defined only for a small subset of S. In this case R is better defined by a collection of such "reaction rules."

can be different from one system to the other. Sometimes the extracted elements (s_1 and s_2) are eliminated from the multiset after they are extracted [830], sometimes not [282]. Sometimes the number of molecules in the population is kept constant [63, 792], sometimes it can vary [790]. Even the operation sometimes does not produce only one result but a whole multiset [391, 790]. All those differences, which make up the diversity of artificial chemistries, make them more difficult to define. For this reason we try to keep things as simple as possible in this chapter and concentrate on the operation R neglecting the dynamics. Thus the formal framework presented now allows a *static* description and analysis of an artificial chemistry.

The following definitions like *organization, closure,* and *self-maintenance* are formulated following [281, 283].

Definition 12.1.2 *(Organization)*
Let O be a set of elements as a subset of S, $O \subseteq S$. Then $\mathbb{O} = \langle O, R \rangle$ is an organization *iff $\mathbb{O}$ is closed and self-maintaining. That is, iff $\forall a_1, \ldots, a_n \in O, R(a_1, \ldots, a_n) = b$, with $b \in O$ (property of* closure) *and $\forall a \in O, \exists b_1, \ldots, b_n \in O$ such that $R(b_1, \ldots, b_n) = a$ (property of* self-maintenance).

To simplify matters the properties "closed" and "self-maintaining" are also used to characterize sets if it is clear what kind of reaction rule R is meant. So, we can also call a *set O* an organization with respect to a reaction rule R, if the properties of closure and self-maintenance of def. (12.1.2) hold.

Given a set of elements and a reaction rule we can, of course, consider the set of all possible organizations on this set. We will indicate this set as $\mathfrak{O}$:

$$\mathfrak{O} \equiv \{\mathbb{O} \subseteq \mathbb{S} : \mathbb{O} \text{ is closed and self-maintaining}\}. \tag{12.1}$$

Just as sets have subsets, we can consider suborganizations of organizations:

Definition 12.1.3 *(Suborganization of an organization)*
Let $\mathbb{A} = \langle A, R \rangle, \mathbb{B} = \langle B, R \rangle \in \mathfrak{O}$ be two organizations ($\mathbb{A}, \mathbb{B} \in \mathfrak{O}$), such that $A \subset B$ ($A \subseteq B$), then $\mathbb{A}$ is a **suborganization** *of $\mathbb{B}$. We will indicate this with $\mathbb{A} \lhd \mathbb{B}$ ($\mathbb{A} \unlhd \mathbb{B}$).*

A natural order among suborganizations can be defined. This will be important later on.

Definition 12.1.4 *(Maximal suborganization)*
Let $\mathbb{A} = \langle A, R \rangle, \mathbb{B} = \langle B, R \rangle \in \mathfrak{O}$ be two organizations ($\mathbb{A}, \mathbb{B} \in \mathfrak{O}$), with $\mathbb{A} \lhd \mathbb{B}$, and such that there is no suborganization, $\mathbb{C}$, with $\mathbb{A} \lhd \mathbb{C} \lhd \mathbb{B}$. Then $\mathbb{A}$ is called the **maximal suborganization** *of $\mathbb{B}$.*

We will now list some simple properties of closed and self-maintaining sets.

Proposition 1 *The intersection of two closed sets is still a closed set.*

Proposition 2 *The union of two self-maintaining sets is still a self-maintaining set.*

This is true not only if we deal with a finite number of sets but also if we deal with an infinite number of sets. Thus, given any set $T \subseteq S$ we can always define C the *smallest closed set that contains T* and M the *biggest self-maintaining set that is contained in T*. C will be defined as the intersection of all the closed sets that contain T. $C \equiv \{x \in S : \forall D \subseteq S, T \subseteq D, \text{ and } D \text{ closed},$ such that $x \in D\}$. M will be defined as the union of all the self-maintaining sets contained in T. $M \equiv \{x \in S : \exists D \subseteq T, \text{ with } D \text{ self-maintaining, such that } x \in D\}$

12.1.1   Organization Generated by a Self-Maintaining Set

Let A be a self-maintaining set. A can be used to uniquely define an organization O. How is this organization generated? We have to note that while each element of A must be generated by at least one interaction of elements of A (self-maintaining property), it is not excluded that some new elements might be generated by the interaction of other elements of A.

Let A^1 be a new set, all of whose elements are directly generated by A:

$$A^1 \equiv \{x \in S \text{ such that for every } a_1, \ldots, a_n \in A, R(a_1, \ldots, a_n) = x\} \tag{12.2}$$

Since A is self-maintaining, the new set can only be larger, $A^1 \supseteq A$. But then, since $A^1 \supseteq A$, A^1 is self-maintaining, too. If $A^1 \neq A$, which is the general case, we can go further and generate another set A^2, that now takes into account the interaction of new elements with each other and with the elements of A, to possibly produce elements not contained in A^1.

In general we can construct sets $A^m \supseteq A^{m-1}$ with

$$A^m \equiv \{x \in S \text{ such that for every } a_1, \ldots, a_n \in A^{m-1} R(a_1, \ldots, a_n) = x\} \tag{12.3}$$

If there exists an m such that $A^m = A^{m-1}$ then $A^m \equiv O$ is closed and an organization.[2] O is well defined, closed and self-maintaining, thus O is an organization. We will call $\mathbb{O} = \langle O, R \rangle$ the *organization generated by the self-maintaining set A*. Notably, O is the smallest closed set that contains A.

12.1.2   Organization Generated by a Closed Set

In a similar way, we can start from a closed set and arrive at an organization: Let A be a closed set, and let B be the largest self-maintaining subset of A, $B \subseteq A$. Then B is both closed and self-maintaining, and therefore, an organization. Thus, $\mathbb{O} = \langle B, R \rangle$ is an organization.

To see this, we need to simply assume this would not be the case, and find a contradiction. So, B is a self-maintaining set by definition. Let $a \in A \setminus B$. If $\exists b_1, \ldots, b_n \in B : R(b_1, \ldots, b_n) = a$ then $B \cup \{a\}$ would be a self-maintaining set in A, bigger than B, which is against our hypothesis of B being the largest self-maintaining set in A. Thus, there cannot be any $a \in A \setminus B$ generated by the elements in B, thus B is closed. We will call $\mathbb{O}$ the *organization generated by the closed set A*.

12.1.3   Organization Generated by a Set

Let A be a set, B the largest self-maintaining set in A, and C the smallest closed set containing A. Thus $B \subseteq A \subseteq C$. C is a closed set, thus it uniquely generates an organization. *Let D be the organization generated by C.* In general $B \subseteq D \subseteq C$. We will define D as the *organization generated by the set A*. Of course, if A is an organization itself then B, C and D are all equal to A. In the most general case $A \nsubseteq D$ nor $D \nsubseteq A$, i.e. there are elements in A that are not in D, and there are elements in D that are not in A. The next section will allow us a closer look at these reactions.

12.1.4   Shadow of an Organization

We have seen how, given any set S, we can associate an organization O. Of course a number of sets might be associated with the same organization. That is to say, that regardless of which set

---

[2] Even if there does not exist a finite n such that $A^m = A^{m-1}$ we could define O, as the union of all the A^i: $O \equiv \bigcup_{i \in} A^i$.

we start with, we shall always end up with the same organization. This leads to the definition of *shadow*. The shadow of an organization is the set of all the possible sets which generate that organization.

Definition 12.1.5 *(Shadow of an Organization)*
Let $\mathbb{S} = \langle S, R \rangle$ *be an algebraic chemistry with* $|S|$ *molecules and* $|R|$ *reaction rules, and* $\mathscr{P}(S)$ *the set of all possible sets of molecules. Suppose that* $\mathbb{O}$ *is an organization in* $\mathbb{S}$, *then we define* $\mathbb{S}_{\mathbb{O}} \equiv \{T \in \mathscr{P}(S) \mid T \sim O\}$. *We call* $\mathbb{S}_{\mathbb{O}}$ *the* shadow *of the organization.*

If we have more than one organization in a system, each one will have its own shadow, which will naturally divide the power set of all possible combinations of molecules into belonging to different shadows.

12.2 Generators

A further concept to understand in connection with organizations is the notion of a *generator*. A generator is an operator that formalizes (and generalizes) the preceding notions: Given a set of molecules, it returns another set of molecules with some specific characteristics. So, given a set S, and an operator "Generate Closed Set" which we shall abbreviate with G_C it holds that

$$G_C(S) = T \qquad (12.4)$$

where T is closed. So G_C symbolizes the process (which might go through a number of iterations) that we described above, whereby a set grows until it is finally closed. Note that G_C does not just associate to each set an arbitrary closed set, but will instead return the smallest closed set which contains S.

When looking at the generation of a closed set this way we can see immediately that

$$G_C(T) = G_C(G_C(S)) = T. \qquad (12.5)$$

In other words, we have reached a state where a further application of the operator does not change the set any more.

Vice versa, we can also formulate an operator for generating a self-maintaining set, given a set S. The operator "Generate Self Maintaining Set," abbreviated G_{SM}, will generate the largest self maintaining set contained in S.

$$G_{SM}(S) = U, \qquad (12.6)$$

with U self-maintaining. Again, applying G_{SM} a second time to the result of its first application, U, does not change the outcome any more:

$$G_{SM}(U) = G_{SM}(G_{SM}(S)) = U. \qquad (12.7)$$

We can see now that a set that is not closed will potentially build the other elements that can be built, thus expanding until it fits the smallest closed superset. Conversely, a set that is not self-maintaining will start to lose elements until it reaches the biggest self-maintaining set. In other words, if we start from a set C which is closed but not self-maintaining, it is natural to expect the system to shrink to its biggest self-maintaining subset. Indeed, it can be proven [791] (under the condition that there is an outflow) that $G_{SM}(C)$ will not just be self-maintaining, but also closed.

Thus, it will be an organization. On the other hand, if we start from a self-maintaining set T that is not closed, we can reasonably expect the system to grow to its smallest closed superset.

The opposite is also true: Given the self-maintaining set T, $G_C(T)$ is not only closed, but also still self-maintaining. Thus we can take a closed set and contract it to its biggest self-maintaining subset which is going to be closed (and thus an organization). And we can take a self-maintaining set and can expand it into its smallest closed superset which is going to be a self-maintaining set, an organization again.

But what will happen if we start from a set that is neither closed nor self-maintaining to generate an organization? Will the set shrink or will it expand? *A priori* we cannot know. One can either expand S to its smallest closed superset $C = G_C(S)$, and then contract it again, by applying the operator G_{SM}. This will generate an organization we call O_1:

$$O_1 = G_{SM}(G_C(S)). \tag{12.8}$$

Or, one could apply the operators in the opposite order, first shrinking to a self-maintaining set, which subsequently gets expanded into a closed set. This way, we obtain another organization, we call O_2

$$O_2 = G_C(G_{SM}(S)). \tag{12.9}$$

O_1 is generally not the same as O_2, in fact it is easy to show that $O_2 \subseteq O_1$.

Generally we shall define an operator $G_O(S)$ that generates an organization as

$$G_O(S) := G_{SM}(G_C(S)). \tag{12.10}$$

We choose this alternative because $G_{SM}(G_C(S))$ tends to produce a bigger organization than its counterpart which results in giving us more information about the possible evolution of a system.

So we know that given any set S, we can always generate an organization by applying the generator G_O. Here again, applying G_O repeatedly, will not change the outcome in terms of the set.

$$G_O(S) := O \tag{12.11}$$

$$G_O(G_O(S)) = O. \tag{12.12}$$

This is what we mean when we say that an organization is a *qualitative fixed point.* It is qualitative, since we look only at the static behavior of the set, not quantitative behavior, something we shall discuss later.

12.3 Bringing Order into Organizations

We have now learned that using a different sequence of the application of generators, we might end up with different organizations. If we start with subsets of the set originally considered, or supersets, we also can reasonably expect different outcomes on the level of resulting organizations. Thus, it makes sense to look at families of organizations, and to try to order them in some systematic way.

12.3.1   Union and Intersection of Two Organizations

First, we need to define what we mean by the union or intersection of two organizations:

Definition 12.3.1 *(union of organizations)*
Let $\mathbb{A} = \langle A, R \rangle, \mathbb{B} = \langle B, R \rangle \in \mathfrak{O}$ *be two organizations. We will define the* union of organizations *as the organization generated by the union of the two sets A and B ($A \cup B$). We will indicate this organization as* $\mathbb{A} \sqcup \mathbb{B}$.

Definition 12.3.2 *(intersection of organizations)*
Let $\mathbb{A} = \langle A, R \rangle, \mathbb{B} = \langle B, R \rangle \in \mathfrak{O}$ *be two organizations. We will define the* intersection of organizations *as the organization generated by the intersection of the two sets A and B ($A \cap B$). We will indicate this organization as* $\mathbb{A} \sqcap \mathbb{B}$.

Note that the term "generated" is important here and should be understood as previously defined. In general, given two organization $\mathbb{A} = \langle A, R \rangle, \mathbb{B} = \langle B, R \rangle \in \mathfrak{O}$ we shall have

$$\mathbb{A} \sqcap \mathbb{B} \subseteq A \cap B \subseteq A, B \subseteq A \cup B \subseteq \mathbb{A} \sqcup \mathbb{B} \tag{12.13}$$

and

$$\mathbb{A} \sqcap \mathbb{B} \trianglelefteq A, B \trianglelefteq \mathbb{A} \sqcup \mathbb{B}. \tag{12.14}$$

12.3.2   Lattice of Organizations

Organizations in an algebraic chemistry nicely relate one to another fitting into a well-known algebraic structure called a *lattice*. The standard definition of a lattice [732] is:

Definition 12.3.3 *(lattice)*
Let L be a set of elements, where two binary operations are defined by $\cup$ *and* $\cap$*. Let those operations be such that* $\forall x, y, z \in L$ *the following properties are held:*

$$
\begin{array}{ll}
x \cup y = y \cup x & \textit{(commutative law on } \cup\textit{)}, \\
x \cap y = y \cap x & \textit{(commutative law on } \cap\textit{)}, \\
x \cup (y \cup z) = (x \cup y) \cup z & \textit{(associative law on } \cup\textit{)}, \\
x \cap (y \cap z) = (x \cap y) \cap z & \textit{(associative law on } \cap\textit{)}, \\
x \cap (x \cup y) = x & \textit{(absorptive law)}, \\
x \cup (x \cap y) = x & \textit{(absorptive law)}.
\end{array}
\tag{12.15}
$$

Then $< L, \cup, \cap >$ *is a* lattice.

According to this standard definition, a lattice is a partially ordered set in which any two elements have a greatest lower bound and a least upper bound. This means that while two elements are not comparable directly, they are comparable via these extreme bounds. For our purposes here, we can identify the variables x, y as different organizations $\mathbb{X}$ and $\mathbb{Y}$, while the operations on them can be identified with the union $\sqcup$ and intersection $\sqcap$ of organizations defined earlier. Thus we can see that organizations form a lattice.

12.4 Novelty and Innovation

Let us recap again what we said about the terms "closure" and "self-maintenance." *Closure* means that every element produced through reactions between a set of objects is in the set of objects, i.e., no element produced will fall outside of the set. *Self-maintenance*, on the other hand, states that every object that is in the set is produced by some or other of the possible reactions between elements of the set.

Novelty, therefore, has a natural definition. Namely, we can say that whatever is produced, that is not in the present set, is new. Recall the definition of the set A^1 in equation 12.2. If $A^1 \neq A$, something we assumed to be generally the case, then A^1 will contain at least one element x that is not part of A, i.e.

$$\{x \in A^1 : x \notin A, x = R(a_1, \dots, a_n) \text{ with} \{a_1, \dots, a_n\} \in A\} \tag{12.16}$$

We can say that x has come about by a reaction of elements in A, but hasn't been in A, and therefore belongs to the difference set $N \equiv A^1 \setminus A$. We call x, or more generally, the set N, a *novelty* with respect to A. Once it appears in set A^1, it will contribute to reactions and might generate further new elements which will then be contained in A^2, as we have discussed. Note that what we said here does not depend on A being a self-maintaining set. It is true for a general set of elements, except for closed sets. We always can define the set of novel elements N according to eq. (12.16), provided A is not closed.

What happens in the case of a closed set A? In that case, by definition, new elements cannot possibly be generated using our reaction system underlying the interaction of elements in A. In other words, for $A \equiv C$, with C being a closed set, $A^1 \setminus A \equiv N = \emptyset$. Thus, closed sets do not allow for novelty. This conclusion can be transferred to organizations that we have learned are closed sets, too. To make matters worse, even for sets which are not closed and self-maintaining, if a novelty is generated, it does not necessarily have to survive for ever. If, for example, a novelty is part of a self-maintaining set, it may disappear again, at least in the systems we consider here — reactive flow systems. This calls for more detailed considerations of the dynamics of the reactions.

Despite the general statement, that novelty cannot be produced by an organization, there are possibilities to consider. For example, if the inflow to the reactive flow system suddenly contains a new element that never before was present, that might cause novelty to be generated. This way, however, the novelty is caused by an outside event. Other outside events might be "mutations," i.e., that an element in the system spontaneously transforms itself into another element. Again, an outside event will have to happen to enable such a transition. But suppose it happens. Then the lattice of organizations gives us a good way to think about novelty. We need to think of it in terms of movements in the lattice. If we move up in the lattice to a more encompassing set, say from a set A^1 to a set A^2, we will observe novelty in the form of the difference set $N = A^2 \setminus A^1$. Note that we have made use here of the ordering feature of the lattice, assuming that $A^2 \supset A^1$, so the difference is "positive." Vice versa, if we move downward in the lattice, that is from an organization encompassing a set $A^3 \supset A^4$ where A^4 has less elements than A^3, we have lost some elements.[3]

Now, a movement in the lattice is not possible really if we stay in the static world of an algebraic chemistry. Since the only fact that counts there is the formal relation between sets, and how this comes about by interactions between elements. However, the lattice structure is just a

---

[3]Interestingly, we have no word for this loss of an element of a system, a kind of negative novelty, except perhaps the word "extinction," borrowed from ecology.

static picture of what is really happening in these kinds of systems: In reality, they are dynamic, there are multiple copies of many (or all) elements in the set, in fact, we have to describe the system with a multiset. In the course of the dynamics of reactions, then, it can indeed happen that a movement occurs from one organization to another, always via the links in the lattice. Movements down to smaller organizations will happen if there are not enough individual elements of a particular type of element to withstand the constant inflow/outflow of the system. In the next chapter we will consider the dynamics of reaction systems and how the dynamics influences the outcome over the long run, and therefore determines the fate of organizations.

Before we go there, however, it is necessary to discuss a topic closely related to novelty, which is frequently mixed conceptually with that term: *innovation.* Innovation requires novelty, but it is not the same as novelty. We have to think of it this way: A system might produce many novelties but only a few of these really will end up being considered innovations. Thus, the term *innovation* refers to novelties that are "selected" and become permanent fixtures of a system. We can think of them as those novelties that will subsequently be produced by the system in sufficient numbers (again, we argue here from the dynamic point of view), so as to not disappear again.

We can now understand why innovations are so much more visible than novelties. They really are a mass phenomenon. If there is enough of a particular novelty staying in the system, we recognize it, and say that the system has changed and has incorporated an innovation. If a novelty, however, is only a blip, and due to unfortunate circumstances it disappears again quickly in the course of the dynamic evolution of the system, it is not measurably present in the system. Sometimes, however, a certain novelty is necessary to cause another novelty to become an innovation. In that sense, it is like a scaffolding that disappears again after the system has moved on. If one knows the reaction laws of the system, one can indirectly infer the presence of this novelty, but it remains invisible to the eye of the observer.

12.5 Examples of the Statics of Organizations

Here we will apply the mathematical concepts described in previous sections. Three different artificial chemistries are investigated by looking at their reaction network and displaying the lattice of organizations. As we will see, each chemistry possesses a characteristic lattice.

12.5.1 Modulo Chemistry

Given two prime numbers p_1, p_2 and their product $m = p_1 p_2$ the *modulo chemistry* is defined as follows: We have as *molecules* the natural numbers smaller than a maximum number m, including zero. $S = \{0, 1, ..., m - 1\}$. The *reaction product* s_3 is the sum of s_1 and s_2 module m: $s_3 = s_1 + s_2 \bmod m$.

Table 12.1 shows the reaction table for this type of system with $p_1 = 3, p_2 = 5$, and the resulting lattice of organizations. Figure 12.1 shows the simple lattice structure resulting. There are only two nontrivial suborganizations with more than one element, $\{0, 5, 10\}$ and $\{0, 3, 6, 9, 12\}$, as can be read off the reaction table. This particular configuration stems from the fact that the divisor m is the product of two prime numbers.

Table 12.1 Reaction table of a module chemistry with $m = 15$.

operator s_1	operand string s_2													
	0	1	2	3	4	5	6	7	8	9	10	11	12	13	14
0	0	1	2	3	4	5	6	7	8	9	10	11	12	13	14
1	1	2	3	4	5	6	7	8	9	10	11	12	13	14	0
2	2	3	4	5	6	7	8	9	10	11	12	13	14	0	1
3	3	4	5	6	7	8	9	10	11	12	13	14	0	1	2
4	4	5	6	7	8	9	10	11	12	13	14	0	1	2	3
5	5	6	7	8	9	10	11	12	13	14	0	1	2	3	4
6	6	7	8	9	10	11	12	13	14	0	1	2	3	4	5
7	7	8	9	10	11	12	13	14	0	1	2	3	4	5	6
8	8	9	10	11	12	13	14	0	1	2	3	4	5	6	7
9	9	10	11	12	13	14	0	1	2	3	4	5	6	7	8
10	10	11	12	13	14	0	1	2	3	4	5	6	7	8	9
11	11	12	13	14	0	1	2	3	4	5	6	7	8	9	10
12	12	13	14	0	1	2	3	4	5	6	7	8	9	10	11
13	13	14	0	1	2	3	4	5	6	7	8	9	10	11	12
14	14	0	1	2	3	4	5	6	7	8	9	10	11	12	13

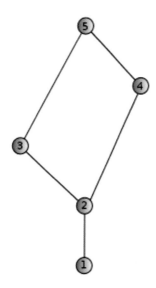

Figure 12.1 Lattice of organizations of modulo chemistry with $m = 15$. (1) The empty organization: $O_1 = \{\}$. (2) Organization containing only the self-replicator: $O_2 = \{0\}$. (3) Organization containing all valid numbers that can be divided by 5: $O_3 = \{0, 5, 10\}$. (4) Organization of all valid numbers that can be divided by 3: $O_4 = \{0, 3, 6, 9, 12\}$. (5) Organization of all molecules: $O_5 = S$.

Table 12.2 Reaction table of a 4-bit matrix-multiplication chemistry (nontopological horizontal folding). A "0" indicates an elastic collision. This means that there is no reaction product defined so that in case of a collision the molecules do not react.

operator s_1	operand string s_2															
	0	1	2	3	4	5	6	7	8	9	10	11	12	13	14	15
0	0	0	0	0	0	0	0	0	0	0	0	0	0	0	0	0
1	0	1	0	1	4	5	4	5	0	1	0	1	4	5	4	5
2	0	0	1	1	0	0	1	1	4	4	5	5	4	4	5	5
3	0	1	1	1	4	5	5	5	4	5	5	5	4	5	5	5
4	0	2	0	2	8	10	8	10	0	2	0	2	8	10	8	10
5	0	3	0	3	12	15	12	15	0	3	0	3	12	15	12	15
6	0	2	1	3	8	10	9	11	4	6	5	7	12	14	13	15
7	0	3	1	3	12	15	13	15	4	7	5	7	12	15	13	15
8	0	0	2	2	0	0	2	2	8	8	10	10	8	8	10	10
9	0	1	2	3	4	5	6	7	8	9	10	11	12	13	14	15
10	0	0	3	3	0	0	3	3	12	12	15	15	12	12	15	15
11	0	1	3	3	4	5	7	7	12	13	15	15	12	13	15	15
12	0	2	2	2	8	10	10	10	8	10	10	10	8	10	10	10
13	0	3	2	3	12	15	14	15	8	11	10	11	12	15	14	15
14	0	2	3	3	8	10	11	11	12	14	15	15	12	14	15	15
15	0	3	3	3	12	15	15	15	12	15	15	15	12	15	15	15

12.5.2   4-Bit Matrix Chemistry

The matrix chemistry that we already have seen in chapter 3 will serve us as a second example. Recall that molecules are binary strings that react with each other through matrix multiplication. An example of this system, with 15 strings, results in the reactions of table 12.2. It shows the reaction table of a 4-bit matrix reaction using the so called nontopological horizontal folding. More details of this system can be found in [63, 64].

The resulting lattice of organizations (figure 12.2) contains a surprisingly large number of organizations and is relatively complex compared to the number of molecules constituting the artificial chemistry. Indeed, the number of strings is 15, the number of closed sets is 70+, and the number of organizations is 54.

12.5.3   Random Reaction Table

Finally we take a look at an artificial chemistry which is randomly generated. Randomness does not refer to the interaction process; the reaction table is deterministic, but what is produced by which reaction is drawn with a random number generator. The *molecules* in our random artificial chemistry are natural numbers. $S = \{1, 2, \ldots, 15\}$. The *reaction table* is generated by drawing for each entry a number from S randomly (not shown here).

Figure 12.3 shows a typical lattice of organizations resulting from such a process. It contains only three organizations: All three organizations are trivial in the sense that they contain either none, only one, or all elements of the set S. Frequently, only two organizations can be encountered in such systems, the empty organization and the organization containing all elements. A larger number of organizations appears with rapidly decreasing probability.

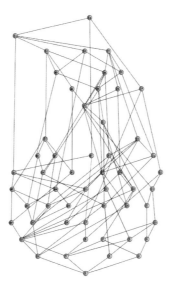

Figure 12.2 Lattice of organizations of the 4-bit matrix-multiplication chemistry shown in table 12.2.

Figure 12.3 Typical lattice of organizations of a random reaction table with 15 molecular types. The random chemistry contains three organizations. (1) The empty organization, (2) one self- replicator, and (3) the organization of all molecules.

12.6 How to Calculate Closed and Self-Maintaining Sets

There are a number of ways in which we can calculate closed and self-maintaining sets. For the sake of clarity, we shall focus on the simplest methods here. One can find more efficient and elaborate methods in the literature.

The most straightforward way to calculate a closed set is to start from a set of molecular types, which we shall denote here $s^{(1)}, ..., s^{(n_0)}$, with n_0 the number of species we start with. The algorithm needs to exhaustively check all combinations of types, running each combination of existing molecular types through the reaction matrix,[4] to find out whether there are novel types produced by interactions. If novelty is found, we denote the new species by $s^{(n_0+1)}, ..., s^{(n_1)}$. Next, these novel molecular types need to be added to the original set of species, and the test for novelty has to be repeated. Suppose in the second test, another set of novel types is discovered, $s^{(n_1+1)}, ..., s^{(n_2)}$. Then we again will have to add the new types and repeat the process, until finally we end up with a set where no novel species can be discovered any more. At that point, we can say that we have found a closed set. This entire process was summarized in the last section as the application of the generating operator G_C onto the initial set of species.

The following segment shows the pseudocode for a simple closure algorithm:

Input: Set of species $I = \{s^{(1)}, ..., s^{(n_0)}\}$

Output: Set of species $C = \{s^{(1)}, ..., s^{(n_k)}\}$ that is closed

```
while ¬terminate do
C ← I;
    foreach reaction of species in C
      if product(reaction) ∉ C
      I ← C∩product
      fi
    end
      if I = C
      terminate
      od
```

It is clear that the search process will end always when we have exhausted all possible different molecular types. So the full set of all molecular types in our reaction system will always be the closed set if none other is discovered before. Effectively, nothing new can be discovered then. However, in implicit AC systems, which are in many ways more interesting from the point of view of novelty and evolution, we don't know the full set of molecular types, and it can be a challenge to find a closed set for certain combinations of initial molecules.

We should keep in mind that with this sort of search we only have determined the closed set stemming from one particular starting condition (the initial set). If we are interested in determining all closed sets of a reaction system, we have to repeat this operation for all different combinations of starting conditions, i.e., for all different combinations of initial molecular types. One can easily imagine that the number of runs is considerable due to the combinatorics of the problem. To be exhaustive, 2^n combinations have to be checked, if n is the number of all possible molecular types.

---

[4] If the reaction matrix is already calculated, otherwise, the matrix will have to be gradually filled "on the go".

Note that there are ways to shortcut calculations, since any intermediate set of molecules that turns up during a closure calculation will have been checked already, and can be excluded from a stand-alone check. One shortcut that can be taken is not to calculate a closed set from every initial condition, but simply to check whether a combination of species is a closed set or not. If we are only interested to list all closed sets accessible from a reaction system, we can apply this alternative strategy. This will save us computations, though we have to go through each initial combination with an exponential effort. So this will also not be feasible for larger sets.

Another possibility is that one might be satisfied with a stochastic answer to the question of closed sets, in the sense that not all combinations are checked, but only a predetermined number. Each of these checks will be done on a random selection of species and this way while we might not end up with a complete list of all closed sets, we will have captured a good part.

If we aim to find organizations, we will have to subject, in a second step, all these closed sets — regardless of how we calculated them in the first place — to a self-maintenance test. That is to say, an algorithm tests whether each of the species in the closed set is also produced by one of the reactions in the closed set. If this is the case, the set is both, closed and self-maintaining. However, in the overwhelming majority of cases, most of these sets will not be self-maintaining, and thus they can be discarded. Only a few will usually pass this second test and therefore stand the criteria for an organization.

The following segment shows the pseudocode for the self-maintenance algorithm, that starts from one particular set $I = \{s^{(1)}, ..., s^{(n_0)}\}$:
Input: Set of species $I = \{s^{(1)}, ..., s^{(n_0)}\}$
Output: States whether I is self-maintaining

```
SM ← I;
   foreach reaction of species in SM
      if product(reaction) ∉ SM
      Test not passed!
      terminate
      fi
   end
   Test passed, set is self-maintaining
```

This test has to be repeated for all closed sets, and only those sets that pass the test will be listed in the list of organizations.

A cleverer method to approach the problem of finding the organizations of a reaction system is to go the other way around: First, seek out all self-maintaining sets, and then use this, usually smaller, number of sets and check them for closure. While in the exhaustive case, a number of 2^m different combinations has to be checked, with m being the number of produced string types under consideration, the resulting number of different sets will be smaller from the outset because $m < n$ or even $m << n$. This is advantageous, since the second pass through these sets involves now expanding the sets to their next closed set. Many unnecessary calculations will have been avoided if we do it this way. Note, however, that we might not find all organizations that we found with the former method. So while the method is way faster, it could overlook a small subset of organizations.

As an example, our previously discussed 4-bit matrix chemistry, which we depicted with its Hasse diagram in figure 12.2 results in table 12.3, if one of the algorithms for calculating organizations is applied. We can see how quickly the combinatorics of organizations can get out of

Table 12.3 List of all organizations of the 4-bit matrix chemistry. Number 0 is the empty set, number 51 is the full set of all elements that naturally close the Hasse diagram. All others will strongly depend on the details of interactions.

ID	Number of Species	Species Set
0	0	{}
1	1	{1}
2	1	{8}
3	1	{9}
4	1	{15}
5	2	{1, 8}
6	2	{1, 9}
7	2	{6, 9}
8	2	{8, 9}
9	2	{9, 15}
10	3	{6, 9, 15}
11	3	{7, 9, 15}
12	3	{9, 14, 15}
13	3	{11, 13, 15}
14	3	{1, 8, 9}
15	4	{1, 2, 4, 8}
16	4	{1, 3, 5, 15}
17	4	{7, 11, 13, 15}
18	4	{8, 10, 12, 15}
19	4	{9, 11, 13, 15}
20	4	{11, 13, 14, 15}
21	5	{1, 2, 4, 8, 9}
22	5	{1, 3, 5, 9, 15}
23	5	{7, 11, 13, 14, 15}
24	5	{7, 9, 11, 13, 15}
25	5	{8, 9, 10, 12, 15}
26	5	{9, 11, 13, 14, 15}
27	6	{1, 2, 4, 6, 8, 9}
28	6	{1, 3, 5, 7, 9, 15}
29	6	{1, 3, 5, 11, 13, 15}
30	6	{7, 9, 11, 13, 14, 15}
31	6	{8, 9, 10, 12, 14, 15}
32	6	{8, 10, 11, 12, 13, 15}
33	7	{1, 3, 5, 7, 11, 13, 15}
34	7	{1, 3, 5, 9, 11, 13, 15}
35	7	{6, 7, 9, 11, 13, 14, 15}
36	7	{8, 9, 10, 11, 12, 13, 15}
37	7	{8, 10, 11, 12, 13, 14, 15}
38	8	{1, 3, 5, 7, 9, 11, 13, 15}
39	8	{8, 9, 10, 11, 12, 13, 14, 15}
40	9	{1, 2, 3, 4, 5, 8, 10, 12, 15}
41	10	{1, 2, 3, 4, 5, 8, 9, 10, 12, 15}
42	11	{1, 2, 3, 4, 5, 6, 8, 9, 10, 12, 15}
43	11	{1, 2, 3, 4, 5, 7, 8, 9, 10, 12, 15}
44	11	{1, 2, 3, 4, 5, 8, 9, 10, 12, 14, 15}
45	11	{1, 2, 3, 4, 5, 8, 10, 11, 12, 13, 15}
46	12	{1, 2, 3, 4, 5, 8, 9, 10, 11, 12, 13, 15}
47	13	{1, 2, 3, 4, 5, 7, 8, 10, 11, 12, 13, 14, 15}
48	13	{1, 2, 3, 4, 5, 7, 8, 9, 10, 11, 12, 13, 15}
49	13	{1, 2, 3, 4, 5, 8, 9, 10, 11, 12, 13, 14, 15}
50	14	{1, 2, 3, 4, 5, 7, 8, 9, 10, 11, 12, 13, 14, 15}
51	15	{1, 2, 3, 4, 5, 6, 7, 8, 9, 10, 11, 12, 13, 14, 15}

hand. It bears repeating, though, that a stochastic approach can be chosen where only a predetermined number of producer sets are checked for fulfilling our conditions.

12.7 Summary

This concludes our discussion of the (static) structure of organizations. We have seen that there are two key features of sets that determine whether they can be considered organizations or not, namely closure and self-maintenance. These criteria can be tested if a reaction matrix is available that determines how elements of the set react with other elements of the set. In many cases, the implicit nature of reaction rules will not allow the formulation of an explicit reaction matrix. Rather, we will need to calculate the reaction matrix "on the go", with its elements filled in over the course of the expansion of the species set. Those types of systems are quite interesting in their own right, and we shall study them in more detail in the chapter on constructive systems.

Next, we shall learn that the organizations that can be found by the structural calculations and tests discussed here are a superset of what is really interesting, namely, dynamically stable organizations. This will form the content of the next chapter.

THE DYNAMICS OF ORGANIZATIONS

I n the last chapter we learned what could constitute an organization and how we might, starting from a reaction table, derive the potential organizations of a reaction system. However,
the situation is a bit more complicated than we dared to explain there, in that a few more
conditions need to be fulfilled for a system to actually possess those structures that we call organizations.

Essentially, we have to consider not only the static connections in the reaction graph but also
the flow of material/energy/information through the graph as a very important criterion when
examining organizations. This is what we shall discuss in the current chapter. We will come
to understand that different organizations might have different stability attached to them, with
unstable organizations in a reaction system never able to form for an extended period of time,
whereas stable organizations having a chance to emerge and stay around a long time before they
sometimes might disappear again.

We first have to turn our attention to flows.

13.1 Flows, Stoichiometry and Kinetic Constants

Suppose, a reactive system consists of two molecular species, A and B, which produce each
other, but can also decay quantified by $k_1, \ldots, k_4$:

$$
\begin{aligned}
r_1 : \quad & A \xrightarrow{k_1} B \\
r_2 : \quad & B \xrightarrow{k_2} A \\
r_3 : \quad & A \xrightarrow{k_3} . \\
r_4 : \quad & B \xrightarrow{k_4} .
\end{aligned}
\tag{13.1}
$$

One could think of a system of radioactive materials that would somehow produce each other, but, by a small amount, slowly decay and disappear out of the reaction system (becoming other molecules inert to any kind of reaction).

We can easily see what happens if such a system is allowed to react: Initial amounts of A and B would react into each other, while at the same time slowly "leaking" out of the system until all material would have disappeared and the reactions came to a standstill for lack of material. There is a reaction sink, where material simply disappears from this system. If we applied our previous notions of closure and self-maintenance, would the reaction system be closed and self-maintaining? Self-maintaining, in a logical sense — yes, because each of the participating molecular species, A, and B, are produced by some reactions, but closed — not if we consider the decay of some of the molecules as a leak out of the system. Only if we included the empty set as part of the system can we still argue that the system is an organization. In the literature, this is done [791] which leads to somewhat strange conclusions. For example, the empty set has to be counted as an organization and, sure enough, this empty organization is part of any organizational lattice one could derive from every reaction system. Still, here the system does not have a stable nonempty equilibrium state, because A, B will disappear. Thus, only the trivial empty organization is stable.

So we have to tighten our definition of organization and include in the definition a reference to flows. We shall tie that to the self-maintenance property of a reaction set A, and, in addition to requesting that any molecule within the set A needs to be produced by a reaction in the set

$$A \equiv \{\forall x \in S : \exists a_1, \ldots, a_n \in A, \text{such that } R(a_1, \ldots, a_n) = x\}, \tag{13.2}$$

we shall request that the flux vector (the vector of reaction rates) of reactions within the set A is positive, i.e. $v_R > 0$ and that the production rate of a molecular species i is positive or at least 0. The latter condition can be derived from the stoichiometric matrix, $\mathbf{M}$, that we have first met in chapter 2, in equation 2.21.

With this more tightened definition, $\{A, B\}$ in the reaction system 13.1 is not an organization any more. We can see that by calculating the flux vectors of the preceding equation system:

$$\vec{v} = \begin{pmatrix} v_1 \\ v_2 \\ v_3 \\ v_4 \end{pmatrix} = \begin{pmatrix} k_1 a \\ k_2 b \\ k_3 a \\ k_4 b \end{pmatrix} \tag{13.3}$$

and then see, whether the production rate of molecules A, B can be ≥ 0. The stoichiometric matrix reads

$$\mathbf{M} = \begin{pmatrix} -1 & 1 & -1 & 0 \\ 1 & -1 & 0 & -1 \end{pmatrix}. \tag{13.4}$$

We now recall equation 2.21, where the left hand side is often written as a production rate vector $\vec{f}$:

$$\frac{d\vec{s}}{dt} = \mathbf{M}\vec{v} = \vec{f} \tag{13.5}$$

In our example,

$$\vec{f} = \begin{pmatrix} -(k_1 + k_3)a + k_2 b \\ k_1 a - (k_2 + k_4)b \end{pmatrix}, \tag{13.6}$$

and we can easily see that these flows cannot be both ≥ 0, given that concentrations a, b and kinetic constants $k_1, \ldots, k_4$ must be larger than 0.

However, a slightly varied system

$$r_1: \quad A \xrightarrow{k_1} 2B \tag{13.7}$$

$$r_2: \quad B \xrightarrow{k_2} 2A \tag{13.8}$$

$$r_3: \quad A \xrightarrow{k_3} . \tag{13.9}$$

$$r_4: \quad B \xrightarrow{k_4} . \tag{13.10}$$

can have all production rates at least 0, depending on the choice of kinetic constants, so that the set $\{A, B\}$ can now be said to be self-maintaining under these circumstances, and therefore an organization (again, counting the empty set as par of this organization).

To test our new notion of self-maintenance, let's now consider a slightly more complicated case, where the disappearance of molecules is not so obvious. Suppose we have three different reaction species, $A, B,$ and C and the following reaction system would be considered:

$$r_1: \quad A \xrightarrow{k_1} B \tag{13.11}$$

$$r_2: \quad B \xrightarrow{k_2} A \tag{13.12}$$

$$r_3: \quad A \xrightarrow{k_3} C. \tag{13.13}$$

What would happen here? Would A be an organization? No, because it is not closed, neither would B, but C would be closed. Would either pair $\{A, B\}$, $\{A, C\}$ or $\{B, C\}$ be an organization? No again, because none of them would be closed. Finally, $\{A, B, C\}$ would be closed, but would it be self-maintaining? From a logical point of view, yes: Each element can be produced from within the set. So the system would form an organization. But here, we again find the same problem, that we encountered earlier — there is a more or less speedy reduction of initial amounts of A and B into C, with C being inert. Because of this reaction into an inert species, the end result of this reaction dynamics will be a maximum amount of C, under any starting conditions, which means that $\{A, B, C\}$ is not a stable organization. This can be seen again from the calculation of the flux vector and the stoichiometric matrix. As before, arbitrary flux vectors will not be able to generate production rates ≥ 0:

$$\vec{f} = \begin{pmatrix} -1 & 1 & -1 \\ 1 & -1 & 0 \\ 0 & 0 & 1 \end{pmatrix} \times \vec{v} \tag{13.14}$$

with at least one component of $\vec{f}$ smaller than 0, $f_l < 0$.

So, clearly, looking at the statics of reaction graphs is not sufficient to determine the fate of organizations. We also have to look at the flow of molecules, restrict the types of reactions allowed for the definition of self-maintenance, and consider the stability of organizations.

13.2 Examples of the Dynamics of Organization

In this section we will look at a number of examples of dynamic behavior of an artificial chemistry. As model system we again employ the matrix chemistry we have already encountered in chapter 3. Here, we shall us a $N = 9$ matrix chemistry, with the possibility to study 511 different types of strings. We shall tie the discussion back to what we have seen previously in this system, but also learn something new. Organization theory will help us to understand the phenomena we are going to see.

	1	2	3	4	5	6	7	8	9	10	11	12	13	14	15	16	17	18	19	20	21	22	23	24	25	26	27
1	1	0	1	0	1	0	1	8	9	8	9	8	9	8	9	0	1	0	1	0	1	0	1	8	9	8	9
2	0	1	1	0	0	1	1	0	0	1	1	0	0	1	1	8	8	9	9	8	8	9	9	8	8	9	9
3	1	1	1	0	1	1	1	8	9	9	9	8	9	9	9	8	9	9	9	8	9	9	9	8	9	9	9
4	0	0	0	1	1	1	1	0	0	0	0	1	1	1	1	0	0	0	0	1	1	1	1	0	0	0	0
5	1	0	1	1	1	1	1	8	9	8	9	9	9	9	9	0	1	0	1	1	1	1	1	8	9	8	9
6	0	1	1	1	1	1	1	0	0	1	1	1	1	1	1	8	8	9	9	9	9	9	9	8	8	9	9
7	1	1	1	1	1	1	1	8	9	9	9	9	9	9	9	8	9	9	9	9	9	9	9	8	9	9	9
8	2	0	2	0	2	0	2	16	18	16	18	16	18	16	18	0	2	0	2	0	2	0	2	16	18	16	18
9	3	0	3	0	3	0	3	24	27	24	27	24	27	24	27	0	3	0	3	0	3	0	3	24	27	24	27
10	2	1	3	0	2	1	3	16	18	17	19	16	18	17	19	8	10	9	11	8	10	9	11	24	26	25	27
11	3	1	3	0	3	1	3	24	27	25	27	24	27	25	27	8	11	9	11	8	11	9	11	24	27	25	27
12	2	0	2	1	3	1	3	16	18	16	18	17	19	17	19	0	2	0	2	1	3	1	3	16	18	16	18
13	3	0	3	1	3	1	3	24	27	24	27	25	27	25	27	0	3	0	3	1	3	1	3	24	27	24	27
14	2	1	3	1	3	1	3	16	18	17	19	17	19	17	19	8	10	9	11	9	11	9	11	24	26	25	27
15	3	1	3	1	3	1	3	24	27	25	27	25	27	25	27	8	11	9	11	9	11	9	11	24	27	25	27
16	0	2	2	0	0	2	2	0	0	2	2	0	0	2	2	16	16	18	18	16	16	18	18	16	16	18	18
17	1	2	3	0	1	2	3	8	9	10	11	8	9	10	11	16	17	18	19	16	17	18	19	24	25	26	27
18	0	3	3	0	0	3	3	0	0	3	3	0	0	3	3	24	24	27	27	24	24	27	27	24	24	27	27
19	1	3	3	0	1	3	3	8	9	11	11	8	9	11	11	24	25	27	27	24	25	27	27	24	25	27	27
20	0	2	2	1	1	3	3	0	0	2	2	1	1	3	3	16	16	18	18	17	17	19	19	16	16	18	18
21	1	2	3	1	1	3	3	8	9	10	11	9	9	11	11	16	17	18	19	17	17	19	19	24	25	26	27
22	0	3	3	1	1	3	3	0	0	3	3	1	1	3	3	24	24	27	27	25	25	27	27	24	24	27	27
23	1	3	3	1	1	3	3	8	9	11	11	9	9	11	11	24	25	27	27	25	25	27	27	24	25	27	27
24	2	2	2	0	2	2	2	16	18	18	18	16	18	18	18	16	18	18	18	16	18	18	18	16	18	18	18
25	3	2	3	0	3	2	3	24	27	26	27	24	27	26	27	16	19	18	19	16	19	18	19	24	27	26	27
26	2	3	3	0	2	3	3	16	18	19	19	16	18	19	19	24	26	27	27	24	26	27	27	24	26	27	27
27	3	3	3	0	3	3	3	24	27	27	27	24	27	27	27	24	27	27	27	24	27	27	27	24	27	27	27

Table 13.1 Extract from reaction table of a 9-bit matrix-multiplication chemistry. A "0" indicates an elastic collision. This means that there is no reaction product defined so that in case of a collision the molecules do not react.

13.2.1 Focusing into an Organization

The first example will demonstrate the phenomenon of focusing from a larger set of types into an organization. Our system is the same as that discussed in chapter 3. Strings are "folded" into matrices, and react with other strings to produce new strings. A population of 1000 string molecules, or the corresponding differential equations (eqs. 3.26—3.27) are used to simulate the behavior of this system.

Figure 13.1 shows the behavior of the deterministically simulated differential equations, if we start the simulation using about equal initial concentrations for $s^{(j)}$, $j = 2,...., 7$. In Figure 13.1 we can see barely a difference between the concentrations of these six species. However, none of them can survive very long, as they all react to produce $s^{(1)}$. A look at the reaction table (see table 13.1) teaches us that in this "boring" part of the reaction table not much is happening,

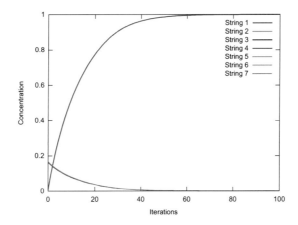

Figure 13.1 System $N = 9$. ODE simulation, starting from a virtually equal distribution of strings $s^{(2)}...s^{(7)}$. Additional string sort $s^{(1)}$ appears and takes over the entire population, focusing all material into this self-replicator.

except a continuous transformation of all initially present sorts into species $s^{(1)}$, which is itself able to self-replicate.

This behavior is typical for reaction systems, it means that the flow of material is channeled into only a few (or in this case one) species, namely those that form an organization. We can speak of a material focusing of the system. It is as if $s^{(2)}, ..., s^{(7)}$ are all raw materials being transformed into something more useful and stable for this system. After some time, only a small subset of the initial distribution of molecules is left over, or, as in this particular case, a subset of the closed set produced by the initial distribution that didn't even exist at the beginning of the simulation but was produced by reactions.

Here, some of the equations for the types involved, that is for $i, j \in \{1, ..., 7\}$ can be read off from table 13.1 as:

$$s^{(i)} + s^{(j)} \longrightarrow s^{(i)} + s^{(j)} + s^{(1)}. \tag{13.15}$$

For other encounters among these molecules collision are elastic, symbolized by a 0 in the reaction table. As a result of these reactions, the only valid organization, besides the empty set, is

$$\mathbb{A} = \langle A, R \rangle \text{ with } A = \{s^{(1)}\} \tag{13.16}$$

and R symbolizing the reactions defined in our matrix chemistry.

We used a deterministic simulation with ODEs, which will always produce the smallest closed superset of the initial distribution of types, before falling back to the largest self-maintaining subset of this superset. Thus, the organization produced by this sort of simulation will be what we termed O_1 after eq. 12.10. That is not necessarily the case for an explicit molecular simulation where the finiteness of the reaction vessel might lead to the disappearance first of certain sorts of molecules and thus ultimately to other results.

Before we look at a case of contingency which this latter behavior embodies, we want to inspect a second example of organization forming in the matrix chemistry described above. Figure 2 shows the same system, now with a different initial condition, all strings being concentrated in type $s^{(12)}$.

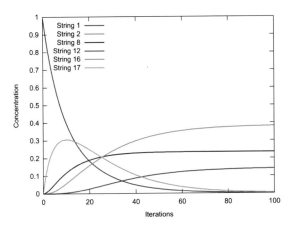

Figure 13.2 System $N = 9$. ODE simulation, starting from string type $s^{(12)}$. Additional string sorts $s^{(1)}, s^{(2)}, s^{(8)}, s^{(16)}, s^{(17)}$ appear, with $s^{(17)}$ an intermediate growing in concentration and then decaying again, as it is not part of the organization that takes over the population in the end. Concentration curve of $s^{(2)}$ is hidden behind that of $s^{(2)}$.

We can see a steep decline in the concentration of $s^{(12)}$ because the reaction with itself produces another string type ($s^{(17)}$) that subsequently gives way to the organization consisting of string types $s^{(1)}, s^{(2)}, s^{(8)}, s^{(16)}$, because according to table 13.1,

$$s^{(17)} + s^{(17)} \longrightarrow s^{(17)} + s^{(17)} + s^{(17)} \tag{13.17}$$

$$s^{(17)} + s^{(12)} \longrightarrow s^{(17)} + s^{(12)} + s^{(8)} \tag{13.18}$$

$$s^{(12)} + s^{(17)} \longrightarrow s^{(12)} + s^{(17)} + s^{(2)} \tag{13.19}$$

$$s^{(8)} + s^{(12)} \longrightarrow s^{(8)} + s^{(12)} + s^{(16)}. \tag{13.20}$$

The story goes as follows: As $s^{(12)}$ reacts with itself initially, it produces $s^{(17)}$ and steeply rises as $s^{(12)}$ declines. Now, $s^{(17)}$ is a self-replicator, which explains its slow fading after being produced. However, as it becomes more frequent and reacts with $s^{(12)}$ a new string appears, $s^{(8)}$, which in turn produces $s^{(16)}$. $s^{(2)}$ appears through another pathway at the same speed as $s^{(8)}$ (these reactions are nonsymmetrical) which helps to dampen and ultimately to reverse the growth of $s^{(17)}$ and to stabilize the emerging organization. Again, we can see that first a closed set is formed, visible around iteration 20, before the organization takes over the reaction vessel completely around iteration 100. Note that $s^{(17)}$ is a self-replicator, but it reacts with other strings as well which ultimately is its undoing.

Generally speaking, different starting conditions (as well as different flows into the reactor, which we haven't discussed in detail), will result in the system to focus its material into different organizations.

13.2.2   A Contingency Thought Experiment

In this example we shall look at another phenomenon that frequently occurs in artificial chemistries. We can best see this by a very simple experiment that we can do with a small AC consisting of two self-replicators.

The system is defined by the following reactions:

$$s^{(1)} + s^{(1)} \longrightarrow s^{(1)} + s^{(1)} + s^{(1)} \tag{13.21}$$

$$s^{(2)} + s^{(2)} \longrightarrow s^{(2)} + s^{(2)} + s^{(2)}. \tag{13.22}$$

All other reactions are elastic. This is a competitive system where an advantage in any of the species could trigger a decision to either of the two sides. Suppose we start an experiment where we set up a reactor with exactly the same number of $s^{(1)}$ and $s^{(2)}$, say 500 copies of each in a pool of 1000 exemplars. What will happen? Due to the stochastic character of collisions, we will have random encounters, producing fluctuations in the concentration of species. However, the situation is highly volatile and, in time, one of the two species will have a large enough advantage to become resistant against fluctuations and win the competition. We cannot predict from the outset which one of the two species will be the long-term winner of this dynamics, it has to play itself out. Figure 13.3 shows a sample simulation of this experiment.

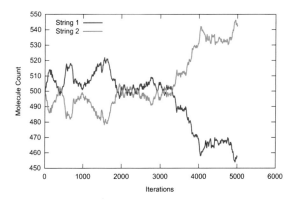

Figure 13.3 A reaction system with two self-replicators. A population of 500 copies of each competes for the reactor. Around iteration 3,500 the competition is decided.

Even under conditions where there is a slight advantage for one or the other of these species, the decision is open, though with a bias to the side that had the original advantage. What this teaches us is that it really depends on all the successive reactions happening in the reactor which self-replicator[1] finally dominates. In other words, we have to run the simulation to know which one will win. Now suppose this is part of a larger system, maybe a system of spatially distributed reactors, that are allowed to interact with each other, perhaps on a slower time-scale. Their interaction might lead to different outcomes, depending on which of the two self-replicators has won. Then a major qualitative difference will result in the overall system from this contingency of collision order.

Modeling the system with (deterministic) differential equations would take away these dependencies and not produce results commensurate with a simulation of single collisions. The existing contingencies in the dynamics of our artificial chemistry are hidden by an ODE approach that again shows the limits of deterministic modeling and the strength of event-by-event simulation.

---

[1]A self-replicator stands for an organization that could, in a larger system, be a larger organization.

13.3   Observing Organizations

Now that we have observed different dynamics, it is time to look at descriptive techniques. We have already seen in the last chapter when we discussed the structure of organizations, that a Hasse diagram could be used to describe the relationship between different organizations in a reaction system. The Hasse diagram depicts the network of relationships between potential focal points of dynamics, which in a dynamical system could be attractors of the dynamics.

However, if we want to gain a better understanding of what is going on during a dynamical reaction simulation we need to be able to watch the presence (or absence) of organizations. To that end, each organization gets assigned a special name. This was already implicitly referred to when we first depicted a Hasse diagram in figure 12.2. There, we had each organization, regardless how many species it had, carry a unique ID; see also table 12.3. We can use this ID as an index and treat each of the organizations as a species-equivalent entity whose presence or absence in the reactor vessel can be observed in the same way as one would measure the concentration of a species. Note that some of the species themselves are organizations, so they naturally have two names, a species name and an organization ID.

Clearly, once an organization (id) has taken over the reactor vessel, its organization has a concentration or strength of $c_{id} = 1.0$. But how can we observe organizations in this way? The first step is to know what a certain organization looks like in terms of concentrations of species. This knowledge is not always available, but if it is, we can make use of that knowledge.

Suppose for the sake of an example, we are interested in the organization of the example of figure 13.2. By using the stoichiometric coefficients often we can calculate the concentration vector corresponding to the particular organization resulting. Note that by simply watching the system we can also measure these concentrations. Let's give this organization an identity, say we call it o_1. In terms of species concentrations, $\vec{o}_1$ is a vector that possesses as many dimensions as there are or can be species in the system and we can write

$$\vec{o}_i = \begin{pmatrix} c_1 \\ c_2 \\ ... \\ c_N \end{pmatrix} \times \frac{1}{\sqrt{c_1^2 + c_2^2 + ... + c_N^2}}, \tag{13.23}$$

where N species can exist in the system in principle. The enabling trick to watch organizations emerge is that we need to realize that we have to normalize this vector $\vec{o}_i$ such that

$$\vec{o}_i \cdot \vec{o}_i = 1. \tag{13.24}$$

This is done by the normalization factor in equations 13.23, which is abbreviated as $\|\vec{c}\|$. We can then use all the resulting vectors $\vec{o}_i$ with $i = 1, ..., K$ assuming we know of K organizations, to watch which one will emerge from the reaction dynamics $\vec{x}(t)$. We only have to also normalize this vector, by dividing it by $\|\vec{x}(t)\|$

$$\vec{x_N}(t) = \frac{\vec{x}(t)}{\|\vec{x}(t)\|} \tag{13.25}$$

at the time of observation, and then multiply this normalized concentration vector with each of the organization vectors:

$$o_i(t) = \vec{x_N}(t) \cdot \vec{o}_i, \tag{13.26}$$

where o_i now refers to the strength of organization $\vec{o}_i(t)$ being present in the reaction vessel.

Let's consider an example: Figure 13.4 shows the dynamics of a 2x2 matrix chemistry, starting from a nearly equal distribution of $s^{(1)}$ and $s^{(8)}$. We know already from the previous discussion that one of the two species will prevail. So the figure in this particular simulation shows what we call organization o_8 as being the strongest at the outset of the simulation.

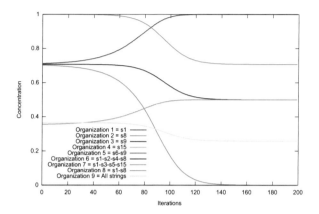

Figure 13.4 Organization prevailing after dynamics has run its course.

This does not astonish, as it is a nearly equal distribution of $s^{(1)}$ and $s^{(8)}$ that makes up the reaction vessel. Some other organizations are present, too, like $s^{(1)} - s^{(2)} - s^{(4)} - s^{(8)}$ (by virtue of some of its members being present) as well as $s^{(1)} - s^{(3)} - s^{(5)} - s^{(15)}$. After a number of iterations of the dynamics, o_1, consisting of $s^{(1)}$ emerges as the winner, while o_2, consisting of $s^{(8)}$ disappears. This in turn has an impact on the strength of other organizations, such as o_6 or o_7 as well. In summary, we can see one organization emerging as the dominant one in the reaction vessel.

There is one last point to understand about the observation of organizations: Some organizations have a near-infinite number of different expressions in terms of concentrations, because they are dependent on initial conditions. Thus, different initial conditions will result in the same species participating, but with different concentrations. In such cases, it is not possible to define organization vectors in a unique way. One could say that the entire system acts like a memory for the initial conditions. While it requires a delicate balance of stoichiometric factors, such systems might be particularly useful for storing information. The very least one can say is that the amount of contingency is increased substantially in those systems.

13.4   Probabilistic Notions of Closure and Self-Maintenance

Until now, we have considered the notions of closure and self-maintenance as mathematical properties that are crisply defined. Either a set was closed or it was not, and either a set was self-maintaining or it was not. However, since we are dealing with physical, chemical, and biological systems that exhibit these properties, the question should be posed whether these notions must be perfectly fulfilled to allow for the class of phenomena we are seeing.

How about if closure and self-maintenance are developed into probabilistic notions? Thus, if we would have only "approximate" closure and "approximate" self-maintenance available, what would that entail?

If the elements of a set are all known, we can examine the set and the interactions of its elements, and clearly identify all closed sets. However, if interactions are defined implicitly, and new elements can appear, the matter is different. There might be a very small chance that something new is produced in one of the many possible reactions of elements. Suddenly it becomes a matter of probability whether something outside the existing set of elements is produced.

This clearly calls for an examination whether our current notions of closure and self-maintenance, which are very rigid in the sense of all-or-nothing cannot be defined more softly so as to allow to classify cases which are in the probabilistic regime. It turns out that even with fully knowing the set of elements and the exact interaction outcomes, we can look at a probabilistic definition of closure and self-maintenance.

We shall start here with a consideration of the well-defined case. Let $S = \{a, ..., z\}$ be a set where we define reactions between elements using a reaction matrix that lists the educts as columns and rows, and the products as entries in the table. We then select a closed subset $S_C \subset S$ with $S_C = \{k, ..., r\}$ and consider its reactions. By definition, these reactions will be producing nothing outside of S_C. Suppose now we select another subset of S, $S_A \subset S$ that partially overlaps with the subset S_C, $S_A = \{c, ..., l\}$ with $S_A \cap S_C \neq \varnothing$ and $S_A \neq S_C$. Can we ascribe a degree of "closure"?

Suppose elements of S_A have r reactions. These can be classified: m_C of r reactions will produce elements of S_A, and $n_C = r - m_C$ reactions will produce elements not in S_A. Thus we could define a degree of closure α_C as

$$0 \leq \alpha_C = \frac{m_C}{r} \leq 1. \tag{13.27}$$

The same considerations applies to the definition of self-maintenance. For any subset $S_B \subset S$ with $S_B = \{k, ..., x\}$ that partially overlaps another subset S_S that is self-maintaining, $S_B \cap S_S \neq \varnothing$, $S_B \neq S_S$, we can classify its reactions into m_S of a total of r reactions that produce an element of S_B, and $n_S = r - m_S$ reactions (out of r) that produce an element not in S_B. Again, the degree of self-maintenance could be defined as

$$0 \leq \alpha_S = \frac{m_S}{r} \leq 1. \tag{13.28}$$

Now that we have these new features, could we overcome the tendency of certain reaction systems (either to open or to decay) by stochastic removal or addition of species? E.g., suppose a species s would not be produced in the system. Could we, by adding it, conserve the degree of self-maintenance? s would have to be input to the system, i.e., "food."

On the other hand, if a species s were produced but would possibly lead the system away from its degree of closure, couldn't we, by removing it, secure the degree of closure of the system? Among other things, this naturally leads to the necessity of filters (and membranes).

Suppose now that we do not have an explicit set of elements $S = \{a, ..., z\}$, among which we could define S_C and S_B. Instead, we have rules for the interactions of members, and a start state from which the system develops. Under many circumstances, the set S will not be finite. What is more, we don't know whether an arbitrary subset S_C will be finite. We are not able to define closure and self-maintenance under those conditions. But can we define a degree of closure and a degree of self-maintenance?

Based on empirical evidence, perhaps we can! This would entail to observe the population of elements in the multiset and their development over time. For each step in time, we could then also determine the degree of closure and self-maintenance of the system. Suppose that a is our start population (it could be one species, or a number of species lumped together), and we see the following reactions taking place:

$$a + a \longrightarrow a + a + b \qquad (13.29)$$

$$a + b \longrightarrow a + b + a \qquad (13.30)$$

$$b + a \longrightarrow b + a + a \qquad (13.31)$$

First, we have 100% a which gradually decays and produces b. But as soon as there are more b, however, another reaction starts to happen,

$$b + b \longrightarrow b + b + c, \qquad (13.32)$$

and the set opens again. So we should be able to monitor, with the help of the observables defined earlier, how the system gradually moves toward a more stable state (with higher degrees of closure and self-maintenance).

13.5 Summary

This chapter should allow the reader to understand the important difference between the structure and dynamics of an organization. We have used the argument of flow of material/energy/ information to recognize this difference. We have then seen different examples of dynamics and explained one way to measure the emergence of organizations. Finally, we have discussed a more relaxed version of closure and self-maintenance, two crucial terms for defining organizations. These relaxed versions allow us to consider open systems as we find living systems that are open systems in the real world.

14

SELF-ORGANIZATION AND EMERGENT PHENOMENA

N ow that we have discussed the theory of organization to some degree, it is time to turn our attention to what has become a core concept of systems science: self-organization. In general terms, this refers to the ability of a class of systems to change their internal structure and/or their function in response to external circumstances. Elements of self-organizing systems are able to manipulate or organize other elements of the same system in a way that stabilizes either structure or function of the whole against external fluctuations. The process of self-organization is often achieved by growing the internal space-time complexity of a system and results in layered or hierarchical structures or behaviors. This process is understood not to be instructed from outside the system and is therefore called *self-organized*.

Related to self-organization (and sometimes mixed with it) is the concept of emergent phenomena, behaviors, structures, or properties of a system that are not entirely determined by the constituents making up that system. The term "phenomena" points to them being observed at a certain (higher) level, and the term "emergent" means that they arise out of the system, possibly by a process of self-organization, carrying novel properties not observed by looking at the system's lower level entities. "Emergent" should be seen in contrast to "resultant," which refers to properties at a higher level that are already observable at the lower level, like a mere summation or aggregation. The actual process by which emergent phenomena are produced is called "emergence" and has spawned a lively literature since it was first introduced into the philosophical literature in the 19th century [124, 510, 584].

According to De Wolf and Holvoet [219], self-organization and emergence "emphasize very different characteristics of a system's behavior,"[1] but they often appear together. One key characteristic of both is nonlinearity in interactions of systems components. We can say that both terms describe processes that happen in dynamical systems over time, and both focus on stability, i.e., temporal constancy, of certain aspects of its behavior. Whereas emergence studies higher-level patterns and their dependency on lower level components, self-organization stud-

---

[1] We, however, dispute their statement that "[they] can exist in isolation" as too strong.

ies these patterns in relation to outside influences (like the flow of energy or material). In other words, studies of emergence look at the inner workings of a system, whereas self-organization looks at environmental conditions and how they impact on the stability of system behavior. It is so difficult to keep these concepts apart because — generally speaking — the environment interacts with the system as a whole, which in turn impacts the interaction of its parts, intertwining both concepts deeply.

In this chapter, we shall discuss self-organization and emergence/emergent phenomena and translate them into the context of artificial chemistries. Let's start with a few examples of self-organization in natural and artificial systems.

14.1   Examples of Self-Organizing Systems

Classical examples of natural self-organizing systems are the formation of Benard convection cells in nonequilibrium thermodynamics, the generation of laser light in nonlinear optics and the Belousov-Zhabotinsky reaction in chemistry [203]. These are examples from the nonliving world, and the complexity of resulting macroscopic space-time patterns is restricted.

Nearly unrestricted complexity through self-organization can be achieved in the living world. For instance, the interaction of species in foodwebs could be looked at from this point of view [786]. Or, consider the idea of self-organization of the Earth's biosphere known as the Gaia hypothesis [524, 525]. This model states that Earth's living and nonliving components self-organize into a single entity called *Gaia*. Gaia can be understood as the whole of the biosphere, which is able to self-stabilize. The model states, in other words, that the biomass of Earth self-regulates to make conditions on the planet habitable for life. In this way, a sort of homeostasis would be sought by the self-organizing geophysical/physiological system of Earth. These ideas have been connected to the Darwinian theory of evolution via natural selection [508, 794], providing a mechanism by which such a stable state can be assumed to have emerged. Other examples of natural self-organizing systems can be found in the upper half of table 14.1.

There are numerous examples of man-made systems or systems involving humans that exhibit self-organization phenomena. Among them are traffic patterns, self-organizing neural networks, cellular phone networks, and the development of web communities.

Take, for example, traffic flow patterns: Macroscopic patterns of traffic jams on highways have been observed and experimentally examined [455]. Their appearance is closely related to traffic density, the model of behavior for drivers, and the traffic flow that this allows [860]. Traffic flow is an open system, and it develops waves of traffic jams (solitons) excited by the density of traffic. Transitions between different traffic flow patterns have been considered as phase transitions, typical products of self-organization in the nonliving world. A number of examples of self-organizing systems from different fields are given in table 14.1, lower section. The last column shows the phenomena emerging as system behaviors from the entities of these systems listed in the third column. Important to produce those patterns are the flows named in column 2. The provide the driving force to self-organization.

Artificial chemistries can also be classified among man-made self-organizing systems. In this case, what emerges and can be observed are organizations. In most of the examples we have seen in the previous chapters, however, there is not much of a flow of molecules into the systems. As a result, usually a single stable organization will be found, a somewhat unsurprising feature of systems that can be seen to seek a stable equilibrium. Things become more interesting, and organizations will become more complex, if we start to increase the flow of molecules into and out of ACs. At the current time, this field is not yet well developed in ACs, but this certainly is an area where more research is warranted.

14.2 Explanatory Concepts of Self-Organization

Despite half a century of inquiry, the theory of self-organization is still in its infancy. There is no "standard model" of SO, only various aspects emphasized by different researchers. Here we will discuss the most important of these.

14.2.1 Nonequilibrium Thermodynamics

Thermodynamics has been concerned with the notion of order and disorder in physical systems for more than a century. The theory of self-organization has to address fundamental issues of this field. The most important question in this regard is, how order can arise through self-organization.

Classical thermodynamics has focused on closed systems, i.e., systems isolated from external influence in the form of matter and energy flow. This allowed to understand the processes involved when a system evolves undisturbed. A key result of this inquiry is the second law of thermodynamics, originally formulated by Carnot and later refined by Clausius in the 19th century. It states that "any physical or chemical process under way in a system will always degrade the energy." Clausius introduced a quantitative measure of this irreversibility by defining entropy:

$$S \equiv \int dQ/T, \tag{14.1}$$

with Q the heat energy at a given temperature T. In any process of a closed system, entropy always rises

$$\frac{dS}{dt} \geq 0. \tag{14.2}$$

According to Eddington, 1928 [245] this universal increase in entropy "draws the arrow of time" in nature.

Boltzmann had reformulated entropy earlier in terms of the energy microstates of matter. In his notion, entropy is a measure of the number of different combinations of microstates in order to form a specific macrostate.

$$S = k_B \, ln(W), \tag{14.3}$$

with k_B Boltzmann's constant and W the thermodynamic probability of a macrostate. He argued that the macrostate with most microstates (with maximum entropy) would be most probable and would therefore develop in a closed system. This is the central tenet of equilibrium thermodynamics.

More interesting phenomena occur if the restrictions for isolation of a system are removed. Nicolis and Prigogine [611] have examined these systems of nonequilibrium thermodynamics, which allow energy and matter to flow across their boundary. Under those conditions, total entropy can be split into two terms, one characterizing internal processes of the system, $d_i S$ and one characterizing entropy flux across the border $d_e S$. In a generalization of the second law of

Table 14.1 Examples of self-organizing systems. Upper part of table: Systems from the natural world; lower part of table: Man-made systems.

System	Flow	Self-organizing entities	Emergent Phenomenon
Atmosphere	Solar energy	Gas molecules	Patterns of atmospheric circulation
Climate	Energy	Precipitation, temperature	Distribution patterns
Liquid between plates	Heat	Particle circulation	Convection patterns
Laser	Excitation energy	Phase of light waves	Phase-locked mode
Reaction vessel	Chemicals for BZ reaction	Chemical reactions	Patterns of reaction fronts
Neural networks	Information	Synapses	Connectivity patterns
Living cells	Nutrients	Metabolic reactions	Metabolic pathways / network patterns
Food webs	Organisms of different species	Species rank relation	Networks of species
Highway traffic	Vehicles	Distance of vehicles	Density waves of traffic
City	Goods, information	Human housing density	Settling patterns
Internet	Information packets	Connections between nodes	Network connection patterns
Web	User requests for website access	Links between websites	Web communities communities
Artificial Chemistries	Molecules	Reactions of molecules	Organizations

thermodynamics, Prigogine and Nicolis postulated the validity of the second law for the internal processes,

$$\frac{d_i S}{dt} \geq 0, \tag{14.4}$$

but explicitly emphasized that nothing can be said about the sign of the entropy flux. Specifically, it could carry a negative sign and it could be larger in size than the internal entropy production. Since the total entropy is the sum of both parts, the sign of the total entropy change of an open system could be negative,

$$\frac{dS}{dt} = \frac{d_i S}{dt} + \frac{d_e S}{dt} < 0, \tag{14.5}$$

a situation impossible in equilibrium thermodynamics. Thus, increasing order of the system considered would be possible through export of entropy. Self-organization of a system, i.e., the increase of order, would not contradict the second law of thermodynamics. Specifically, the nonequilibrium status of the system could be considered a source of order.

Even in the distance from thermodynamic equilibrium, however, certain stable states will occur: the *stationary* states. These states assume the form of *dissipative structures* if the system is far enough from thermodynamic equilibrium and dominated by nonlinear interactions. The preconditions for dissipative structures can be formulated as follows:

1. The system is open.

2. The inner dynamics is mainly nonlinear.

3. There are cooperative microscopic processes.

4. A sufficient distance from equilibrium is assumed, e.g., through flows exceeding critical parameter values.

5. Appropriate fluctuations appear.

If those conditions are fulfilled, the classical thermodynamic branch of stationary solutions becomes unstable and dissipative structures become stable system solutions.

Artificial chemistries could easily be fulfilling all these conditions. What is required is that there is a supercritical flow of molecules into the system, and that the simulation allows to capture stochastic effects necessary for fluctuations to develop. It would also be interesting to study the entropy and energy content of ACs and base their examination on concepts of nonequilibrium thermodynamics.

14.2.2 Synergetics

Prigogine's description of dissipative structures is formally limited to the neighborhood of equilibrium states. As Haken pointed out, this is a severe restriction on its application and in particular precludes its formal application to living systems. Instead, Haken proposed *order parameters* and the *slaving principle* as key concepts for systems far from equilibrium [358]. Let the time evolution of a continuous dynamical system is described by

$$\frac{dq}{dt} = N(q, \alpha) + F(t), \tag{14.6}$$

where $q(t) = [q_1(t), q_2(t), ...q_N(t)]$ is the system's state vector and N is the deterministic part of the system's interaction whereas F represent fluctuating forces, and α are the so-called control parameters. Then the stable and unstable parts of the solution can be separated by linear stability analysis, as can the time-dependent and time-independent parts. As a result, the solution can be written as

$$q(t) = q_0 + \sum_u \xi_u(t) v_u + \sum_s \xi_s(t) v_s. \tag{14.7}$$

v_u, v_s are the unstable and stable modes, respectively, and $\xi_u(t), \xi_s(t)$ are their amplitudes. These amplitudes obey the following equations

$$\frac{d\xi_u}{dt} = \lambda_u \xi_u + N_u(\xi_u, \xi_s) + F_u(t) \tag{14.8}$$

$$\frac{d\xi_s}{dt} = \lambda_s \xi_s + N_s(\xi_u, \xi_s) + F_s(t), \tag{14.9}$$

with λ_u, λ_s characterizing the linear part of the equations and function N_u, N_s summarizing the nonlinear deterministic components. The slaving principle formulated by Haken now allows to eliminate the stable mode development by expressing them as a function of unstable modes

$$\xi_s(t) = f_s[\xi_u(t), t]. \tag{14.10}$$

Thus, the unstable modes (order parameters) enslave the stable modes and determine the development of the system's dynamics. This result is useful both to describe *phase transitions* and *pattern formation* in systems far from equilibrium.

Synergetic concepts have been applied in a variety of disciplines [359]. In chapter 13 we have seen how organizations in ACs can be observed. From the viewpoint of synergetics, AC organizations could be modes v of the artificial chemistry system dynamics, and their strength could be the equivalent to amplitudes ξ.

14.2.3 Chaos and Complexity

The treatment of *chaotic systems* has been derived from nonlinear system theory. Chaotic systems are usually low-dimensional systems that are unpredictable, despite being deterministic. The phenomenon was originally discovered by the meteorologist E. Lorenz in 1963 [523], although H. Poincare in 1909 was aware already of the possibility of certain systems to be sensitive to initial conditions [668]. The reason for the difficulty in predicting their behavior stems from the fact that initially infinitesimal differences in trajectories can be amplified by nonlinear interactions in the system. These instabilities, together with the lack of methods for solving even one-dimensional nonlinear equations analytically, produce the difficulties for predictions. Modern theory of deterministic chaos came into being with the publication of a seminal article by May in 1976 [561].

Complex systems, on the other hand, have many degrees of freedom, mostly interacting in complicated ways — that is, they are high-dimensional. All the more astonishing is the fact that our world is not totally chaotic in the sense that nothing can be predicted with any degree of certainty. It became apparent, that chaotic behavior is but one of the ways nonlinear dynamical systems behave, with other modes being complex attractors of a different kind.

Complexity itself can be measured, notably there exist a number of complexity measures in computer science, but describing or measuring complexity is not enough to understand complex systems. Turning again to artificial chemistries, we can ask what might be proper measures of complexity for an AC, and what does it teach us about the behavior of such a system?

14.2.4 Self-Organized Criticality

For particular high-dimensional systems, Bak et al. [52] have suggested a dynamic system approach toward the formation of fractal structures, which are found to be widespread both in natural and artificial environments. Their canonical example was a pile of sand. They examined the size and frequency of avalanches under certain well-prepared conditions, notably that grains of sand would fall on the pile one by one. This is an open system with the forces of gravity and friction acting on the possibly small fluctuations that are caused by deviations in the hitting position of each grain of sand. He observed how the grains would increase the slope of the sand pile until more or less catastrophic avalanches developed.

Bak suggested the notion of *self-organized criticality* (SOC) as a key concept which states that large dissipative systems drive themselves to a critical state with a wide range of length and time scales. This idea provided a unifying framework for the large-scale behavior in systems with many degrees of freedom. It has been applied to a diverse set of phenomena, e.g., in economic dynamics and biological evolution. SOC serves as an explanation for many power-law distributions observed in natural, social and technical systems, like earthquakes, forest fires, evolutionary extinction events, and wars. As opposed to the widely studied low-dimensional chaotic systems, SOC systems have a large number of degrees of freedom, and still exhibit fractal structures as are found in the extended space-time systems in nature.

We can observe SOC in action in ACs, when, upon the introduction of a new type of molecule, the emergence of a whole new organization is caused through the interaction of that molecular type with other already present types and the results of their mutual interaction.

14.2.5 The Hypercycle

In a series of contributions since 1971, Eigen and Schuster have discussed particular chemical reaction systems responsible for the origin, self-organization and evolution of life [249–251]. By considering *autocatalytic sets* of reactions they arrived at the most simple form of organization, the *hypercycle*, which is able to explain certain aspects of the origin of life. They have considered a chemical reaction system composed of a variety of self-reproductive macromolecules and energy-rich monomers required to synthesize those macromolecules. The system is open and maintained in a nonequilibrium state by a continuous flux of energy-rich monomers. Under further assumptions they succeeded in deriving Darwinian selection processes at the molecular level. Eigen and Schuster have proposed rate equations to describe the system.

The simplest system realizing the previously mentioned conditions can be described by the following rate equations

$$\frac{dx_i}{dt} = (A_i Q_i - D_i)x_i + \sum_{k \neq i} w_{ik}x_k + \Phi_i(\vec{x}), \tag{14.11}$$

where i enumerates the individual self-reproducing units and x_i measures their respective concentrations. Metabolism is quantified by the formation and decomposition terms $A_i Q_i x_i$ and $D_i x_i$. The ability of the self-reproducing entities to mutate into each other is summarized by the quality factor for reproduction, Q_i, and the term $w_{ik}x_k$, which takes into account all catalytic productions of one sort using the other. A_i, D_i are rate constants for self-reproduction and decay respectively. The flow term Φ_i finally balances the production / destruction in this open system in order to achieve $\sum_k x_k = const.$

By introducing a new feature called excess production,

$$E_i \equiv A_i - D_i \qquad (14.12)$$

and its weighted average

$$\bar{E}(t) = \sum_k E_k x_k / \sum_k x_k \qquad (14.13)$$

and symbolizing the "intrinsic selection value" of a sort i by

$$W_{ii} = A_i Q_i - D_i, \qquad (14.14)$$

one arrives at reduced rate equations

$$\frac{dx_i}{dt} = (W_{ii} - \bar{E}) x_i + \sum_{k \neq i} w_{ik} x_k. \qquad (14.15)$$

These equations can be solved under certain simplifying assumptions and notably yield the concept of a *quasispecies* and the following extremum principle: A quasispecies y_i is a transformed self-replicating entity with the feature that it can be considered as a cloud of sorts x_i whose average or consensus sequence it is. The extremum principle reads: Darwinian selection in the system of quasispecies will favor the quasispecies that possesses the largest eigenvalue of the rate equation system above.

14.2.6   The Origin of Order

In the 1990s, Kauffman [447] pointed out one of the weaknesses of Darwinian theory of evolution by natural selection: It cannot explain the '*origin* of species' but rather only their subsequent development. Kauffman instead emphasized the tendency of nature to constrain developments along certain paths, due to restrictions in the type of interaction and the constraints of limited resources available to evolution. In particular, he upheld the view that processes of spontaneous order formation conspire with the Darwinian selection process to create the diversity and richness of life on Earth.

Previously, Kauffman had formulated and extensively studied [445] the NK fitness landscapes formed by random networks of N Boolean logic elements with K inputs each. Kauffman observed the existence of cyclic attractor states whose emergence depended on the relation between N and K, and the absolute value of K. In the case of large K ($K \approx N$), the landscape is very rugged, and behavior of the network appears stochastic. The state sequence is sensitive to minimal disturbances and to slight changes of the network. The attractor length is very large, $\approx N/2$, and there are many attractors. In the other extremal case, $K = 2$, the network is not very sensitive to disturbances. Changes of the network do not have strong and important consequences for the behavior of the system.

Kauffman proposed NK networks as a model of regulatory systems of living cells. He further developed the notion of a *canalizing function* that is a Boolean function in which at least one variable in at least one state can completely determine the output of the function. He proposed that canalizing functions are an essential part of regulatory genetic networks.

14.3 The Emergence of Phenomena

Interestingly, one of the first examples of emergent phenomena were those connected to chemical reactions. Two reactants with certain features come together and produce a third chemical with other, novel features, not contained in the reactants present earlier. These features are called emergent phenomena of the chemical reaction system. For example, there might be types of chemical bonds in the new molecule that did not exist in either of the atoms before the reaction. Thus, qualitatively new types of interactions came into being through the reaction which might convey some hitherto unobserved properties to the molecule.

It is important to keep apart epistemological aspects of emergent phenomena, those that have to do with their explanation, from ontological aspects, those that speak about the objects themselves. It is frequently said that emergence refers to phenomena unpredictable by looking at the lower-level entities, yet this language refers to epistemological aspects of emergence of not being able to explain the qualitatively new phenomena on the higher level. While these are a legitimate aspects, they need to be seen separately from ontological aspects [145], which refer to the occurrence of qualitative novelty in a system over time.

14.3.1 Observation

Observation of the properties and behaviors of objects is a key to grasping the concept of novelty. Whenever components of a system — first kept apart, then brought together — start to interact with each other, these interactions change the behavior of the whole.[2] While prior to them interacting, we have a mere collection of objects, once they start to interact and influence each others behavior, which is usually accompanied by energy exchange if material is involved, we have a system with its dependencies and potential emergent phenomena.

Now, in order to see whether there are new properties of the system as a whole, and whether these new properties change the properties of its constituents, we first have to measure them with some dedicated apparatus. Properties of objects do not exist in the empty space, they are the measurable quantities of observable qualities. Thus, both on the level of the constituent objects and the system as a whole it only makes sense to discuss qualities that are observable, and of quantities that are measurable. If, at the level of the system as a whole, new qualities "emerge," we speak of an emergent phenomenon. A change in the system has occurred, on the other hand, if we can measure certain quantities in the system as a whole to be different from the collection of the same measurable quantities in the constituent objects (the parts). In the latter case, we have measured quantities which are associated with qualities observable on both levels.

How would we be able to detect a new observable quality? It is here that the discussion of emergence often becomes tangled in a web of philosophical argument and counterargument. Qualitative differences are something we know when we see it, but we have a hard time defining it in the abstract. So let's take an example from physics: states of matter as associated with bulk properties. A solid is a state in which matter does not change its volume or shape depending on its environment (at least not easily detectable). However, when applying heat energy, a solid will sooner or later melt and transit into another state of matter, a liquid. The liquid state is defined by matter adapting its shape to the environment (its container) but maintaining its volume (at least in first approximation). In physics, one speaks of phases of matter and a phase transition at the point where energy suddenly changes the qualitative properties of matter. We are so used to

---

[2]We prefer the term "interaction" over the term "bonding" of Bunge [145].

Table 14.2 Hierarchy of subject levels

LEVEL	SUBJECT
8	Sociology/Economics/Politics
7	Psychology
6	Physiology
5	Cell Biology
4	Biochemistry
3	Chemistry
2	Atomic Physics
1	Particle Physics

this process in daily life that we don't think about it, but at a phase transition, a new property of matter can be seen emerging.

By describing matter with phases and the qualitative change with phase transitions, we have in no way explained how this qualitative change comes about. For that to be possible, we need to look at a lower level of matter, i.e., at its constituent parts (atoms and molecules) and how these parts interact with each other. Nevertheless, a qualitative difference in behavior can be observed at the bulk or macro level of organization of matter, which then calls for an explanation at the lower level of organization of matter. Here, as in many cases of phase transitions, two levels of description are used: in our example, the bulk macrolevel and the atomic microlevel. Changes in the behavior at the macrolevel are explained by changes in the behavior of constituents at the microlevel. Once we have found the causal effects of behavioral changes on the microlevel at the macrolevel, we usually are satisfied with the explanation.

This brings us to the question of levels and scale.

14.3.2 Scale

We have learned that we can make best sense of the world by partitioning it in layers of a hierarchy. Thus, in the area of the sciences, for example, a division of subjects has taken place that allows us to consider different subjects at their relevant respective level of resolution. Following Ellis [254] we can list a number of subjects in a hierarchy of levels as in table 14.2. These levels come about by us focusing on certain details of the world around us, and by trying to explain things at that level of detail by observation and theorizing.

Our attempt to observe and model reality, however, will at all times be bound by the means of our observational tools and the limitations of our intellectual power to explain. Thus, we assume these layers of entities to be characterized by interactions at the same level and isolation between levels. However, this is an approximation to a reality where the difference is less crisp, with strongly interacting entities within levels and weakly interacting entities between levels.

The hierarchy of layers is ordered by the scale we apply to observation. The body of work known under the term "hierarchy theory" addresses the issues of levels in connection with scale of observation [26] in a general complex systems context. Other terms often used in connection with scale, mainly in the area of observation and description of hierarchies, are "resolution" and "scope" [733]. Essentially, in order to make use of a certain scale while measuring the properties of entities, this scale needs to be applicable multiple times. However, neither very few nor very many multiples of scale should be required. In the former case, each individual entity takes on a more important role, whereas in the latter case, individual entities and their properties "dissolve" in the statistics of their features [676]. Thus, a reciprocal relationship exists between the scope[3] and resolution. The larger the scope of observation, the lower the resolution must be in order to arrive at a reasonable multiple of scale. Resolution characterizes the finest details that can be resolved through observation at that particular level.

To move from a more elementary level (at the bottom of the hierarchy) to a higher level in the hierarchy, we need to apply some coarse-graining techniques that allow us to lump together many smaller entities at the lower level into fewer larger entities entities at the higher level. This means we have to lower the resolution. Often this is done by some sort of averaging (coarse-graining) of the properties carried by a collection of lower-level entities. Moving away for a moment from physical systems, coarse-graining or averaging in social systems could be achieved by voting. In this case, we can picture individuals on a lower level belonging to groups on a higher level. Group formation and interaction is thus the interesting emergent phenomenon when moving to a higher level in a social system.

One of the consequences of this averaging is that the process of coarse-graining produces equivalence classes. That is to say, the individual lower-level entity has influence only on the behavior of its higher level counterpart insofar as it can determine a change in its behavior, which will often nonlinearly depend on the averages of properties of the lower level. In addition, there might be influences exerted from higher-level entities down to lower-level entities, known as top-down causation (more on this later).

In recent years, so-called multiscale methods have been developed [913] that aim at modeling systems at different scales simultaneously. If the world is indeed organized in hierarchies of layers, as we previously stated, then multiscale methods naturally should be expected to produce better models than single-scale methods. We can imagine multiscale methods to model a system at different scales, small scales in regions where high resolution is necessary to understand the system's behavior, coarser scales where less detailed models are sufficient. Thus, locally, the scope of a model is increased to deal with details that would otherwise disturb the relation between the model and the system whose behavior it tries to capture.

Figure 14.1, left, shows an example of a system where a clear hierarchy between two levels can be seen forming. Level I entities are nicely separated into different groups at level II, which in turn form groups on an even higher level, while themselves possibly formed from even lower level entities, with bottom-up or top-down information flow. This is, of course, a modelers dream system, with a clear-cut formation, as in figure 14.1, right, called a hierarchical or tree data structure in computer science.

Yet, reality is not as crisp and clear-cut. A more realistic example of what we find in the messy real world is depicted in figure 14.2, left. In this example, some entities of the lower level are part of more than one group/entity at the higher level, and therefore will influence the entities at a higher level in more than one way. Such a system has a more complicated structure; see for example figure 14.2, right. The information flow in such a system includes lateral exchange, by virtue of overlapping membership of individuals in different groups.

As well as overlapping memberships, there could be entities whose influence is felt at a level higher than the next level in the hierarchy, thus defying the simple averaging and coarse-graining procedure we described previously. Again, group formation in social systems is an example. Such a system is better described as the particular graph data structure of a network; see for example figure 14.2, right. A more general case would be a system consisting of entities connected

---

[3]Scope is used mostly in a spatial dimension, but sometimes also in formal or even a temporal dimension.

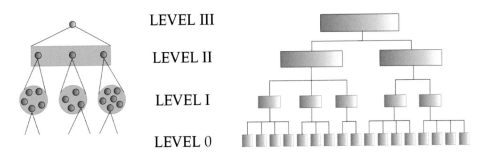

Figure 14.1 Hierarchical system with entities clearly separated into different groups. Left: Entities; Right: Structure.

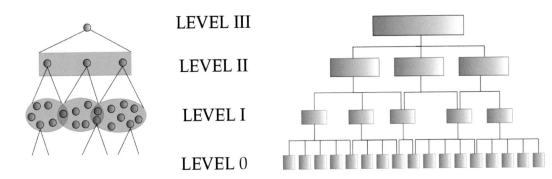

Figure 14.2 Hierarchical system with entities of overlapping group participation. Left: Entities; Right: Structure.

in arbitrarily connected networks of entities. The connection pattern, connectivity of nodes, and strength of connections in particular could then be used to construct partial hierarchies; see figure 14.3 a, b for an example.

A number of algorithms have been published in recent years that aim at the discovery of "community structure" in graphs, which would be useful to discover these hierarchical layers [183, 289]. In particular when a large number of nodes is involved (like in genetic networks), a probabilistic approach can be safely assumed. The hierarchy is only a reflection of the probability that two nodes are connected. It is clear that starting from a hierarchy like in figure 14.3 d, one could construct a network like in figure 14.3 c. However, even the backward route of identification of an approximate hierarchy can be gone, using a maximum likelihood method combined with a Monte Carlo sampling algorithm.

14.4 Explanatory Concepts of Emergence

The term "emergence" has a long history. A thorough discussion would not be possible within the context of this section, for entire books have been written about this subject [123].

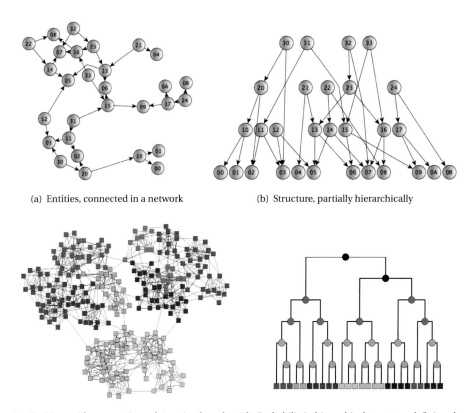

(a) Entities, connected in a network

(b) Structure, partially hierarchically

(c) Entities, with connections determined proba-
bilistically based on neighborhood.

(d) Probabilistic hierarchical structure defining the
neighborhood.

Figure 14.3 More general systems of hierarchization. (For c, d see [183], with permission from the author.)

Emergence has been variously defined. Here we would like to follow the notion of Mario Bunge, the famous Argentinian philosopher and physicist, who in the introduction to his book *Emergence and Convergence: Qualitative Novelty and the Unity of Knowledge* [145] states concisely: "The term 'emergence' refers to the origin of novelties." Emergence is the explanatory concept that is supposed to allow to explain novelties like molecules, life, mind, social norms, and the state, to name a few of the entities that beg for explanation with classical (reductionist) scientific methods. Bunge sees the relation between parts and wholes as key to understanding emergence.

14.4.1 Weak Emergence and Strong Emergence

Mark Bedau explains that what makes emergent phenomena so mind-boggling is the apparent metaphysical unacceptability of the assumption that they are both, dependent on underlying processes (at the micro level) and, at the same time, autonomous from underlying processes. In other words, instead of parts and wholes, Bedau discusses emergence in terms of levels (macro and micro), and states the problem of emergence to explain this apparent contradiction [89]. Bedau discerns nominal emergence, weak emergence and strong emergence. In the definition of these three types of emergence different aspects of the term are taken into account. The common feature of all three types is the notion that there are properties that can be possessed by entities on the macro level, but not by entities on the micro level. This sort of minimal common denominator he calls "nominal emergence." This term, being minimal, is not able to explain how and why emergence occurs, it merely describes the fact of macro properties that are not found on the micro level.

Macro properties come in different flavors, however, so Bunge restricts the realm of emergence to those properties of the macro level that are global, and not distributive. By that he means properties that are not simply the sum over all the elements, of which there could be many. Examples of relevant "global" properties mentioned by Bunge [145, p12] are, for instance, the synchrony of a neural assembly, the self-regulation of an organism or machine, the validity of an argument, the consistency of a theory, etc. The more traditional name for global properties among philosophers of emergence is emergent as opposed to "resultant" properties. Distributive properties as mentioned before, therefore, can be identified as resultant properties. Nominal emergence, according to Bedau, does not differentiate between global and distributive or emergent and resultant properties, which makes it a very broad definition.

What, now, are the stricter concepts of weak and strong emergence, according to Bedau? The most restrictive concept, strong emergence, adds the following further requirements: A strongly emergent property is *supervenient* and has *irreducible causal power*. Supervenience is a concept that explains the bottom-up formation of emergent properties. However, in addition to this feature, strongly emergent properties have irreducible causal power, i.e., causal power that cannot be explained by the causal power of lower-level entities. What is meant by causal power? Well, it is a process by which the higher level of a system (say the macro level, the global level) influences and determines behavior of lower-level (microlevel) components. It is important to understand the term "irreducible causal power." This means that there is no explanation for its effects on the lower level, by merely taking into account the lower-level entities and their behavior, be it simple or complicated (by way of interactions). Kim, in particular, points out a problem with this irreducible causal powers [459, 460]: If there is irreducible causal power acting from a macro level to determine the behavior of components at the micro level, this causal effect would possibly counteract causal effects from micro-level interactions of components themselves. In other words, there would be a causal overdetermination of the behavior of micro-level entities that would lead to contradictions. Bedau therefore dismisses strong emergence as a scientifically implausible if not irrelevant concept and instead proposes weak emergence as a less radical, intermediate notion with sufficient scientific credibility. We disagree with him on this for reasons explained later in this chapter.

Weak emergence requires supervenience, i.e., the dependence of macro-level properties on micro-level properties, but dismisses irreducible causal powers of the macro-level entities. Instead, "the central idea behind weak emergence is that the emergent causal powers can be derived from micro-level information, but only in a certain complex way" [89]. In other words, among macro-level properties emergent from micro-level interactions is in particular their causal effects on the micro level. These causal effects can therefore be explained without recurrence to irreducible causal powers at the macro level.

Kim is less skeptical than Bedau about the possibility that an account of irreducible causal powers can be found, but he emphasizes that without such a consistent account the entire value of the notion of emergence breaks down. By proposing weak emergence Bedau, on the other hand, emphasizes that a scientific account of irreducible causal powers is neither required nor possible. This discussion requires some more elucidation, and we will devote the next section to top-down causation.

14.4.2 Hyperstructures and Dynamical Hierarchies

Another explanatory concept in regard to emergence is the idea of hyperstructures, originally introduced by Baas with follow-up work on the dynamics of hierarchies by Rasmussen et al. Baas introduced hyperstructures as a framework to conceptualize emergence and he formally connected emergent entities to category theory [45, 46]. Baas considers the formation of hyper-structures as a process consisting of the repeated application of three ingredients: Objects S_i, with index i characterizing a particular object in a family, a mechanism to observe properties, O, with properties being formalized as $P \in O(S_i)$ and interactions among objects, I, which use the properties of objects. Together, these ingredients form a constructive process that leads to higher-order objects, when stabilized. This is formalized as $R(S_i, O(S_i), I)$. The result of this process could be a family of new objects, S_j^2, which would have their own set of observations, O^2. These could be either completely disjoint, partially overlapping, or equal to the set of observations O at the lower level which we now call first-order objects. Observations of second-order objects would now lead to second-order properties, $P^2 \in O^2(S_j^2)$ which would lead to second-order interactions I^2. In turn the interaction among the second-order objects could lead to even higher-order objects. Iterating these operations (object formation and interactions) will produce what Baas calls a hyperstructure, $S^N = R(S^{N-1}, S^{N-2}, ..., S^1)$. The situation is depicted in figure 14.4.

Where does all that leave us in regard to emergence? The constructive process leading to higher-order objects with possibly new properties allows to define emergent properties as follows: *Emergent* properties $P_e \in O^2(S_j^2)$ would be properties that are not found among the components of a second-order object S_j^2, i.e., $P_e \notin O^2(S_k)$ where k runs over the components of S_j^2 thus containing all first-order components of this second-order object. Note that emergent properties can only exist with higher-order objects, beginning at the second order.

The formation of higher-order objects is the subject of influential work on dynamical hierarchies [689]. With two examples, the formation of hierarchies of subsequently more complex objects is simulated and studied. These two examples are taken from the area of molecular dynamics (a 2D and a 3D simulation, respectively) where the formation of vesicles from monomers with certain properties is studied. The process of interaction is constituted by aggregation, whereby single molecules and polymers are driven to cluster together through physical forces (first from monomers into polymers, then from polymers into vesicles). Aggregation is, however, only one of a number of possible mechanisms for constructing higher-level objects. In social systems, this mechanism would be better characterized as group formation. Another completely different mechanism is network formation. In networks, interactions happen on edges between nodes (objects), and the relation between level of an object is constituted by its connectivity in the network.

In dynamical hierarchies, once a new level of objects is created, the interactions of objects are usually a new means of communication. The communication could happen within the new level (i.e., between the objects of the new level), or between new and old levels. Generally, the complexity of objects will need to increase in the process, in order to be able to handle additional communication/interaction. While this happens automatically in a natural system, in a

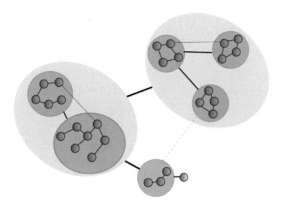

Figure 14.4 Hyperstructures according to Baas [46]. Nodes represent objects (dark gray = first order, light gray = second order, ellipsoid = third order). Links represent interactions (one line = first order; two lines = second order; 3 lines = third order; dashed lines = interactions between objects of the same order in different higher-level objects.

Table 14.3 Physicochemical dynamical hierarchy, according to [689]

Level of description	Molecular structure	Observed functionality
Level 3	Micelles	inside/outside, permeability, self-reproduction
Level 2	Polymers and water	elasticity, radius of gyration
Level 1	Monomers and water	phase separation, pair distribution

simulation provisions need to be made that new properties can be included in higher-level objects. Thus, for example, the data structures simulating objects need to be enhanced to hold new properties. Rasmussen et al. criticize a widespread belief in the complex systems community that they call the complex systems dogma, which states that simple objects plus their interactions can generate complex phenomena. At least as far as dynamical hierarchies of more than two levels is concerned, their idea is the following ansatz: "Given an appropriate simulation framework, an appropriate increase in object complexity of the primitives is necessary and sufficient for generation of successively higher-order emergent properties through aggregation."

Going back to the examples of monomers forming polymers forming micelles under the presence of water, the properties of objects of higher order are interactions that are increasingly constrained by the objects. For instance, polymers restrict the dynamics of participating monomers, and micelles restrict the dynamics of participating polymers. Thus, top-down or downward causation is produced by higher-order structures and we have strong emergence. The situation can be depicted as in table 14.3.

14.4.3 Other Concepts

De Haan [217] discriminates between three different types of emergence: emergence as discovery, mechanistic emergence, and reflective emergence. He introduces what he calls the "conjugate," a joint appearance of a high-level emergent phenomenon with an emergent property of the lower-level entities. In other words, what happens at the higher level has an influence on the entities at the lower level by providing them with an emergent property. He cites order parameters in nonlinear physics as typical examples of this conjugate.

Ellis [255] emphasizes that the lower-level entities provide a variety of behaviors from which the higher-level emergent system selects what it prefers. "Higher levels of organization constrain and channel lower level interactions, paradoxically thereby increasing higher level possibilities. A key feature here is *multiple realizability* of higher level functions, and consequent existence of *equivalence classes* of lower level variables as far as higher level actions is concerned." As an example of an equivalence class, Ellis mentions the micro-state of the molecules of a gas. There are many billions of states that will lead to the same higher-level state, such as the temperature of that gas. Thus, there is a certain independence of the higher-level state, i.e., "the same higher level state leads to the same higher level outcome, independent of which lower level states instantiate the higher level state."

Bishop and Atmanspacher [120] base a characterization of emergence on the existence of context. Their perspective is that of an observer who describes the relationship between different levels of a system in terms of necessary and sufficient conditions for interlevel interactions. *Conceptual emergence* can offer a framework for explanation of certain emergent phenomena. They define it this way: "Properties and behaviors of a system in a particular domain (including its laws) offer necessary but not sufficient conditions for properties and behaviors in another domain." Domain here refers to levels. In later work [119] Bishop offers an account of physical phenomena like the Rayleigh-Benard convection as explainable by contextual emergence. They write: "Although the fluid molecules are necessary for the existence and dynamics of Benard cells, they are not sufficient to determine that dynamics, nor are they sufficient to fully determine their own motions. Rather, the large-scale structure supplies a shaping or structuring influence constraining the local dynamics of the fluid molecules. This is contextual emergence in action [119, p.7]."

Bar-Yam [77,78] also emphasizes the existence of strong emergence in nature. He argues that it is possible to develop a scientifically meaningful notion of strong emergence using the mathematical characterization of multiscale variety. A sign of the presence of strong emergence would be, in his definition, oscillations of multiscale variety with negative values. He demonstrates his ideas with the example of a parity bit of a bitstring. While the parity bit (calculating the parity of the entire string) is determined by all bits, no bit in itself can determine its value. Notice that the property we are talking about (the value of a bit) can be measured in the same units for the entire system (i.e., the parity bit) as for its components (the bits that produce the parity bit). He then lists a large number of examples where strong emergence appears in a system and connects strong emergence to selection. He points out that because selection is a constitutive part of strong emergence, systems working under selection (like living organisms) should be expected to demonstrate strong emergent behaviors.

14.5   Emergence and Top-Down Causation

We now enter the "holy grail" of the discussion about emergence. In the last section we have seen that emergence is intimately connected to the notion of top-down causation, sometimes also called downward causation. There is a metaphorical dimension to this use of language, assuming "higher" levels at the macro scale where we can observe, and "lower" levels at the micro scale where we think entities are located that constitute macro phenomena. The language used to describe the difference between levels and the directionality of cause-effect relationships is perhaps not exact, but it refers to causes for bottom-up causation on the lower level, causes for top-down causation on the higher level. One other caveat: Usually discussion centers on neighboring levels; however, it is well known that this focus is a gross oversimplification, since we generally have to deal with hierarchical systems with multiple levels.

With that in mind, we can now look at causation. As Kim has pointed out, it is the irreducible causal power of the higher level which is a necessary condition for emergence. He further poses the conundrum whether this could happen given that there are low-level causes determining the entities at the lower level.[4] Wouldn't there be a conflict from the top-down causation intervening with what is already going on in causal relationships at the lower level? Wouldn't the primary cause for behavior of entities at the lower level be their low-level interactions and causes? And where would that leave top-down causation to act on?

This question brings us to the nature of causation and to the necessity to introduce time. Without time, without what we might tentatively call a cause that is appearing at a given moment, and the effect observed some time later, there could not be established a notion of cause-effect relationship. A cause always has to precede its effect. With the notion of time, however, we come into the realm of dynamics, and it is in this realm that the apparent conflict between micro-level causes and top-down causes (from a higher, emergent level) can be resolved.

The gist of the argument is that "emergence" is a process whereby novelty is generated and becomes part of a system. Emergence does not refer to static appearance of entities, but to a dynamic process that possibly requires many steps. It is accompanied by a destabilization of the status of a system, a subsequent instability phase, which is usually an exponential growth process, with a final stabilization of the new entities. Thus, there are interactions in the system that go from negative feedback (stable) to positive feedback (unstable, growth), and then later back from positive feedback to negative feedback (stable). In other words, emergence is an iterative process whereby bottom-up causation and top-down causation interact in a way that allows them to gradually align their effects.

Explaining this in other words: The "conflicting" causes from different levels are not different from potentially conflicting causes on a single level. These causes will compete for influence on the entities, and their effects will gradually change such that the overall effect minimizes any conflicts. It is the effects whose causes are in minimal conflict which have the largest growth rate in behavior. Coupled with the fact that top-down causes are themselves present only to the degree that the emergent entities are present, this allows the process of emergence to be described in the language of selection: The higher-level (emergent) entities will cause changes to the lower level through selecting the most appropriate behaviors for their own (i.e higher-level) survival. Other behaviors are still being probed by the lower level, but they are becoming gradually downgraded to fluctuations as the emergent entities succeed in establishing themselves.

---

[4]Think of interactions of objects of the lower level as these low-level causes.

Now this sounds somewhat teleological, and indeed a number of authors have discussed emergence in a teleological context. But teleology is an interpretation of an otherwise quite automatic process. It has close similarity to Darwinian selection in the sense that the behaviors with the most "fitness" (read least conflict) are selected and stabilized.

In a similar vein, a recent contribution by Noble [618] emphasizes the fact that hierarchical systems in biology have layers from which causal influences point to other layers. Why would there be a "privileged" direction of causation only going upward, from each level to the next higher? Influences could be acting in both directions, up and down. Using the case of a dynamical model of the heart, Noble makes a convincing case for strong emergence. He points out that dynamical systems models using differential equations allow for certain types of specifications (constraints) to appear in the course of simulations. That is, the system's behavior influences and gradually changes its boundary conditions. The situation is depicted in figure 14.5 .

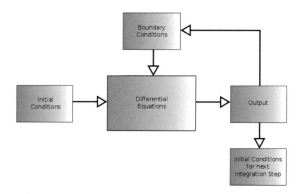

Figure 14.5 The way differential equation systems can constrain their own behavior: A system of differential equations is in and of itself not able to provide reasonable solutions. In addition to the equations proper, a set of initial conditions and boundary conditions needs to be specified for specific solutions. Once they are supplied (from the outside), output can be produced. But this output can influence the boundary conditions as well as initial conditions for a later time step. In this way, the solutions as a whole ("the system") can influence its parts (its component equations). (Adopted from [618])

The boundary conditions influenced by the output of the dynamical system at a given time will in turn influence the dynamics at a later time. This way, the dynamical system self-selects over time what dynamical solutions (stable attractors) it approaches. A lucid discussion of the example of Rayleigh-Benard convection cells — a prime example of self-organization phenomena and emergence — is given in [119]. Bishop writes: "Cases such as fluid convection where the molecular domain provides necessary but insufficient conditions for the behavior of convection cells, and where the convection cells constrain the behaviors of fluid molecules, are rather typical in complexity studies [119, p. 9]."

14.6 Summary

In this chapter we have reviewed self-organizing systems and how they relate to emergent phenomena. We have looked at the different notions of emergence and discussed top-down causation as the key mechanism to bring about emergence.

The importance of the notion of emergence derives from the fact that it can generate something new, either at the current level of interactions, but more importantly at a higher level. This is all the more powerful, as it is not a one-off phenomenon, but can happen repeatedly, thus opening the way to an arbitrary growth in complexity and power of the underlying system.

Constraint, selection, differential growth rates of potential (collective) behavior are the key players when it comes to produce emergent phenomena. The neighboring levels at which these influences are observed, however, are only part of a potentially large hierarchy of levels that come into existence through repeated transitions where a new level suddenly appears and gets populated with entities. It is certainly worth taking a closer look at the details of novelty and how it can be established in an upwardly open hierarchical system. We shall turn to this topic now under the heading of "construction."

CHAPTER

15

CONSTRUCTIVE DYNAMICAL SYSTEMS

I t has often been said that Charles Darwin, in his book on the *Origin of Species*, aimed at one
theme (the origin) but delivered on another (the mechanism of selection). Fontana and Buss,
in their article [282] discuss this at greater length and choose the title of their contribution
accordingly. The question not clearly answered by Darwin is how it is that new species emerge
out of older ones, in a way similar to the process of new cells emerging out of previous ones in the
cell division process of cell biology. Before the important event, only one cell was present, but
after the event of cell division, suddenly two cellular entities are present, both of which develop
independent from each other. One could say that something new has come into the world with
this simple process of cell division.

15.1 Novelty, Innovation, Emergence

Many biological systems work like that, and the reach of this phenomenon goes far beyond biology. Within biology, the appearance of new species (Darwin's question) the appearance of new
cells and individuals, the appearance of new genes within genome evolution, or the appearance
of new functions in evolution are examples of the same kind of process. Figure 15.1 depicts these
structurally similar processes. In chemistry, new compounds are regularly created by chemical
scientists, but they were also created in the history of the universe by a process of repeated collisions of previously existing compounds that happen to stick together and form different compounds after their chance encounter.

Similarly, in the Economy, new inventions are made by economic agents, and some of these
are successful as innovations in that they sweep the economy and become used widely. In psychology, concepts that are learned are similarly formed anew in the neural substrate of individuals. In a sense, these new entities take a life of their own, while prior to their formation, no clear
representation was distinguishable in the network of cells.

Astronomy studies the birth of new stars and galaxies, determined by the large-scale laws of
physics that govern at astronomical scales of space and time. New entities emerge on an ongoing

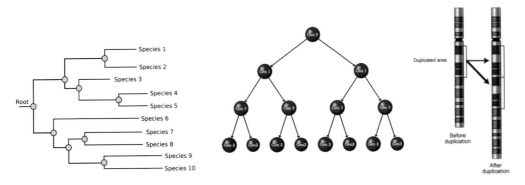

Figure 15.1 Three events in evolution: Speciation (left), cell division (middle) and gene duplication (right, courtesy National Genome Research Institute).

basis in the wider universe. We can even go back to the beginning of the cosmos and its subsequent early development. There were clear moments in time when something new happened, like the "condensation" of atoms out of the plasma, the decoupling of light from matter, etc.

In all of these cases, new things appear and new sorts of interactions take place that from that moment on do not disappear again, and indeed, not only do they appear, but they also start to dominate the subsequent dynamics of the system considered, and in this way influence the entire process under study in a nonlinear, i.e., history-dependent way.

A typical example of the process, yet again from biology, is the development of a multicellular organism out of the fertilized egg cell, the zygote (see figure 15.2). An organism, possibly consisting of trillions of cells, is constructed from those humble beginnings. It is not only the question where the information can reside to determine the fate of all these different cells in their respective morphology, but also how this construction is orchestrated in such a way that it proceeds orderly despite the massive amount of material that is needed (and not produced by the zygote itself).

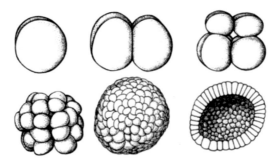

Figure 15.2 Zygote division: Developmental process of a multicellular organism [716]. Reproduced with permission from Elsevier.

There are serious mathematical problems with a description of constructivity as provided by the preceding examples. In order to get this problem sorted out, we have to clearly discern among the terms *novelty, innovation,* and *emergence.* Müller and Wagner define a novelty as a "de novo structure or a derived modification of an existing one which significantly alters its function" [603]. We can see here that Müller and Wagner consider two possibilities, de novo and derived. The former can be divided into two different subclasses: (1) a de novo structure at the same hierarchical level as the ones it is compared with (a duplication, birth, or speciation event as described above) , and, (2) a de novo structure at a higher level coming about through integration of lower-level structures (group formation, division of labor, coherent pattern formation). This latter case (2) underlies emergent phenomena as discussed in the last chapter. The second possibility mentioned by Müller and Wagner, derived modification, is also interesting, though not so much in connection with constructive systems.   For example, a case where an existing gene assumes a new function is covered by this second possibility. Gould and Vrba have termed such a process an *exaptation* [344].

Innovation can apply to all novelties. While a novelty could simply be a blip in the course of events of a system's dynamics, and could disappear again, an innovation happens when the new entity (regardless whether appearing on the same hierarchical level or on another hierarchical level) or modification establishes itself for a sufficiently long period of time, and possibly triggers further downstream events, like diversification or radiation.

Under these definitions, *emergence* is strictly reserved for events when the novelty appears on a higher level, due to the collaboration of entities of a lower level. It is thus the outcome of some sort of integration activity. Note that emergence could refer to a first-of-a-kind appearance of an entity on a new hierarchical level, as we suggest, or to the appearance of further entities on a level already existent and populated by entities, a use of the term by some authors in the literature, which we, however, discourage.

Constructive dynamical systems are intended to model and simulate innovation and emergence, i.e., the dynamics of novel appearance of entities on the same level as well as on a higher level.

15.2 Birth Processes at the Same Level

Let us focus first on the conceptually simpler case of a novelty in the form of a birth process at the same hierarchical level where entities already existed (and were modeled) before. Processes that generate this type of change are, for instance; mutation, recombination, duplication and subsequent divergence, chance encounters, and change in the environment. We will look later at the second case, in which new entities emerge on a higher level, a level that has previously not existed.

The two phenomena require different formal treatment. Take an ODE approach. While rate equations are usually deterministic, so that there really is no room for surprise and innovation, we might be able to model the occurrence of a novelty in the following way. Let

$$\dot{x}_i(t) = f(x_i(t), x_{j \neq i}(t)) \text{ with } 1 \leq i \leq N \tag{15.1}$$

be the general set of equations describing a system with N entities, where we have separated terms of interaction with other entities into the second dependence variable, $x_{j \neq i}(t)$, on the right hand side. For example we could write ($1 \leq i \leq N$ with N the number of types):

$$\dot{x}_i(t) = a_i x_i + x_i \Sigma b_{ij} x_j + x_i \Sigma c_{ij}^k x_j x_k, \tag{15.2}$$

which allows the reaction of molecules i with j or of i with j under the presence of k. The parameters a, b, c determine the possible trajectories of the system. As Ratze et al. [699] emphasize, the ODE system really contains all possible trajectories of the system. In other words, given a realization by initial conditions, $x_i(t)$ is determined and will develop toward its asymptotic state, given a_i, b_{ij}, c_{ij}^k.

What happens in a birth event? As we said before, a new component x_{N+1} will appear and has to be added to the system of equations. That means a new equation appears, and an entire set of new constants has to be added, too. That is, for $i, j, k = 1 \ldots N$, the parameters $a_{N+1}, b_{N+1,j}, b_{i,N+1}, b_{N+1,N+1}, c_{i,j}^{N+1}, c_{N+1,j}^k, c_{i,N+1}^k, c_{N+1,N+1}^{N+1}, c_{i,N+1}^{N+1}, c_{N+1,j}^{N+1}, c_{N+1,N+1}^k$ have to be determined and added to the model. Once in place, the evaluation of the equations (by calculation or simulation) might take a completely different course and might lead to another, previously inaccessible state. Through addition of the new entity, the dimensionality of the state space has changed, which allows qualitatively different outcomes of the dynamics.

As we can easily see, this provides a formidable problem for a mathematical description. While the dynamics can be extracted from the equations — although already this might be difficult due to their nonlinear character — the occasional addition of equations and terms in equations is not something a mathematical formalism is able to handle. Thus, we either have the choice of foreseeing all potential mathematical equations from the outset, a task that is impossible to achieve in combinatorial spaces, or we have to resort to a higher-level metamathematical description, a description that has as its content the number and nature of the variables under consideration in our dynamical system model and the way of their interaction.

15.2.1 A Simple Example

Let us discuss a simple example, this time with differential equations that are less complicated than equations 15.2. Let's start with a system with just one component. Suppose we have a dynamical system with the following rate equation:

$$\dot{x}_1(t) = -ax_1(t), \tag{15.3}$$

where a is a positive constant, say $a = 0.1$. This equation can be easily solved, Figure 15.3 depicts the functional behavior calculated through numerical solution (upper curve). Now let's suppose a new entity appears, say x_2, also following its own dynamic rate equation. Initially, let's assume it is independent of the previous entity x_1. Let's further assume that the equation for x_2 can be written similarly as

$$\dot{x}_2(t) = -bx_2(t), \tag{15.4}$$

with $b = 0.4$. This is again an exponentially decaying quantity, depicted in figure 15.3 as the lower curve.

So far, nothing really surprising has happened. The introduction of a new entity has enlarged the system, yet not changed its behavior, though we could say that enlarging the system *is* a change in behavior. This is what was called resultant effects, i.e., an aggregation of behaviors, in the previous chapter 14. For example, we could try to represent the entire system through the average of x_1 and x_2, and we would not detect a qualitative change in behavior.

However, if we now introduce an interaction between the two entities, which we shall model in its simplest form as a coupling between the variables (see equation 15.5),

$$\begin{aligned} \dot{x}_1(t) &= -ax_1(t) - cx_2(t) \\ \dot{x}_2(t) &= -bx_2(t) + cx_1(t), \end{aligned} \tag{15.5}$$

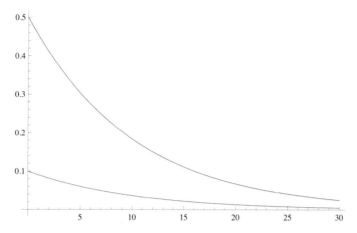

Figure 15.3 Upper curve: Numerical solution of equation 15.3: Exponential decay from initial condition $x_1 = 0.5$. Lower curve: Numerical solution of equation 15.4: Exponential decay from initial condition $x_2 = 0.1$.

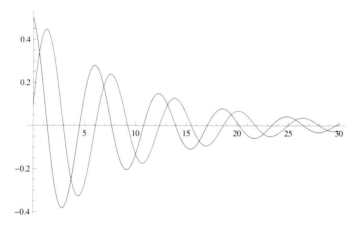

Figure 15.4 Numerical solutions for both variables in equation 15.5: From initial conditions $x_1 = 0.5, x_2 = 0.1$ an exponentially decaying oscillation has developed.

with a coupling strength $c = 1$, we see a very interesting effect depicted in figure 15.4: An oscillator has formed by virtue of the coupling of the two variables, showing dampened (exponentially decaying) sinusoidal oscillations of both quantities. This qualitatively new behavior came into being after we had chosen a nonzero coupling constant between the two variables of our system. Thus — as we probably suspected all along — it is the interaction between entities that allows for qualitatively new (emergent) behavior of a system. So we can add as many new entities to a system as we want, as long as we do not arrange for interactions between the components, no qualitatively new behavior will be generated. Yet, if we arrange for interactions, even an extremely simple system like the two-component system x_1, x_2 will show novel behavior.

So far, we have merely looked at a manually established change in a system. We could afford to act manually because the system was low-dimensional, and setting up coupling constants for one or two quantities can still be done by hand. But how can we treat systems with many entities where interactions (coupling constants) between them are changing on a regular basis? Our next example will explain just such a system.

15.2.2   A Model of Autocatalytic Sets

In a series of papers since 1998, Jain and Krishna have examined simple network models of interacting molecular species in autocatalytic sets. These systems possess two types of variables, fast-changing dynamic variables characterizing the population counts of species, and slow-changing variables characterizing the interaction graph (network) of species [425–428]. Autocatalytic sets have been proposed as key stepping stones to life, and examined previously in [49, 447]. We shall discuss the Jain and Krishna model here because it captures important aspects of constructive systems.

Species in this system are symbolized by nodes of a graph, with their interactions ("catalytic function") being symbolized by edges in the graph. This is similar to our reactions graphs from previous chapters, except simplified in that each directed edge between nodes A and B means that sort A catalytically produces sort B.

Figure 15.5 The two sorts A and B interact by A being able to catalytically produce B.

So, in this system, we find n different species $S = \{s_1, ..., s_m\}$ with their interactions noted in an adjacency matrix $C \equiv (c_{ij})$. Each entry c_{ij} could be either "1" or "0," depending on whether a catalytic function of species i on species j exists or not. The nodes and their edges constitute the structure of the network, which additionally has dynamical variables, symbolized by population variables lumped together into a vector

$$\vec{x} = \{(x_1, ..., x_m) | 0 \le x_i \le 1, \sum_{i=1}^{m} x_i = 1\}. \tag{15.6}$$

While population variables $\vec{x}$ change quickly over time, the structural variables s_i and c_{ij} also change, albeit at a much slower pace. A simulation of the system will essentially let the fast variables achieve equilibrium before the slow variables change. Assuming the slow variables are constant, the dynamics of a fast variable is described by the following rate equation:

$$\dot{x}_i = \sum_{j=1}^{m} c_{ij} x_j - x_i \sum_{k,j=1}^{m} c_{kj} x_j \tag{15.7}$$

which reaches a fixed-point attractor after some time, starting from an arbitrary concentration $\vec{x}_0$. The adjacency matrix C is initially given by a sparse random matrix, where each element c_{ij} is set to "1" with a low probability p. It remains "0" with probability $1 - p$. c_{ii} are always set to "0" so that no self-replicators are allowed. The first term in equation 15.7 is the interaction term, while the second term is the well-known flow term, which normalizes the overall concentration to

$$\sum_{i=1}^{m} x_i = 1. \tag{15.8}$$

Now, in order to get the slow moving dynamics, the species with the smallest concentration x_{min} are selected and one of them, call it x_r, is removed from the system. With it, naturally, the connections c_{ri}, c_{ir} disappear. Instead, a new node, $\bar{x}_r$ is substituted in its place, with its connections $\bar{c}_{ri}, \bar{c}_{ir}$ being reassigned values based on the probability p mentioned before. Thus, in effect this negatively selected node has its interaction wiring in the network being replaced. Once the structure has changed, x_r is set to a constant, x_0, and the remaining species concentrations are perturbed around their current value, with the normalization condition 15.8 enforced. A new round of fast-moving dynamics is simulated with the modified system.

The model in this form was intended to capture the formation of autocatalytic sets in a primordial pond in which chemical species accumulate and interact, occasionally being washed out by a flood from rain or waves, with new species brought in by such an event as well.

So, if the model is started in a sample run with $m = 100$ nodes and the probability of connections $p = 0.0025$, at first the connectivity of nodes is rather low, with linear chains for connections; see figure 15.6, left. However, at some point, due to random replacement of a node, a small autocatalytic set can arise, as in figure 15.6, right.

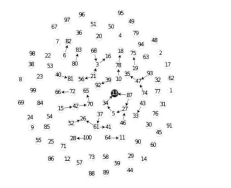 

Figure 15.6 Left: Initial connectivity of the network with $m = 100, p = 0.0025$. Right: Emergence of smallest autocatalytic set with two member species at iteration 2854. From [427], with permission.

This completely changes the dynamics of the network. Suddenly, a strongly growing component of the network is present that — through selective forces that push out nonparticipating nodes — forces a complete reorganization of the network, until all nodes participate in the autocatalytic set to a larger or smaller degree. Figure 15.7, left, shows this state of the network. While the first phase of network development is called *random phase* by Jain and Krishna, this second phase is called *growth phase*, which ends when all existing nodes first have become members of the autocatalytic set.

After the full network is part of the autocatalytic set, changes in connectivity might simply refine the structure of the network (what JK call the *organizing phase* of the network) or result in new stronger autocatalytic sets that are able to out-compete the existing set, and therefore displace the original ACS through a crash of the system; see figure 15.7, right. This might happen repeatedly, with the system entering another growth phase after a crash. Setbacks might even occur that lead the system back to a state similar to one of its earlier states, as in figure 15.8, left and right.

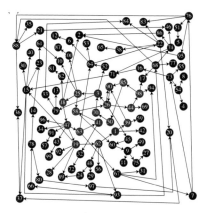 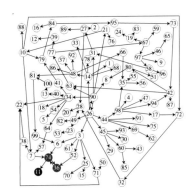

Figure 15.7 Left: Growth phase of network ending in all nodes being connected to the core. Right: Organization phase during which competition improves the stability of the autocatalytic set through rewiring and crash and regrowth cycles. From [427], with permission.

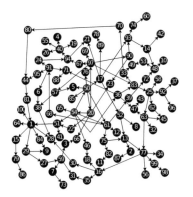 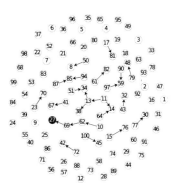

Figure 15.8 "Boom and bust" cycles lead to repeated crashes and regrowth of an ACS. From [427], with permission.

We can see here clearly that there are different types of novelties entering the system. Some of these novelties (new nodes with their new random connectivity patterns) are able to replace part of the network through the very nature of their connectivity patters, whereas others are not able to survive very long without being replaced. One might argue that such a model is not very realistic, since in the real world a novelty doesn't substitute for an existing entity right away. That might be the result of a lengthy development, but usually doesn't happen by construction as in this simplified model. Yet, it turns out that the substitution rule in the JK model is just accelerating the development without changing it much. This is similar to what we discussed in earlier chapters with the bit string chemistry, where it was possible to also leave existing strings present and simply add new strings. The net effect of substitution is simply an amplification of selection pressure.

Coming back to the question at the start of this subsection: How can we treat systems where there is novelty being generated on a constant basis? Well, we have seen a single run now of such a more complex system. It is easily imaginable that other runs would result in quantitatively different outcomes, though qualitatively the same type of phases of development will be observed. And while nodes are substituted, it is really their connections that make the difference in network dynamics, not the nodes themselves. By simply rewiring a node, without replacing it, we can achieve the same effect. So, again, as in the simple example of the oscillator, it is the interactions terms (here symbolized by the adjacency matrix c_{ij}) that allow the system to exhibit new behavior.

What are the consequences for state space? In state space, a node corresponds to a dimension, and thus if we add a node to the network, which replaces a node already existing, the net effect is that the dimensionality of state space will not change. What that means is that novelty can be captured even in the case when the state space does not change dimensions. A novelty will merely find its expression in a discontinuity in the change in state space, when states start to suddenly change in a different way, upon the introduction of a novelty. So the state trajectories will develop differently from that point on. Once new equilibrium values have been reached, the system will again be ready to experience a discontinuity. So, in other words, a treatment using differential equations is not possible across these discontinuities.

15.2.3 Classification of Innovation

The work of Jain and Krishna led them to propose a classification system for innovations [428]. In order to set this in context, we have to discuss the connection between their concept of autocatalytic sets and our concept of organizations as self-maintaining and closed sets. It turns out that there is a close relation: Autocatalytic sets are self-maintaining sets, i.e., for each member of the autocatalytic set, there is a reaction of elements within the set that produces that member. So, autocatalytic sets are self-maintaining, but not necessarily closed. While closure is possible for autocatalytic sets, this is usually not the case. However, if we look at the dynamics of ACS systems, we will find that only closed autocatalytic sets survive in the long run.

Based on the simple reaction equation (15.7), which contains just a linear interaction term, we can expect that graph theory will have something to say about the behavior of those systems. And indeed, the linear algebra of these systems is such that the interaction (adjacency) matrix C can be analyzed and predictions made about the dynamical attractor behavior of a system based on these equations. In fact, the Perron-Frobenius eigenvalue λ_1 of matrix C can be calculated. It is real and nonnegative, and the attractor of the system is the eigenvector corresponding to this eigenvalue of C. In more general systems, organizations are the equivalent of the eigenvectors of these linear systems.

Returning to innovations: As we have seen, innovations can be generated by replacing nodes in the ACS network with a new rewiring. Thus, without changing the dimension of the state space, new behavior can potentially be generated. Each replacement is called a novelty, but only a small percentage of novelties catches on and changes the behavior of the system. Jain and Krishna define such a change in system behavior as the feature that the new species has a nonzero concentration in the attractor following its introduction to the system. The actual classification system of Jain and Krishna differs here from ours.

Type A: Short-lived innovations
Case 1: Blip
Short-lived innovations are those that are generated by noncyclical subgraphs in the overall network of interacting species. Thus, no new self-maintaining sets are generated by the rewiring of

the network. In the case of no preexisting organization, the addition of a new pathway does not add much to the system, and it survived only by chance. Such changes are short-lived due to the brittleness of the feeder to the innovation.

Case 2: Incremental

What could also happen, though, is that a parasitic interaction is added to an already existing self-maintaining set, which allows the system to draw some material flow into the new species. But even if the attractor of the system is perturbed somewhat by such an addition, the new species will mostly show up in peripheral branches of the interaction network and thus be of not much consequence to the self-maintaining set at its core.

Type B: Long-term innovations

The second major type of innovation is able to add a new self-maintaining subset (an irreducible subgraph) to the network of interactions or to modify an existing one.

Case 1: Independent organization

A new self-maintaining subset is added, but independent of an existing self-maintaining set. In this case, it depends on the eigenvalue λ_1 of the subgraphs what will happen.

Subcase a: Core-shift

If the eigenvalue of the new subgraph is larger than the eigenvalue of the previously existing ACS, the new ACS will take over the system. This is called a core shift.

Subcase b: Birth of an organization

If prior to the addition of the new ACS there did not exist an ACS, then the new node and its interactions will allow, for the first time, the emergence of an ACS in the system. This case is similar to *a*, but technically should be kept apart and receive its own category.

Subcase c: Dormant

If the eigenvalue of the new subgraph is smaller than the previously existing ACS, the new ACS effectively becomes a part of the periphery of the of the existing core. It will extract some material flow from the system but will not be able (at least not without further modifications) to generate a major change in the behavior of the system.

Case 2: Modification to an existing organization

If the newly added node with its interactions becomes part of the existing ACS, again two subcases can be discerned:

Subcase a: Core enhancing

The new node receives input from the existing core and provides output to the existing core and thus enlarges the core, which increases its eigenvalue λ_1. This is the by far most frequent subcase.

Subcase b: Neutral

If all nodes are already part of the ACS before the innovation arrives, a new node can also, in principle, be replaced without any effect on the organization. This is technically possible, but it happens only under very restricted conditions.

This classification of innovation, is still tied somewhat to the specific algorithm applied in the JK system. For example, in principle a core destroying subcase could imagined, but based on the selective forces in the JK system, this is virtually not happening there. As soon as one relaxes the deterministic selection, such a case becomes plausible. We have to leave it to the future to define a more generally applicable classification of innovations.

15.2.4   An Expanding State Space

A constructive system with an expanding state space is given, for instance, by an organism that grows from a single cell (the zygote) into a multicellular being through the process of ontogenesis. This is an extremely sophisticated and highly orchestrated process that adds complexity through an existing growing system and occasionally reduces the complexity again in a process of pruning. Figure 15.2 shows the early stages of such a process.

By adding new elements, such systems are increasing their state space and thus allow more complex tasks to be performed once the new elements are integrated into the system. As before, the fact that a new component is born is not in itself decisive, but the fact that the new component will start to interact with previously existing components and thus begins to influence the subsequent development of the entire system. As we have stated elsewhere, it is difficult to capture these processes in differential equations, unless one adds new equations every time a new entities is added to the system. However, in a simulation system, it might be possible to capture the event through the allocation of new memory to the state of a new variable being added to the system. We are reminded here of the statements early in this book that mathematical and algorithmic modeling of a system could differ substantially if we consider the processes of transformation that allow novelty to be introduced.

In principle, it should be possible through the type of growth process discussed here to reach a system complexity such that an arbitrarily complex tasks can be performed. Notably, if principles of hierarchical organization are applied which allow logarithmic scaling of complexity, a system will expand its reach. This ability, however, naturally leads to the necessity to consider new levels in a system, being formed or populated for the first time: A new type of novelty.

15.3   The Emergence of Entities on a Higher Level

In the previous section, we discussed novelty emerging on the same level as already existed (referred to as de-novo structure at the same hierarchical level, subclass (1), by Müller and Wagner [603]). For instance, in the case of our low-dimensional oscillator example, one variable already existed, when we added a second one and provided for couplings with the first. Similarly, in ACS systems, new species nodes are added while old ones are destroyed,with the new nodes again on the same level as the previously existing ones.

What, however, happens when a new entity emerges for the first time? This is subclass (2) in Müller and Wagner, a de-novo structure at a higher level resulting from an integration of lower-level structures. Formally, a collective group of equations can be lumped together, since they form a coherent pattern. This could happen in a system consisting of a fixed number N of components, it does not require the addition of new components, $N + 1, \ldots$. Rather, what is required is a specific combination of entities that interact in a very specific way, which becomes visibly integrated and therefore perceivable over time.

We have seen examples in chapter 14 where we discussed this phenomenon from the perspective of a definition of emergence. What is important is that the emergence over time of a particular pattern establishes an equivalence class of states that all lead to the same pattern. Thus, it is not necessary to always have the same concentrations of individual lower-level entities, the connectivity of low-level entities will ultimately produce the pattern regardless. This connectivity comes about by the interaction of the components, which allows them to compensate each others' behavior. It is this network of interactions that provides a buffer between the possibly fluctuating behavior of the low-level entities and the higher-level behavior. If there is a dynamical mapping process between the levels, as is regularly the case in natural systems, we speak of a dynamical organizing process and a resulting attractor. The attractor can be considered a pattern in its own right, with multiple different micro states leading to it as its equivalence class. However, we can also achieve a similar equivalence class in static mapping systems, as exemplified by genetic programming. In that case, the mapping is required to be a many-to-one mapping from lower level entities to higher level entities, creating equivalence classes via neutral variation [408]. Generally, the pressure to produce robust entities produces these kinds of relationships [903].

For example, such a robust attractor pattern might be the combination of x_1 and x_2 in equal parts. To be an attractor, we would require it to be the outcome of a dynamical process. Technically, the pattern could be expressed by a concentration vector $\vec{x}_p = 1/Z(1,1,0,....,0)$ where Z a normalization factor assuring $\vec{x}_p \times \vec{x}_p = 1$. Thus, it would be observable by measuring the concentrations of elementary components. The pattern $\vec{x}_p$, though consisting of these basic components $x_1, x_2, ..., x_N$ with a particular value, could, however, be interpreted as a new entity and formally given a separate scalar concentration y_1, on the longer time-scale where the dynamics has settled. Keep in mind that y_1 really is an equivalence class of an entire field of x vectors: All those x-vectors that lead to it once we let the dynamics of the system play itself out. Thus, we have given it a new symbol y_1 signifying the entire attractor basin.

The difference between measuring the concentration (or presence in the system) of y_1 and the measuring of low-level concentrations is important. First, the time development of $y_1(T)$, while dependent on the time development of $x_i(t)$, is generally slower, and its values develop on a longer time scale, symbolized here by writing the time variable as T. This is a reflection of the fact that y_1 is an equivalence class of x variables, which means its change should be somewhat dampened compared to changes of the latter variables. In other words, x variables are allowed to compensate each others' changes and still result in the same y. We can express this in a different language by saying that x variables form a group and cooperate to produce y. Group formation is therefore a very important step in the emergence of new levels of organization.

In biology, many examples exist of novelties emerging from the interplay of previously independent components. Szathmáry and Maynard Smith [565] have termed them *evolutionary transitions*. Table 15.1 lists the examples they gave in their book. Evolutionary transitions are characterized by cooperative interactions between originally individual and competing entities that gradually form more cohesive wholes through cooperation until they are themselves (on the higher level of groups) able to sustain an individuality and the accompanying drive to survive/become stable. This is accompanied by a reduction in the independence (and therefore individuality) of its component parts, which give up sovereignty for the common good. The work of Szathmáry and Maynard Smith, Okasha [636], Michod [582], and Wilson [920] all point to this direction.

Returning to the formal treatment of y, while this new variable captures the development of lower-level variables x cooperating, it does not participate in the competition among molecular species instated by the flow term. This can be understood from the point of view of reproductive strength of y: There is no reproductive strength of y allowing it to multiply. Rather, it is tied to the activity of its components, which compete among themselves.

Things would be more interesting if there were more than one type of y active in the reactor. Generally, this is impossible in a well-mixed reactor, only one of the potentially alternative groups will be dominant. However, if we look at the higher level of a *set of reactors*, we can consider a higher-level selection process among reactors. McMullin and co-workers offer the clearest example of such a procedure to date: Multilevel selection among reactors [223].

Table 15.1 Major transitions in biological evolution, according to [565]

Before transition	After transition
Replicating molecules	Populations of molecules in protocells
Independent replicators	Chromosomes
RNA as gene and enzyme	DNA genes, protein enzymes
Bacterial cells (prokaryotes)	Cells with nuclei and organelles (eukaryotes)
Asexual clones	Sexual populations
Single-celled organisms	Animals, plants, fungi
Solitary individuals	Colonies with nonreproductive castes
Primate societies	Human societies

Emergence of a new level starts with group formation. Okasha summarizes the steps for the emergence of new entities in such a system in the following steps:

Stage 1 Collective fitness is defined as average component fitness. Cooperation spreads among components.

Stage 2 Collective fitness is not defined any more as average component fitness but is still proportional to it. Groups start to emerge as entities in their own right.

Stage 3 Collective fitness is neither defined nor proportionally follows average component fitness. Groups have fully emerged, with their fitness becoming independent of component fitness.

Fitness here refers to the ability to reproduce and survive selection events. McMullin's model, also discussed in chapter 11, simplifies matters by jumping over the second stage and implementing competition among reactors as the higher-level entities. While the weakness of such an approach is that it passes the key stage of emergence of a new level, it gives a means to model the multilevel selection process in a system consisting of more than one level. Perhaps it can serve as an inspiration for how to algorithmically model the entire process.

15.4 Summary

We have seen that one should discern between two processes:

(1) The appearance of a new entity on the same hierarchical level as previously existing entities, e.g., a new gene could appear that has a new function.

(2) The appearance of a new entity on a different new level of the hierarchy of entities, as, for example, in the cooperation of a number of genes to produce a particular function, or other types of higher level interactions.

While the former can be conceptually grasped as a branching process, whereby one entity gives birth to two, the latter can be conceptually understood as the process of group formation that ultimately leads to independent and competing units on the level of groups. Wagner has pointed out that neutral mappings are a key factor in producing the equivalence classes of lower-level entities that lead to higher-level identity (either dynamically or statically). He emphasizes that the selective pressure for robustness on the higher level ultimately leads to these neutral mappings. Haken explains that a time-scale separation between lower-level entities that change on a fast time-scale and higher-level entities that change on a slow time scale will lead, through the application of the "slaving principle" to a constraining of lower-level entities through higher-level entities. This, and Ellis's notion of causal slack in equivalence classes of lower-level entities are the hallmarks of emergence.

Part V

Applications

16

APPLICATIONS OF ARTIFICIAL CHEMISTRIES

A rtificial Chemistries have been applied to a variety of domains, ranging from wet chemical computing to automated music composition. The present chapter is the first of part V of this book, covering various application domains of artificial chemistries. The current chapter discusses a selection of examples intended to be at the same time simple enough and illustrative of a broad range of AC application possibilities.

The first AC example has taken inspiration from the world of bacteria, namely their chemical composition and behavior. Bacteria must be able to detect rapid changes in their environment and to adapt their metabolism to external fluctuations. How do they achieve such a feat? Bacteria monitor their surroundings with membrane-bound sensors and intracellular signaling cascades. As a result of this chemical signal processing, the metabolic and genetic systems of a bacterium change, leading to a change in its behavior. These regulatory mechanisms of bacteria can act as a model for the design of robust control systems based on artificial chemistries. We shall look into this in more detail in section 16.1.

Because of its natural parallelism, chemistry has inspired many AC algorithms for communication in computer networks, applications where distributed operation and decentralized control are of paramount importance. Section 16.2 will provide a discussion of some of these algorithms, with more discussion about distributed algorithms using ACs to appear in chapter 17.

Artificial chemistries have also been applied to study the emergence and evolution of communication codes in general, ranging from the genetic code to codes used by the immune system, and those used in natural languages. Indeed, because ACs capture the dynamics of interacting objects in an intuitive way, they are well suited to model the dynamics and evolution of natural and artificial languages in general. In section 16.3, we will examine how this can be done.

Music can be seen as a creative form of communication, and the musical language is an intricate form of artistic expression whose underlying cognitive processes are still poorly understood. There as well, artificial chemistries can provide a contribution. Indeed, music fragments have been composed with the help of artificial chemistries. Section 16.4 will provide a brief overview of such explorations.

Another example of where artificial chemistries have been applied is in the implementation of mathematical proof systems. The analogy comes from the idea that transforming a mathematical statement into another is similar to a chemical reaction. And while many transformations might be possible to a given mathematical statement, the proof of a statement will ultimately result in the most simple statement possible: "True" (or "False"). Thus, a proof could be considered to be a search for a sequence of transformations of a statement until this sort of simple statement is reached [620]. This search could be conducted in parallel, by subjecting the statement to different kinds of transformations (reactions), and further to transform the resulting statements in parallel, until a simple statement of the form mentioned before is reached. A number of scholars have claimed that artificial chemistries and mathematical proof systems have something in common [148,284], and we will look at the possibility in section 16.5.

The final example we intend to discuss here is some of our own work in merging artificial chemistries with genetic programming. The idea here is to consider the data flow in a program (considered as a sequence of instructions manipulating the state of a number of processor registers) as chemical reactions. More to the point, if an instruction makes use of the content of a register, there is some connection akin to a chemical reaction between this register and the register receiving the result of the instruction. Thus, a program could be considered as analogous to a chemical reaction network, with nodes being the registers, and edges being the instructions connecting these nodes. Technically, one would work with a bipartite graph, where both registers and instructions are nodes (of different kinds), with the rule that only nodes of different kinds can be connected. A more detailed discussion will comprise section 16.6.

The examples discussed here are a small selection of applications that have been used in the literature. The following chapters of part V will discuss more examples from more focused application areas. The reader should keep in mind, however, that there are some subjective elements in the selection of these applications, and we do not claim to have exhausted the list of interesting applications that have been studied using an AC approach.

16.1   Robots Controlled by Artificial Chemistries

Single-cell living organisms need the capability to process information in order to keep them autonomous and surviving in unknown and often unpredictable environments. However, their need for adaptive (intelligent) behavior cannot be fulfilled by using a nervous system. After all, they simply consist of one single cell. Nevertheless, single-celled organisms possess a tool in the form of other networks (possibly equivalent to multicellular organism's nervous systems): Reaction networks of various types — signaling networks, metabolic networks, and genetic networks. These networks are recruited to suit the needs of behavioral control. Would it be possible to learn from these methods and apply them to the control of robots?

Before we discuss some examples where these ideas have been used with robot control, it is a good idea to look at examples of natural systems in which these phenomena can be observed. We focus on two such systems here: bacterial chemotaxis controlled by metabolisms, and bacterial stress response to environmental changes.

It should be emphasized, however, that single-celled autonomous organisms are not the only source of inspiration for chemical control of robots. Another source is the endocrine system, one of the four adaptive systems active in higher-order multicellular organisms. Yet another source is the chemistry of neural systems, which is in itself a complex system overlaying and coupled to the electrical system of signal processing in nervous systems.

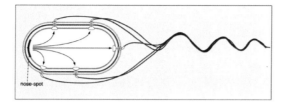

Figure 16.1 Schematic depiction of an *E. coli* bacterium with its chemoreceptors and flagella motors, [653]. © Science 1993, with permission.

Generally speaking, the information processing system of an agent needs to possess the ability to form an internal complex system of some type, capable in some way of capturing features of the external world. All these examples allow the formation of complex networks, which seems about what is needed to provide a reservoir of behavior for control.

16.1.1 Chemotaxis and Behavior

Chemotaxis is the movement of cells or biological organisms in response to chemical gradients. It might involve the flight from concentrations of toxic chemicals or the approach to areas where nutrient chemicals are available for the organism. In our discussion here, we will focus on the simplest mechanisms to explain such processes. How do bacteria achieve this? Figure 16.1 shows schematically the example of an *E. coli* bacterium: A localized set of chemoreceptors connects (via chemical pathways) to the control of bacterial flagella, which, collectively, form a type of motor, driving the bacterium.

The general control strategy of such a bacterium is to seek out areas where "beneficial" substances can be found. It does so by controlling its movement based on a gradient of chemicals. The measurement of a gradient is performed through detecting the temporal concentration changes of a chemical while swimming. Typical swimming speeds allow a bacterium to compare concentrations over ten to a hundred body lengths in a matter of seconds. Based on the chemical comparison signal, the flagella rotors are directed either clockwise or counterclockwise. Thus, a very simple threshold-like decision can occur between the two directions. Furthermore, the geometry of the flagella rotors is such that a counterclockwise rotation will lead to them forming a bundle, which results in a straight swimming movement ,whereas a clockwise rotation of the flagella leads to a breaking up of the flagella bundle. Since the flagella are distributed around the cell, this means that the movement is now incoherent and leads to a "tumbling" of the bacterium in place. Again, the transformation of control signals into actual movement is rather simple, based on a geometric principle.

Note that this is only the simplest of all cases. What happens if both an attractant (nutrient) and a repellent (toxin) are detected by chemical receptors has been described in a classical paper by Adler and Tso [25]: Bacteria are able to process both inputs and come up with a reasonable response, based on the strength of both signals.

Artificial chemistries have been used to apply these inspirations and devise simple control systems for simulated and real robots.

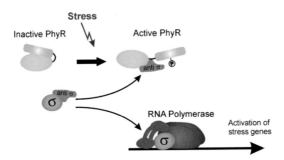

Figure 16.2 General mechanism of stress response of bacterial cells. Sigma factors in passive state in unstressed cells. In response to stress, the sigma factor is released and attaches to DNA enabling transcription by polymerase. From [294], with permission.

16.1.2 Bacterial Stress Response

The second example of simple control being exerted by the environment on behavior is bacterial stress response. This is related to the ability of bacteria to adapt to their environment through changes to their genetic and metabolic networks. For the sake of the argument, let us focus here on the genetic network.

Sudden changes of the environment, such as the intrusion of a disease-causing bacterium into its host organism, require a coordinated switch of a whole set of genes. Similarly, bacteria that grow in nutrients exponentially will sooner or later reach a stage where they have exhausted their resources. Again, a coordinated switch of activity of many genes at the same time has to happen, to transit into what is called the stationary phase, where a bacterial populations does not grow any further. Both of these phenomena have been found to be related to a change in gene expression controlled by so-called sigma factors, proteins that attach to DNA and the polymerase molecule that reads off DNA [350,494]. Sigma factor proteins have multiple regions effectively enabling the transcription initialization at certain promoter sites. Their action is such that they are only present during initialization of transcription. As soon as transcription gets going, they disassociate again from the polymerase molecule to provide the same service to another polymerase molecule. A change in environmental conditions causing the stress associated with it activates certain sigma factors (see, e.g., Figure 16.2) which in turn then activate the reading off of certain stress response genes leading to a changed expression of proteins in response to the stress. For this to happen, the environmental stress impacts on the equilibrium between a free (and reactive) form of a sigma factor and its passive form.

16.1.3 A Metabolic Robot Control System

In a number of contributions, one of the authors of this book has addressed the issue of designing simple robot controllers that use various forms of artificial chemistries. The basic idea is to set up a network of reactions (which could consist of signaling, metabolic, and/or genetic reactions) to connect a simplified input stream of simulated molecules to a simplified out stream of control signals translated into robot behavior. The network will then act as the "computer" or control system determining under which input conditions which behavior should be activated. Figure 16.3 depicts a simple schema of the system examined [959] with the substrate referring to the input stimulus and products being behavioral signals.

In our first exploration we focused on handwritten reaction networks built from an artificial polymer chemistry and from an artificial enzyme-substrate chemistry. The latter was chosen for an implementation test and showed the anticipated behavior. We had constructed the system such that input from a number of light-sensitive sensors and from infrared powered distance sensors would be fed into the network, with a requirement that the system should seek light and avoid short distances to objects (walls, obstacles). The output of the network was a stream of "molecule signals" controlling the two motors of a wheeled Khepera robot. A schematic view of the robot is shown in figure 16.4.

Later experiments focused on evolving reaction networks rather than hand-designing them. To this end, a bipartite representation of chemical reactions was introduced. A bipartite graph contains two sorts of nodes, (1) one type representing the various chemical substances participating in the reaction, and (2) another type representing the chemical reaction itself. In this way, metabolic pathways can be modeled, provided the bipartite graphs obey additional constraints coming from chemistry, like material balance. This constraint fixes that no material can be created *ex nihilo*. Further, this graph is directed, with edges representing the flow of molecules, and weighted, with the weight representing the catalytic or inhibitive activity of a substance to a reaction.

Formally, we have a directed graph G, $G = (V, E)$ with vertices V, consisting of two types $V = R \cup S$, with $R = \{v_{r_1}, ..., v_{r_n}\}$ the set of reactions and $S = \{v_{s_1}, ..., v_{s_m}\}$ the set of molecules. Each of these entities carries a unique name. Edges E connect molecules to reactions and also form a set $E \rightarrow V \times V \times R$ with each edge $E = (e_1, ..., e_l)$ consisting of a 3-tuple $e_i = (v_i, w_i, k_i)$ with the following condition:

$$v_i \in R \Rightarrow w_i \in S \qquad \text{or} \qquad v_i \in S \Rightarrow w_i \in R. \qquad (16.1)$$

In addition, there are special edges that symbolize catalytic/inhibiting effects on reactions. These are always of the type $v_i \in S \Rightarrow w_i \in R$ pointing from a molecular to a reaction. Both

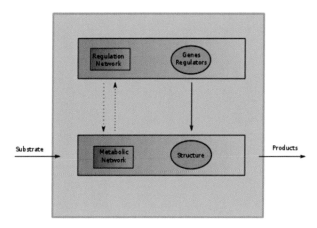

Figure 16.3 Schematic reaction network, adapted from [959].

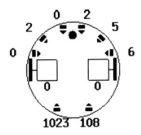

Figure 16.4 Khepera robot schema. The robot is equipped with six proximity sensors at the front, two on the side, and two at the back. They are based on infrared reflection measurement. In the situation depicted, there is an obstacle left in the back of the robot. In addition, there are two motors for wheeled motion, both currently receiving no signal. From [958], ©MIT Press 2001, with permission.

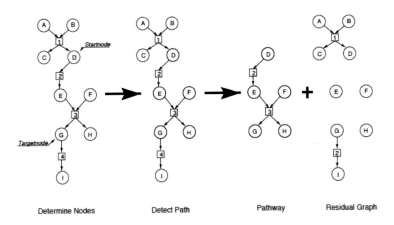

Figure 16.5 How to detect a pathway in a reaction graph. (a) Search for start and target substance, represented by nodes D and G. (b) Depth-first search is used to determine shortest path from start to target node. (c) The pathway is isolated from the rest of the graph, leaving a residual graph with partially unconnected nodes. From [958], ©MIT Press 2001, with permission.

types of reactions have reaction rates, and therefore their edges carry appropriate weights, the former of a default size k_0, the latter in a range $k_i \in k_{min}, k_{max}$.

A metabolic pathway is a subgraph $P(v, w) = (V', E')$ with $v, w \in S$; $v, w \in V'$; $V' \subseteq V$; $E' \subseteq E$. P only contains nodes that are encountered on the shortest path from v to w. In addition, every reaction r in P must be participating in the path. There are more definitions, but we refer the interested reader to the literature [958]. Figure 16.5 shows which steps are necessary to detect a pathway from one substance (here D) to another (here G).

Once we have established the representation for a reaction system, we can now consider its variation under evolution. This is done in a typical fashion of a genetic programming system, an evolutionary algorithm applied to a size-changing representation using the operations of mutation and crossover [76]. We can imagine mutations to take one reaction graph and randomly modifying it by a choice of the following operations:

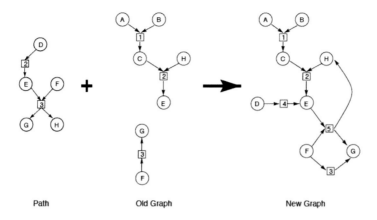

| Path | Old Graph | New Graph |

Figure 16.6 How to insert a pathway into a reaction graph. (a) The extracted path from a graph as in figure 16.5 is connected. Missing node D is created for this purpose. All existing molecular nodes and reaction nodes remain unchanged. From [958], ©MIT Press 2001, with permission.

1. Addition or deletion of a molecule node (start and target nodes stay fixed).

2. Addition or deletion of a reaction node (as far as it doesn't remove target nodes from the graph).

3. Addition or deletion of catalysts/inhibitors.

4. Changing the type of reaction.

5. Changing the catalytic/inhibitory rates of reactions.

Crossover operations work by merging two graphs at a start and target node. We refer again to figure 16.5. Suppose it was determined that the crossover should combine two different reaction graphs starting at node D and ending at node G. We take the path from figure 16.5 and insert it into another reaction graph from which we have removed the path from D to G. Figure 16.6 shows how the potentially dissected parts of the second graph (called "old graph") are connected with the previously extracted path from the other graph to form a full reaction graph.

Clearly, these operations will end up changing the reaction graph in various ways, in general varying its length as well as its constituent parts. Notably, the generation of new nodes and reactions means that the system is constructive, i.e., the total number of constituents and reactions present at the beginning is not conserved over the course of an evolutionary run. Technically speaking, the set of terminals in the GP system varies during the course of a run, and the size of reaction graphs varies as well.

Details of evolutionary system can be studied elsewhere [958], but we shall end our discussion with the presentation of the results of a sample run. In figure 16.7 we can see on the left the reaction network resulting from a run, cut down to the simple essential reaction pathways. With this behavioral program equipped, the Khepera robot behaved as shown in figure 16.8. The robot is shown in a series of snapshots in a 2D landscape with obstacles. Once it approaches a wall, it turns around in its preferred direction (left) and moves in the opposite direction. The evolution of the artificial reaction network took place in real hardware, and we were able to show that fundamentally the same network (up to scale factors) would evolve in a simulated and the real world, at least for this simple task.

Other important work in the area of robot control using artificial chemistries or the simulation of (simplified) bacterial behavior can be found in [660]. There, Penner et al. have formulated a simplified AC that models actual bacterial chemotactic behavior. Their approach was, however, slightly different in that they did not consider mass action directly. Instead, their AC seeks to model the diffusion of signaling molecules through the cytoplasm in an implicit way: each reaction rule has a firing probability reflecting the probability that the involved educt molecules are in sufficient close proximity to each other for the reaction to take place.

More recently, a generalized class of systems has been explored in the context of control of complex, even chaotic, dynamical systems. This combines metabolic, signaling, and genetic networks into one system [519,520]. The AC of the metabolic network is under control of an artificial genetic regulatory network [68], and both are influenced by the signaling network transmitting signals from the environment (see figure 16.9 for a rough sketch).

16.2 ACs for Networking

In this section we look at applications of ACs to the computer networking domain. The difference between computer networking and distributed systems in general is that the former focuses on low-level communication protocols that must interface with the machine and the transmission lines, whereas the latter have a wider scope, including high-level applications that interface with the end user. In computer networks, efficiency and speed are major concerns, concerns that are even more prevalent in recent times due to the enormous demand for fast Internet access. In contrast, distributed systems generally prioritize usability, availability, and fault tolerance above other requirements, in order to offer a comfortable and reliable application experience to the end user. In spite of the differences, both areas share many common principles, such as the quest for parallel and decentralized algorithms that should be able to achieve a desired global behavior without a central information storage, relying instead on local information exchanges. This is where artificial chemistries become interesting for these domains.

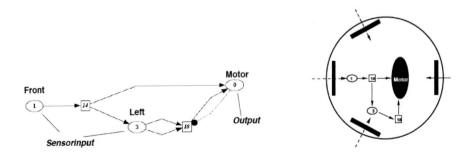

Figure 16.7 Reaction network evolved for a robot to avoid obstacles. Left: Sensor input is fed from front and left side and react to provide a signal to the motor. As a result, the robot turns around in a preferred direction when encountering an obstacle. Right: Configuration on Khepera robot. From [958], ©MIT Press 2001, with permission.

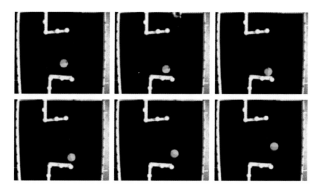

Figure 16.8 Snapshots of the evolved behavior of the Khepera robot in a 2D landscape with obstacles. From [958], ©MIT Press 2001, with permission.

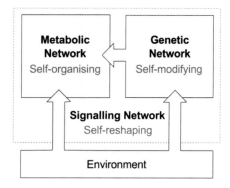

Figure 16.9 Sketch of an artificial biochemical network, a combination of an artificial genetic network, an artificial metabolic network, and an artificial signaling network.

We start this section with an overview of approaches with ACs to networking. Then we show an example of a communication protocol developed with the *Fraglet* language, a programming language inspired by chemistry and based on string rewriting. We shall postpone a general discussion about AC applications to distributed computing to chapter 17.

One branch of artificial chemistry research for networking takes inspiration from spatial chemistries such as reaction-diffusion systems. Pattern formation with reaction-diffusion has been used to build distributed algorithms for wireless networks [244, 527, 608, 946], and autonomous surveillance systems [414, 945]. The typical spot patterns resulting from activator-inhibitor systems have been used to regulate channel access in ad hoc wireless networks [244], and to make decentralized but coordinated decisions on the placement of node functions in sensor networks [608]. Alternative sets of underlying chemical reactions to solve this coordination problem were explored in [941], and their automatic evolution was reported in [939]. Reaction-diffusion has been applied to prevent and control congestion in wireless mesh networks [946]. Venation patterns obtained by anisotropic diffusion in an activator-inhibitor mechanism were used by [526, 527] to build "data highways" to conduct information in dense wireless sensor networks. Activation-inhibitors have also been applied to the autonomous coordination of a distributed camera surveillance system [945], in order to decrease blind spots in the surveillance area. Also in the surveillance domain, an algorithm inspired by reaction-diffusion is proposed [414] to track the movement of targets and adjust the video coding rate accordingly.

Another branch of artificial chemistry research for networking builds upon chemical metaphors and multiset rewriting. The Fraglets programming language discussed below falls into this category. With respect to related approaches such as Gamma, CHAM, MGS, and P systems, the rewriting rules in Fraglets have the peculiarity of operating only at the front symbols of a molecule: this constraint makes their processing efficient, as only a few symbols have to be inspected and modified at each execution step. The approach is actually similar to the way data packets are processed in the current Internet. As will become apparent later (section 19.3.1) such an operating mode resembles the DNA Automaton [768], where the enzyme-DNA complex progressively reads and digests the input molecule by its front side.

16.2.1 Chemical Communication Protocols with Fraglets

A chemically inspired programming language called Fraglets was proposed by Tschudin [862] for the design and evolution of communication protocols. In this language, a molecule is a string that carries data or instructions (program fragments), and can be seamlessly transported as a data packet over a communication infrastructure such as the Internet. Network nodes in the infrastructure (computers, routers, sensor/smart devices) can be regarded as chemical vessels where molecules collide and react. Chemical reactions apply string rewriting rules that consume the reactant molecules and produce new molecules in the process. Since molecules may carry instructions, the overall distributed program may get rewritten in the process.

Table 16.1 Subset of the Fraglets instruction set from [862]

Unimolecular reactions (transformations)		
Operator	*Input*	*Output*
dup	[dup A B *tail*]	[A B B *tail*]
exch	[exch A B C *tail*]	[A C B *tail*]
nop	[nop *tail*]	[*tail*]
split	[split *string₁* * *tail*]	[*string₁*], [*tail*]
send	$_A$[send B *tail*]	$_B$[*tail*]
Bimolecular reactions (string concatenation)		
Operator	*Input*	*Output*
match	[match A *tail₁*], [A *tail₂*]	[*tail₁* *tail₂*]
matchp	[matchp A *tail₁*], [A *tail₂*]	[matchp A *tail₁*], [*tail₁* *tail₂*]

Table 16.1 shows some basic instructions of the Fraglets language. The instructions are divided into two categories: unimolecular reactions or transformations, and bimolecular reactions that concatenate two strings. Fraglet rules always operate at the front part of the molecule. The first few symbols of each string determine the rewriting action, and the operator is consumed after execution, which is analogous to the way packet headers are processed in the protocol stacks of computer networks. The transformation dup duplicates an input symbol, exch swaps two input symbols, nop consumes a symbol, split cleaves the fraglet at the first occurrence of the

'*' symbol, and send transmits the tail of the fraglet to the specified machine. Since the operator symbol is consumed during the reaction, the program is degraded as it is executed. The bimolecular reaction rules basically merge the tails of two fraglets under the condition that their head symbols match exactly. The matchp rule is a persistent version of match in which the rule itself is not consumed during the reaction (analogous to a catalyst in chemistry). This rule is useful as a persistent program storage.

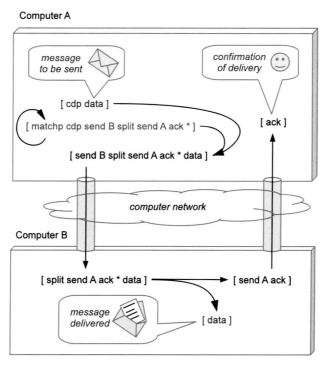

Figure 16.10 Confirmed Delivery Protocol (CDP) in Fraglets [862]: this is a simple protocol that sends a message from machine A to machine B, returning an acknowledgment to the user after the message has been successfully delivered to its destination. The persistent (seed) code is shown in red; data items are shown in blue; intermittent code fragments generated during execution are shown in black, and result from the reaction between the seed code and the available data items. Adapted from [862].

Figure 16.10 shows the execution flow for a simple communication protocol written in Fraglets: The confirmed delivery protocol (CDP) transmits a message from machine A to machine B and returns an acknowledgment to the user when the transmission is completed. The program (red molecule) reacts with an incoming data molecule, generating a send fraglet that ships a split fraglet from machine A to machine B: upon arrival at node B, the split fraglet cleaves a portion of the message that is delivered to the destination, and returns another portion to machine A as an acknowledgment of correct delivery. Note that machine B initially had no CDP program: it executes the code fragments as they arrive. Therefore, this implementation of CDP falls into the category of active protocols, that is, protocols that transport their own code across the network, in contrast with traditional protocols that require the full program to be installed beforehand in all the machines where it executes. This simple example shows how a chemical metaphor can lead to unconventional ways to implement distributed algorithms.

The full implementation of a more complicated protocol, called flow control with credits, can be found in [862]. The flow control with credits protocol keeps track of packet sequence numbers and reorders them in case they are received in the incorrect order. A number of subsequent protocols were later designed [576, 581] or evolved [914, 943] for Fraglets. Chapter 17 will come back to Fraglets, including an example from the distributed computing domain.

16.3   Language Dynamics and Evolution

Language is one of the most remarkable traits that distinguish humans from other animals. Although some animals communicate using protolanguages that associate meaning to signs such as pheromone trails and sounds, humans are the only extant species capable of complex languages that feature elaborate grammatical rules and enable the expression of abstract concepts and ideas. Human languages are far from static, and have been postulated to evolve in a Darwinian sense. Several computational models of language dynamics have been proposed, including mathematical models based on differential equations [4], and agent-based models focusing on the interactions among individuals in a population [806, 812].

The methods used for modeling language dynamics have many aspects in common with artificial chemistries, evolutionary dynamics, and the modeling of biological systems. For instance, the process of learning one or more languages can be modeled using dynamical systems [375], and the evolution of languages can be modeled using evolutionary games [624]. However only a few recent proposals have explicitly used ACs to model language processes. ACs have been used to model the mapping of language to meaning [342], as well as the convergence to a common language in a population of agents [212]. These studies illustrate the potential of ACs for the modeling and simulation of language dynamics, although much remains to be explored in this field, as pointed out by the referenced works and by [166]. One of the advantages of ACs in this area is their ability to capture the properties of both ODE- and agent-based models, an integration that is currently lacking in the field [165]. There is thus plenty of room for more research in this direction, and it is this untapped potential that we would like to highlight in this section.

The study of natural language dynamics is highly interdisciplinary, spanning linguistics, evolutionary biology, and complex systems science, among other areas. It covers at least three aspects [166]:

- Language evolution studies how the structure of a language — its syntactic and semantic elements — may evolve out of a primitive protolanguage.

- Language competition looks at the dynamics of two or more languages as they compete for speakers in a society, possibly leading to the survival, death, or coexistence of some of those languages.

- Language cognition seeks to understand how language is interpreted and produced by the brain.

We now discuss these topics in more detail, with focus on the methods used in language research that share common aspects with methods used in ACs. Beware that our perspective is highly biased toward these commonalities and is not intended to reflect the language research field as a whole. We intentionally leave out any studies that do not involve computational modeling, such as anthropological, psychological or statistical studies, since these are outside the scope of this book.

16.3.1 Language Evolution

A model of language evolution based on evolutionary game theory has been developed by Nowak and colleagues in a series of papers. In [624] the authors showed how a population of individuals can converge to a protolanguage that associates sounds to objects. They showed that when each object is associated with a specific sound, errors (misunderstandings) limit the number of objects that a protolanguage is able to describe. This "linguistic error limit" can be compared to Eigen's "error threshold" [249] in primitive replicators under the RNA world hypothesis (see chapter 6). They showed how words composed of multiple sounds, and later a grammar, can be used to overcome this limitation. They further argued that a grammar could have evolved as a means to minimize communication errors in the protolanguage, causing it to transition from a set of word associations to a full grammatical language. The evolutionary dynamics of such transition is further explored in [626], showing that syntax offers a selective advantage only when a sufficiently high amount of signals need to be expressed.

The evolution of a Universal Grammar (UG) is investigated in [623]. The hypothesis of a UG was introduced by Chomsky to refer to the biological ability of any human child to learn any human language. UG is supposed to be a rule system that is able to generate the search space of all possible candidate grammars. The process of language learning (language acquisition in linguistic terms) can then be described as a search in the UG space using the UG rules. The UG itself is believed to be innate, and to have evolved at some point in human history, however this issue is controversial [375, 808]. A good UG would allow children to learn the language of their parents in a short period of time. For this purpose, it would have to allow the exploration of a large but relatively constrained search space, using only a limited amount of sample sentences as input. From this perspective, language learning resembles an evolutionary search process. In this context, a formulation of language acquisition is presented in [622] using evolutionary game theory. Language dynamics is described by a system of ODEs that can be seen as an extension of the quasi-species equation and a special case of the replicator equation.

16.3.2 Language Competition

Language competition can be modeled using deterministic Lotka-Volterra equations or probabilistic models, in well mixed or spatial scenarios (for instance, using reaction-diffusion) [4, 166, 656]. In such studies, internal structure of languages is usually not taken into account, neither are the different learning abilities of speakers. Some of the results in this area reveal phenomena that are strikingly similar to well-known findings from theoretical biology: For instance, dominance of one language (with extinction of all others) is observed in systems without spatial considerations [4], whereas a phenomenon of resilience or coexistence is observed in spatial systems [656] (reminding us of the hypercycle in space [125]). The formation of spatial patterns similar to Turing patterns has also been observed, with a segregation of the geographical space into different language regions [33, 656]. Furthermore, social connectivity among individuals can be seen to have an influence on language spread and competition: spatial patterns formed under a small world social structure [165] differ from those on regular lattices, with hub individuals playing a major role in language propagation. These results are consistent with the theory of evolutionary dynamics on graphs [621].

16.3.3   Language Cognition

So far we have discussed language dynamics mostly at the level of populations of individuals. The studies of language cognition focus on the inner language-processing mechanisms that occur inside the brain, such as language acquisition, usage, and loss. Since language is a communication means, it does not make sense to study it for an isolated individual; therefore, the population aspects are still present. From a computational perspective, individual speakers are modeled as agents in a multiagent system, and we must look at the dynamics of the system at two levels: inside the agents' minds, and at the population level.

Undoubtedly the most famous work in this field is the one by Luc Steels [804, 806, 812]: During the late 90s, he demonstrated the emergence of a common vocabulary (lexicon) in a group of software agents, then in a group of robotic agents. He then set out to investigate how grammars can emerge and evolve, using increasingly sophisticated robots. Steels's framework is a distributed multiagent system where the agents are autonomous and have no access to global information. The system is open, permitting the dynamic the influx and outflux of agents and meanings [804]. Agents interact via language games, among which naming games that lead to the emergence of a shared lexicon [805] via positive reinforcement of successful word-meaning associations.

According to Steels [807], the spontaneous emergence and complexification of natural languages may be explained by a combination of the following ingredients: evolution, coevolution, self-organization, and level formation. The evolution of languages occurs via cultural transmission of language constructs with some variation over the generations, combined with a selective pressure to maximize communication success while minimizing cognitive effort. Coevolution between language and meaning occurs, since a rich language offers the means to express more complex concepts, that in turn cause the language itself to become more complex. Self-organization explains the convergence to a shared language in a population of agents, without central control, and in spite of many equally fit alternative language structures. Level formation might explain the origin of syntax by composition of schemas, leading to the emergence of hierarchically structured grammars.

Below we shall examine two aspects of Steels' extensive work: a formulation of naming games with artificial chemistries, and the more prospective investigations around the emergence of grammars.

Naming Games

A formulation of the naming game as an artificial chemistry is presented in [212]. In their system, a set of meaning molecules $M_i \in M$ and a set of sign (word or name) molecules $S_j \in S$ float in a chemical reactor where they interact with adaptor (or code) molecules $C_{ij} \in C$ that define the mapping from word to meaning. In the simplest form of the naming game, two agents must agree on a name out of a set of N_s names for a given concept or meaning M. The game then operates according to the following set of chemical reactions:

$$M + C_j \xrightarrow{k_s} S_j \tag{16.2}$$

$$M + S_j \xrightarrow{k_c} C_j \tag{16.3}$$

$$M + S_j + C_j \xrightarrow{k_a} 2S_j \tag{16.4}$$

$$M + S_j + C_{j'} \xrightarrow{\kappa_1} M + S_{j'} + C_{j'} \tag{16.5}$$

$$M + S_j + C_{j'} \xrightarrow{\kappa_2} M + S_j + C_j. \tag{16.6}$$

Reaction 16.2 generates a word S_j for meaning M using a code C_j. Reaction 16.3 produces the code C_j able to generate the M-S_j mapping. Reaction 16.4 leads to the replication of a successful word S_j: a word is successful when it matches its corresponding meaning through the corresponding code. Reactions 16.5 and 16.6 eliminate mismatching words (resp. adaptors) by replacing them with matching ones. The authors then convert the system of reactions 16.2—16.6 to an ODE and show that it achieves the same lateral inhibition behavior that had been reported in previous work [809], converging to a shared lexicon.

The Origin of Grammars

Steels also applied his framework to explore the origins of grammar in populations of robotic agents [806,808]: whereas the lexicon is a set of associations between word and meaning, a grammar is a set of associations between a syntactic schema (defining the structure of a sentence) and a semantic schema (embedding the meaning of a sentence). Steels refuses the hypothesis of an innate and static UG, in favor of a complex adaptive systems view [808], according to which languages and their grammars change continually; any universal tendencies we might observe are actually emergent global properties stemming from local agent interactions. Following this view, Steels and colleagues have proposed Fluid Construction Grammar (FCG) [811], a formalism that seeks to capture the flexibility and variability of natural languages in a computational context. FCG is based on hierarchical grammatical constructions, that are divided into semantic and syntactic poles. When these two poles match to some degree, they are merged in order to map sentences to meaning and vice versa. Multiple constructions may compete due to synonyms or ambiguities in the sentences, leading to alternative matching poles. A search method is then used to assist in the match and merge process [810]. An FCG implementation in Lisp is offered to the research community as open source software.[1] Since its inception, numerous experiments have been carried out using FCG, including the evolution of specific grammatical elements such as German agreement systems [116].

Note the similarities between the FCG and artificial chemistries: Construction poles in FCGs could be regarded as molecules that match by lock-and-key binding and merge by chemical reactions. In [212], the combination of FCG and artificial chemistries is indeed regarded as a promising approach to investigate the still unknown mechanisms of formation of grammatical codes. However, it is not clear whether or how FCGs could really occur in the human brain. Building upon a previous hypothesis of natural selection in the brain [274], Chrisantha Fernando [272] proposes a chemical metaphor combined with a spiking neural network as a hypothesis for the

---

[1] http://www.fcg-net.org

Figure 16.11 Examples of music generated by artificial chemistries. Left: Three homophonic phrases [592]; Right: Three polyphonic phrases [854], ©Springer 2008, with permission.

implementation of FCGs in the brain. According to his hypothesis, the physical symbol system needed to implement FCGs in the brain could take the form of spiking neurons that recognize certain temporal spike patterns as symbols (akin to words), which may then bind together through chemical reactions to form more complex symbol structures (akin to grammars). Some ideas about how learning, matching, and merging operations might be performed on such neural structures are then sketched. Much remains to be done however, and we are only beginning to scratch the surface of what might be possible at the intersection of language dynamics and ACs.

16.4   Music Composition Using Algorithmic Chemistries

Next, let us discuss an example from the arts, from music in particular. The composition of music has long been the realm of humans alone. However, it has been noted that some animals produce music, a field studied under the name "zoomusicology" [546]. It turns out that what we call music, i.e., the aesthetic use of sound, is used among animals for communication. Think of the songs of birds or whales, both in completely different environments. Machines have also been used not only to play music, but to compose, both according to classical composition principles [612] and with the help of randomness and selection (in an interactive mode) [717].

Artificial chemistries can also be used for music composition. The approach starts with the observation that the sequence of musical notes comprising a melody could be considered a pattern. Following rules of composition could be interpreted as the application of combination rules between patterns, and the creative part of the composition could be captured in random factors that allow certain combinations to take place and not others. In order to avoid random sounds, chords could be part of the patterns that are combined, and there might even be some stricter rules for sequences filtering out disharmonic combinations.

Kazuto Tominaga has published some stunning examples of algorithmically generated music based on these principles on his website [2]. One study he performed with collaborators was to look at a method to produce simple phrases in a homophonic environment [592]. He later also worked on the composition, with a similar system, of polyphonic phrases [854]. Figure 16.11 shows a number of music examples generated by this method.

It seems that algorithmic chemistries provide the right mix of randomness and structural filtering abilities to allow for special constructs to emerge. The experiments reported were executed in a well-stirred reactor tank.

16.5 Proof Systems

Fontana and Buss have observed [284] that artificial chemistries could be applied as logical proof systems. The reasoning is based on an equivalence between λ-calculus, the system in which Fontana formulated his artificial chemistries, and proof theory. The equivalence is called the Curry-Howard isomorphism [406] and states that a type σ in λ-calculus corresponds to a logical formula or proposition/statement σ. Thus, a proof could be constructed by finding an appropriate λ-term of type σ. Fontana and Buss then state that the mathematical notion of truth (as conveyed by the proof of a statement) could be fairly mapped into the construction of a molecule: Existence of the molecule would mean truth of the proof. This provides a very interesting link between the "static" world of mathematics and the dynamic world of physical or, rather, chemical objects. Such a link is of enormous importance when we start thinking about the nature of time, as it is reflected in the dynamics of physicochemical interactions, as opposed to the notion of timelessness that governs most parts of mathematics [79].

We will look at this problem in the context of automatic theorem provers, a classical branch of Artificial Intelligence. An automatic theorem prover is fed with a theory $A = \{A_1, \ldots, A_n\}$ of expressions. These expressions might be the axioms of that theory, or they might include derived expressions that have been shown to be true. In addition, another expression P is given, and the question to be answered is whether P is consistent with the theory A, symbolized by $(A \models P)$. The automatic theorem prover takes the expressions of A and combines them into new expressions in a process called "inference." If P is producible by a series of inferences from A, P can be called a theorem of A. The sequence of steps that lead to this construction is said to be the proof $(A \models P)$. Interestingly, we can see here that in this way A is constructive, in that it has now lead to a theory $A' = \{A_1, \ldots, A_n, A_{n+1} \equiv P\}$ incorporating the new theorem.

In [147, 148] the idea of generating proofs was studied using first-order predicate calculus, a logical representation of statements that is simple enough to calculate inferences. First-order calculus makes statements about objects and their properties and relations ("clauses"). It is called first-order because the objects of statements cannot be relations of other objects. We used "binary resolution" as a method for inference. Thus, the molecules are are given by clauses and reaction rules are considered to be the inference rules transforming those clauses. We can imagine that a proof might take a substantial amount of inferences, i.e., reactions, to complete. The advantage of looking at such a method as an artificial chemistry is the parallel character of the system: A traditional proof is a well defined sequence of steps, whereas a reaction system has stochastic reactions taking place, which might or might not succeed in a proof. Thus, the AC is really a proof search system, a tool that possibly arrives at a proof if sufficient "reactions," that is, inferences, have taken place.

Once the molecule of the proof is constructed — which requires possibly substantial computation time, since many of the inferences might lead to dead-ends in the form of molecules that cannot be further transformed — the sequence can simply be read off from its construction sequence. Thus, all dead-ends naturally will not show up in the sequence of reactions, though there might be subsets of the reaction sequence that cancel each other out, and can therefore be removed from the sequence.

There is a deep relation between the application of inference rules and chemical reactions, because inference rules have to match the patterns of clauses to be able to unify/resolve the clauses and produce a new clause, while in chemistry the 3-dimensional shapes of molecules must correspond to each other in order for the reaction to take place. Thus the shape of a molecule and the "shape" of a clause are equivalent.

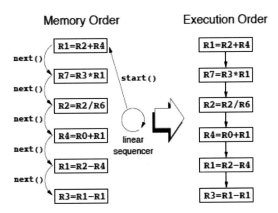

Figure 16.12 Execution of a regular program in a von Neumann computer. The von Neumann machine is a stored program computer that holds both data and program instructions in memory. Two aspects of this execution are therefore memory order ("space") and execution order ("time"). From [74], ⓒSpringer 2005, with permission.

16.6 Artificial Chemistry and Genetic Programming

In this section, we introduce another way of looking at programming a classical register-machine computer. The idea here is to consider the simplest instructions of a computer in a register-machine language as equivalent to chemical reactions. Thus, for instance, suppose you have a computer that applies the following data manipulation to the content of registers r_1, r_2, r_3:

$$r_3 = r_1 + r_2 \tag{16.7}$$

What we are saying with this operation is that we take the data content of register r_1, interpret it as a number, and add it to the data content of register r_2, again interpreted as a number. The result is stored in register r_3. Often in computer science, this is alternatively written as:

$$r_3 \leftarrow r_1 + r_2 \tag{16.8}$$

With that notation, not much is missing to realize that this is somewhat similar to what could be seen as a chemical reaction:

$$r_1 + r_2 \rightarrow r_3 \tag{16.9}$$

Indeed, the operation of addition performed in this instruction could as well be a chemical reaction. Where we need to be careful is when we look at the content of these registers, which will be different depending on what we have determined as the initial conditions of the system. The instructions will work for every content (at least if precautions have been taken to interpret the content as numbers). Contrast that with a chemical reaction, where r_1, r_2, r_3 stand for chemical species and the only thing that might change is the concentration or count of each of these types in a reaction. In that sense, a chemical reaction is a much simpler process than the execution of an instruction.

The idea that a collection of instructions of a register-machine computer language could be used for an artificial chemistry has many facets that have not been explored fully yet. In what

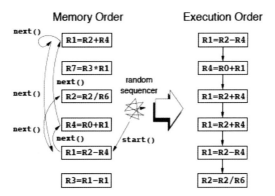

Figure 16.13 Execution of a program akin to an AC in a von Neumann computer. The sequence generator returns a random order for the execution of instructions, in this way decoupling memory location and time. From [74], ©Springer 2005, with permission.

follows, we shall restrict ourselves to a small subset of what could be done with this idea, and we encourage the reader to come up with alternative scenarios. What we are going to focus on is the higher-level notion that a set of instructions could be executed in a different way from the traditional approach of algorithmics where a strict sequence has to be obeyed. Figure 16.12 shows an example of a program and its implementation in a computing register machine, looked at from the traditional perspective. Each instruction will be executed step by step, which is done by moving through the spatial/memory locations where a program is stored, and, moving one step at a time, translate that set of instructions into a sequence of instruction that is executed one by one. A call to the start() and next() function will allow to step through the memory space, and produce the sequential order of execution usually considered the only one produceable.

If, however, we use the AC framework, we see that it is not necessary to have the sequence of reactions (instructions) in the same order every time the program is executed. In fact, in chemistry, due to the random processes of Brownian motion, there will be many different ways to order different reactions. Transported into the realm of programs, a more realistic scenario of execution is depicted in figure 16.13. Here, start() will start the execution at an arbitrary, randomly determined instruction, and next() will allow to step through the program in a randomly determined order. Notably, a second execution of the same program will results in a different order of execution of the same instructions laying in memory space.

The interesting point about this random execution of a program is that the instructions have not changed, it is just the sequence that has been broken up. Thus, any dependencies between instructions will be dissolved by such an approach. If the execution of the same program is repeated multiple times, the behavior will tend toward an average behavior . We have examined this approach in a series of contributions [74, 75, 497, 498].

It is natural to ask what a program in a random sequence could be applied to. Given that the instructions remain the same, would it not, on average, always produce the same output? This depends critically on the decision where to feed input and to read off output. If we fix those two locations by determining from the outset which registers are used as input registers (read-only) and which are to be used as output registers (read-write), then we have a way to observe —

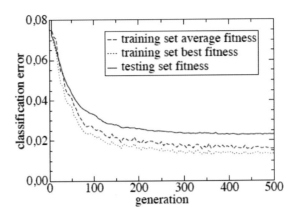

Figure 16.14 A typical run of the ACGP system. From [74], ©Springer 2005, with permission.

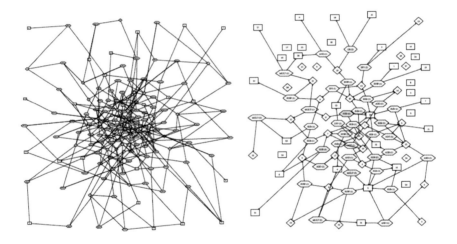

Figure 16.15 Graph structure of the resulting program, run on the Thyroid program. Left: Initial best program, consisting of 100 randomly assembled instructions; Right: Best program of generation 465. Instructions are symbolized by hexagons, whereas the registers they feed their data to (and from) are symbolized by diamonds. From [74], ©Springer 2005, with permission.

regardless of the execution order — what a particular set of instructions can do. We suspected that certain soft computing tasks, like the recognition of patterns, might be useful applications of this approach. In order to improve on the performance of a particular program, we would define an error function which could be improved upon by evolving the set of instructions. This is what we mean by "genetic programming" in this context. Thus, as in other GP systems [76], we would introduce variation operations like mutation of instructions and crossover between programs, and evaluate a population of programs constructed as a sequenceless (multi)set of instructions.

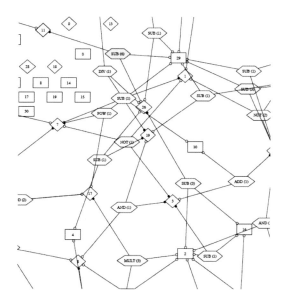

Figure 16.16 Magnified graph structure of the resulting program, run on the Thyroid program. Register 11, upper left, contains the output of the program. Darker color indicates how frequent this instruction is represented in the multiset of this program. The numbers in parentheses indicate how often a particular instruction is executed. From [74], ©Springer 2005, with permission.

As an example, we show in figure 16.14 how a system of this type behaves when applied to a classical pattern classification problem (thyroid cancer) from the UCI dataset [47]. In this case, a population of 100 individual programs has been run for 500 generations. In order to make sure that the maximum number of instructions has been reached in each evaluation, five times the number of instructions has been executed. The system shows a behavior well known from GP systems. The testing set fitness is always worse than the training case fitness, fitness generally declines quickly at the outset until it grows slower in declining later on. Some fluctuations occur in the behavior, based on the method of sampling from our training set.

The resulting programs can be depicted in the graphs shown in figure 16.15. These graphs show the best "program" in different generations. Registers are symbolized by diamonds, instructions by hexagons. What these graphs capture is the data flow through a network of registers, and its transformation via functions (instruction operations). The left part of figure 16.15 shows a much larger network than the right part. This is because a number of registers and instructions are not used any more in the course of evolution. Even the right part is somewhat larger than necessary. Figure 16.16 shows the most important part of the program.

Why would one want a system to process information in this form? This approach lends itself well to parallel execution. Since there is no coordination between the execution of instructions, any subset of instructions could be executed on a different process, in principle. Thus, the system would not have a requirement for coordination or communication, and the system would be scalable at run-time: New processors could be recruited to start processing instructions if it turns out to process information too slowly. A program would not necessarily need to reside in contiguous memory space. Finally, a program would be fault tolerant since the number of instructions in the multiset corresponds to their importance in arriving at the correct result of the process.

16.7 Summary

In this chapter we have discussed a few examples where Artificial Chemistries have been applied to solve a wide range of problems. In the upcoming chapters we will delve deeper into specific different classes of AC applications.

The systems covered in this chapter were all examples of *in silico* artificial chemistries. Chapter 17 will cover computation with *in silico* ACs in more detail. Before moving to the wet lab, Chapter 18 provides an overview of how ACs can be used in the computational modeling of real biological systems. For wet applications of artificial chemistries, e.g., *in vitro* DNA computing or *in vivo* bacterial computing, the reader may consult chapter 19.

Finally, chapter 20 goes beyond chemistry and biology, discussing mechanical, physical, and social systems where ACs have been successfully applied.

COMPUTING WITH ARTIFICIAL CHEMISTRIES

A chemical reaction entails a transformation from one state (the collective state of the educts) to another (the products of the reaction). Computation also transforms states by processing input data into output results. Hence, chemistry can be seen as a natural model of computation and can be recruited for computational purposes. Throughout this book we have already seen some examples of algorithms implemented with artificial chemistries, such as decomposing numbers into primes (Chapter 2), or sorting numbers in parallel using Gamma or MGS (Chapter 9). In this chapter we will delve more deeply into the techniques for programming inspired by chemistry.

The term *chemical computing* encompasses both computation inspired by chemistry [231] and computation with chemical and biochemical substrates [793]. The former will be the focus of this chapter, and refers to computation models that take inspiration from chemistry but run *in silico*, that is, as software on traditional electronic hardware. The latter has also been called molecular computing, biomolecular computing, molecular information technology, or wetware [132] for short, and covers computation performed at the level of single molecules in the context of bioelectronics, *in vitro* using molecules in the wet lab, or *in vivo* with living cells or small organisms [10, 99, 194, 356, 444, 776, 951]. Some of these systems will be discussed together with other wet artificial chemistries in chapter 19.

In a more general context, we can say that chemical computing belongs to the broader field of natural computing, which in turn is a subfield of the area of unconventional computation. Natural computing covers a number of nature-inspired computation on various substrates, including quantum computing, wetware, and biologically inspired computing.

While the construction of "wet" chemical computers *in vitro* and *in vivo* still poses major research and technological challenges, much can still be learned from the construction of chemical computers *in silico*, and the use of ideas from chemical computing in everyday algorithmic practice. Chemistry can be a rich source of inspiration for the design of new algorithms running on conventional hardware, notably for distributed and parallel algorithms. Algorithms that have to deal with uncertain environments or nondeterministic behavior can also benefit from an inspiration from chemistry.

In silico chemical computing also poses challenges of its own: Programming in a chemical way departs from conventional programming because it requires "thinking parallel" and "thinking chemically." Being still a young research domain, general techniques and guidelines for *in silico* chemical computing are hardly available, and chemically inspired algorithms are currently not widespread through mainstream computer science. On the other hand, these challenges can stimulate many of us to propose new unconventional algorithms that might solve problems in creative ways, and might be more robust than conventional solutions due to their inherent parallelism and self-organizing properties. Some of these algorithms may also work as initial models for future wet chemical computers, simulating the wet environment on conventional hardware.

This chapter reviews some of the current efforts toward techniques and algorithms for *in silico* chemical computing. We start by uncovering some tentative principles of implementation of chemical programs, such as how to structure these programs, the levels of organization of program structures, the interplay between emergence and programmability, and the thermodynamic aspects of computation, including reversible computing. A discussion of algorithmic examples that have been proposed in the literature then follows, focusing on two application fields: the use of chemistry as a metaphor for parallel, distributed and unconventional algorithms, and the simulation of wet chemical computing *in silico*. For the first application field, we draw examples from search and optimization in section 17.2, and from distributed systems in section 17.3. The second application field appears in section 17.4, and is a preparation for the topic of computation with wet chemistries that will be covered in chapter 19.

17.1 Principles of implementation

The design of computer programs is the domain of software engineering. In the early days of computer science, programs were written directly in machine language, using binary or hexadecimal numbers that coded for machine instructions. Assembly language then followed, in which programmers could input the machine instructions using mnemonic alphanumeric codes instead of raw numbers. Higher-level programming languages appeared some time later, starting with structured languages such as Ada, C, and Pascal, then moving to object-oriented languages such as C++ and Java, scripting languages such as Perl and Python, and culminating with graphical software design languages such as UML.[1] Other languages departed from the mainstream imperative languages by capturing alternative ways to express programs, such as via logic inference (Prolog), via function composition (Lisp), or via formal specifications (VDM,[2] Lotos[3]). Programming languages reflect software engineering best practices on how to design and structure programs such that they are at the same time clear, efficient and easy to maintain.

In the case of chemical computing *in silico*, two extreme situations are often encountered: whereas many systems are based on machine and assembly language (such as the automata reaction chemistry and the assembler automata from chapter 10), equally many others are based on formal calculi such as Gamma and P systems. The former tend to express programs that are difficult to understand for humans, and lend they themselves mostly to automatic programming. The latter can express programs that are in some cases very intuitive for humans (like the

---

[1] Unified Modeling Language
[2] Vienna Development Method
[3] Language of Temporal Ordering Specification

Gamma examples from Section 9.2), but tend to be far from chemical reality and to resist automatic programming. There is currently a lack of chemical programming languages that would be clear enough for humans, and at the same time close enough to chemistry, enabling the design of efficient programs. Programs in that language would need to evolve by themselves and to scale up in complexity, and would be properly implementable on wet chemical computing devices of the future. We are not aware of any specific guidelines on how to design chemical programs for efficiency and clarity. Some general strategies for chemical programming have been put together in [231], some principles of programming for Gamma have been proposed in [56], a set of design principles for organization-oriented chemical programming have been outlined in [552], and some advice on how to design assembly-level chemical programming languages for evolvability has been laid down in the context of Avida [628].

In this section, we have attempted to put together some preliminary design principles based on the few available guidelines from literature, combined with our own experience. We focus on common principles valid for chemical computing on top of any substrate, and on specific principles applicable to *in silico* chemical computing.

17.1.1 Structuring Chemical Programs

The first step in the design of a conventional computer program is to define the data structures and basic algorithms operating on them. In object-oriented programming, for instance, this amounts to defining classes of objects and methods (procedures) operating on them. Correspondingly, the first step in designing a chemical program is to define the AC triple (S, R, A) as explained in chapter 2. We now discuss this triple for the case of chemical computing.

The Set of Molecules S

The set of molecules S now encode the chemical program together with the information that it needs to process. In some chemistries, only the data items are explicitly represented as molecules, whereas the program itself is immutable. The original Gamma formalism is an example of this type of chemistry: The multiset rewriting rules are not explicitly encoded as molecules; therefore, they cannot be rewritten. Other chemistries allow program and data to inhabit the same space and have a similar molecular structure. In these chemistries, it is possible to operate on the programs just as we would operate on data items; hence, it is possible to modify or rewrite the programs as they run, a feature that can bring both benefits and risks and that still remains poorly explored. An example is the higher-order γ-calculus, where the rules themselves are expressed as molecules.

There are typically two orthogonal ways to store information in a chemical program: within the structure of the molecules, and within their concentrations. In the former case, molecules may represent data structures such as integer numbers, strings or graphs. In the latter case, the information is in the concentration vector holding the current amount of each type of molecule in the system. These two storage alternatives may be combined in the same program, and their interplay can result in interesting behavior not commonly found in conventional programs, such as resilience to instruction deletion [575] and program mutations [74].

The input and output molecules must be defined according to the way information is stored:

- In a chemistry that basically works with concentration information, the concentration of some initial molecules will impact the dynamics of the concentrations of some other molecules. The steady state of the resulting dynamical system can be considered as result of the computation, or the dynamics itself can be considered as result (for instance, expressing an oscillatory behavior), in the case of chemistries that run continuously.

- If the chemistry stores information primarily within the molecular structure, the injection of initial molecules encoding the initial state of the system triggers the computation of the output result, that must be expelled from the system, or extracted from the final "soup" of molecules by pattern matching with some recognition pattern(s) present only in these output molecules.

In the case of spatial ACs, the position of each molecule or the compartment where it resides also bears information that can have an impact on the outcome of the computation; patterns may form in space due to the accumulation of chemicals in certain regions, and different compartments may respond differently to the molecules present due to a different mix of internal chemicals and reactions.

The Set of Reaction Rules R

The set of reaction rules R corresponds to the actual computer program. In the case of a higher-order chemistry, the set R contains more generic rules able to operate on both data and program molecules, typically the instruction set that dictates how program molecules operate on data molecules.

According to the way the set R is defined, we can distinguish two approaches to chemical programming: design by a human programmer, and automatic programming. Designing chemical programs by hand can be difficult, especially as programs grow complex and are not intuitive. It can also be difficult to predict the macroscopic behavior resulting from the microscopic reaction rules. For this reason, automatic programming is often employed to generate chemical programs using optimization techniques such as evolutionary computation [74, 92, 206, 220, 506, 519, 958]. Programs evolved in this way tend to look quite different from those designed by humans.

The Reactor Algorithm A

The reactor algorithm A now establishes the program execution flow: Starting from an initial set of molecules (that may also contain a seed program), the algorithm A will determine how the information in these molecules is processed and how the results can be obtained. The dynamics defined by A also includes the geometry of the reaction vessel [231]: a well-mixed setting such as in basic Gamma, a compartmentalized setting such as in P systems, or a spatial setting such as in cellular automata.

Recall from chapter 2 that the algorithm A may be deterministic or stochastic. Therefore, chemical programs may also be executed in a deterministic or stochastic way in contrast with traditional programs, which are mostly deterministic. The outcome of the computation may also be deterministic or stochastic. A deterministic outcome occurs when the output is always the same given the same input. A stochastic or nondeterministic outcome occurs when for the same input values, one may obtain different output results every time the program is executed. Such a stochastic behavior is often undesired in the context of *in silico* applications of chemical computing.

Caution must be exerted here not to "blame" this outcome directly on the deterministic or stochastic nature of the reactor algorithm A itself: although a nondeterministic outcome is most often associated with a nondeterministic algorithm, it is also possible to have a deterministic outcome with a stochastic algorithm. The prime number chemistry of chapter 2 and the Gamma examples of chapter 9 illustrate this case well: The computation produces the same result whatever the order we choose to pick the molecules.

It is also possible to reduce the amount of stochasticity by increasing the amount of molecules in the system, or by devising clever reaction networks such that the outcome is predictable even in the face of a stochastic execution [74].

Note that defining programs as a set of molecules that react is a model that is in stark contrast with today's computer architectures based on von Neumann machines (see section 10.3). In the AC computing model, conceptually there is no central memory storage, and no program counter (although the underlying implementation on a traditional computer must still abide to these rules). Instead, the execution order is dictated by random molecular collisions and by the applicable reaction rules. High-order ACs share an important common point with von Neumann machines, though. Both treat the program as data, allowing programs to be operated upon.

17.1.2 Dealing with Emergence

Computation with a chemistry is a dynamic process embedded in a complex system [519, 814]. Thus, chemical computation must inherently face emergent phenomena: properties at the higher levels emerge from interactions at the lower levels. Following the emergent computation premise [72, 286], when programming with a chemistries, one often wishes to benefit from the emergent properties of the system and to engineer the system at the microscopic level to achieve the desired macroscopic behavior. This can be seen in contrast with traditional computer science, where typically every possible effort is made to produce controllable and predictable results from well-defined algorithms. This emergent aspect also makes programming with chemistries much less intuitive than traditional programming, at least for computer scientists. One must "think chemically," which entails parallelism at the microscopic scale (as multiple reactions happen independently) and a dynamic system behavior at the macroscopic scale (as changes in concentrations have an impact on the computation flow). Thus, parallelism is primary, with seriality achievable at the collective level, a very interesting switch from the traditional point of view of computer science. Moreover, multiple levels of emergence may arise as the system is structured into several levels of organization.

Emergence versus Programmability Trade-off

Back in the 1980s Michael Conrad, one of the pioneers of molecular computing, identified the following trade-off principle [193, 194] inherent in the design of any computer: "a system cannot at the same time be effectively programmable, amenable to evolution by variation and selection, and computationally efficient."

By "effectively programmable," he meant programmable by humans in the traditional way, typically by giving instructions with the help of a programming language that can be compiled into machine code in order to be executed. Several decades later, his trade-off principle still seems to hold. Progress in genetic programming and related techniques for the automatic evolution of computer programs have revealed that automatically generated programs do not resemble those written by humans [493]. Instead, they tend to exhibit some features that are typical of biological genomes, such as the accumulation of dormant code akin to introns, redundancy of functions, complex interactions among elements, and emergent behavior, among others [71].

Conrad argued that molecular computers could be programmed by learning and evolution, instead of by using traditional means, and this could make them more flexible and adaptable, at the cost of some efficiency and ease of control by humans. Indeed today, the automatic evolution of chemical programs is a promising research direction in chemical computing [74,206,220,506, 519], since the programming of chemical algorithms by traditional means quickly becomes too cumbersome as the systems grow large and complex.

By making chemical programs more easily evolvable, we lose hold of their programmability and obtain programs that elude human understanding. Moreover, these programs tend to be less focused in performing their intended task, because part of their structure is molded by evolution to withstand mutations and to survive to the next generation. On the other hand, this emergent design can also make these programs more robust and adaptive.

At the other side of the spectrum, currently there are several ongoing efforts to mimic electronics in wet chemical and biochemical computers, such as the implementation of logic gates with biochemical reactions [386], in reaction-diffusion computers [813, 857] or with bioengineered genetic regulatory networks in cells [35, 908]. These efforts could be running the risk of losing the valuable properties of chemistry and biology that some computer scientists long to obtain: self-organization, robustness, and evolvability. Conrad's trade-off principle remains valid in this context: by making chemical computers programmable like electronic computers, their evolvability will likely be lost, and they could end up with the same brittleness that characterizes present-day computer hardware: a change in a single bit of instruction causes a disruption in the entire computation flow.

We currently witness what looks like a paradoxical situation: while some chemists and biologists are seeking to program their systems like conventional computers, some computer scientists seek to program their systems like chemical mixtures and biological organisms. As with any trade-off situation, a compromise must be reached, and it is likely that the proper balance lies somewhere in between, at a point yet to be discovered.

Levels of Organization

In conventional software engineering practice, it is recommended to structure the programs in a hierarchical way in order to encapsulate complexity. The same practice is also desirable if we wish chemical programs to scale up and solve complex problems. However, it is not straightforward to obtain such scalability in chemical computing, due to the dynamic nature of chemical processes, and the lack of support from the underlying chemical programming languages. P systems are naturally hierarchical; however, the hierarchies in P systems do not naturally map to the hierarchies in conventional programs. Although approaches to structured chemical programming have been proposed (such as a structured version of Gamma [293]), there is still a contrast between the structuring degrees of freedom coming from chemistry and the typically rigid hierarchies in classical programming.

It is helpful to organize chemical computation according to the material substrate it involves (or takes inspiration from), and by increasing levels of complexity. Considering molecules and their reactions as one class, we discern complex chemical reaction networks as another category, and further on, spatial and compartmentalized computation, that is, computation with protocells and unicellular organisms, with the last category also comprising computation with multicellular organisms including growth and morphogenesis. Hence, as the complexity increases, chemical computation tends to look biological.

Chemical programs are typically structured into one or more out of three basic levels of organization:

- Molecules: at this level, virtual molecules represent data structures that store information. A commonly used structure is a string of letters from an alphabet, akin to a biological polymer sequence. The information in these structured molecules can sometimes be interpreted as a sequence of "machine" instructions that — when interpreted — give rise to the execution of a program that transforms a molecule into a different one, assembles new molecules from given instructions or from templates provided, or dismantles unneeded molecules into decaying waste products.

- Reaction networks: At the level of the dynamics of multiple reactions forming a range of reaction networks from simple to complex, information is encoded in the vector of concentrations of each species in the system, and the computation proceeds as these concentrations change in time. Thus, the interaction laws produced by the chemical reaction networks constitute the program of this computation. This is a dynamic systems perspective to computation, an early notion [727] that is recently gaining more prominence [814].

- Spatial organization: Due to side reactions and other limitations, well-mixed systems forming reaction networks do not scale to large sizes, therefore it is necessary to segregate reactions into separate compartments or different regions in space. This leads to the third component of chemical computation, space, which is key to implementing realistic chemical programs.

These three levels are complementary to each other, and a chemical computation method may make use of a combination of these three elements in different weighting. A balanced approach is computation with artificial genetic regulatory networks (GRNs) [68, 130, 479, 846], in which the genome (at the molecular level) influences the dynamics of chemical reactions that may lead to spatial pattern formation and behavior.

Some open issues in this area include how to choose the appropriate level(s) for each problem, how to compose levels in a coherent and functional way, and ultimately how to design the system such that it is able to choose the appropriate levels by itself, to construct higher levels that display the desired emergent properties dynamically as needed [689], and to transition between levels automatically [609].

17.1.3 Thermodynamics and Reversible Computing

Traditional computer processors typically execute their programs in the "forward" direction: machine instructions are executed in sequence from the first to the last one, with occasional jumps and interruptions that break the default sequential mode. Even parallel computers are generally built from arrays of sequential processors. Usually operations cannot be undone, unless the corresponding reverse operation is explicitly included in the sequence of instructions. In contrast, a *reversible computer* [108, 295, 374, 849] would be based on reversible instructions that could be executed in both forward and reverse directions, like a reversible chemical reaction. Reversibility would not only render these computers able to backtrack from potentially wrong computation paths, but could also lead to significant energy savings.

In 1961, Landauer [491] stated a principle that for every physically irreversible operation performed on one bit of data, an amount of energy equivalent to at least $k_B T \ln 2$ Joule (where k_B is the Boltzmann constant and T is the absolute temperature of the system) ends up dissipated as heat and is lost to the environment. He distinguished between physical and logical reversibility. A physically reversible operation could in theory be performed without energy consumption: the kinetic energy expended in switching state (remember figure 2.4) is reclaimed when the transition state is stabilized. In contrast, a physically irreversible operation erases one bit of data, losing information about its previous state. This loss of information generates entropy, leading to heat dissipation.

An operation is logically reversible if its output uniquely defines its input [491]. No information is lost in the process, and the operation can be reverted by running the inverse function on the output to recover the original input values. In contrast, a logically irreversible operation could be converted into a physically reversible one by storing the previous state of the computation such that it could be undone. Such a reversible computer could run without heat dissipation, provided that no information would ever be erased. However, the machine would need to be restored to its initial state after it had finished its intended task, and it would require an enormous amount of memory to keep track of the history of the computations in order to reverse them.

In 1973, Charles Bennett [107] proposed to improve this reversibility technique by copying the output before running the machine in the reverse direction, and he showed that it was possible to decrease the amount of storage necessary by breaking the computation down into several steps. He showed that this method could be adjusted to dissipate as little energy as needed, at the expense of an equivalent decrease in computation speed. In that same paper [107], Bennett also identified the biological process of DNA transcription into RNA as an example of naturally occurring thermodynamically reversible computation: the RNA polymerase transcribes DNA to mRNA in a process that is physically and logically reversible, comparable to a tape copying machine. A few years later (1982) [108] he hypothesized how a chemical Turing machine might be built out of enzyme-catalyzed operations on a DNA "tape." Such a Turing machine could operate in a logically reversible way, that is, it would be able to undo computation steps, since chemical reactions may also proceed in the reverse direction.

Bennett's enzymatic Turing machine is depicted in figure 17.1. Table 17.1 shows the corresponding transition rules. The Turing tape takes the form of a string of beads with two types of attached molecules representing bits 0 and 1. The head of the Turing machine is implemented by molecules that encode the possible states of the machine (A and B in this example). At the top of the figure, the machine is in state A, reading symbol 0. Enzyme E_{A0} matches this current state and symbol combination, binding to both A and 0 molecules. The binding causes molecule 1 to replace 0 at the current tape position. At the same time the enzyme detaches molecule A on one side and releases molecule B on the other side. The latter binds to the right of the string, driving the machine to a new position, where B becomes the new head. This design requires one type of enzyme for every possible combination of current state and symbol read from the tape. The transition (symbol to be written to the tape and movement to the left or to the right) is encoded in the configuration of the enzyme and its binding sites. Enzymes, tape molecules, and head molecules of all required types are present at several copies in the machine's environment. The correct operation of the machine relies on their specific bindings to the tape location where the computation is taking place.

The enzymatic Turing machine is reversible because the "memory" of past transitions is "stored" in the enzyme in the form of occupied or free binding sites: a "regretful" enzyme could run backward by binding to the head and state that it has just inserted, chopping them off and reinserting the molecules from the previous state that it had grabbed before. In order to drive

Table 17.1 Transition rules for Bennett's enzymatic Turing machine [109]

head state	bit read	change bit to	change state to	move to
A	1	0	A	left
A	0	1	B	right
B	1	1	A	left
B	0	0	B	right

Table 17.2 Reversible logical gates by Fredkin and Toffoli [295, 852]: In the Toffoli Gate, the output is $y_0 = x_1$ AND x_2 if the control line c is zero, otherwise it is $y_0 = x_1$ NAND x_2; the variables y_1 and y_2 are used to store a copy of the input values such that the operation can be reversed. In the Fredkin gate, the output is $y_1 = x_2$ and $y_2 = x_1$ (crossover) if $c = 0$, and a copy of the input otherwise; the control line c remains unchanged in the output (c'). In both gates, there is a unique mapping from input to output and vice versa, such that the operation can be reversed.

Toffoli Gate					
input			output		
c	x_1	x_2	y_0	y_1	y_2
0	0	0	0	0	0
0	0	1	0	0	1
0	1	0	0	1	0
0	1	1	1	1	1
1	0	0	1	0	0
1	0	1	1	0	1
1	1	0	1	1	0
1	1	1	0	1	1

Fredkin Gate					
input			output		
c	x_1	x_2	c'	y_1	y_2
0	0	0	0	0	0
0	0	1	0	1	0
0	1	0	0	0	1
0	1	1	0	1	1
1	0	0	1	0	0
1	0	1	1	0	1
1	1	0	1	1	0
1	1	1	1	1	1

computations in the forward direction, this system must be maintained out of equilibrium by a constant supply of new enzyme molecules with free current state sites (and bound next sites), and removal of "used up" enzymes with bound current state sites (and free next state sites).

Around the same time that Bennett proposed his enzymatic Turing machine, Fredkin and Toffoli [295, 852] showed how reversible logic gates could be designed, and how these could be combined into *conservative logic circuits* able to implement any Boolean function. Table 17.2 depicts the Toffoli gate and the Fredkin gate, two alternative implementations of a reversible logic gate with three inputs and three outputs. The Toffoli gate implements a reversible AND/NAND function of the inputs x_1 and x_2, where the result is stored in y_0 and the two other outputs y_1 and y_2 are used to store a copy of the input state such that the gate can operate in reverse mode. A change in the control line c causes the output y_0 to be inverted, such that the total amount of zeros and ones remains balanced, a requirement for the reversibility of a gate. The Fredkin gate implements a crossover (x_1 goes to y_2 and x_2 to y_1) when the control line c is zero, and simply copies the input to the output otherwise. Both gates have a unique mapping from input to output and vice versa, thereby ensuring reversibility.

Reversibility as a design property could lead to potentially significant energy savings with respect to traditional computers. Current electronic computers still dissipate orders of magnitude more heat per operation than Landauer's theoretical lower bound for irreversible computing, however technology is improving quickly. Once this lower bound is reached, reversible com-

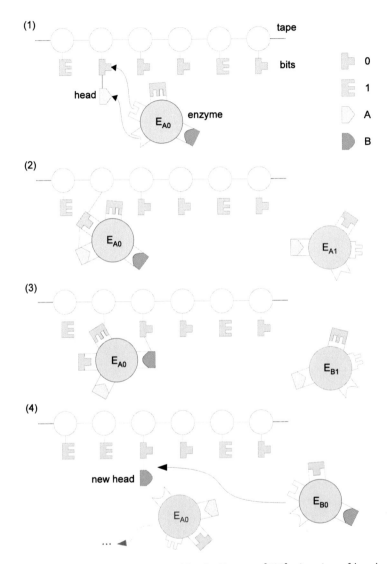

Figure 17.1 Hypothetical enzymatic Turing machine by Bennett [109]: A string of beads with attached 0 and 1 molecules represents the tape. Molecules A and B (possible machine states) act as the head of the Turing machine. An enzyme with specificity for the current state and symbol realizes the state transition by binding and removing the molecules from the current tape position and releasing new molecules that bind to the next state position.

puting will become an interesting alternative in order to reach further energy savings per bit of computation.

Even if the theoretical and physical difficulties with reversible computing can be surmounted, achieving reversible computing in practice is far from trivial: It requires a whole new range of programming techniques and tools, starting with reversible programming languages [44, 944]. While physical reversibility cannot be attained with conventional hardware, logical reversibility can still offer some advantages from a software engineering perspective; it has been used to implement rollback mechanisms in fault recovery [138], backtracking from incorrect computations in optimistic parallel simulation [160], and reversible debuggers able to step through programs in both directions [174], among other applications.

Chemical reactions offer a natural model of reversible computing, although models based on physics and quantum computing are also possible. For a long time, reversible computing remained mostly an academic curiosity. Recently, however, renewed interest in reversible computing was sparked, motivated by current limitations in electronic technology, which is quickly reaching a barrier beyond which heat dissipation becomes a major obstacle for the design of faster processor chips [189]. Thus, reversible computing has attracted increasing attention in recent years [218, 662, 849], dominated by solutions based on adiabatic digital electronics and quantum computing. There is still room for research into reversible computers based on chemistry and biochemistry, realizing Bennett's vision of an enzymatic computer.

In the case of *in silico* chemical computing however, energy savings do not apply, since programs must still run on conventional hardware. New reversible chemical computing algorithms could still be beneficial, though, in order to contribute to the related software engineering efforts mentioned earlier, and to the design of future wet reversible computers.

17.2 Search and Optimization Algorithms Inspired by Chemistry

Artificial chemistries have been used to design optimization algorithms that make use of a chemical metaphor to search for solutions to difficult problems [62,213,337,438,489,614,665,912,937]. Such chemically-inspired algorithms belong to the class of heuristic optimization approaches, together with evolutionary algorithms (EAs) [339,590], ant colony optimization (ACO) [239], and particle swarm optimization (PSO) [454,669], among others.

Heuristic optimization algorithms start with a set of suboptimal (typically random) solutions, and progressively improve upon them, until solutions that are considered to be sufficiently close to the optimal ones are found. These algorithms target complex search and optimization problems that typically require the search through a vast space of possible solutions. The problems addressed by theses heuristic schemes are often NP-hard, that is, the amount of computation steps needed to find an exact solution increases very fast with the size of the solution, making any attempt to find an exact solution prohibitive for all but modest sized problems. For this reason, NP-hard problems often call for heuristics able to find approximate solutions that are close enough to the optimum, but achievable within a feasible amount of computation steps.

Two examples of well-known NP-hard problems are the traveling salesman problem (TSP) and the graph coloring problem. The TSP consists in finding the tour of minimum cost (in terms of distance traveled or fuel consumed, for instance) that visits all the cities on a map, using only the available roads, and visiting each city only once. After visiting all the cities, the salesman usually goes back to the city where he started, closing the tour. The TSP is known to be NP-hard: in the worst case, the amount of computation steps needed to find an exact solution increases exponentially with the number of cities on the map. The graph coloring problem consists in assigning colors to nodes on a graph such that no two adjacent nodes have the same color. Each node receives one color out of k available colors. It has been shown that this problem is NP-hard for $k \geq 3$.

One of the advantages of using artificial chemistries for optimization is that by looking at the optimization process as a dynamical system [814], it is possible to think of such an optimization process as "falling toward the solution" [194] as the dynamical system tends to move away from unstable configuration states toward a more stable attractor.

In this section we will examine two early algorithms [62, 438] that illustrate the general idea of how ACs can be used for search and optimization. More recent algorithms in the domain include [213, 337, 489, 614, 665, 912, 937].

17.2.1 The Molecular Traveling Salesman

The molecular traveling salesman [62] is an optimization algorithm aimed at solving NP-hard problems with an artificial chemistry, with the TSP as an application case study. In the molecular TSP, candidate tours are encoded as strings (analogous to biomolecules such as DNA) that carry a quality signal (or fitness value). Machines (analogous to enzymes) float in the reactor and operate on these strings to change and select them according to a fitness function embedded within the machine's knowledge.

Data strings representing candidate solutions take the form $\vec{s} = (s_0, s_1, ..., s_N)$, where N is the number of cities in the tour, $s_i \in \{1, ..., N\}$ for $i = 1, ..., N$ is the tour sequence, and $s_0 \in \mathbb{R}^+$ is the fitness of the tour. Hence, the first element of the string contains its quality signal, whereas the remaining elements represent the candidate tour. For example, the string $\vec{s} = (l, 3, 1, 4, 2)$ says that the tour starting with city 3, then visiting cities 1, 4, and 2 in this order, has fitness l.

The fitness l of a string $\vec{s}$ is the sum of the euclidean distances between the cities. In a two-dimensional map:

$$l(\vec{s}) = \sum_{i=1}^{N} \sqrt{(x_{i+1} - x_i)^2 + (y_{i+1} - y_i)^2} \tag{17.1}$$

where (x_i, y_i) are the coordinates of city s_i in space, and $s_{N+1} \equiv s_1$ (i.e., the tour is closed).

Each machine picks a number n_{op} of strings from the "soup," modifies them producing new strings, and evaluates the new strings according to their fitness criterion, which in the TSP case is given by function l in equation 17.1. The n_{op} best strings are released into the "soup" and the others are discarded.

The string modification operation is hardwired in the machine. In the case of the TSP, four sorts of machines are present in the reactor: an exchange machine (E-), a cutting machine (C-), an inversion machine (I-), and a recombination machine (R-). These machines operate on a single string, except the R-machine that operates on a pair of strings: the E-machine exchanges two random cities in the tour; the C-machine cuts a segment from the tour and inserts it at another location in the same string; the I-machine acts like the C-machine but also inverts the segment before moving it; the R-machine performs a recombination between the tours of two strings. These operations are performed with safeguards to ensure valid tours. After the operation (E-, C-, I-, or R-) is applied, the fitness of the new string(s) is computed, and the n_{op} best strings are selected for survival, where $n_{op} = 1$ for E-, C- and I-machines, and $n_{op} = 2$ for R-machines.

In order to solve a TSP instance with N cities, the reactor starts with a population of M data strings containing random candidate tours of length N, and one or more machines of each of the four possible kinds. The machines randomly collide with the strings present in the reactor, and operate on them in parallel. The result is an optimization process that gradually converges to a population of tours with shorter total distances approaching the optimum. This optimization algorithm makes use of the intrinsic parallelism of an AC, where the local action of machines on strings leads to a global optimization effect without the use of global information.

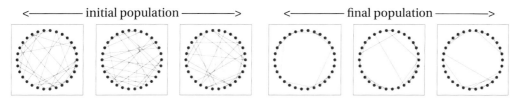

<———— initial population ————>   <———— final population ————>

Figure 17.2 Example of optimization by the Molecular TSP on a toy problem where the optimum tour is a ring: starting with random tours (left), after 1000 generations the system has found solutions that are close to the optimum (right). For conciseness, here only 3 individuals from the population are shown for each case. Results obtained with our reimplementation of the algorithm in Python, on top of the PyCellChemistry package, see appendix. See [62] for more results.

Figure 17.2 shows an example of run where the cities are placed on a circle, such that the optimum topology (ring of cities) can be easily visualized. We can see that starting from random solutions, the system is able to find solutions that are close to the optimum. The simulations also reveal that the recombination operator accelerates the optimization process, at the cost of a loss in the diversity of the population, a trade-off that is well known today in the domain of evolutionary algorithms.

By changing the content of the strings, their fitness function, and the operators working on them, it is possible to solve other problems using this chemical algorithm.

17.2.2 The Chemical Casting Model

The chemical casting model (CCM) introduced by Kanada in [438, 439] is another production/rewriting system inspired by the chemical metaphor. The CCM can also be can be regarded as a model for emergent computation. The motivation was to develop a new method for searching for solutions to constraint satisfaction and optimization problems, a family of NP-hard problems encompassing the TSP and the graph coloring problem.

The molecules in CCM are composed of atoms and their reaction rules depend on the type of search problem being examined. In general, atoms are components of a potential solution of the problem. In the examples shown in [438, 439], the whole population (working memory) represents one point in search space. Each atom has a type and a state. Atoms can be connected by links which may be used to specify a concrete problem like in the graph coloring example. In this case links are fixed during a simulation. For a traveling salesman problem (TSP), links can represent a solution to the problem and thus have to change during the search process.

The reaction rules have the general form

$$LHS \rightarrow RHS \tag{17.2}$$

where the left-hand side (LHS) is a pattern. If a subset of atoms matches the LHS it may be replaced by atoms which match the right hand side (RHS).

The quality of a solution is measured by a local evaluation function called local order degree (LOD). The LOD is used to decide whether a rule can be applied or not. There are two types of LODs: The first type maps one atom to a number, and the second one maps two atoms to a number. A rule may be applied if the sum of the LODs of the participating atoms on the left-hand side is smaller than the sum of the LODs on right hand side. We can see that the quality of the solution is only measured locally by this method.

In [438] Kanada has also introduced an annealing method by assigning an additional energy state variable, called frustration, to each atom. Frustration is similar to kinetic energy. It is added to the LOD so that they may react even if the previous LOD criterion is not fulfilled, provided frustration is high. In case of an elastic collision, frustration of colliding atoms is increased by a factor c, otherwise (i.e., when a reaction happens), frustration is reset to its initial value.

The authors have successfully demonstrated the application of the CCM to different constraint satisfaction problems, e.g., graph coloring, the N-queens problem, and the TSP. The simulations show that even if the current solution is evaluated only locally, the system can converge to the global optimum. They have also shown how the locality of the reaction rules can be tuned by adding a catalyst. Adding a catalyst reduces the locality of a rule, which roughly leads to faster convergence speed but increases the probability that the system gets stuck in a local optimum.

The advantage of this approach is that many processors may operate on one instance of a problem solution in parallel with minimal communication because every evaluation of the quality and every variation is performed locally. In some cases (like graph coloring), even synchronization and access conflict to a shared memory can be ignored, because they only generate a harmless noise of low intensity.

17.3 Distributed Algorithms Using Chemical Computing

Due to their parallel nature, artificial chemistries have inspired numerous distributed algorithms and communication protocols for computer networks. Some typical algorithms include routing [527, 581], load balancing [576, 580], congestion control [371, 578, 946], and distributed coordination [467, 468, 551, 554, 608, 945]. Some of these algorithms have been designed by humans, others have been obtained by artificial evolution.

In this context, artificial chemistries offer a natural metaphor for applications in which a number of autonomous entities must interact by local communication only, in order to achieve a desired global behavior. Moreover ACs enable a robust algorithmic design with proof of properties such as convergence and stability, by applying analysis techniques from dynamical systems [520, 863] and from chemical organization theory [554].

In this section we illustrate how distributed algorithms can benefit from the natural parallelism in ACs, through some examples from the literature: the first example (sec. 17.3.1) shows a distributed load balancing protocol written in the Fraglets programming language that had been introduced in chapter 16; the second example (sec. 17.3.2) depicts the use of the Avida platform for the evolution of distributed algorithms, with focus on decentralized synchronization and desynchronization algorithms; finally (sec. 17.3.3) we show how chemical organization theory can be used for computation.

17.3.1 Distributed Computing Based on Chemical Dynamics with Fraglets

The Fraglets language is a string rewriting AC proposed by Tschudin [862] for the design and evolution of communication protocols for computer networks (see chapter 16). It is able to express self-modifying and self-replicating code [575, 942], and showed some promising robustness properties, such as tolerance to instruction loss [864], and the autocatalytic recovery from code deletion attacks [575]. These properties have been harnessed to design a self-healing multipath routing protocol [581], and a resilient link load balancing protocol [576, 577]. One of the original goals of the language was the automatic synthesis and evolution of communication protocols for computer networks. A number of simple protocols and algorithms could indeed be evolved for Fraglets [914, 943]. When attempting to evolve more complex Fraglet programs automatically however, some evolvability concerns were detected such as code bloat, an elongation tendency, and frequently broken chemical reaction pathways [579, 943]. Countermeasures to mitigate these problems included a gene expression mechanism to control the production of valid code [935], a new language based on circular molecules that avoided the characteristic tail buildup in Fraglets [936], a mechanism to enforce energy constraints [579], and an artificial immune system that would detect and eliminate harmful code [753]. Despite these efforts, a successful strategy for the automatic evolution of distributed communication protocols using Fraglets remains mostly an open issue.

A more successful research branch derived from Fraglets is the engineering of communication protocols based on chemical dynamics [576, 594]. Similar to membrane computing and related chemical paradigms, the Fraglets interpreter ran originally in a maximally parallel way, executing as many reaction rules as possible, until the reactor reached an inert state, waiting for the next incoming packet. Soon enough, however, (as it happened to membrane computing as well [661, 718]), it became clear that reactions had to be scheduled according to the law of mass action, taking specific reaction rates into account, in order to capture phenomena such as autocatalytic growth, decay, competition of replicators, diffusion speed [575, 580], which turned out to have a crucial impact on the behavior and performance of the distributed algorithms considered. Stochastic simulation algorithms such as Gillespie SSA [328] and the next reaction method [324] were used for this purpose [580]. These algorithms had to be extended to deal with multiple compartments (as in [661, 718], albeit with a limited hierarchy [580]), since the Fraglets simulator had to consider arbitrary network topologies made up of multiple interconnected reactors. The results in [575, 576, 580] deeply rely on such mass action scheduling, which was later used to design a packet scheduler acting on the queues used to ship packets between nodes in a network [578]. A "chemical congestion control" protocol was then proposed on top of such a packet scheduler [578], and it was shown that the protocol behaves similarly to TCP, the standard and most widely used congestion-control protocol on the Internet today. A distributed and adaptive rate controller based on enzyme kinetics then followed [595], supported by a thorough stability analysis [594] together with experiments on a real network [594, 595]. Whereas a congestion controller dynamically adapts the transmission rate to the current network load, a rate controller seeks to regulate the network traffic to match externally defined rate limits. A distributed rate controller is useful to regulate access to shared resources such as cloud computing infrastructures.

Next we illustrate the Fraglets approach to engineering communication protocols based on chemical dynamics, through the chemical disperser protocol presented in [576].

A Chemical Disperser Algorithm for Load Balancing

The goal of a load balancing algorithm is to share the workload of processors evenly among tasks. The chemical disperser algorithm in [576, 580] solves this problem by diffusing "job molecules" (representing the tasks to be executed) to the nodes of the network. Starting from a random configuration where each node has an arbitrary amount of job molecules, the system converges to a stable configuration where all nodes have an equal amount of job molecules. It is straightforward to prove that this final amount corresponds to the average of the total amount of job molecules initially present in the network, and moreover, that this distributed equilibrium is stable: when adding or removing job molecules at arbitrary places, the system again settles to the equilibrium where tasks are equally distributed over processors, as desired.

The full protocol for the disperser has been implemented in Fraglets, and is depicted in [580], whereas an analysis of the core averaging part can be found in [576]. Such core is represented by the following reaction, executed at each node in the network:

$$C_{ij} + X_i \xrightarrow{k_i} C_{ij} + X_j \quad , \quad \forall (i, j) \in E \qquad (17.3)$$

where E is the set of edges (links) in the network, X_i is a job molecule currently situated in node i, and C_{ij} is a catalyst molecule found in node i and responsible for sending job molecules X from node i to its neighbor node j. The actual sending process of X from i to j is represented by the conversion of molecule X_i to X_j.

The concentration of each molecule species C_{ij} is the same everywhere and is kept constant. Moreover all the reactions have the same rate coefficient k_i. Therefore, the more neighbors a node has (that is, the higher its degree in network terminology), the higher the chance of a molecule X to encounter a catalyst C and to be shipped to a neighboring node. This means that for a given concentration of X, nodes with a higher degree have a faster outflow of X. It also means that for nodes with the same degree, those with a higher concentration of X will also be faster at draining the excess molecules. As a result, the system settles to an equilibrium where each node contains an amount $\bar{x}$ of X molecules equal to the average concentration of X throughout the network: $\bar{x} = \sum_{i=1}^{n} x_i(t = 0)/n$, where n is the total amount of nodes in the network. The details of the proof can be found in [576].

Figure 17.3 shows a sample run of our independent implementation of the disperser algorithm in a network of four nodes. This implementation can be found in the Python package described in the appendix. Gillespie SSA is used for the stochastic simulation of the set of reactions (17.3) for the four nodes. At $t = 0$, 1,000 job molecules X are injected in node 4. After some time these molecules get redistributed throughout the network, and each node ends up with an average of around 250 molecules. The molecule counts still fluctuate due to the stochastic nature of the system. At $t = 10s$, 600 X molecules are injected in node 2, raising the global average to around 400 molecules per node. At $t = 20s$, 300 molecules of X are expelled from node 2 and destroyed, lowering the global average as expected. Note that the nodes that are further away from the point of injection or deletion (nodes 2 and 3 in case of the initial injection in node 4; and node 4 in case of the subsequent injection and deletion events in node 2) converge more slowly, but eventually all the nodes settle to the same average values.

This simple algorithm provides a decentralized way to calculate average values in a network. It is a chemical implementation of a discrete mass-conserving diffusion process on an amorphous topology, as discussed in [941]. It is well known from physics that plain diffusion converges to the average overall concentration, and this idea has been applied to load balancing since the late 1980s [152, 205, 257]. The advantage of the disperser algorithm lies in its simplicity and consequent provability: by representing tasks as molecules and selecting reactions according to the mass action law, the algorithm never has to check for inconsistencies such as a negative load; moreover, the chemical reactions can be directly converted to the corresponding differential equations that can then be checked for convergence and stability. Note that the algorithm

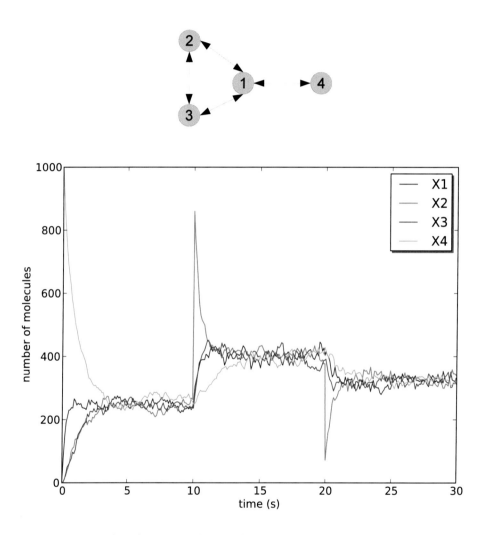

Figure 17.3 Disperser algorithm in a small network. Top: network topology with 4 nodes connected via bidirectional links. Bottom: Plot of the amount of job molecules at each node as a function of time, for a sample run of the algorithm.

leaves out many issues of practical relevance, such as variable processor speeds and capacities, transmission delays, the dynamic in- and outflow of jobs as users submit new tasks for execution and existing tasks are completed, and uncooperative nodes that would "steal" or ignore jobs assigned to them. However, it suffices to illustrate the concept of chemical programming for such a class of distributed algorithms.

17.3.2   Evolving Distributed Algorithms with Avida

In Avida (chapter 10), digital organisms take the form of assembly-like programs that evolve in a virtual Petri dish where they interact as an ecology. The Avida platform has been successfully used in numerous experiments ranging from models of biology to distributed systems [18, 567].

If left unattended, the organisms in Avida compete with each other for computation resources, finding increasingly clever strategies to survive and reproduce, leading to a digital ecosystem where co-evolution takes place. In such digital ecologies, organisms evolve to spend their precious CPU cycles in pursuing their own survival agenda. Such a setting is suitable for experiments aimed at studying biological evolution, such as [18].

In order to evolve digital organisms containing algorithms able to solve a user-defined problem, a strategy to provide incentives to complete on user-defined tasks must be devised. For this purpose, it is necessary to introduce an externally defined fitness function that rewards the desired behavior, similar to what is done in evolutionary algorithms [339, 590]. In Avida this is achieved by introducing the notion of "merit" [468]: each digital organism is endowed with a *merit* attribute that determines how many instructions it is allowed to execute relative to other organisms. The higher its merit, the more instructions the organism is allowed to execute. Consequently, for two organisms with the same size and the same replication code, the one with the highest merit replicates more quickly, giving it a selective advantage. Organisms earn merit by performing user-defined tasks such as implementing a desired algorithm. Merit may also decrease as a form of punishment for bad behavior.

Avida has been applied to automatically evolve a variety of software engineering and computer network algorithms, such as distributed leader election [468], cooperation for energy saving in mobile devices [567], synchronization and desynchronization in a group of processors [467], and flood monitoring with sensor networks [337], among others. We illustrate this digital evolution approach with the example of distributed synchronization and desynchronization from [467].

Example: Evolution of Synchronization and Desynchronization

Synchronization and desynchronization algorithms play an important role in distributed systems. An example of synchronization algorithm in distributed systems is heartbeat synchronization, a technique to synchronize periodic signals generated by different nodes in a network. The opposite algorithm, namely desynchronization, aims at spreading out signals from different nodes, such that they do not access the same resource at the same time, avoiding conflicts.

The evolution of distributed synchronization and desynchronization algorithms was shown in [467] using the Avida platform. The idea was to evolve behaviors analogous to the way fireflies synchronize their flashes. In analogy to the natural evolution of firefly synchronization in nature, the digital evolution abilities of Avida were harnessed in [467] to automatically generate the desired synchronization and desynchronization behaviors. For this purpose, the Avida instruction set was enhanced with a new set of instructions allowing flashes to be sent out and captured by other organisms, such as the flash instruction (broadcast a flash message to the neighborhood), and if-recvd-flash (to check if a flash was received from one of the neighbors).

Since synchronization and desynchronization algorithms require the cooperation between nodes of the system in order to coordinate their behaviors, an evolutionary algorithm able to perform group selection was used. In this framework, digital organisms are divided into sub-populations called *demes*, and the fitness function rewards how close the behavior of the entire deme is to the target one. In the case of the synchronization task, the ideal behavior is for all nodes to send their flash at the same time; in the case of the desynchronization task, the ideal behavior is that each node sends its flash as far apart in time from the others as possible.

Demes compete with each other in the digital world of Avida, such that a fitter deme has a higher chance of replicating than a less fit one. When a deme replicates, the entire deme replaces another deme in the Avida world. Inside a deme, the subpopulation evolves as the organisms suffer mutations. However, these local mutations inside the deme are not passed onto the next generation. Instead, the original germline used to seed this deme is inherited, in a process called germline replication. Germline replication had been proved successful in evolving cooperative behaviors in earlier Avida studies; therefore, it was chosen also for the evolution of synchronization and desynchronization tasks.

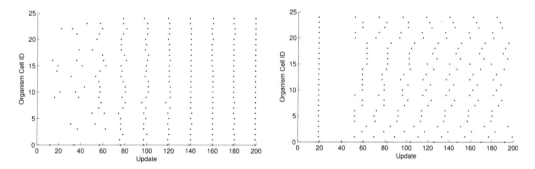

Figure 17.4 Evolved synchronization (left) and desynchronization (right) behaviors on top of Avida, from [467]. The y-axis indicates the ID of the Avida cell where the digital organism resides. The x-axis indicates the simulation time, measured in updates, the Avida simulation time unit. Each dot corresponds to a flash sent out by a given digital organism at a given time. Reprinted from Knoester and McKinley, Artificial Life 17(1):1-20 [467], © 2011, with permission from MIT Press.

Figure 17.4 shows examples of evolved synchronization and desynchronization behaviors. The plots show the behavior of a single subpopulation (deme) containing 25 copies of the most commonly evolved genome during the experiments. Initially, the deme starts at a state that is far from the optimum. With time, the individuals within the deme gradually converge to the desired behavior.

An analysis of the code for the best individuals evolved reveals that the synchronization behavior is implemented by a combination of instructions that either advance or delay the execution of the flash instruction, such that the frequency of flashes is adjusted upon reception of an external flash message. A large number of the instructions in the evolved code have no apparent purpose other than spending CPU cycles, contributing to the frequency adjustment process. The evolved desynchronization code operates in a similar manner, with the difference that in this case the frequencies are adjusted such that the organisms flash at different times. In both cases, the evolved algorithms achieve the desired behavior in a decentralized way, using only local information.

17.3.3 Computing with Chemical Organizations

Chemical organization theory (chapter 12) allows us to identify the organization structure of a chemical program: Given a chemical reaction network that is doing some computation, as chemical reactions transform the set of molecular species present in system, such computation can be regarded as a transition between chemical organizations [550,552]. With such perspective in mind, it is possible to design and analyze chemical programs using chemical organization theory: A chemical algorithm to compute the maximal independent set problem is presented in [551,553], and chemical implementations of logic circuit elements such as a XOR gate, a flip-flop, and an oscillator are presented in [550]. One of the motivations for using organizations for computation is to obtain robust algorithms because organizations do not vanish. A recent review of organization-oriented chemical programming can be found in [552].

Recalling from chapter 12 that an organization is a set of molecular species that is closed and self-maintaining. Closed means that it produces only molecules that are already part of the system, that is, it does not create new molecules. Self-maintaining means that the system is able to regenerate any molecules that are used up in reactions, so the system as a whole does not decay, although the concentrations of the molecules in the system may fluctuate dynamically. Since no new molecular species are created and the existing ones maintain themselves, in principle the group of species in the organization remains stable in time, unless some external event intervenes to change things. These properties confer robustness to a system when it is designed to converge to an organization state.

When computing with chemical organizations, information is primarily encoded by the presence or absence of a given molecular species at a given moment. This is in contrast to other approaches to chemical computing, which store information within strings, or in the concentrations of substances. Since the organization structure changes by adding or removing species from the system, it is from this perspective that computation can be regarded as a movement between organizations. As such, organization-oriented chemical computing offers a coarse-grain, macroscopic view of the underlying chemical processes: Since the organization as group "survives" in spite of dynamic fluctuations in the concentrations of its members, what actually matters in the long term is the set of species that are members of the organization, not their precise amounts. In a way, information is digitized as concentrations are reduced to zeros (species absent) or ones (species present). The examples below illustrate this idea more clearly.

Example 1: A Logic XOR Gate Using Chemical Organization Theory

In [550] a recipe to design logic circuits using chemical organization theory is proposed. A "chemical XOR" gate is shown as a case study. This XOR gate works as follows: Two molecular species are assigned to each input and output variable, each corresponding to one possible state of the boolean variable. For a XOR gate with inputs a and b and output c, the following molecular species are used: $\mathcal{M} = \{a^0, a^1, b^0, b^1, c^0, c^1\}$ where a^0 represents $a = 0$ and a^1 represents $a = 1$, and likewise for b and c.

The truth table for the XOR gate is presented on the left side of table 17.3. The output is true when only one of the inputs is true. The reaction rules stem directly from this truth table, and are shown on the right side of table 17.3. Furthermore, since x^0 and x^1 (for $x = a, b, c$) represent

Table 17.3 Truth table for a XOR gate (left) and corresponding reactions within an organization-oriented implementation (right)

input		output	reaction
a	b	c	
0	0	0	$a^0 + b^0 \longrightarrow c^0$
0	1	1	$a^0 + b^1 \longrightarrow c^1$
1	0	1	$a^1 + b^0 \longrightarrow c^1$
1	1	0	$a^1 + b^1 \longrightarrow c^0$

mutually exclusive states, a "particle vs. anti-particle annihilation" rule is added for each such pair of species:

$$a^0 + a^1 \longrightarrow \emptyset \quad , \quad b^0 + b^1 \longrightarrow \emptyset \quad , \quad c^0 + c^1 \longrightarrow \emptyset \qquad (17.4)$$

Running this chemical program consists in throwing input molecules in the reactor, letting them react according to the preceding rules, and looking at the output when the system reaches a stable state. The authors show that the Hasse diagram for this chemical program contains 15 organizations and that the organization that survives at the end of the run corresponds to the answer to the computation task, according to the truth table.

The authors also show how this simple case study can be extended to more complex circuits involving multiple interconnected gates. A chemical flip-flop consisting of two NAND gates, an oscillator, and other cases are also demonstrated using the same principle, including a distributed algorithm that will be discussed below.

As with other chemical computing approaches, algorithms based on chemical organization may be engineered by humans or evolved automatically. The emergence of evolution in a chemical computing system is investigated in [555] using organization theory. The evolution of chemical flip-flops is studied in [506] with the help of chemical organization theory. It is shown that computations can be seen as transitions between organizations, and that evolutionary improvements are mostly correlated with changes in organizational structure.

A shortcoming of organization-oriented chemical computing is that one species is needed per state of each variable: the absence of a substance must be represented explicitly by the presence of another substance. Thus it requires even more species than other methods, such as computation with concentration dynamics (section 17.4.3), for which the number of different species needed is already an issue.

Example 2: Solving the Maximal Independent Set Problem Using Chemical Organization Theory

The maximal independent set (MIS) problem is a graph problem that encounters applicability in algorithms for communication networks, such as determining cluster heads in wireless networks [554]. The goal is to partition a set of network nodes into two classes: those that are servers (or masters, or leaders, or cluster heads), and those that are slaves or clients. The partition should be such that each slave can find at least one master within its neighborhood, and masters are surrounded only by slaves, not by other masters.

More formally, given an undirected graph $G = \{V, E\}$ with vertices $v_i \in V$ and edges $(v_i, v_j) \in E$, a set of vertices $I \subset V$ is independent when it contains no adjacent vertices, that is, $\forall v_i, v_j \in I : (v_i, v_j) \notin E$. An independent set I is maximal (I_{max}) if no vertex can be added to it without violating its independence property. In this way, the nodes within the MIS can be considered as master nodes, and the others as slave nodes.

A chemical algorithm to compute the maximum independent set problem was proposed in [554]. It works as follows: two molecular species s_j^0 and s_j^1 are assigned to each vertex $v_j \in V$. A high concentration of s_j^1 indicates that vertex v_j is in the MIS ($v_j \in I_{max}$), whereas a high concentration of s_j^0 indicates its absence from the MIS ($v_j \notin I_{max}$).

Since species s_j^0 and s_j^1 represent mutually exclusive states, the first reaction rule to add is one that makes sure that, upon collision, molecules s_j^0 and s_j^1 annihilate each other:

$$s_j^0 + s_j^1 \longrightarrow \emptyset \tag{17.5}$$

Next, for each vertex v_i, molecules s_i^1 get produced when no neighbor of v_i is in I, such that v_i can then be included in the MIS:

$$s_j^0 + s_k^0 + ... + s_l^0 \longrightarrow n_i s_i^1 \tag{17.6}$$

for all n_i neighbors of v_i, that is, linked to v_i via edges $(v_j, v_i), (v_k, v_i), ..., (v_l, v_i) \in E$.

Finally, a vertex that considers itself in I attempts to conquer its position by destroying any competing neighboring molecules:

$$s_j^1 \longrightarrow s_i^0 \quad , \quad \forall (v_j, v_i) \in E \tag{17.7}$$

By using chemical organization theory analysis, the authors are able to prove that this simple set of rules is able to solve the MIS problem in a decentralized way, and moreover that the solutions obtained are instances of organizations. Such a feature renders the solution robust to perturbations: When the system is disrupted by adding or removing s_i^0 or s_i^1 molecules, it will tend to fall back into the MIS solution again, since this solution is an attractor of the underlying dynamical system. The authors also point out that the system should be able to reconfigure itself when nodes are added or deleted.

17.4   *In Silico* Simulation of Wet Chemical Computing

Another application of *in silico* chemical computing is the computational simulation of algorithms for wet chemical computers. Since it is currently not possible or too expensive to implement many wet computers in the wet laboratory, an initial idea about their behavior can be gained via computer simulations. Indeed, the simulation of chemical and biological systems is another important application of artificial chemistries, to be covered in chapter 18. Such systems can be extended to cover man-made synthetic chemical and biological compounds, as will be exposed in chapter 19. In the present section we prepare the ground for these upcoming chapters by providing an initial flavor of some early and simple models of wet computers.

17.4.1   Biochemical Logic and Switching

In the early 1970s, Seelig and Rössler proposed to make use of bistable chemical reactions to implement chemical flip-flops [727, 758]. A bistable reaction "switches" from one concentration level to another when the concentration of some "input" chemical exceeds a threshold. Therefore it can be seen as a biochemical switch. In this way, it is possible to compute with chemistry by using analog concentration values to encode digital information. Other examples of bistable chemical switches include [516, 685].

Another early attempt toward chemical computers came in the 1970s as well, when Tibor Gánti postulated the existence of "fluid machines" or "soft automata" [305, 306] made of chemicals that react mostly in an autocatalytic way. He showed that cyclic reactions could form the basis for an abstract protocell model that he called the "chemoton" or "chemical automaton," which is a "program-directed, self-reproducing fluid automaton" (see also Chapter 6). Gánti sketched elementary operations using chemistry, such as AND- and OR-branching in chemical reaction networks, cyclic coupling, and autocatalysis. He showed how such elementary operations could be combined to produce several important building blocks of fluid machines: chemical sensors built out of reversible reactions, chemical switches made of OR-branching reactions, chemical oscillators made of coupled cyclic reactions, self-reproducing cycles to maintain the system alive, and self-consuming cycles to decompose waste materials. Gánti explained that fluid machines are able to perform useful work in a thermodynamic sense, analogous to steam engines, and that their simplest form is a cyclic chemical reaction network.

In the 1980s, based on earlier ideas by Robert Rosen [721, 722], Masahiro Okamoto and colleagues proposed an enzymatic model of a biochemical switch based on a cyclic coupling between an excitatory and an inhibitory factor [632]. They later showed how this simple switch could be extended to play the role of a "chemical diode" [633], a McCulloch-Pitts neuron enhanced with memory storage [635], and finally how biochemical neurons could be composed to form a prototype of a biochemical computer called an "artificial neuronic device" [631] akin to a neural network. The possibility to build logic gates using such biochemical switches was briefly evoked in [634], but not demonstrated at that time.

Later, Hjelmfelt and colleagues [386] showed how to build logic gates using interconnected chemical neurons. Their chemical neurons were based on a modified version of Okamoto's neurons including only reversible reactions and controlled by an enzyme. In [386] it is shown how to compose these chemical neurons into neural networks, and how to implement logic gates such as AND, OR, NOT, NAND, etc., using these interconnected neurons. In a subsequent paper [387] the same authors show how to build finite state machines out of similar chemical neurons. Such machines are synchronized using a chemical clock, and include a binary decoder, a binary adder, and a stack memory, all made of interconnected chemical neurons. The authors show that it is in principle possible to implement a universal Turing machine using such chemical elements. The informal universal computation proof shown in [387] was extended in [539] to include comprehensive qualitative results. It was therefore established that chemical computation can indeed be made Turing Universal.

Next we shall examine Okamoto's switch [633] as an example of such early ideas on the design of components for wet chemical computers.

Example: Okamoto's Biochemical Switch

Okamoto's switch [633] can be described by the following set of chemical reactions:

$$I_1 \longrightarrow X_1 \tag{17.8}$$
$$I_2 \longrightarrow X_3 \tag{17.9}$$

$$A + X_3 \xrightarrow{k_2} B + X_4 \tag{17.10}$$

$$B + X_1 \xrightarrow{k_1} A + X_2 \tag{17.11}$$

$$X_2 \xrightarrow{k_3} \emptyset \tag{17.12}$$

$$X_4 \xrightarrow{k_4} \emptyset \tag{17.13}$$

Substances I_1 and I_2 represent the input to the system, acting as precursors to the production of substrates X_1 and X_3, respectively. The output of the switch is read from the concentrations of enzymes A and B. These enzymes are cofactors that convert X_3 to X_4 and X_1 to X_2, respectively (see reactions 17.10 and 17.11). Both X_2 and X_4 decay as shown in reactions 17.12 and 17.13. The concentrations of input substances I_1 and I_2 are assumed to be controlled externally in order to maintain the input to the system at the desired level, so their concentrations are not affected by reactions 17.8 and 17.9, although they are consumed in those reactions.

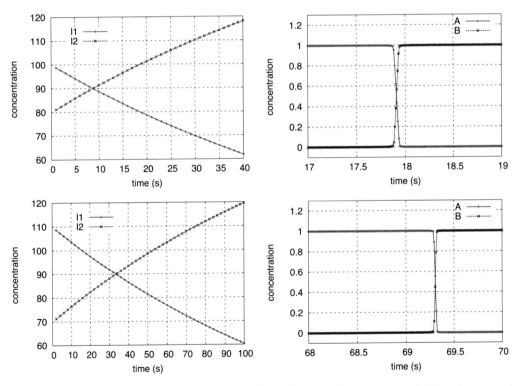

Figure 17.5 Biochemical switch by Okamoto et al. [633]. Top: original setup from [633]: a few seconds after the two inputs reach equal concentration, the concentration of A flips from one to zero, while B flips in the reverse direction. Bottom: the switching time can be adjusted by acting on the inputs I_1 and I_2; here, the switching point is delayed by slowing down the rate of conversion from I_1 to I_2.

Figure 17.5 (top) shows the behavior of the switch using an ODE simulation for the parameters from [633], namely $k_1 = k_2 = 5 \times 10^4$, $k_3 = k_4 = 10.0$, with initial concentrations

$X_1(0) = X_3(0) = 0$, $X_2(0) = X_4(0) = 8.0$, $A(0) = 1.0$, and $B(0) = 0$. The input I_1 decreases in time while I_2 increases. We simulate this by starting with an initial concentration of $I_1(0) = 100$ and $I_2 = 80$, and then converting I_1 to I_2 in an irreversible reaction $I_1 \to I_2$.

Figure 17.5 (bottom) shows what happens when we slow down the rate of conversion from I_1 to I_2, causing the encounter $I_1 = I_2$ to happen later in the simulation. As a result, the switching point for A and B is shifted to a later time. In this way, the switching time can be adjusted by acting on the inputs I_1 and I_2.

This is just a very rough sketch of how the switch works. A detailed mathematical analysis can be found in the original paper [633], including a prediction of the switching time given the input signals. They also show how the switch flips from on to off and vice versa in response to a sinusoidal input signal.

Chemical switches are generally based on the principle of bistability: A dynamic system is bistable when it has two stable steady states, with an unstable steady state in the middle. The unstable one causes the switch to instantaneously flip to one or other stable state to the left or to the right, given a slight perturbation. The existence of bistability in a chemical system can be detected and proven through the mathematical analysis of the ODE resulting from the set of chemical reactions. Such knowledge can also be used to design new switches. Similar analysis techniques can be applied to design other circuit elements, such as oscillators. A detailed description of analysis techniques for nonlinear chemical systems can be found in [258], including guidelines for their design.

17.4.2 Biochemical Oscillators

Oscillators are important in biology because they provide clock signals to pace the activities of the cells, such as the circadian clock that drives our 24-hour circadian rhythm. Similarly, they are also important design components in biochemical computers, because they provide clock signals that allow synthetic biochemical circuits to function in a synchronized fashion, in analogy to electronic circuits. An early model of a biochemical oscillator is the Brusselator [675].

Example: The Brusselator Model of a Biochemical Oscillator

The Brusselator [675] is expressed by the following chemical reactions:

$$A \xrightarrow{k_1} X \tag{17.14}$$

$$B + X \xrightarrow{k_2} Y + D \tag{17.15}$$

$$2X + Y \xrightarrow{k_3} 3X \tag{17.16}$$

$$X \xrightarrow{k_4} E \tag{17.17}$$

In this model, an autocatalytic enzyme X produces more of itself from substrate Y (reaction 17.16); however, this conversion is partially reversed by the presence of chemical B, which forces enzyme X to release Y in reaction 17.15, generating waste product D in the process. X is regularly injected (reaction 17.14) and also decays to waste product E (reaction 17.17). The interconversion between X and Y generates oscillations in the concentrations of these two substances, under some parameter ranges. The concentrations of substances A and B are assumed to remain constant throughout the process (for instance, due to a buffering mechanism that compensates for their depletion in reactions 17.14 and 17.15).

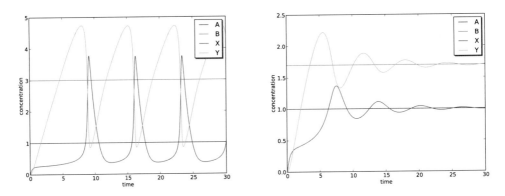

Figure 17.6 Brusselator in the unstable (oscillatory) regime (left) and in the stable regime (right).

Figure 17.6 shows the behavior of the Brusselator in two cases: the first (left side) shows the Brusselator under an unstable parameter region that leads to sustained oscillations; the right side shows a stable regime when the oscillations are dampened and end up disappearing as the system converges to a steady state. The relation between the concentrations of A and B determines the resulting behavior: By calculating the equilibrium of the system and analyzing its stability, it can be demonstrated that sustained oscillations occur for $b > 1 + a^2$ (where $b = [B]$ and $a = [A]$), in the simplified case where $k_i = 1$, $i = 1..4$.

The plots on figure 17.6 were obtained via the numeric integration of the ODEs resulting from reactions 17.14 to 17.17, for this simplified case (that is, with all the k_i constants set to one), and with $a = 1$, thus the oscillatory region occurs for $b > 2$. Indeed, on the left side the value $b = 3$ was used, while on the right side we had $b = 1.7$. The initial concentrations of the other substances were all set to zero.

The Brusselator is an idealized theoretical model of a chemical system. The authors themselves point out that reaction 17.16 is unrealistic since it requires the simultaneous collision of three molecules, which is a very unlikely event in real chemistry. However, such idealized models are widespread in the literature, since they simplify the mathematical analysis of the resulting dynamical system, while capturing the essential qualitative phenomena of interest.

Other well-known biochemical oscillators include the BZ reaction [276] and the Repressilator [256] which both appear in chapter 19.

17.4.3   Analog Chemical Computation of Algebraic Functions

Another stream of research shows that it is possible to compute in an analog way with a chemistry, using the information encoded within the dynamics of concentrations: it is possible to implement simple mathematical functions with chemical reaction networks and then compose them to build more complex functions [141,220,511]. Both manual design and automatic evolution have been used to obtain reaction networks able to compute a target function. We illustrate this approach with a simple example of computing a square root function, and highlight some of its shortcomings.

Example: Computing the Square Root of a Number Using Concentrations

As an example of "analog" computation of functions using a chemistry, let us imagine that we would like to compute the function $y = \sqrt{(x)}$. The following pair of chemical reactions solve this problem, as will be shown below:

$$X \xrightarrow{k_1} X + Y \tag{17.18}$$

$$2Y \xrightarrow{k_2} \varnothing \tag{17.19}$$

According to the law of mass action, the speed of each reaction above can be written respectively as:

$$v_1 = k_1 x \tag{17.20}$$

$$v_2 = k_2 y^2 \tag{17.21}$$

and the change in concentration of Y can be expressed as:

$$\dot{y} = v_1 - 2v_2 = k_1 x - 2k_2 y^2 \tag{17.22}$$

The result of the computation is read after the system reaches equilibrium, when the concentration of Y no longer changes:

$$\dot{y} = 0 \Rightarrow k_1 x - 2k_2 y^2 = 0 \Rightarrow y = \sqrt{\frac{k_1 x}{2k_2}} \tag{17.23}$$

In order to obtain $y = \sqrt{x}$ we must set:

$$\frac{k_1}{2k_2} = 1 \Rightarrow k_1 = 2k_2 \tag{17.24}$$

Figure 17.7 shows the result of this computation for various values of x, using ODE integration of equations 17.22 for $k_1 = 2$ and $k_2 = 1$, which satisfy condition 17.24. This is a simple way to perform a chemical computation in which information is provided in the form of concentration of substances. Other similar solutions for the square root computation exist, and various arithmetic functions can be encoded in a similar way. More complex functions can be composed out of elementary ones, by feeding the product of one reaction network as input for another one, and so on, forming complex webs of reactions performing such chemical computations.

This is a simple way to perform computation with a chemistry, but it has many shortcomings that limit its applicability: First of all, it is difficult to encode negative numbers with such a representation based on concentrations. Second, we must wait for the system to reach steady state before reading the result, which can take sometimes too long. Moreover, not all systems reach a steady state, some oscillate; some grow. Finally, and more important, the composition of complex functions out of simple ones requires different chemicals in every step, in order to avoid interference among the multiple building blocks. This second shortcoming severely limits the scalability of this approach, especially for implementation with real chemistry. In [386] the authors already admit that the number of required chemical species is a practical shortcoming of their approach. However, they point out that this might not be a big obstacle if the species are polymer chains like proteins (which is the case anyway for enzymes), due to their virtually unlimited variety of possible types. They also point to spatially distributed or compartmentalized systems as another potential solution to reduce the number of needed species. Such systems will be covered in chapter 19.

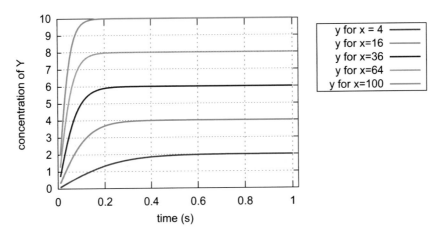

Figure 17.7 Chemical computation of $y = \sqrt{x}$, example of solution using reactions 17.18 and 17.19: the concentration of substance X encodes the value of x, while the result y is read from the concentration of substance Y after the system reaches a steady state.

17.5 Summary

In this chapter, various techniques for computing with artificial chemistries have been outlined. Some of the proposed approaches are still in preliminary stages, and there is a lack of solid programming principles and guidelines. In spite of these challenges, *in silico* chemical computing brings a different perspective to conventional computer science: parallelism as opposed to sequentiality, instruction "soups" as opposed to neat program counters, and systems that "fall" toward the solution as part of their dynamics. Some ideas discussed in this chapter will recur in chapters 18 and 19 in the form of biological models or wet computation ideas. These common ideas might one day coalesce into common principles of implementation valid across a wide range of biochemical computing substrates.

All models are wrong, some are useful.

GEORGE BOX

CHAPTER

18

MODELING BIOLOGICAL SYSTEMS

A s abstract chemical systems, artificial chemistries find natural applicability in the modeling of biological systems. In this chapter we examine the role of artificial chemistries in this context. One of the strengths of ACs for modeling biology is their ability to abstract away from the biochemical details while retaining the constructive potential to deal with combinatorially complex models of biomolecules and their interactions. Such a level of resolution lies somewhere between fine-grain molecular modeling [499, 751] and more macroscopic models such as those based on differential equations and agent-based modeling [83, 711, 907].

In chapter 2 we discussed modeling and simulation in general, but here it is worth to be a bit more specific about the role of modeling in biology. Biological models attempt to explain biological phenomena using mathematics and computer simulations of biochemical process occurring at the level of cells, tissues, and organs. The computational modeling of biological systems is situated between mathematical modeling and experimental research, and is used to test how well the different hypotheses match existing experimental data, complementing other computational techniques devoted to hypothesis generation, the mining of large data sets, and the analysis and optimization of system parameters [465].

Biological phenomena most often result from complex interactions among different elements at multiple scales of resolution, and reductionist models are often not sufficiently powerful in biology. The first step to build a good biological model is to identify the level of detail one wishes to observe in order to capture phenomena of interest. This then has consequences for other levels and their interaction with this level. The main difficulty is to strike a proper balance between simplistic models that do not hold sufficient explanatory power, and exhaustive models that become intractable. In more recent times, many projects strive to model complete systems at multiple scales, such as the Human Brain Project [1] and others.

We start our discussion at the molecular level, with algorithms for RNA and protein folding. Folding has been extensively used as a metaphor in artificial chemistries, due to the mapping between structure and function that it entails. An introduction to enzymatic reactions and the binding of proteins to genes then follows. These two types of reactions are recurrent in the literature related to modeling of the dynamics of biochemical pathways, which is discussed next. We then devote a section to genetic regulatory networks (GRNs), since several ACs modeling GRNs exist and can also be applied to other domains such as robotics and engineering. A treatment of cell differentiation and multicellularity follows, which naturally calls for spatial ACs before we close with morphogenesis.

We focus on extant biological organisms in this chapter. Other biological models appear elsewhere in this book; see chapter 6 for an overview of models of early life and potential synthetic organisms, chapter 19 for the engineering of new chemical reaction mechanisms applicable to various biological problems, and chapter 7 for models of biological evolution and ecosystems.

18.1   Folding Algorithms

As discussed in chapter 5, biopolymers such as DNA, RNA and proteins have a linear chain primary structure, and can fold into intricate 3D shapes. The properties of a molecule change with the shape it adopts. In this way, DNA can only be read after unwinding its coiled structure and separating its double strands; RNA can form loops and hairpins, among other structures, that may allow it to act as a specific binding domain or as a catalyst; and proteins fold into complex shapes with sophisticated enzymatic and structural properties.

The 3D shapes of biomolecules can be observed experimentally with the help of laboratory techniques such as X-ray crystallography and nuclear magnetic resonance as well as chemical and enzymatic modification assays. However, these techniques are laborious, slow, and expensive. Since the folded 3D structure is crucial for their biochemical function, many algorithms have been proposed to predict 3D shape from polymer sequence. These algorithms typically seek to obtain the minimum energy configuration of the molecule, which in most cases corresponds to the stable folded form. In what follows we will discuss some of these algorithms, focusing on their relation to ACs. Indeed, many ACs seek to bring interesting properties of folding into artificial systems. The matrix chemistry of chapter 3 is one example, other examples will appear throughout this section.

18.1.1   RNA Folding

RNA molecules play several important roles in the cell, some of which have only recently been discovered. The most frequently occurring type of RNA is the ribosomal RNA (rRNA), which is the predominant compound of ribosomes and accounts for about 80 percent of the total RNA in living cells. Other RNA types include messenger RNA (mRNA), which carries DNA information to the ribosomes for translation into proteins; transfer RNA (tRNA), which helps translating mRNAs to proteins; and various recently discovered RNAs that may have various catalytic and regulatory roles.

RNAs are typically single-stranded molecules that may fold into various three-dimensional shapes and allow them to carry out specific biological functions. The folding of an RNA macromolecule is mainly caused by complementary base-pairing between segments of RNA, with loops formed where no base-pairing occurs. The folded RNA structure is often represented in dot-bracket notation; the original sequence is represented as a string of identical length composed of nested parenthesis representing base pairs, and dots representing unpaired nucleotides. Table 18.1 shows two examples, whose resulting shapes can be seen in figure 18.1. These shapes were obtained using the RNAfold server from the Vienna suite of RNA processing tools [349], an extensive software package for RNA research that is also well known in the

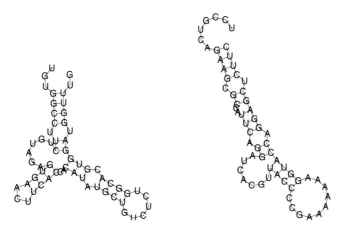

Figure 18.1 RNA folding examples, generated automatically using the Vienna RNAfold server (http://rna.tbi.univie.ac.at), from the two RNA sequences of table 18.1.

AC community. RNA folding algorithms typically seek to minimize the free energy of the RNA structure using dynamic programming [209,766]. More information about the particular folding algorithm used in the Vienna suite can be found in [389, 390], and an overview of the various tools available in the Vienna suite can be found in [349].

Table 18.1 Two randomly generated RNA sequences of 60 nucleotides each (generated using the Random DNA Sequence Generator at http://www.faculty.ucr.edu/~mmaduro/random.htm), together with their corresponding structure in dot-bracket notation, obtained with the Vienna RNAfold server (http://rna.tbi.univie.ac.at). Their two-dimensional shapes can be visualized in figure 18.1.

	First RNA (left side of Fig. 18.1):		
Sequence:	UGUGGCCUUCUGUAGAGUGAACUUCACCACAUAUGCUGUCUCUGGCACGUGGAUGGUUUG		
Structure:	...(((((.((......((((...))))..(((.((((......)))).))))).))))..		
	Second RNA (right side of Fig. 18.1):		
Sequence:	UCCGUCAGAAGCGCCAUUCAGGAUCACGUUACCCCGAAAAAAAGGUACCAGGAGCUCUUC		
Structure:	.......(((((.((..(((.((.......((((..........)))))).))))).))))		

Although the basic folding principle for RNA is free energy minimization [549], in reality RNA structures are influenced by various other factors [759] such as the formation of pseudo-knots, noncanonical base-pairing, non-nearest neighbor and sequence-specific effects, the potential influence of enzymes called RNA chaperones [378,850], and the existence of groups of alternatively folded structures for the same sequence. Such complications often lead bioinformaticians to compare homologous RNA sequences (sharing a common ancestry) from different organisms in order to infer their structure, using a human-guided method called comparative sequence analysis [311,353]. This method consists of inferring the structure of a family of similar homologous macromolecules by assuming that their structure is similar to the structure of a macromolecule of the same family whose structure has been already determined experimentally. To date, comparative sequence analysis is still considered the most accurate method to predict RNA secondary structure in bioinformatics, and is usually the benchmark against which new algorithms are tested [759]. Its shortcomings are that it is not fully automated and that it is only applicable to naturally occurring structures for which multiple homologs are available for comparison. For synthetic *de novo* RNAs, or for rare RNAs with few or no known homologs, free energy minimization remains the method of choice. Statistical learning methods have also been proposed in order to predict more complex RNA structures by relying on fewer parameters [237]. See [731,759] for more details about the biophysical process behind RNA folding and the various computational methods proposed to predict it.

RNA Folding in Artificial Chemistries

The matrix chemistry of chapter 3 is an example of an artificial chemistry where RNA folding is used as a metaphor to obtain a 2-dimensional matrix operator from a one-dimensional vector of numbers. The matrix chemistry owes much of its interesting behavior to folding, such as the ability to form organizations. Other ACs built upon an RNA folding metaphor include [481,824, 873].

RNA folding algorithms based on free energy minimization have been used in artificial chemistries to understand the emergence and evolution of biochemical reaction networks where reactions are catalyzed by ribozymes. For this purpose, an RNA artificial chemistry was constructed [873], where molecules are RNA sequences that fold into secondary structures using the Vienna package [349,389,390]. The folded RNAs may act as ribozymes in chemical reactions. The catalytic function of a given RNA molecule is assumed to lie within its longest loop, and is determined with the help of Fujita's imaginary transition structures [300], a hierarchical classification scheme for chemical reactions. Resulting chemical reactions are performed via the graph rewriting scheme from ToyChem [103] (covered in section 11.3 of this book). This RNA AC therefore implements a two-step genotype-phenotype map that first maps the genotype (RNA sequence) to a corresponding structure (folded shape), and then maps the structure to a function (catalytic activity) of the molecule.

Using the RNA AC proposed in [873], several aspects of ribozyme-catalyzed artificial metabolisms have been studied. Evolution experiments are reported, showing that small-world connectivity emerges in large networks containing highly connected metabolites. In [874] the neutrality of the genotype-phenotype map was studied through random walks in genotype space, revealing a large number of neutral variants, as well as a variety of alternative functions that can be reached within a short distance from a given sequence. These findings indicate that such systems can be simultaneously robust to harmful mutations and evolvable (able to promptly discover useful novelty). Subsequent evolution experiments [280, 875, 876] showed a partition of the evolutionary process into qualitatively different phases: at the beginning, metabolic pathways are extended in the forward direction, through the emergence of enzymes able to extract energy from food molecules by breaking them into increasingly simpler components. In later stages of evolution, complex metabolic pathways are formed as existing enzymes are involved in new reactions and further specialize their function.

Other examples of ACs in which molecules are RNAs that fold into secondary structures include [481,824]. In [481] binary strings are "folded" into a planar graph structure using a dynamic programming approach similar to algorithms used in regular RNA folding. The purpose of this study is to examine the role of neutral mutations in escaping from local optima during evolution. A "mathematical folding" model is proposed in [824], where molecules are represented as strings that can "fold" into a graph structure where edges are chemical bonds with covalent, ionic, hydrogen, or van der Waals strength. The folded "enzyme" obtained by this process can then function as a control-flow computer, performing operations on a set of input tokens. As an example, a "splitase" enzyme is constructed, which is able to split a set of molecules into hydrophilic and hydrophobic clusters.

A prominent research topic of wet ACs and synthetic biology (see also chapter 19) is the engineering of artificial RNAs for various biological purposes, including novel therapies in medicine. In this context, algorithms that solve the inverse folding problem [349, 389], providing a family of neutral sequences that map to a desired shape, which is important for the design of novel RNAs with specific functions.

18.1.2 Protein Folding

The protein folding problem, or protein structure prediction problem, consists of devising algorithms able to derive the 3D shape of a protein from its sequence. Chapter 5 depicted some typical folded protein structures (see figure 5.6). The final shape of a protein is influenced by several factors, and knowledge in this area is still far from complete. Predicting protein structure is of utmost importance in the medical domain, since misfolded proteins can cause severe illnesses such as Alzheimer's, Parkinson's, and Creutzfeld Jakob diseases. Current research effort is also in progress in designing new proteins that can help fighting widespread human diseases such as AIDS, cancer, or malaria.

The protein folding problem is more complicated than the RNA folding problem, for several reasons. First of all, there are many types of amino acids, each with their own properties, including hydrophobicity, charge, and size. A protein folding algorithm must take all relevant properties into account. Typically, in an aqueous environment, the hydrophobic amino acids will find refuge in the interior of the molecule, while the hydrophilic amino acids will tend to stick out on the surface. Moreover, many proteins incorporate additional elements such as metal ions, which have an active role in their operation, for instance, the iron of hemoglobin binds to oxygen, enabling the molecule to transport oxygen through the bloodstream. The shape of a protein may also change depending on its surrounding environment, which will vary from organism to organism. Moreover, the folding process is often assisted by chaperones [369, 734], specialized enzymes that can aid in the folding process, or may prevent a protein from folding prematurely, or recognize misfolded stretches.

Due to the combinatorial nature of protein sequences, the protein folding problem can be formulated as a (combinatorial) optimization problem. Particular instances of the protein folding problem have been shown to be NP-hard [110]. Many protein folding algorithms therefore rely on heuristic optimization techniques to navigate the vast amount of possible configurations in solving the energy minimization problem. Folding algorithms based on various optimization techniques have been proposed, including genetic algorithms [473, 659, 684], simulated annealing [735], estimation of distribution algorithms [740], ant colony optimization [181, 775], to name a few.

Another approach to the protein folding problem, often used in bioinformatics, is to infer protein structures by comparing related homologous sequences, sometimes with the help of statistical machine learning algorithms [354]. However, this approach does not help to shed light

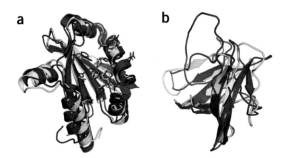

Figure 18.2 Examples of protein structure predictions, aiming at different targets, (a) and (b), set for the CASP9 contest http://predictioncenter.org/casp9/. Starting from predictions by Rosetta@Home (red), the predictions by the FoldIt players (yellow) get closer to the experimentally observed crystal structure of the target protein (blue). Images produced by the FoldIt Void Crushers Group [457] using the PyMOL software package (http://www.pymol.org). Reprinted by permission from Macmillan Publishers Ltd: *Nature Structural & Molecular Biology*, vol. 18 pp. 1175–1177, © 2011.

on the physicochemical mechanisms responsible for the folded shapes, nor does it help to infer the shapes of *de novo* synthetic proteins.

The difficulty of protein folding has led researchers to propose *FoldIt* [197, 457], an online multiplayer game in which humans can help in the folding effort by manually interacting with a protein structure and by attempting to obtain an optimal folding. FoldId players have succeeded to solve concrete folding problems with biological relevance, such as discovering the crystal structure of a retroviral protease [457].

Other initiatives to involve the general public in the folding effort include *Folding@Home* and *Rosetta@Home*. Both count on the help of volunteers around the world, who install and run the folding software on their computers, donating computational resources to compute large protein foldings. *Folding@Home* [86, 492, 774] uses a Markov model to predict the more probable folding paths starting from the protein sequence. It is applied to purposes such as drug design and molecular dynamics simulations. Rosetta@Home [210, 686, 778] uses a Monte Carlo free energy minimization procedure based on realistic physical models, in combination with comparisons with known protein structures retrieved from available bioinformatics databases. Rosetta has also been used to design new proteins and to solve related molecular modeling problems such as the docking of small molecules to proteins, protein-protein and protein-DNA interactions, receptor-ligand bindings, and RNA folding.

International competitions are organized periodically to determine the best folding methods and to assess the current state of the art in the field [602]. Figure 18.2 displays an example of result, in which predictions from Rosetta@Home were refined by a FoldIt team to yield models that more closely resemble the experimentally observed structures. So far no solution to the protein folding problem has become sufficiently well established, and the efforts continue to be pursued.

Since folding real proteins is a complex and computationally expensive problem, few artificial chemistry approaches have been directly based on it. Perhaps the closest examples in this context are Suzuki's Network AC (NAC, see section 11.3) and the *Aevol* platform by Beslon and co-workers [654]. In NAC, node chains akin to biopolymers fold into cluster structures akin to globular proteins, that can have a catalytic function. In the Aevol platform, an artificial chemistry is defined that includes biopolymers akin to DNA, RNA, and proteins. In this chemistry, genes are binary strings that code for proteins using a codon translation table similar to the codon to amino acid translation mechanism found in nature. The resulting protein is then "folded" into a mathematical function that expresses "metabolic" activity of the protein in terms of a fuzzy set formalism. Proteins in this formalism can range from polyvalent proteins with low specificity and low efficiency, to highly efficient and specialized proteins. Using the Aevol platform, the authors studied the effect of homologous gene rearrangements on the evolution of genetic regulatory networks, showing that such rearrangements are important to the evolvability of these networks.

18.2 Basic Kinetics of Biomolecular Interactions

18.2.1 Enzyme Kinetics

Enzymes are substances that act as catalysts in biological systems. Most enzymes are proteins, and they act by binding weakly to a set of substrates, accelerating their reaction without actually being consumed in the reaction. As introduced in chapter 2, enzymes accelerate reactions by decreasing the activation energy necessary for the reaction to occur. This occurs in both directions for a reversible reaction, therefore the equilibrium of the system does not change under the presence of the enzyme, it is just reached more quickly. Such a decrease in activation energy is possible because the reactants (substrates for the enzyme) fit into a sort of "pocket" in the enzyme, which is the enzyme's active site. They can then come into contact more easily, hence requiring less energy for the reaction to occur.

An enzymatic reaction can be represented in a compact way as:

$$S \underset{E}{\rightleftharpoons} P \tag{18.1}$$

where E is the enzyme, S is the substrate, and P is the product of the reaction. This reaction actually occurs in multiple steps:

$$E + S \rightleftharpoons ES \rightleftharpoons EP \rightleftharpoons E + P \tag{18.2}$$

where ES is an enzyme-substrate complex, and EP an enzyme-product complex.

Figure 18.3 depicts the energy levels for the various steps of reaction 18.2. The enzyme binds to the transition state more strongly than to the substrates or products. The transition state then becomes more stable than in the uncatalyzed case: The energy barrier is lower, facilitating the reaction.

A widely used model of enzyme kinetics is the Michaelis-Menten model [42]. It is a simplification of reactions 18.2, in which the product formation step is assumed to be irreversible:

$$E + S \underset{k_{a'}}{\overset{k_a}{\rightleftharpoons}} ES \overset{k_b}{\longrightarrow} E + P \tag{18.3}$$

By writing the rate equations for this system of reactions (applying mass action in each reaction step), and considering that the net rate of formation of ES should be zero (due to mass conservation), it is easy to obtain the Michaelis-Menten equation:

$$\frac{d[P]}{dt} = \frac{v_m[S]}{k_m + [S]} \tag{18.4}$$

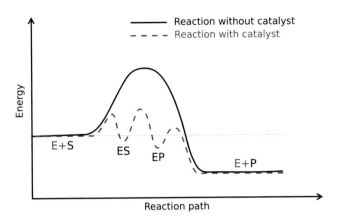

Figure 18.3 Schematic representation of how enzymes decrease the activation energy necessary for the reaction to occur. Black curve: uncatalyzed reaction. Red curve: catalyzed reaction featuring two intermediate steps: the production of the enzyme-substrate complex (ES), and of the enzyme-product complex (EP), each step requiring a much lower activation energy barrier than the uncatalyzed case.

where:

- $v_m = k_b[E_0]$,

- $[E_0] = [E] + [ES]$ is the total amount of enzyme (free plus bound to substrate) available in the system

- and $k_m = (k_{a'} + k_b)/k_a$ is the Michaelis-Menten constant.

Equation 18.4 expresses the rate of product formation as a function of the amount of substrate S. When S is so abundant that $S \gg k_m$, the rate tends to v_m, which is the maximum rate of the reaction. Thus the reaction saturates at some point where there is insufficient amount of enzyme to convert all the substrate. This is depicted in figure 18.4.

18.2.2 Hill Kinetics

An important characteristic of GRNs is that the reaction rate laws are often assumed to follow Hill kinetics, as opposed to mass action. Hill kinetics assumes that two or more ligands (proteins in the GRN context) bind cooperatively to a single receptor (a gene in the GRN context). That is, a group of proteins act in cooperation in order to activate or repress a given gene. A single protein alone is not enough.

Hill kinetics can be obtained with the following reversible chemical reaction [741, 916]:

$$G + nP \underset{k_r}{\overset{k_f}{\rightleftharpoons}} C \tag{18.5}$$

where G is a gene, P is a protein, n is the number of proteins that bind to the gene in order to activate/repress it, and C is the complex formed by the association of n proteins P with gene G. The coefficient n is called the Hill coefficient.

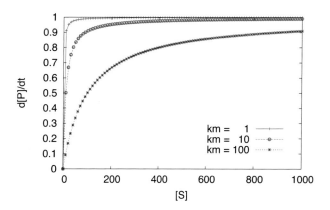

Figure 18.4 Plot of the rate of an enzymatic reaction as a function of the amount of substrate S, following the Michaelis-Menten equation.

Note that this scheme is a rather idealized model of reality, since it assumes that n molecules bind simultaneously to G, which is unrealistic in practice. However it is still useful as a qualitative model: it has been shown [916] that the Hill coefficient n is better interpreted as the level of cooperativity among multiple ligands that bind non-simultaneously to a receptor, rather than the actual amount of ligands that bind simultaneously to the receptor.

At chemical equilibrium we have:

$$k_f GP^n = k_r C \implies K_d = \frac{k_r}{k_f} = \frac{GP^n}{C} \tag{18.6}$$

where K_d is the reaction dissociation constant. The amount of gene does not change, hence:

$$G + C = G_0 \tag{18.7}$$

By defining $K^n = K_d$, and substituting Eq. 18.7 in Eq. 18.6, the fraction of occupied and free gene molecules can be obtained:

$$H_1 = \frac{C}{G_0} = \frac{P^n}{K^n + P^n} \tag{18.8}$$

$$H_2 = \frac{G}{G_0} = \frac{K^n}{K^n + P^n} \tag{18.9}$$

Functions H_1 and H_2 are both known as Hill functions. They can be normalized by defining $x = \frac{P}{K}$, hence:

$$H_1(x) = \frac{x^n}{1 + x^n} \tag{18.10}$$

$$H_2(x) = \frac{1}{1 + x^n} \tag{18.11}$$

Figure 18.5 (top) shows the shape of the Hill functions for various values of the Hill coefficient n. The bottom plots show the ODE integration of reaction 18.5 for the same coefficients, starting

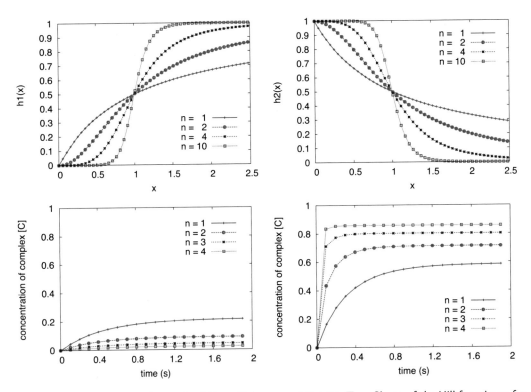

Figure 18.5 Hill kinetics under varying hill coefficients $n = 1, 2, 4, 10$. Top: Shape of the Hill functions: for $n > 1$, the Hill functions have a sigmoid shape. Bottom: Change in concentrations of the gene-protein complex C, for the case where the equilibrium concentration of free protein P is around $P \approx 0.5$ (left) and $P \approx 1.5$ (right).

with initial concentrations $G(0) = P(0) = 1$, $C(0) = 0$. When the equilibrium concentration of protein is low (bottom left plot, $P < 1$) higher values of hill coefficient lead to a slower conversion of gene G to gene complex C, and the conversion stagnates at a low proportion of C. Whereas when $P > 1$ (bottom right plot), higher coefficients lead to a steeper conversion, leading to a much higher proportion of C at the end of the simulation. This is consistent with the Hill curves shown in the top plots, where higher coefficients lead to a more pronounced sigmoid shape, with a higher proportion of occupied genes for $P > 1$.

Assume now that protein P is an activator, that is, it promotes the expression of a gene product X_1 (typically an mRNA that can be later translated into a protein):

$$C \xrightarrow{k_1} C + X_1 \tag{18.12}$$

In this case, the rate of formation of X_1 can be written as:

$$\dot{x}_1 = k_1 C \tag{18.13}$$

By substituting eq. 18.8 in eq. 18.13 we obtain:

$$\dot{x}_1 = k_1 G_0 \frac{P^n}{K^n + P^n} \tag{18.14}$$

Therefore the rate of production of X_1 obeys a Hill function, increasing with the amount of activator P present at a given moment, similar to figure 18.5 (top left).

A similar line of reasoning can be applied to obtain the rate of production of a gene product X_2 when protein P is a repressor protein. In this case, only the free gene G produces X_2:

$$G \xrightarrow{k_2} G + X_2 \tag{18.15}$$

and the rate of formation of X_2 can be obtained with the help of eq. 18.9:

$$\dot{x}_2 = k_2 G_0 \frac{K^n}{K^n + P^n} \tag{18.16}$$

Therefore, the rate of production of X_2 also obeys a Hill function, this time decreasing with the concentration of P, similar to figure 18.5 (top right).

The sigmoid shape of the Hill functions can be regarded as a switching behavior, which has two implications. First, biochemical switches can be designed using GRN dynamics; second, by assuming steep Hill functions, GRNs can be modeled as binary circuits, with genes as binary components that can be switched on and off. However this is clearly a simplification: when only one protein activates or represses the gene, the kinetics does not follow a sigmoid shape, as can be seen in figure 18.5 (top) for $n = 1$. Moreover, usually only one copy of the gene is present, and few copies of the proteins; with such low copy numbers, stochastic effects cannot be neglected, and an ODE is a poor approximation for the behavior of the system. Nevertheless, Hill kinetics is pervasive in the GRN literature, since it enables the deterministic modeling of GRNs in a simple and elegant way. Moreover, in the case of algorithms inspired by GRNs, it is often not necessary to mimic real biological GRNs very accurately; therefore, sigmoid models are applicable in such contexts.

18.3 Biochemical Pathways

A biochemical pathway is a sequence of chemical reactions occurring in a cell. Each pathway has a specific function, such as the transformation of some chemical substance or the communication of a chemical signal. Within each cell, numerous pathways operate in parallel, and many of them are interconnected into intricate webs of chemical reactions, forming complex reaction networks.

The biochemical reaction networks found in living cells can be divided into three major groups: metabolic networks, genetic regulatory networks (GRNs), and signaling networks. These networks are linked to each other: typically, biochemical signals outside the cell activate receptors on the cell membrane, that trigger signaling transduction pathways inside the cell, which culminate with the activation or inhibition of the expression of some target gene. The product of the expression of such gene can be a transcription factor for the GRN, an enzyme with a metabolic role, or a protein that participates in a signaling pathway. In this way, the various biochemical networks are regulated by the dynamics of their interactions.

We start with the evolution of metabolic networks using ACs [384, 638]. We then review some other approaches in this area, involving more complex ACs, as well as formal ACs that have been used to model biochemical reaction networks, among them P systems already discussed in chapter 9. Finally, we give an overview of the extensions needed in AC algorithms such that they can be used to model real biological systems.

18.3.1 Studying the Evolution of Metabolic Networks with ACs

Metabolic networks can be represented as graphs where the various chemicals and reactions are represented as nodes that are interconnected to each other via links. Recent studies to characterize the topology of metabolic networks found in living organisms revealed that these networks have a tendency to exhibit scale-free or small-world structures [430, 904] and are often modular [655, 700]. Modularity in metabolic networks refers to the occurrence of clusters of highly interconnected groups of chemicals and their corresponding reactions, with few links traversing different clusters. These findings have motivated many subsequent investigations to explain the reason behind the formation of such topologies [186, 442, 906]. In this section, we highlight two examples of ACs [384, 638] that have contributed to such efforts.

The String Metabolic Network (SMN)

The string metabolic network (SMN) of Ono and colleagues [638] is an AC where chemicals are strings of letters and reactions perform the ligation, cleavage or recombination of strings in a reversible way. Strings of arbitrary length may be formed out of an alphabet of eight letters $\{a, b, c, ..., h\}$, in two types of reversible chemical reactions:

$$A + B \quad \rightleftarrows \quad AB \tag{18.17}$$

$$AB + CD \quad \rightleftarrows \quad AD + CB \tag{18.18}$$

The first reaction is a ligation with corresponding cleavage in the reverse direction, and the second reaction is a reversible recombination between two molecules. Examples of the first and second kind of reaction are:

$$adbg + ef \quad \rightleftarrows \quad adbgef \tag{18.19}$$

$$facbaha + eefg \quad \rightleftarrows \quad facbfg + eeaha \tag{18.20}$$

Each reaction is catalyzed by a corresponding enzyme. The metabolic system of an organism is represented by its set of enzymes (representing the reactions that they catalyze) plus the set of other chemicals (metabolites) involved in the reactions. The organism possesses a genome that encodes the enzymes for its metabolism. At the beginning of the "life" of an organism, an initial set of metabolites are present as resources for the metabolic network. The organism synthesizes other compounds from this initial set, using the enzymes expressed by its genome. A regulated inflow/outflow of substances ensures that the total amount of substances inside the cell is kept stable. The kinetic coefficients for the reactions are set in such a way that the synthesis of longer molecules is more difficult than that of smaller ones.

Metabolic networks are then evolved by mutating the genomes of the individuals, running the resulting metabolic network, and evaluating the mass of each organism (in number of letters) after a fixed number of timesteps.

The mutation operation works as follows. One enzyme (here equivalent to a reaction) is chosen at random; a random compound from its educt side is replaced with another compound already appearing in the organism's compound set; a new point of recombination is then chosen, where applicable; finally, the resulting enzyme is added to the genome. This mutation operation is similar to a gene duplication followed by a mutation.

Since the amount of chemicals inside the organism is kept constant, a larger mass indicates that the organism is able to synthesize longer compounds, which is a difficult task due to the

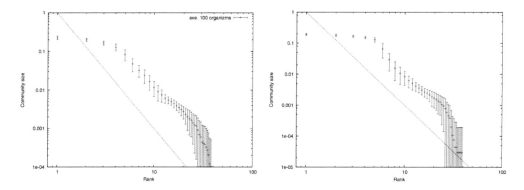

Figure 18.6 Distribution of community sizes in natural (left) and in artificial (right) metabolic networks [638]. The left plot is the average over 100 organisms that possess large metabolic networks, according to a database of genomes and metabolic pathways. The right plot results from the evolution of organisms based on the string metabolic network AC, and shows the average over 30 runs of the SMN evolutionary algorithm. We can see that both natural and artificial networks display very similar curves. Reprinted from Ono et al., LNAI 3630, pp. 716–724, [638], © 2005, with permission from Springer.

higher energy barrier toward such synthesis. A larger mass is therefore considered as an indication of higher metabolic efficiency, and chosen as a fitness criterion to select individuals for the next generation.

Using such an evolutionary approach, the authors show the emergence of modular networks similar to natural metabolic networks, characterized by a power-law distribution of their community sizes. A community is a module, and modules are identified by an algorithm that partitions the graph into subgroups or modules in such a way that the amount of links between modules is minimized. It then suffices to classify these modules by number of nodes (community size) in order to obtain the community size distribution. Figure 18.3.1 compares the distribution of community sizes for natural (left) and artificial (right) metabolic networks. Whereas such modularity is observed in both natural and artificially evolved networks, the authors also show that such modularity does not occur in networks created by random mutations, without selection pressure.

Metabolic Networks Based on Bonded Numbers

In the earlier SMN example, organisms were subject to a fixed environment, defined mainly by the initial amount of metabolites and the inflow and outflow of substances to control the internal concentration of chemicals. In contrast, Hintze and Adami [384] study the evolution of metabolic networks in a context where the environment might change dynamically during the lifetime of an organism. The impact of such changes on the structure of the evolved metabolic networks is then analyzed.

In the artificial chemistry in [384] (we call it the "Bond Number Chemistry," or BNC for short), molecules are strings made of three types of atoms: 1, 2, and 3. The number of the atom represents the number of bonds that it forms. That is, atom 1 may form at most one bond with another atom, atom 2 may form up to two bonds, and atom 3 may form up to three bonds. A molecule is valid when all its atoms are at their maximum number of bonds. Examples of valid molecules in-

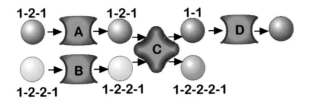

Figure 18.7 Example of metabolic pathway in the BNC-based cell model from [384]: Import proteins A and B carry the precursors 1-2-1 and 1-2-2-1 into the cell, where they are transformed by catalyst C into metabolic byproducts 1-1 and 1-2-2-2-1. Molecule 1-1 is excreted out of the cell with the help of export protein D. Reprinted from figure 1A of Hintze and Adami, PLoS Comp. Bio. 4(2) [384], © 2008 Hintze and Adami, available under the terms of the Creative Commons Attribution 2.5 Generic (CC BY 2.5) license.

clude 1-2-1, 1-2-3=3-1, or 1-3=2, where '-' stands for a single bond and '=' for a double bond. Using such bond rule, 608 valid molecules with length up to 12 can be constructed. Unlike real chemistry, these molecules are linear, thus bonds do not cause the formation of branches or rings.

Reactions in the BNC are similar to the recombination reactions in the SMN. Given two reactants, a cleavage point is chosen within each, and their tails are swapped. Due to the bond rule, however, only a fraction of all possible reactions lead to valid molecules. Even under these conditions, the authors show that there are more than 5 million valid reactions. As an example of valid reaction, in the reaction below the broken bonds at the cleavage points are indicated by '|' in the intermediate stage, and both products are valid molecules at the final stage:

$$1\text{-}2\text{-}2\text{-}1 \ + \ 2\text{=}3\text{-}3\text{=}2 \ \longrightarrow \ 1\text{-}2|2\text{-}1 \ + \ 2\text{=}3|3\text{=}2 \ \longrightarrow \ 1\text{-}2\text{-}3\text{=}2 \ + \ 2\text{=}3\text{-}2\text{-}1 \qquad (18.21)$$

An organism is a cell that contains BNC molecules, a genome, and proteins. The genome contains a set of genes that code for proteins that may have transport or metabolic functions. Genes are strings from the alphabet {0,1,2,3}, analogous to the four DNA bases A, C, T, and G. Each gene encodes the expression level of the protein that it produces, the affinity of this protein to its target molecules, and the specificity of the reaction that it catalyzes. Unlike in nature, proteins are also encoded using the alphabet {0,1,2,3}, and are translated almost one-to-one from their corresponding genes. For example, the import protein 12321000000 transports a molecule 1-2-3=2-1 into the cell. An additional affinity encoding scheme models the binding strength of a protein to target BNC molecules according to an affinity score (see [384] for details). Using such a protein-metabolite interaction model, metabolic pathways can be unfolded. Figure 18.7 shows an example of pathway, including the transport of precursor molecules into the cell, their transformation inside the cell, and the excretion of metabolic byproducts into the extracellular environment.

Cells live in a 2D virtual world, where the smallest 53 of the 608 possible molecules are provided as food resources for the cells. Food molecules diffuse from certain locations on the grid. A population of cells evolve on a grid using a genetic algorithm in which cells compete for the available food resources. In order to survive, cells must be able to transport proteins in and out of the cell, and to synthesize larger molecules (metabolites) from the 53 types of food molecules available in the environment. Organisms are evaluated for their ability to synthesize complex metabolites. A cell divides when it contains a sufficient amount of such complex metabolites.

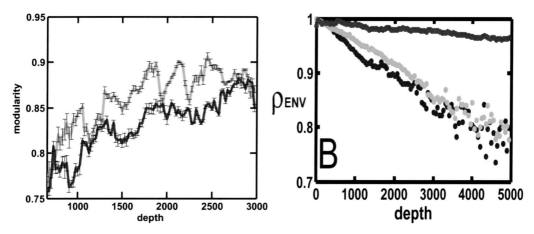

Figure 18.8 Evolution of network modularity (left) and environmental robustness (ρ_{ENV}, right) for the BNC-based cell model from [384], under three different environments: static (green), quasi-static (blue), and dynamic (red). On both plots the x-axis (depth) indicates the descent along the ancestor line (from the first, oldest individual with depth zero, to its most recent descendant with depth 5000). Reprinted from figs. 6 and 7B of Hintze and Adami, PLoS Comp. Bio. 4(2) [384], © 2008 Hintze and Adami, available under the terms of the Creative Commons Attribution 2.5 Generic (CC BY 2.5) license.

Mutations in the genome may occur upon cell division, leading to new variants with potentially different metabolic activities.

Using this model, the authors compared the evolved metabolic networks under three different environments: a static environment where the location of the food sources is fixed; a quasi-static environment where a single food source is moved at each update; and a dynamic environment where all food sources may move at every update. Their simulation results show that metabolic networks evolve more slowly in dynamic environments; moreover, they appear to be less modular. On the other hand, they are more robust to environmental noise, since they evolve alternative metabolic pathways in order to survive using different food sources. Figure 18.8 summarizes these findings.

The fact that the evolved metabolic networks under the BNC model are less modular in dynamic environments contradicts related work [441] that had indicated that a changing environment would, on the contrary, foster the emergence of modularity. As discussed in [384], the reason for this discrepancy is most probably the nature of the environmental changes in [384], which is highly random, as opposed to the more systematic changes introduced in [441]. Biological reality lies perhaps somewhere in the middle.

18.3.2 Other AC Approaches to the Modeling of Biochemical Pathways

The two examples described before are based on simple chemistries where molecules are strings of symbols. We now briefly review some more complex and realistic ACs that have been proposed for the modeling of biochemical reaction networks, as well as some formal models aimed at capturing the essential aspects of biological processes with strong analytical methods.

Tominaga and colleagues [855, 856] propose a string-based AC to model biochemical pathways in cells. In contrast with related string-based ACs, Tominaga's AC allows strings to be stacked, forming 2-dimensional patterns that enable them to form double strands or other more complex structures similar to those occurring in natural biomolecules. The reaction rules in this chemistry perform pattern matching and recombination operations that mimic the lock-and-key binding interactions among biomolecules, and the constructive power of biopolymers. In [855] the authors show that a variety of biochemical processes can be implemented using this AC, ranging from metabolic pathways such as the oxidation of fatty acids, to more complex operations on nucleic acids such as DNA replication, transcription from DNA to mRNA, and the translation from mRNA to protein. In [856] they show that their chemistry is computationally universal, and then show how it can be used to implement well-known DNA computing algorithms such as the Adleman-Lipton paradigm and the DNA automaton by Benenson and co-workers (see section 19.3.1).

In chapter 11 we had described the Toy Chemistry by Benkö and colleagues as a example of graph-based chemistry that represents molecules and their reactions in a way that is much closer to real chemistry than the average AC. The Synthon Artificial Chemistry proposed by Lenaerts and Bersini [505] is a more recent graph-based chemistry that has also been proposed to get closer to real chemistry. As such, it also represents molecules as graphs of atoms, and reactions as transformations on graphs. Chemical aspects such as charged molecules, covalent bonds, and ionic bonds are explicitly modeled. Chemical reactions are modeled by extending the notion of isomerism (compounds that have the same chemical composition but with atoms placed at different positions) to ensembles of molecules. From this perspective, the educt and product sides of a reaction can be regarded as isomers of each other at the ensemble level, and reactions are isomerizations between two ensembles. In contrast with the Toy Chemistry, Synthon focuses on the dynamical aspects of the system, including stochastic chemical kinetics and the coevolutionary dynamics of the system. Experiments are reported in [505] that show the partition of the molecule sets into two categories of chemical reaction networks, a phenomenon that could help to explain the origin of homochirality (the fact that in nature, one of two mirror-image molecules is often selected for, in spite of the fact that both molecules have equal functions in principle).

Oohashi and colleagues [643] propose an AC embedded in a virtual ecosystem to study the emergence of biochemical pathways for the degradation and reuse of biomolecular compounds. Their approach aims at validating two models: the hierarchical biomolecular covalent bond model (HBCB) assumes that natural biomolecules adopt a hierarchical structure following the strength of their covalent bonds; the programmed self-decomposition model (PSD) assumes that the degradation of biopolymers (BPs) into biomonomers (BMs, that can be later reused) was selected by evolution as an efficient strategy for recycling biomolecular compounds. Their validation approach combines wet lab experiments with computer simulations. In the simulation part, different self-decomposition processes are expressed with an AC and tested in a virtual ecosystem. Their results show that indeed the virtual organisms able to break BP into reusable BMs have an evolutionary advantage over other organisms that degrade BPs into smaller units that are harder to recycle.

Formal Models of Biochemical Reaction Networks

Chemical organization theory (chapter 12) has been proposed in [168] as a method to analyze biological networks; indeed, important chemical reaction networks such as metabolic and gene regulatory networks can be decomposed into organizations that form a hierarchical structure representing the feasible states and steady states of the system.

Various biological processes have been modeled with the help of MGS [320] (introduced in chapter 9). The ability of MGS to model spatially distributed biochemical reactions is illustrated in [321] through several examples: a simple model of growth expressed as a single rewriting rule; a more complex model for the growth and formation of filaments in cyanobacterium *Anabaena catenula*; the action of restriction enzymes to split DNA strands at a point of cleavage upstream from a specified recognition sequence; a model of the cAMP cell signaling pathway.

P systems (discussed in chapter 9) have also been extensively used to model biology [112, 613, 718, 781, 828, 832]. Traditionally, rewriting rules operated on the objects of the P system in a nondeterministic and maximally parallel manner: at each time step, rules were chosen in a nondeterministic way, and applied until the system would become inert (no further rules could be applied to any region). However, such an algorithm would be inconsistent with biological systems, where chemical reactions happening at different speeds pace the rhythm of the cell operations. Recently [718, 781], P systems have been endowed with variants of stochastic simulation algorithms, enabling them to accurately model real biochemical reaction systems.

In [832] the ARMS system (a special kind of P system which also appeared in chapter 9) was used to model the signaling pathways of the p53 protein , a widely studied transcription factor which plays an important role in stress response to damage in mammalian cells. Mutations in the p53 gene are the cause of many cancers. The authors express the core of the p53 pathways in ARMS, and run the resulting dynamic system. The obtained simulation results show the formation of regulatory feedback loops that agree with the known biological literature.

Another formalism for modeling biochemical pathways with ACs is the chemical graph transformation approach by Valiente et al. [266, 267, 719]. In their approach, molecules are represented as graphs, and chemical reactions operate on these graphs such that the total number of bonds that are broken and created during a reaction is minimal. A set of biochemical reactions applying such minimal transformations is defined as an optimal artificial pathway, or optimal artificial chemistry. The algorithms for constructing such optimal ACs are introduced in [266], and in a subsequent paper [267] these algorithms are applied to validate large databases of metabolic reactions. Using their method, they succeeded to uncover several inconsistencies in a large database, such as unbalanced reactions, and proposed the corresponding fixes also using the same method.

18.3.3 Algorithms for Simulating Large-Scale Reaction Networks

The simulation of large-scale biochemical reaction networks is computationally costly. Algorithms for making such simulations more efficient are subject of current research. One of the problems is the large amount of possible reactions, due to the different conformations of enzymes and other biological molecules, and their different possibilities and speeds of binding to their substrates, and the way multiple proteins may form large complexes, each different complex with special properties and participating in a given set of reactions given by their shape that determines the activity.

As we have seen in chapter 4, Gillespie's SSA is one of the most widely used algorithms for simulating chemical reactions. Its computational cost per time step grows linearly with the number of possible reactions; hence, it is inefficient when this number is large. Alternatives such as the next reaction method (whose cost scales logarithmically on the number of reactions) or Tau-Leap (which is able to execute several reactions in a single, coarse-grained time step) are more efficient for large numbers of reactions, but also require complete knowledge of possible chemical reactions in order to choose which reaction or set of reactions to execute at each time step.

A general overview of methods to deal with very large reaction networks can be found in section 17.4 of [757]. A more specific review of various stochastic simulation methods aiming at modeling chemical reaction networks of real biological systems can be found in [870].

Further improvements targeting specifically large reaction networks have been devised. In [780] a constant-time algorithm is proposed, with a cost that is independent of the number of reactions. It is based on the idea of partitioning the set of chemical reactions into groups of reactions with similar propensities. The price to pay in order to obtain a constant-time cost is a finite probability of reactions being rejected, which would waste computational cycles. Similar to the other algorithms, this algorithm requires complete knowledge of possible reactions in order to assign them to groups.

Is it possible to have a stochastic reaction algorithm that takes into account the different kinetic rates of different reactions, without having to calculate the propensities of all reactions that might possibly occur? One such algorithm is a simple extension of the naive bimolecular collision algorithm described in chapter 2. In order to take into account different reaction speeds, we could calculate the probability of a collision to be effective as a function of the activation energy that needs to be overcome for the reaction to occur. This is known from the Arrhenius equation (eq. 2.32), and the simple algorithm would then become: Pick two random molecules for collision; if their kinetic energy is higher than the activation energy for the reaction, then perform the reaction (effective collision), otherwise return the molecules to the reaction unchanged (elastic collision). Such a simple algorithm is feasible if kinetic energies can be computed for each molecule, and activation energies for each reaction, given the colliding reactants. It is effective if the amount of elastic collisions is much smaller than the amount of effective collisions. This is the case when most molecules have a high probability of reacting with any other molecule. However, systems of this kind, with an almost fully meshed network configuration, would most probably be of little practical use, since the amount of cross-reactions would be overwhelming.

Given the combinatorially large amount of molecular compounds and their possible interactions, it would be more effective to specify reaction rules implicitly based on their shape and chemical properties, and compute their reactions on the fly from these general rules. Hence, instead of exhaustively enumerating all possible compounds and their reactions, one would start with a handful of simple compounds (such as the food set in autocatalytic sets of section 6.3.1), and calculate the rules as they are required by reactions, in order to determine the corresponding products. Such a model could support the emergence of an unlimited amount of new compounds, becoming an instance of a constructive dynamical system (chapter 15). In the bioinformatics literature, such an approach is referred to as *rule-based modeling* of biological systems. A review of rule-based modeling techniques with focus on their application to protein-protein interactions in cell signaling can be found in [388].

In [351] a spatial simulator of complex reacting biomolecules molecules based on rule-based modeling is proposed. The simulator, called SRSim, was initially used to model complex self-assemblies and transportation phenomena such as the movement of molecular walkers on top of microtubules. It was then applied to model more complex biological phenomena, such as the mitotic spindle checkpoint, a complex network of molecular regulatory interactions responsible for correct cell division [415, 861]. SRSim takes into account the detailed spatial conformation of each protein and the complex that it may form, and positions each molecule in space such that only neighboring molecules may react with each other.

18.4   Modeling Genetic Regulatory Networks

Some notions of biological genetic regulatory networks (GRNs) appeared in chapter 5 (section 5.3.3), including the well-known lac operon example that behaves strikingly like a Boolean logical function. Biological GRNs are highly nonlinear dynamical systems, and their behavior is still poorly understood in general. The complexity of GRNs calls for computational methods able to model specific aspects of these networks with a flexibility and level of detail that would be hard or expensive to achieve in wet experiments. At the same time, methods able to explore large and intricate GRNs are also needed, in order to increase the current understanding of their behavior. Finally, engineering GRNs with novel functions could enable new therapies and biological applications with programmed cell functions, as will be further discussed in chapter 19.

Two categories of GRN models can be distinguished: static and dynamic models. Static models aim at analyzing the topology of a GRN, which is represented as a graph with nodes representing genes and links between them representing regulatory interactions. Sometimes the intermediate mRNAs and the proteins involved in the GRN are explicitly represented, sometimes only genes used, with the influence of mRNAs and proteins lumped together into the interactions among genes. Dynamical models aim at investigating the dynamic properties of GRNs as the concentrations of different transcription factors (TFs) rise and fall with time. Both static and dynamic perspectives must be combined for a complete picture of a GRN.

The dynamics of GRNs can be captured at different levels of resolution, from fine to coarse grained. At a fine-grained level, molecular events such as the binding of a TF to a gene, its subsequent dissociation from the gene, and the production of mRNA and proteins, are explicitly represented as chemical reactions that occur with some probability. At a coarse-grained level, genes and their products often assume a binary representation: genes are either active or inactive, and the proteins they produce are either present or absent. To make the model even coarser, only genes may be represented, and their interactions via mRNAs and proteins are subsumed as direct interactions between genes.

Two straightforward ways to model the dynamics of GRNs have already been mentioned. One could formulate the chemical reactions involved, assign kinetic laws with their associated coefficients to them (for instance, using Hill kinetics for the protein-gene interactions, enzyme kinetics for enzymatic protein-protein interactions, and mass action for the rest); and then simulate the resulting system either with deterministic ODEs or with stochastic algorithms such as Gillespie's SSA (chapter 4). The latter captures the noisy and stochastic features of GRNs, and is therefore appropriate to a microscopic simulation where the behavior of each cell can be simulated individually. However, the computational expense is high and biologists still have difficulties to provide the kinetic parameters for the various reactions from experiments. ODE models capture the average behavior of large numbers of identical cells and are currently widely used in biology, since average kinetic parameters over large populations of cells are much easier to measure in the laboratory than parameters for individual cells [126].

A simplified way of modeling GRNs comes with binary models. Binary models originate from the observation that some GRNs behave as Boolean functions, such as the lac operon. The function is active when one substance is present and the other is absent, which can be easily expressed in Boolean logic. One possible binary representation of a GRN is to consider genes as taking ON-OFF (binary) values, whereas the concentration of proteins may assume continuous values: when a gene is on, the protein that is produced by it increases in concentration; when the gene is off, the protein decreases in concentration, because it is not produced any more and suffers slow degradation. For further simplification, proteins may also take binary values, or may be left out altogether. Usually the larger the GRN being modeled, the more simplified its model must be in order to cope not only with computational constraints but also with limitations in the ability to analyze large amounts of data.

We now focus on the use of ACs for GRN modeling. We start with two binary models proposed by Stuart Kauffman, which are often encountered in the AC literature: The NK model, and random Boolean networks (RBNs). We shall then move to artificial regulatory networks (ARNs) which are ACs inspired by GRNs. ARNs have been applied to biology and computer science. Some of the biological applications of ARNs will be discussed in this section, whereas applications in computing are covered in chapter 11.

18.4.1   Kauffman's NK Model of Random Epistatic Interactions

The NK Model [447] aims to model epistatic interactions among groups of genes, that is among genes that influence the expression of other genes. In the NK model, N is the number of genes in the system, and K is the number of epistatic interactions, that is, the number of other genes that influence the expression of any given gene. In graph terms, the GRN is represented as a directed graph with N nodes, each node with K incoming links ($0 \leq K < N$). Each gene can take one of a discrete set of values, typically a binary value.

The NK model is mostly used to create rugged fitness landscapes in evolutionary processes. In that context, each gene is assigned a fitness value that depends on its own state (binary value) and on the state of K other genes that connect to it. Since each gene takes a binary value, and interacts with a fixed number K of other genes, there are 2^K possible combinations of the states of the genes that have an epistatic influence upon any given gene. Together with a gene's own state, this makes 2^{K+1} possible state combinations. Imagine that the GRN is so complex that the effects of each of these possible combinations of states are unknown. Kauffman then proposed to assign a random fitness value (in the form of a real number within the interval 0 and 1) to each of these possible 2^{K+1} combinations. This results in a matrix $M(i, j)$ of epistatic interactions, where $0 \leq i < N$ is the gene number, and $0 \leq j \leq 2^{K+1}$ is the list of possible epistatic state combinations.[1] Given matrix M, the GRN graph of epistatic interactions, and a genome of N genes, a straightforward procedure to calculate the fitness w_i of a single gene i in this model is: first, obtain the states of the K genes that interact with i, together with gene i's own state (g_i), then concatenate these states into a binary string: the corresponding integer number is the position j in matrix $M(i, j)$; a simple lookup in the matrix $M(i, j)$ then determines the fitness of gene i under state j.

The fitness of the whole genome is simply the average of the fitness values of its genes:

$$W = \frac{1}{N} \sum_{i=0}^{N-1} w_i \qquad (18.22)$$

Figure 18.4.1 shows two examples of fitness computation using epistatic matrices for $N = 3$. The first example is the trivial case of $K = 0$ (no epistatic interactions). The fitness of each gene in this case is simply determined by its value: $w_i = M(i, g_i)$. In the second example ($K = 1$) we denote by $j = g_k g_i$ the binary number indicating the state of gene i and of gene k with which it

---

[1]It goes without saying that storing the matrix M explicitly is extremely inefficient in terms of memory usage, even for modest values of K. Therefore alternative algorithms to compute the fitness of genomes in the NK model have been proposed [401].

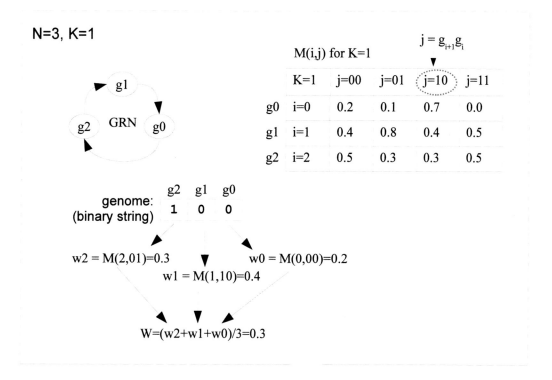

Figure 18.9 Kauffman's NK model: example of fitness computation for $N = 3$ and varying K. Top: $K = 0$ (no epistatic interactions). Bottom: $K = 1$ (gene g_k, for $k = i+1$ has an impact on gene g_i, in a circular fashion.

interacts. In this case, the graph of interactions is very simple: gene $k = i + 1$ has an impact on gene i, in a circular fashion. The fitness of a gene i in this case then becomes $w_i = M(i, g_k g_i)$.

Kauffman [447] showed that by adjusting N and K, various types of fitness landscapes can be created. For $K = 0$, a unimodal landscape is obtained, with a single peak that can be easily attained via hill climbing. If we increase K, the landscape becomes more rugged, and single point mutations may lead to drastic changes in fitness values. At $K = N - 1$, the maximum value of K, the fitness landscape is random, and there is no correlation between the fitness of a given genotype and that of its neighbors.

The NK model is an example of a highly idealized static GRN model. Because the actual effect of epistatic interactions is not known in general, Kauffman simply assumed that they could be considered as random. Below we will see how it can be transformed into a dynamic model of a binary GRN that in some ways resembles cellular automata.

18.4.2   Random Boolean Networks (RBNs)

In 1969 Kauffman proposed a "formal genetic net" [445] as a binary model of a GRN that later became known as a random Boolean network (RBN) [243, 348, 447, 591]. An RBN is a graph with N nodes, each with K incoming links ($1 \le K \le N$) that can be seen as input values. Each node has a Boolean value, and is assigned one Boolean function at random out of the set of 2^{2^K} possible Boolean functions with K inputs and one output. Whereas the Boolean value may change in time, the Boolean function remains fixed once chosen.

An RBN can be seen as an extension of a cellular automaton (CA; see chapter 10) with a more general graph topology instead of a regular grid topology. As such, we can run the system in a way that is very similar to the procedure to run a CA: starting with an initial state where each node has an initial Boolean value (for instance, randomly assigned, or set to specific initial values), the next state of each node is calculated as the output of the Boolean function given the values of its input nodes. The computation then proceeds until some termination condition is met. As in a CA, the system will eventually repeat a configuration producing cycles, because the number of possible configurations is finite and state updates are deterministic. These cycles are the *attractors* of the RBN dynamics. The period of such cycles and the state transitions within cycles produce patterns that characterize the RBN, in a similar way as the patterns produced by CAs characterize the cellular automaton.

Using this model, Kauffman showed that as K is increased from 1 to N, RBNs move from simple to increasingly complex dynamics until they degenerate into a chaotic regime. RBNs with $K = 1$ produce no interesting behavior, those with very large K values behave erratically, whereas interesting behavior occurs for intermediate values of K, such as $K = 2$ or $K = 3$. A particularly interesting case is $K = 2$, where complex behavior emerges. Analogous to its CA counterpart [496], Kauffman suggested that these RBNs displayed dynamics at the "edge of chaos": they are not too rigid to settle into a "dead" fix point, and not too chaotic to look random, hence the right amount of complexity to look "alive."

In order to bring RBNs closer to biological reality, the constant parameter K can be replaced by various probability distributions with mean K. If K follows a power law distribution, GRNs with scale-free topology are formed [291]. These scale-free GRNs display more ordered dynamics and match biological data more closely than the original model with constant K.

Beyond biological GRNs, Kauffman's RBNs turned out to be applicable to a variety of other phenomena, including cell differentiation, neural networks, and complex systems in general. In section 18.5 we will look at how RBNs provided a pioneering explanation for the cell differentiation process.

Recently, RBNs have also been applied to artificial chemistries: The RBN World [264] is a subsymbolic artificial chemistry in which atoms are RBNs extended with bonding sites (called bRBNs, or bonding RBNs), and molecules are networks of bRBNs interconnected via their bonds. The chemical reactions in the RBN World include the formation of bonds between two RBNs, and the cleavage of existing bonds. The aim of the RBN World is to construct ACs able to exhibit rich and complex dynamics, with the emergence of higher-level structure able to transition to higher orders of organization. For this purposes, one of the features of this chemistry is the ability to compose RBNs into hierarchical structures. So far the authors have proposed a set of criteria and metrics in order to identify suitable parameter ranges in RBN chemistries [263], which could help detecting the right chemistries able to produce autocatalytic sets, hypercycles and heteropolymers, and to promote the self-organization of these elements into more complex structures resembling life.

18.4.3 Artificial Regulatory Networks (ARNs)

In order to confer more reality to GRN models, finer-grained models are needed, beyond Boolean models. Artificial regulatory networks (ARNs) [67, 68] model GRNs as networks of interacting bitstrings, enabling not only GRN modeling per se but also modeling their evolution, in order to understand how natural GRNs achieve the intricate level of organization that they usually present.

ARNs have been introduced in chapter 11, in the context of biologically inspired network ACs. Recall that an ARN is composed of genes and proteins, both represented as binary strings. Genes are portions of a linear genome preceded by a promoter region and two regulatory sites, one for the binding of activator transcription factors, and the other for the binding of inhibitory factors. The degree of complementarity between the bit strings of the TF and its binding site determine the rate of expression of the corresponding gene.

Although originally designed to enrich artificial evolution algorithms in a computer science context, ARNs are able to reproduce specific aspects of natural GRNs and their evolution. As an example, heterochrony can emerge in the evolution of ARNs as shown in [68]. Heterochrony is a variation in the timing of onset or offset of certain genes, and is known to play a key role in the formation of particular structures during embryonic development. In an ARN heterochrony may arise as a result of a small change in the degree of matching between regulatory sites and proteins.

Perhaps the most striking resemblance between ARNs and natural GRNs is the emergence of scale-free and small-world GRN topologies through gene duplication and divergence processes in ARNs. Such topologies were initially reported in [67,68] and then more extensively elaborated in [478,479]. Furthermore, ARNs exhibit network motifs resembling those found in natural GRNs [73, 504], and the recurrent use of these motifs can also be explained by a gene duplication and divergence process [504]. Another example of proximity between ARNs and natural GRNs is the evolution of typical concentration dynamics found in natural GRNs, such as sigmoids, decaying exponentials, and oscillations [479, 480].

18.4.4   Open Issues and Perspectives

None of the current models available so far seem fully able to deal with the complex nature of real biological GRNs: most models assume well-stirred vessels, and drastic simplifications are adopted to make the models computationally tractable. In reality, however, the space within cells is very crowded, proteins tend to aggregate into large complexes of several proteins with specific shape, and DNA molecules may bend upon binding of some transcription factors. These circumstances can affect gene expression levels, sometimes by orders of magnitude [126]. New GRN models will have to take these more realistic considerations into account.

18.5   Cell Differentiation and Multicellularity

The process of cell differentiation in multicellular organisms is driven by changes in patterns of gene expression, that lead to different cell types. Such differentiation process has been briefly outlined in chapter 5. The modeling of cell differentiation is therefore closely related to GRN modeling. In the AC domain, research related to cell differentiation mainly focuses on dynamical system processes that can explain the emergence of specialized cell types, and on the origin of cell differentiation as an outcome of GRN evolution in multicellular organisms. A further research interest is the emergence of multicellularity out of communicating independent cells that initially adopt a symbiotic relationship, and then become interdependent until they end up forming a single organism.

Kauffman's seminal paper on RBNs [445] offered a pioneering explanation for the dynamical system mechanisms underlying cell differentiation in biology. Kauffman identified different types of cycles as attractors in the "edge of chaos" regime of RBNs, and conjectured that such attractors could be regarded as different cell types [445]. The cell differentiation process in multicellular organisms would be the result of a convergence to such an attractor in the dynamics of a GRN. The state of the art in biology at that time (1969) did not provide any means to verify his hypothesis.

Furusawa and Kaneko [301] have proposed a model of cell differentiation called *isologous diversification*, that can be considered as a spatial AC. According to their model, cells can differentiate due to instabilities in their internal chemical reaction network, which is a nonlinear dynamical system. Such instabilities are analogous to those responsible for the formation of Turing patterns in reaction-diffusion systems [868]. In [301], cells occupy slots in a two-dimensional grid, where they take up nutrients, grow, and divide. These cells communicate via the diffusion of chemicals to the extracellular medium. Nearby cells may also attach to each other, forming clusters of various shapes and sizes. Each cell contains a set of chemicals that form an autocatalytic network with feedback, such that it often leads to oscillations in internal chemical concentrations. These oscillations are attractors of the reaction network dynamics. The characteristics of the concentration curves of the attractor define a cell type, as illustrated in figure 18.10. When a cell divides, roughly half of the chemicals go to each daughter cell, leading to small differences in chemical composition between the two daughter cells. These differences tend to be amplified, potentially leading to cells that stabilize at different attractors, thus assuming different cell types.

Computer simulations of the multicellular model of Furusawa and Kaneko show that, starting from a single cell of type-0, a cluster of identical cells is initially formed. Once this cluster exceeds some threshold size, some cells in the interior of the cluster start to differentiate into type-1 and type-2 cells. The latter further differentiate into type-3 cells, and a ring pattern emerges (see figure 18.11). Under some conditions of cell adhesion, small clusters may detach from the

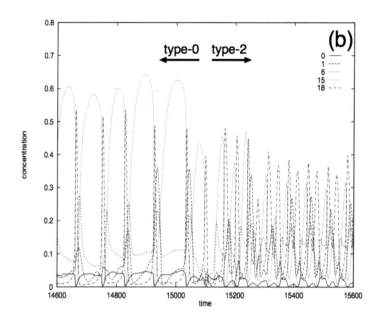

Figure 18.10 An example of cell differentiation in the model by Furusawa and Kaneko [301]. The concentration dynamics characterizes the cell type. Instabilities in such dynamics cause changes in cell type. In the above plot, a type-0 cell differentiates into a type-2 cell: its period of oscillation becomes shorter, and the amplitude of the oscillations decreases for some of its chemicals. Each curve corresponds to one out of 20 chemicals present in the cell. Reprinted from Furusawa & Kaneko, Artificial Life 4(1):79-93 [301], © 1998, with permission from MIT Press.

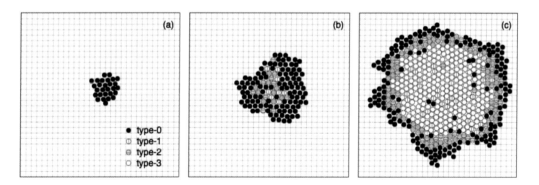

Figure 18.11 Emergence of multiple cell types in a cluster of cells held together by adhesion forces. Reprinted from Furusawa & Kaneko, Artificial Life 4(1):79-93 [301], © 1998, with permission from MIT Press.

multicellular cluster and move away to found new colonies elsewhere on the grid. The original cluster eventually reaches a state where it stops growing and "dies." A life cycle of replicating multicellular "organisms" then has emerged, without being programmed explicitly.

Further refinements of this model [302] show that a ring pattern is the most commonly observed, but the formation of other multicellular patterns such as stripes is also possible. Moreover, the differentiated cell types are robust to small perturbations caused by fluctuations in chemical concentrations. The overall multicellular pattern is also robust to the death or damage of some of its constituent cells.

In a subsequent paper [840], Takagi and Kaneko show that predefined compartments are not necessary for the emergence of cell differentiation; it can also happen in a reaction-diffusion system, which is an extension of the well-known Gray-Scott model that leads to self-replicating spot patterns [345,658]. In [840], compartments able to self-replicate are formed spontaneously, as the self-replicating spots in the original Gray-Scott model. In contrast to the latter however, where the spots are homogeneous, in [840] the internal chemical states may also vary from one spot to another: some may contain oscillatory dynamics, others may settle to a fixed point with constant concentrations. As in [301], each of these different characteristics of the concentration dynamics may be regarded as a specific cell type.

In a more recent survey [304], the biological plausibility of the isologous diversification model by Kaneko and colleagues is discussed. The authors point out that their hypothesis is consistent with recent experiments about inducing pluripotent stem cells in the laboratory. Their model is compared to simpler models based on multistable genetic switches, according to which different stable states could correspond to different cell types. Currently there is still no firm answer concerning the actual mechanism of cell differentiation at such fine-grain level of detail, and further research on stem cells is needed in order to close the debate on this topic.

18.6 Morphogenesis

Morphogenesis is a complex process resulting from the interplay between cell signaling and the corresponding changes in gene expression patterns, leading to a decision between alternative cell fates, such as cell growth, division, differentiation, or death. An introduction to morphogenesis was presented in chapter 5. In this section, we provide some examples of AC models used to study biological morphogenesis.

The father of the mathematical and computational modeling of morphogenesis is undoubtedly Alan Turing. In 1952, he proposed that the patterns observed on the skin of animals (such as zebra stripes and cow spots) could be the result of a reaction-diffusion process involving two interacting chemicals, that he called *morphogens* [868]. These patterns became later known as *Turing patterns*. The nonlinear nature of the reaction-diffusion equations, and the dynamical instabilities that drive pattern formation have attracted the attention of mathematicians, physicists, complex system scientists, and many other researchers over the world. Over the years, a huge body of work has been devoted to the further exploration of Turing patterns and more complex patterns derived from reaction-diffusion systems, including self-replicating spots [345,658], and several animal and plant structures [469,572,573], such as patterns on the surface of seashells, and leaf venation patterns. However, the impact of Turing's theories on real biology remains modest: although there is some evidence that the diffusion of chemical morphogens actually occur in some cases, reaction-diffusion alone is not sufficient to drive the whole morphogenetic process.

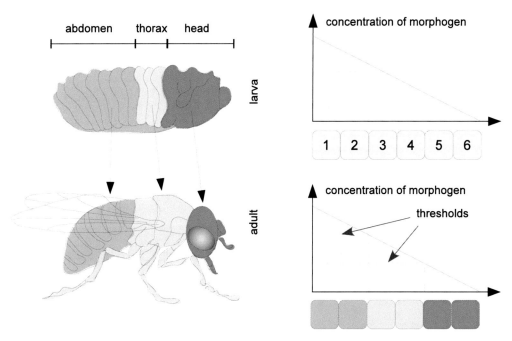

Figure 18.12 Body segmentation in the fruit fly (*Drosophila melanogaster*, left), and its abstract French flag model by Lewis Wolpert [927] (right): the concentration of a morphogen provides a coordinate system to indicate the position of the cells. Each cell is assumed to be able to read and interpret threshold concentrations of morphogen, and to respond to such signal by differentiating into the proper cell type. Left side adapted from [326, p. 268] © 2000 Sinauer Associates, Inc.; right side adapted from [930, p. 27] © 2011 Oxford University Press. Modified and reproduced with permission from the publishers.

By the end of the 1960s, Lewis Wolpert [927] proposed that *positional information* could be provided to the cells by morphogen concentration gradients, and that cells would be able to "read" such gradients in order to make decisions on their fate, typically causing a cell to differentiate into a certain cell type in response to a precise threshold concentration of a given morphogen.

During early embryonic development, the body of insects such as the Drosophila fruit fly is segmented into distinct parts that will later become the head, thorax and abdominal region. As an abstraction for body segmentation in morphogenesis, Wolpert proposed the *French flag model*, illustrated in fig. 18.12. In this model, the French flag represents the three body segments of an organism such as the fruit fly, and the color of each segment indicates the target type of its constituent cells. Starting with a single cell that has an internal (genetic) program, and one morphogen that forms a concentration gradient, the cell should divide and differentiate in order to develop the full organism featuring the three segments, by using only the gradient information and the internal program. For instance, concerning differentiation, a high concentration threshold of the morphogen could activate a gene that turns the cell blue, an intermediate threshold would result in a white cell, and a low concentration, below the other two thresholds, would result in a red cell.

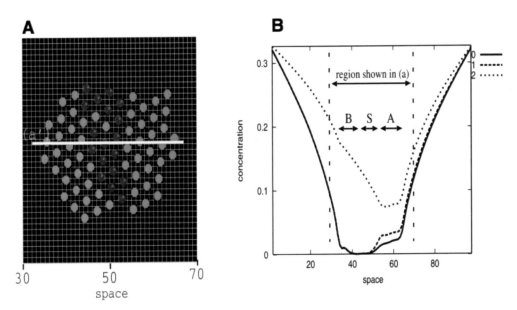

Figure 18.13 Emergence of chemical gradients under the isologous diversification model [303]. The example shows the formation of a stripe pattern (left), and the resulting gradients of nutrient molecules (right). Note that these gradients are not linear, in contrast to Wolpert's model (Fig. 18.12 (right)). Cell types: S = stem cell (red); B = type B cell (blue); A = type A cell (green). Reproduced with permission from The International Journal of Developmental Biology (Int. J. Dev. Biol.) (2006) 50:223-232 [303].

The original French flag model was one-dimensional. Its straightforward extension to two dimensions is still widely used as a benchmark to test developmental techniques [172,366,466,585, 586]. Further extensions including flags from other countries [265,715,928] and arbitrary two- or three-dimensional shapes [34] have also been used in order to capture more complex developmental patterns. Wolpert himself published several updates of his work as more information became available on the biological mechanisms behind morphogenesis [928,929]. Although positional information must indeed be available to cells in some form, it is generally agreed that other processes beyond passive diffusion of molecules contribute to provide positional information to cells, such as active cell-to-cell signal relaying mechanisms, and vesicular transport of signaling particles [456].

Wolpert's model and its variants do not say anything about how concentration gradients are actually constructed and maintained. Some users of these models assume that the gradients are available from the extracellular matrix as a maternal effect. For biological realism however, ultimately the production of morphogens and the maintenance of positional information must also be encoded in the genome, which poses extra challenges for discovering the right genetic program.

A possible explanation for the origin of positional information is offered by [303], in the context of the isologous diversification theory, discussed in section 18.5 in the context of the emergence of multicellularity. In this model, differences in the rates of nutrient uptake by different cell types lead to concentration gradients of nutrient molecules in the space surrounding the

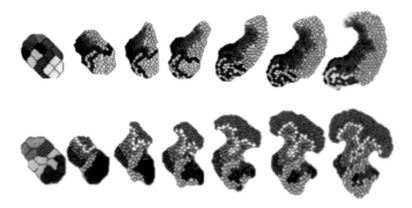

Figure 18.14 Developmental stages of two evolved morphogenetic patterns from [392] (top and bottom). Time flows from left to right. Although the fitness criterion used to evaluate the patterns was simply the number of distinct cell types, relatively complex patterns emerge. Reprinted from P. Hogeweg, Artificial Life 6(1):85-101 [392], © 2000, with permission from MIT Press.

cells. Figure 18.13 shows an example: as the stripe pattern on the left is formed, chemical gradients emerge (right). Cells can use these gradients as positional information cues. Further experiments reported in [303] show that such emergent gradient fields also help stabilizing the morphogenetic patterns, guide pattern formation, and help regenerating the pattern after cell damage. On the other hand, it is also shown that positional information alone is not sufficient for pattern regeneration: the internal state of the remaining cells must also be preserved, such that they can still react to the gradients in a consistent way in order to regenerate the pattern.

18.6.1   Evolution of Morphogenesis

Understanding the mechanisms of pattern formation also includes understanding how these mechanisms evolve in the first place. The combination of evolution and development (evo-devo) has been subject to active research recently [366, 392, 466, 586]. In this section we focus on some research work situated at the intersection between ACs and the evolution of biological morphogenesis.

An early evo-devo model was presented in [464]. It used a genetic algorithm to evolve the genetic rules determining the metabolism of cells. The model also captures intercellular communication, diffusion, and the active transport of chemicals. As a result, morphogenesis arises as an emergent phenomenon.

The evolution of morphogenesis is further investigated in [392] using a two-dimensional model of development controlled by a GRN. Cell adhesion properties are simulated with the help of the cellular Potts model (CPM [335]), in which each cell occupies a number of slots in a cellular automaton (CA). A state change in the CA may represent the movement of a cell or its change in shape. The CA update rules follow a surface energy minimization process that captures differences in cell adhesion that may lead to cell movement, cell growth and division, or cell death. Different GRN expression profiles lead to different cell types. The GRN is modeled with a 24-node Boolean network, of which two nodes are dedicated to intercellular signaling,

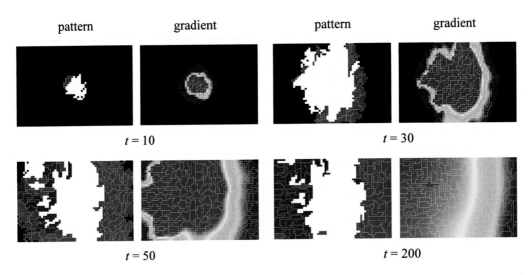

Figure 18.15 Formation of a French flag pattern under the evo-devo model by [466]. Each pair of snapshots shows the cell pattern on the left side and the morphogen gradient on the right side (where the colors indicate the concentration of morphogen, increasing from red to blue). Starting with a single zygote cell at the center of the grid (not shown), a blob containing of a few undifferentiated cells forms ($t = 10$). The magenta region observed in this blob and subsequent images tracks the fate of the original zygote. The cell colony grows and then generates a few red cells ($t = 30$). The blue cells emerge later ($t = 50$), and the pattern becomes progressively more clear ($t = 200$). Screenshots from the interactive Java applet by J. F. Knabe, available at http://panmental.de/ALifeXIflag/ as part of the support and supplementary information to [466].

and 10 other nodes determine cell adhesion properties. The genome that unfolds into the GRN may suffer point mutations. The fitness criterion is the number of different cell types expressed after a fixed number of time steps. Figure 18.14 shows some examples of patterns evolved under this model.

In [466], the evolution of GRNs for morphogenesis is illustrated on the French flag pattern. In this model, a GRN is made of genes that can be on or off, and proteins with varying concentration levels. The amount of proteins present inside a cell control its growth or shrinkage, which could end up with cell division or apoptosis. Among the proteins expressed, some determine the cell's color, others influence its adhesion properties. The genome is represented as a string of base-4 digits, analogous to the four DNA bases. Cells are placed on a CPM grid as in [392], and may secrete morphogens that diffuse in the surroundings.

In [172] the ARN model discussed in section 18.4.3 was extended for pattern formation, by endowing some of the genes in the ARN with a structural function that encodes for a given cell type. At any given point in time, at most one structural gene in each cell is active, and this gene is activated when the concentration of its regulating TF exceeds a threshold. Once differentiated, the cell type remains locked. However the locking mechanism is predefined in the system, not a consequence of the ARN circuit itself. Morphogenetic patterns form in this system due to a sequential activation of the genes corresponding to each target cell type. In this way, early cells lock into the first type whose protein becomes active, later cells lock into the second type, and so forth, until a pattern is formed. Using this system, the authors were able to evolve ARNs to produce concentric squares and French flags. The extension of this system to form arbitrary patterns could be hindered by the requirement of synchronized cellular operation, by the number of proteins required to trigger the differentiation process, and by the inability of cells to further differentiate until they reach their final type.

The evolution of the ability to regenerate tissues after damage has been studied in [34], using a three-dimensional model of artificial organisms whose GRNs closely resemble their biological counterparts. Another interesting piece of work in this area is the evo-devo model in [41], which combines ACs, GRNs, and genetic algorithms to look at how artificial neurons can grow and develop into something that could look like a nervous system.

The common point among all these approaches is the use of ACs combined with spatial considerations and physical constraints necessary for pattern and shape formation. Much research remains to be done. For instance, the fitness function typically used is the distance to a target pattern. Such a fitness function is not biologically plausible, since in nature a shape is selected for when it confers the organism with a survival advantage.

18.7 Summary

The area of modeling of biological systems is vast, and the methods used are very diverse. The niche for artificial chemistries is this context is still not entirely well defined. Many researchers working on models of biology are unaware of the possibilities offered by ACs. This chapter showed a few examples where ACs have contributed to the construction of models of real biological systems, ranging from models of simple biomolecular interactions to more complex models of multicellularity and morphogenesis. A distinctive hallmark of ACs can be perceived in many of these examples: the fact of not being content with merely understanding biology. The mindset of Artificial Life researchers is deeply imprinted to AC; hence, the inherent desire to build and create something novel, ranging from new yet to be proven hypotheses for the origin of certain biological processes to the synthesis of new biological or biochemical components. Reflecting this mindset, chapter 19 will discuss the synthesis of new forms of life, as well as the programming of wet ACs and living organisms to perform desired tasks.

What I cannot build, I cannot understand.

RICHARD FEYNMAN

WET ARTIFICIAL CHEMISTRIES

In this chapter we will attempt an excursion into "wetland:" the world of synthetic biology and wet computers *in vitro* and *in vivo*. The nascent discipline of synthetic biology explores the idea that cells and even entire organisms can be reengineered and reprogrammed to perform specific functions. They might be even created from scratch out of laboratory chemicals. Natural cells and their components are also candidates for computing devices. In his book *Wetware: A Computer in Every Living Cell*, Bray defends the thesis that "living cells perform computations" and shows several examples, from proteins that behave like switches via genetic circuits that implement oscillators to cells that differentiate to form organisms [132]. Not all wet ACs will be alive, though: some will simply have some lifelike characteristics or build upon self-organizing chemicals in a test tube. Yet, they represent novel ways to look at life and its building blocks and to harness their power to create useful applications.

We will first examine some possible ways to build wet artificial cells and their components. We start by looking at alternative nucleic acids and amino acids (section 19.1). From a chemical point of view, these molecules are perfectly plausible, but for some reason they do not occur in biological organisms. There are also alternatives to the existing natural genetic code. Synthetic life forms or living technological artifacts might soon be constructed using these synthetic building blocks. Next we will look at how artificial cells may be concocted in the laboratory, out of naturally occurring compounds, synthetic compounds, or a mix of both (section 19.2).

In the second part of the chapter we shall discuss wet chemical computing, including computation with DNA and other biomolecules, reaction-diffusion computers that operate on Petri dishes (section 19.3), and computation with bacteria and other living beings (section 19.4). The chapter concludes with a discussion of some essential ethical considerations that inevitably come forth when attempting to manipulate or even recreate life in the laboratory (section 19.5).

19.1 Artificial Building Blocks of Life

Our "excursion into wetland" begins with an overview of alternative building blocks for artificial cells, focusing on some possibilities of synthetic nucleic acids and amino acids, and alternatives

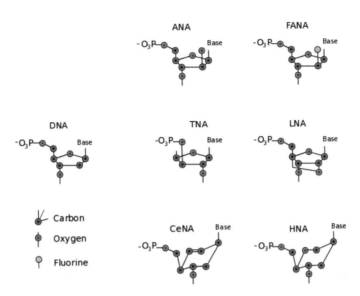

Figure 19.1 Different xeno nucleotides. Sugar molecule is replaced by arabinose in ANA, 2'-fluoroarabinose in FANA, threose in TNA, a locked ribose in LNA, cyclohexene in CeNA and anhydohexitol in HNA. Adapted from [435].

to the "standard" genetic code. It is exciting (and perhaps also a bit scary) to imagine the endless possibilities of life forms that could have been based on these compounds, or that could be synthesized in the future. For instance, the Los Alamos Bug that will be discussed in section 19.2.2 is a model of a cell "inside out," in which PNA, a form of Xeno nucleic acid, is envisaged as a possible genetic material, to be placed not inside the cell, but on its surface.

19.1.1 Xeno Nucleic Acids

DNA and RNA are the basis of life's information processing capabilities because they store and copy the information essential to assemble and construct living organisms. Researchers have examined in recent years which of the properties of DNA and RNA are essential to conveying the functions of storage and inheritance to these macromolecules. As we mentioned in chapter 5, nucleic acids are built from nucleotides composed of a nucleobase, a backbone of a sugar connected through phosphates.

Synthetic chemists have started to modify these elements systematically, to possibly come up with other systems capable of information storage and inheritance. One modification examined is the variation of the sugar that carries the nucleobases. So instead of the deoxyribose of DNA or the ribose of RNA, other sugar molecules are used. Figure 19.1 shows a few of the variants examined to date: ANA, FANA, LNA, CeNA, HNA, TNA, collectively called XNA (for xenonucleic acids).

Some of these have been successfully copied through a process of reverse transcription into DNA, then DNA copy via a laboratory procedure called PCR amplification, and finally reverse transcription into XNA. This requires the development of synthetic kinds of polymerase enzymes able to read off from X/DNA and transcribe into X/DNA and backward [666].

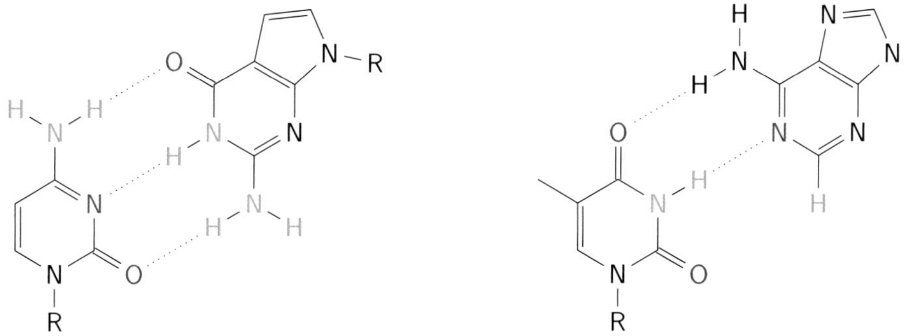

Figure 19.2 The interaction of standard nucleobases via hydrogen bonds. A pair is formed by a smaller (py for pyrimidines C & T) and a larger (pu for purines G & A) partner through hydrogen bonding. Hydrogens are either donated (green, D) or accepted (red, A). Patterns are formed as pyDAA (C) — puADD (G) and pyADA (T) — puDAD (A). Adapted from [106].

Another direction of research takes aim at the nucleobases AGCT/U of DNA/RNA nucleic acids and seeks to replace or augment them with "unnatural" nucleobases. In fact, there are even two methods to approach this goal. The more conservative is to stay close to the natural nucleobases and just slightly modify the principle by which molecular base-pairing occurs in nature [106]. Figure 19.2 shows the interactions among the standard nucleobases. Patterns are formed by pyDAA — puADD and pyADA — puDAD combinations, which provide stability. However, some other combinations of these factors could also be used to stabilize hydrogen bonds, for instance, pyDAD — puADA, or pyADD — puDAA. In total, there are four other combinations, as depicted in figure 19.3, that should be able to perform the same base pairing function [316].

A more radical approach will scan possible artificial nucleobases. Interestingly, there exist more than 60 possible nucleobases to choose from, so criteria for their selection needed to be applied. In [501] the main criterion used was the ability to form base-pairs, which are the fundamental principle by which information held in DNA/RNA can be copied and multiplied. A screen of all possible base pairs from these 3,600 combinations yielded a new base pair candidate, based on the MMO2 and 5SICS molecules (see figure 19.4 for a few variations). When subjected to a second test, namely the efficiency of extension of a strand using such nucleobases, this pair of bases performed very well, confirming the earlier result.

19.1.2 Extension of Amino Acid Alphabet

As we saw in chapter 5, amino acids are the components from which proteins are made. Just to quickly recall the features of these chemical structures: Amino acids have a common structure consisting of an amino group (NH_2) and a carboxyl group ($COOH$), but differ in their side chains. Side chains can be very simple, for instance just a hydrogen atom, or as complex as two carbon rings. In natural proteins, 20 amino acids (for simplicity abbreviated by capital letters, see table 5.1) are most common, which differ in their chemical features. Side chains influence, for example, acidity, basicity, charge, hydrophobicity, and size of the molecule.

Since proteins are the machinery of life, they require a high sophistication, which can be seen when they are produced in biosynthesis: Often, there are additional posttranslational modifica-

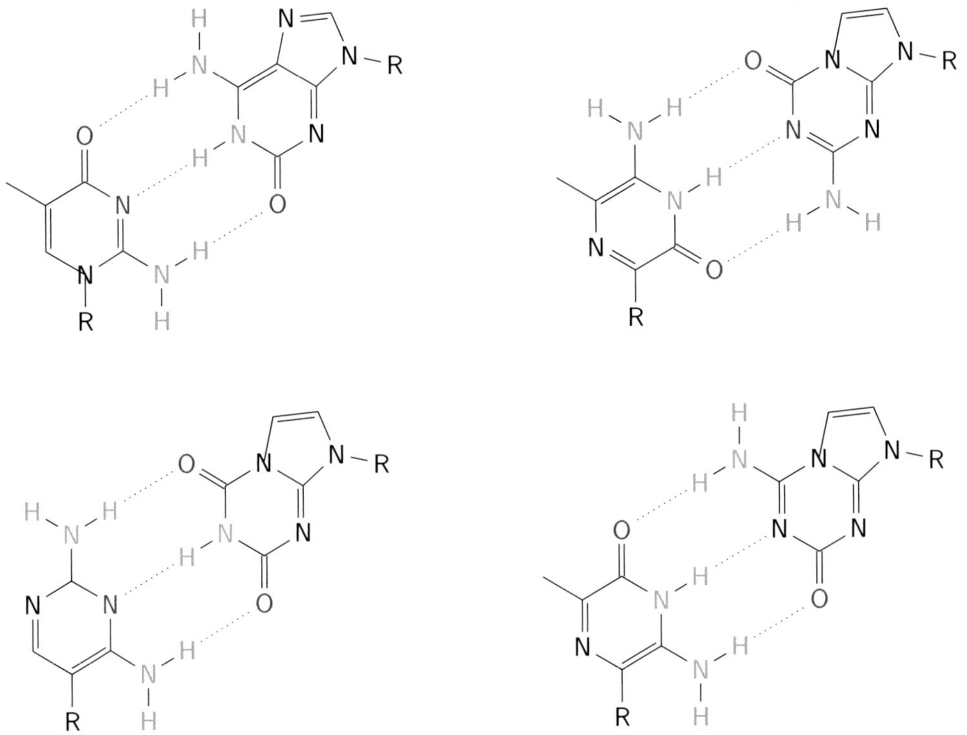

Figure 19.3 The interaction of synthetic nucleobases. New pairing patterns can be formed by hydrogen bonds between pyAAD — puDAA, pyDDA — puAAD, pyDAD — puADA, pyADD — puDAA, provided the geometry of synthetic nucleobases fits. Adapted from [106].

tions necessary, or combinations of proteins with other molecules (cofactors) are required for their proper function. Chemists have asked themselves for quite some time whether the addition of new, not naturally occurring amino acids would not help to expand the range of functions proteins could perform. According to a recent review [909], more than 30 not naturally occurring amino acids have been incorporated into proteins of *E. coli*, yeast or mammalian cells. Figure 19.5 shows examples from the literature.

Clearly, all these and other molecules can be produced via synthetic chemistry, but there are two important questions to answer for progress toward incorporating unnatural amino acids into living cells:

1. How can unnatural amino acids be synthesized with the mechanisms available to a living cell?

2. How can the genetic code be expanded to instruct the incorporation of unnatural amino acids into living cells?

The first question is mostly answered through modification of existing biosynthetic pathways. The second question is the topic of the next subsection.

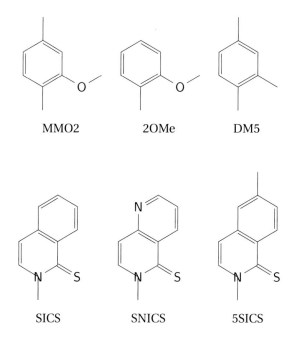

Figure 19.4 A selection of unnatural nucleobases used in [501]. Upper row: Pyrimidine bases; Lower row: Purine bases.

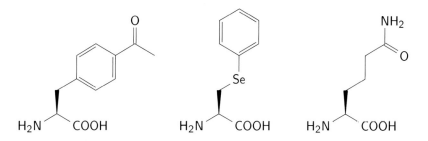

Figure 19.5 A selection of 3 out of 30 unnatural amino acids incorporated into cells of *E. coli*, yeast or mammals, [909].

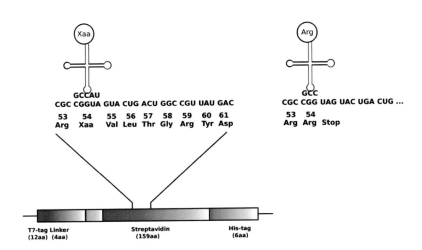

Figure 19.6 Replacement of three letter codon by five letter codon in region 53 — 61 of a streptavidin gene. A: A tRNA$_U$ACCG is aminoacylated with a nonnatural amino acid (Xaa) and correctly decodes the sequence CGGUA. B: In the absence of the five-letter codon, reading proceeds by decoding CGG in position 54 as the natural amino acid arginine, followed by the UAG stop codon. Adapted from [393].

19.1.3   New Genetic Code

The problem of modifying the genetic code is that one must be very careful not to disrupt its regular function. Since the code is redundant, with 61 codons available for the common 20 amino acids, there is room however, to adjust it slightly. For example, the natural genetic code has three extra codons (TAG, TAA, TGA) which code for a "stop reading" signal, generally called *nonsense codons*. These nonsense codons have been used to code new amino acids *in vitro* and to explore the chemical properties that result from the modified proteins so constructed [200]. Chemically, the respective codon tRNA needs to be loaded with the corresponding amino acid, a process called *aminoacylation*.

Since the three stop codons are under heavy use in protein translation as they prohibit the elongation of a peptide, modifications of assigned genetic codes (other than the nonsense codes) has also been performed.

Another method to have control over the specific incorporation of nonnatural amino acids is the expansion of the code to provide four-letter or even five-letter codons [393, 394]. In one particular example, a five-letter codon GCCAU was used to code for a nonnatural amino acid, Xaa (see figure 19.6). The purpose of this operation was to allow a modified version of a protein called streptavidin to be produced by the *E. coli* bacterium in whose genome the code was inserted. Experiments confirmed that the protein was produced correctly.

This is but one example of the power of code modification to produce modified versions of proteins by incorporating variations into the genetic code. While earlier work was depending on *in vitro* methods more recent work has modified the ribosome reading machinery of cells, to allow the natural reading of four-letter sequences and their decoding [610]. So far, no full four-letter or five-letter alphabet has been developed. Molecular scientists are still optimizing single codon readout, increasing efficiency and specificity of the participating reactions [617].

19.2 Synthetic Life and Protocells

In chapter 6 we discussed abstract models of minimal cells or protocells. Recall that these minimal cells were composed of three basic subsystems: a membrane, an information processing subsystem, and a metabolic subsystem. Abstract protocell models such as the chemoton and autopoietic protocells have inspired researchers to create minimal cells in the laboratory using natural or synthetic biomolecules.

Since Gánti's Chemoton [305, 306], the autopoietic units by Varela, Maturana, and Uribe [557, 881] and related early protocell models, many other possible protocell designs have been proposed [534, 691, 694, 754, 787].

Two approaches to build minimal wet artificial cells can be distinguished: top down and bottom up. In the top-down approach, researchers take the genome of a simple extant organism such as a bacterium (parasitic bacteria are typically chosen, due to their small genome), and strip out everything that is purported not to be essential for its survival [707]. This is usually done by gene knockout experiments: One removes a gene and if the cell can still survive and reproduce, one removes another gene, and so on, until the cell so manipulated can no longer survive. A minimal cell is then obtained by backtracking one step. This minimal cell is still able to take up nutrients, maintain itself and reproduce mostly like a normal cell. However it might not be able to survive outside a laboratory environment, which in some cases might also be desirable for safety reasons. Top-down synthetic biology makes extensive use of this approach, as exemplified by the recent achievements of Venter and his team [323], who in 2010 succeeded in obtaining the first artificial cell capable of surviving and reproducing after bootstrapping from a synthetic genome.

In spite of the indisputable merits of this achievement, some researchers wish for even more: they would like to see artificial cells forming entirely out of inanimate matter, as it once happened at the origin of life. By throwing a few biomolecular building blocks into water, and letting self-organization and self-assembly do the remaining work from the bottom up, wet artificial cells should emerge and eventually become capable of reproduction and evolution. This is the realm of synthetic protocell biology.

Note that artificially designed protocells will not necessarily be alive: some protocells might exhibit only a few lifelike properties, but still be useful for some practical purposes. For instance, for safety reasons some protocells might be designed not to reproduce, or to self-destruct after their task is completed.

Below we review some of the approaches to synthetic life and protocells. Our perspective is necessarily constrained from the AC viewpoint. The reader is invited to consult more specialized literature such as [121, 691, 787] for a broader view of the area.

19.2.1 Venter's Artificial Cell

In 2010, Venter and his team [323] successfully bootstrapped a living bacterial cell from a synthetic genome. In very simplified terms, Venter's artificial cell was produced as follows: First of all, the genomes of two laboratory strains of the bacterium *Mycoplasma mycoides* were retrieved from a genomic database and combined into a single donor genome. The donor genome was further engineered to include watermarks in noncoding regions, and other modifications such as gene deletions and mutations, which clearly distinguish the synthetic genome from its natural counterparts. The donor genome sequence was then synthesized, cloned inside yeast and transplanted into a recipient cell. The recipient cell was an "empty shell" of a similar species,

Mycoplasma capricolum, which had its own DNA disabled by a laboratory procedure. Following successful transplantation, the new hybrid cell started to grow and divide, exhibiting the phenotype of an ordinary *M. mycoides* cell, except for a marking blue dye, as shown in figure 19.7. The new synthetic cell was named *Mycoplasma laboratorium* [334, 707].

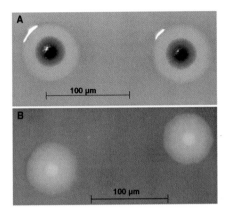

Figure 19.7 The synthetic bacterial cell obtained at the Craig Venter Institute [323]. Colonies of synthetic cells are stained blue (A) due to a gene that the converts a chemical in the medium to a blue compound, distinguishing them from the natural, wild type cells in (B), which contain no such gene and remain white. The phenotype of the synthetic cell behaves otherwise the same as the natural one and is able to replicate, in spite of the artificial genomic modifications that have been embedded in its design. From D. G. Gibson et al., Science 2010 [323]. Reprinted with permission from AAAS.

This new synthetic cell is the first organism ever to carry its creators' names, a few quotations, an e-mail address, and even a web site link in the form of watermarks embedded in its genome. The watermark encoding was engineered in a clever way to make sure that it will not be accidentally decoded into RNA or proteins during the lifecycle of the cell.

The creation of this synthetic cell attracted a lot of attention from the scientific community and from the press. It was often overstated as the first synthetic life form. Although the genome of *M. laboratorium* is indeed synthetic, the full organism is not entirely synthetic, since it used many essential components from the recipient cell. Moreover the synthetic genome is mostly a copy of the natural one. Overall there seems to be some consensus that although this achievement represents a major landmark in synthetic biology, and an important contribution to the advancement of biotechnology and its applications, it is only a starting point on the road towards synthesizing life. Some opinions on this matter have been expressed in [87].

19.2.2   The Los Alamos Bug

One of the most minimalistic examples in the bottom-up category is the Los Alamos Bug by Rasmussen and colleagues [270, 694, 695], shown in figure 19.8. It consists of a self-assembling lipid micelle that harbors metabolic elements in its interior and genetic material anchored to its surface. This "inside out" structure of the Los Alamos bug is its distinctive design hallmark. This design makes it possible to build extremely small protocells, orders of magnitude smaller than the smallest natural cells in existence [694].

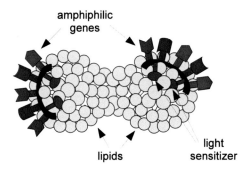

amphiphilic
genes

light
sensitizer

lipids

Figure 19.8 The Los Alamos Bug [270, 694, 695] is a protocell design based on a lipid micelle with anchored genes and a sensitizer chemical. The sensitizer captures energy from light in order to drive the chemical reactions needed to produce lipids for the micelle and to polymerize the genomic sequences. The figure depicts the "bug" in the middle of a cell division process. Adapted from [3].

An initially proposed candidate genetic material for the Los Alamos Bug was peptide nucleic acid (PNA), a type of xenonucleic acid that has a hydrophobic backbone and can be coupled with hydrophilic bases capable of hydrogen bonding, such that complementary double strands can be formed like in DNA or RNA. In contrast with DNA or RNA however (which have a hydrophilic backbone), the hydrophobic backbone of PNA makes it naturally inclined to stick to the lipid micelle: single-stranded PNA will "lie on its back" with the hydrophilic bases facing the water medium, ready to base-pair with floating monomers. The hybridized PNAs formed in this way will tend to sink into the micelle structure due to their hydrophobicity. The absence of water inside the micelle facilitates the ligation of the monomers, completing the formation of the double-stranded PNA structure [695]. Some of these double strands might end up dissociating, and the two single strands then migrate to the surface due to the exposed hydrophilic bases. Once on the surface they can again capture monomers, resuming the replication cycle.

Energy to drive the Los Alamos Bug is provided by light, captured by photosensitizer chemicals that are also hydrophobic and tend to sink into the micelle, where they help to build lipid amphiphiles for the micelle and PNA sequences for the genetic subsystem. This process can be regarded as a minimal metabolism for the protocell. As the metabolic reactions proceed, the protocell's surface-to-volume ratio changes in a way that it ends up dividing into two daughter cells. The viability of this model has been explored in the wet laboratory [694], mathematically [695], and in simulations [270]. Mathematical analysis and computer simulations reveal operational conditions under which the protocell is able to survive and evolve. Wet laboratory experiments have validated separate parts of the design. To date however, to the best of our knowledge, a full and continuous operation of the Los Alamos Bug has not yet been achieved.

19.2.3 Szostak's Heterotrophic Protocells

In contrast to the Los Alamos Bug which is truly unconventional and unlike anything that is alive today, Szostak and coworkers propose an artificial protocell model that can also serve as a plausible hypothesis for the origin of life. In their model [754, 839], depicted in figure 19.9, self-assembling fatty acid vesicles are permeable to the small monomers used to build larger polymers holding genetic information. The information polymers are typically RNA molecules or similar types, which are too large to leave the vesicle and therefore remain trapped inside.

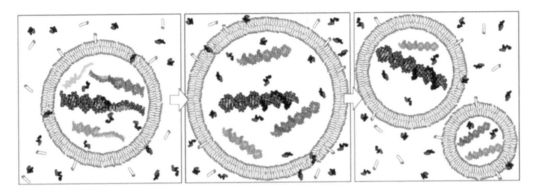

Figure 19.9 In the heterotrophic protocell model proposed by Szostak and his team [541,754,839], fatty acid vesicles enclose genetic material. These vesicles are permeable to small monomers that form the genomic polymers. The genetic material is either capable of nonenzymatic template replication [541] or could fold itself into a ribozyme able to copy the genome [839]. Vesicle membranes grow as new fatty acids are captured from the medium, until they eventually divide by physical forces. Reprinted by permission from Macmillan Publishers Ltd: *Nature*, vol. 454, pp. 122—125 [541], © 2008.

Fatty acids are the amphiphiles of choice for these protocell membranes, since vesicles made of fatty acids are much more dynamic and permeable than those made of phospholipids, facilitating their growth and the passage of small molecules in and out of the cell. Vesicles grow as new fatty acids are incorporated from the medium, and later divide by mechanical forces. The nucleic acid polymers inside these cells can reproduce via ribozyme-catalyzed replication or via nonenzymatic template-directed replication.

This model protocell is theoretically able to display some first signs of evolution: sequences that get copied faster (because they use more abundant monomers, or separate more easily, or display some primitive enzymatic activity) become more abundant, and from early random sequences one would move to sequences that carry more and more information, installing an evolutionary process that builds up organized complexity as a consequence.

The most minimalistic version of Szostak's protocell has no explicit metabolic subsystem; it relies on spontaneous, downhill reactions that need no extra input of energy. Vesicle growth is induced by the osmotic pressure exerted by RNA molecules inside fatty acid vesicles [173], hence vesicles that contain RNA grow at the expense of those that do not contain RNA. More recent variants include some metabolic components such as a ribozyme that synthesizes a small hydrophobic peptide capable of accelerating membrane growth [6]. As alternatives to RNA, various types of xeno nucleic acids such as TNA, PNA, HNA, ANA, and another form called 3'-NP-DNA are being investigated as possible genetic materials for these protocells [121]. Modified base pairs and nonstandard base-pairing have also been considered in order to improve copying fidelity. As for the compartment composition, multilamellar vesicles made of fatty acids have been shown to elongate into filamentous shapes that can grow and divide more reliably than spherical vesicles.

Although significant progress has been made, Szostak's protocells still face obstacles on the way toward their full realization [121]: On one hand, nonenzymatic template replication is still slow and inefficient due to strand separation difficulties, inactivation of monomers and other

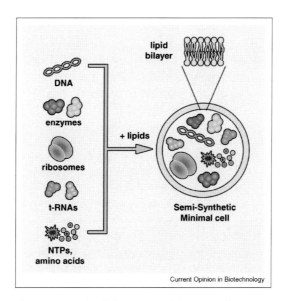

Figure 19.10 The semisynthetic minimal cell by Luisi and his team [176, 177, 535, 801]: essential components of today's cells, such as nucleic acids, enzymes, ribosomes, and amino acids are enclosed in a lipid bilayer membrane. Reprinted from *Current Opinion in Biotechnology*, vol. 20, C. Chiarabelli, P. Stano, and P. L. Luisi, Chemical approaches to synthetic biology, pages 492-497 [177], © 2009, with permission from Elsevier.

problems. On the other hand, ribozyme-catalyzed replication is also hard: the best ribozyme obtained by *in vitro* evolution so far is still too large and complex to be able to self-replicate; moreover it requires ions that end up destroying the ribozyme itself. Ions also cause fatty acid vesicles to precipitate, destroying the cellular compartments. Recently a promising solution for this problem of ion self-poisoning has been proposed [7]. Last but not least, the integrated and coordinated replication of whole protocells in a sustained manner, as well as their evolution, remains among the biggest challenges currently under investigation.

19.2.4 Semisynthetic Minimal Cells

A third stream of thought within the artificial cell research community advocates hybrid approaches that combine the best of both top-down and bottom-up worlds [416, 463, 535, 801]. These protocells are called semisynthetic minimal cells (SSMCs) [176, 177, 535, 801]: liposomes based on self-assembling lipid containers hold several components of existing cells, including DNA, RNA, enzymes, ribosomes, and various monomers, as depicted in figure 19.10. One of the major proponents of such hybrid approach is the school of Pier Luigi Luisi [534, 535, 801], who seeks to realize autopoiesis in the wet lab [531, 533, 536]. Another prominent research group in this area is led by Tetsuya Yomo [417, 463, 948].

SSMCs contain the minimum necessary elements needed for their survival, which is cast in terms of the maintenance of their autopoietic organization. In contrast with Szostak's protocells which are mainly based on fatty acids, SSMC containers are based on lipid bilayers composed of phospholipids, like in today's biological cells. Such vesicles have the advantage of being more stable and robust than those made of fatty acids. However their lack of permeability is one of their biggest drawbacks. The presence of transcription and translation machineries greatly facilitates the replication of genetic material and the production of cell constituents. However, it also makes these protocells much more difficult to maintain, since ultimately the full machinery must also be replicated, which is a big challenge, especially for the translation machinery, composed of ribosomes, tRNAs, and complex enzymes.

Several steps toward liposome-based protocells have been undertaken by Tetsuya Yomo and his team [416, 417, 463, 819, 948]. In the self-encoding replicase system presented in [463], liposomes enclose a template RNA sequence that encodes a replicase able to copy its own template RNA. A translation machinery is also included in the liposomes, in order to produce proteins from RNA. The whole liposome-encapsulated system contains 144 gene products, which is quite large when compared to minimalistic systems such as the Los Alamos Bug or Szostak's heterotrophic protocells, but still small compared to naturally occurring biological cells. The replicase is a catalytic subunit of the $Q\beta$ replicase protein, a phage enzyme that has been extensively used to catalyze RNA replication *in vitro*. The replicase has a high error rate, hence it introduces mutations that can create genetic diversity and potentially lead to evolution. However continuous replication turns out to be hampered by substrate depletion, unexpected molecular interactions, inactivation of the translation machinery, and the inherent competition between translation and replication [416]. In spite of these challenges, subsequent improvements of this system revealed interesting evolutionary dynamics, such as the emergence of small parasitic RNAs, and the evolution of competitive replicators able to stand up against these selfish parasites [61, 417]. Other recent studies include a deeper analysis of the RNA replication mechanism by $Q\beta$ replicase [877], and the use of liposomes as microreactors for protein synthesis and their evolution *in vitro* [615].

Due to the complexity involved in replicating all the constituents of SSMCs in an autonomous and continuous way, we can consider that SSMCs have not yet been fully demonstrated in the wet lab, although several aspects of their design have been achieved [802, 803]. Methods for controlled self-assembly, fusion and division of liposomes have been devised [848], as well as methods for encapsulating enzymes and the DNA transcription/translation machinery inside liposomes. DNA replication via PCR, transcription of DNA to RNA, and lipid synthesis were achieved by enzymes encapsulated inside liposomes. Proteins have also been synthesized inside liposomes, using a commercial kit containing tRNAs, ribosomes and other components necessary for the translation machinery to operate. However, a mechanism for the regulation of transcription is still missing, and the low permeability of the membrane is an obstacle to the input of energy molecules and building blocks to the cell. Without a regulation mechanism and without an efficient transport mechanism in and out of the cell, the SSMCs sooner or later end up exhausting their resources and are unable to maintain themselves. One solution to the transport problem is to express a protein in the liposome that creates a pore in the lipid membrane, allowing the passage of small molecules [619]. Another solution is to use smaller vesicles to carry biomolecules, and to deliver their payload to the SSMCs via vesicle fusion [163]. In summary, although not all the problems have been fully solved so far, steady progress is being made.

19.2.5   Droplet-Based Protocells

Another interesting approach to protocell design that has been receiving considerable attention lately is the use of droplets to encapsulate the protocell contents. Two types of droplets are typically considered: water-in-oil emulsion droplets, and oil-in-water droplets. Both make use of the spontaneous formation of droplets due to the natural repulsion between hydrophobic and hydrophilic particles when placing a drop of liquid into a solvent in which it is immiscible.

Hanczyc and colleagues have devised an oil-in-water droplet system that is capable of motility and sensor-motor feedback coupling [364, 365]. An oil phase containing a fatty acid precursor and a few other chemicals is placed in an alkaline aqueous environment containing surfactants (fatty acids). Upon contact with water, the precursor hydrolyzes into more surfactants. This reaction triggers a symmetry-breaking instability that causes a convection flow that propels the droplet through the water. The droplet moves directionally, expelling lipids behind it as it goes. Figure 19.11 shows a snapshot of a moving droplet. It is also able to follow a pH gradient, moving towards the highest pH, a behavior that can be regarded as a simple form of chemotaxis. The authors argue that such ability to "sense" an external gradient, and to "respond" to it by changing its internal flow state and direction of movement can be seen as an elementary form of cognition. The movement continues until the droplet is depleted of the necessary chemicals. Externally supplied "fuel" surfactants can keep the droplets moving for as long as desired [858].

Figure 19.11 The self-propelling oil droplet by Hanczyc and colleagues [364]. The droplet, shown in red, is moving in an aqueous solvent (blue), leaving a low pH trail behind (white). The diameter of the liquid recipient is 27 mm. Courtesy of M. Hanczyc.

Further achievements in this area include the analysis of self-propulsion mechanisms [54, 424, 556]; the demonstration of division and fusion of droplets [162, 227]; the study of interactions among droplets and their collective behavior [403]; the anchoring of single-stranded DNA molecules to the droplet surface in order to build droplet structures [357]; and perhaps most interestingly, the design of droplet applications such as solving mazes using chemotactic droplets [483, 885], or extracting rare-earth metal ions [53]. See [363] for a recent review.

Self-propelled droplets are protocells without an explicit information subsystem. Yet information is present in these systems in the form of sets of chemicals and their locations, and is "processed" by physicochemical processes causing motion. Following a gradient or solving a maze can be seen as computation tasks. Other different droplet systems have been explicitly designed as wet computers [15]. The potential of protocell-based computation remains largely unexplored. This brings us to the next section on computation with wet chemistries.

19.3 Chemical and Biochemical Computation

The success of modern electronic computers makes us almost forget that it is possible to compute with many substrates other than silicon-based hardware. When Alan Turing devised the computation model now known as the Turing machine (see chapter 10), the word "computer" did not refer to a machine but to a person assigned to the task of "computing" (calculating with paper and pencil) [199]. Turing then imaged a device able to automate the work of such human "computer", reading symbols from a tape, calculating with them, and writing the result back on the tape. The Turing machine is an abstract model of computation based on this operator-tape metaphor, and as such it is not tied to a specific physical realization. The first digital computers were built from vacuum tubes, roughly a decade after Turing's theoretical work [151].

Turing died in 1954, barely one year after Watson and Crick made their fundamental discovery of the DNA structure. But computation with DNA molecules was first achieved in the laboratory about 40 years later [24]. One of Turing's last scientific contributions was a mathematical model describing reaction-diffusion processes as potential explanation for morphogenesis, such as the formation of spot and stripe patterns on the skin of animals [868]. Unfortunately, he did not live long enough to connect his late work on morphogenesis to his earlier work on the Turing machine. Chemical computers based on these ideas had to wait until much later to be invented. Prototypes of reaction-diffusion computers [13] have only recently been built in the laboratory. Many attempts to build general-purpose chemical computers are in progress [59, 768, 955], and it is at least theoretically possible to build Turing universal chemical computers [386, 539].

As highlighted in chapter 17, chemical computing is inherently parallel, since many reactions may happen simultaneously. Models of chemical computation already exist since the 1970s [107, 192, 193, 306, 356, 486], but only recently has technology reached a sufficiently mature stage such that the first prototypes of molecular computers based on real chemistry become possible [10, 13, 99, 444, 767, 776]. Moreover, since chemistry is the basis of life, computation using chemistry can occur at various levels of organization of living organisms, from organic molecules to single-cell and multicellular organisms. Indeed, the information processing capabilities of living beings have been repeatedly highlighted in the literature [132, 249, 306, 776].

Future wet chemical computers are not intended to replace today's electronic technology but to complement it in domains where conventional approaches fall short of providing a satisfactory solution, such as nanoscale health monitors, customized drugs, or the operation of miniaturized robots in hazardous or otherwise inaccessible environments. One probable exception is worth mentioning however: Arguably one of the most promising bio-molecules for a technical application able to compete with classical electrical computers is bacteriorhodopsin. Bacteriorhodopsin is a molecule found in the membrane of *Halobacterium halobium*, and it is responsible for the conversion of solar light into chemical energy. In the mid-1980s it was found that it can be used also as a holographic recording medium [361]. Important properties of this molecule are that its conformational state can be changed by photons of specific wave lengths and that it is very robust and not destroyed when stimulated many times. In addition the current conformational state can be easily detected by using light because the adsorption spectra of bacteriorhodopsin depend on its conformational state [84]. Potential applications of bacteriorhodopsin are high-density memories [118, 543], hybrid electro-optical neural networks [367], and optical pattern recognition [362].

In this section we give an overview of some prominent approaches to wet chemical computing at the chemical and biochemical level. We starting with the popular DNA Computing, then discuss the synthesis of artificial Genetic Regulatory Networks for biological computation, and finally look at chemical computation in space using reaction-diffusion systems. More biological approaches to wet computation will be discussed in section 19.4.

19.3.1 DNA Computing

The DNA molecule has a number of properties that make it very interesting for computing purposes, including its information storage and transmission capacity, and its complementary base-pairing that facilitates error correction. In the field of DNA computing, the computational potential of DNA and related biomolecules is exploited to design new "wet computers," with various biomedical and engineering applications in mind. After earlier theoretical work [108], DNA computing was demonstrated for the first time in a wet laboratory by Adleman, who in his seminal work demonstrated a solution in principle to the Hamiltonian Path Problem using DNA strands [24].

To this day, DNA computing consists of two branches, theoretical and wet DNA computing. The theoretical branch investigates the possibilities of DNA with regard to well-known problems of computer science (e.g., universal computing [85, 129], cryptography [128], or acceleration of computation [515]). A DNA computer model can be regarded as a constructive, analogous artificial chemistry. It is analogous because it should model a real biochemical system accurately such that a molecule in simulation can be identified with a real molecule and a reaction going on in simulation should also have its counterpart in reality. One expects results predicting the computational power of DNA.

Wet DNA computing took off with Adleman's experiment, and is now a well-established branch of wetware. In this section we illustrate this wet approach through Adleman's experiment together with another very different approach: the DNA automaton by Shapiro et al. We will see that these two examples have remarkably many points in common with ACs: The DNA molecules can be regarded as strings of information that bind to each other and perform operations based on the contents of the matching strings, producing other strings in the process. In the interest of conciseness and focus, we leave out many other approaches to DNA computing. The interested reader can find more details in specialized books such as [30, 418, 682], and in recent reviews on the topic [31, 98, 99, 169, 175].

Basic Operations on DNA Molecules

How to compute with DNA? First of all, we have to look at the basic operations that can be done on a DNA molecule. These operations are usually performed by various biological enzymes, discussed in detail in [728]. The hybridization property of DNA can be exploited to obtain binding and recognition operations; restriction enzymes are able to cut DNA strands at specific positions, while other enzymes can glue them back together, or "walk" along the DNA sequence. Similar operations were also identified in [767] aiming at the construction of a molecular Turing machine. In summary, the basic operations are: recognition (binding by an enzyme, with potential conformational changes), ligation or condensation (join), cleavage (split), hybridization (form a double helix by binding to complementary strands or nucleotides), dehybridization (separate a double helix into single strands), and movement along the polymer (enzyme walks on DNA).

The duplication of DNA strands is another frequently used operation in DNA computing, and is achieved via dehybridization and hybridization with new nucleotides, usually with the help of polymerase chain reaction (PCR). PCR is a widely used procedure to replicate DNA in laboratory. It works by thermal cycling: When heated, the DNA double helix separates into two single strands. The mixture is then cooled down, and each single strand hybridizes with free nucleotides in the solution, forming new double strands. The formation of new strands relies on a special DNA polymerase that resists high temperatures. The cycle is then repeated by raising the temperature again, resulting in an exponential amplification in the number of DNA strands.

We now look at how these operations can be applied to perform computations with DNA.

Solving NP-Hard Problems Using DNA

As we mentioned, Adleman's experiment was the first successful wet-lab computation using DNA strands [24]. Adleman proposed a DNA-based algorithm for solving the directed Hamiltonian path problem (HPP), a simplified version of the popular traveling salesman problem (TSP). A short description of the TSP can be found in section 17.2. The HPP is a special case of the TSP where the distances between directly connected cities are all the same, thus reducing the problem to that of finding a path that visits each city only once (any feasible path is equally good). Although simpler in terms of parameters, it has been proven that the HPP is also NP-hard in terms of computational complexity.

More formally, in the directed HPP a graph $G = \{V, E\}$ represents the map, where each vertex $v_i \in V$ ($i = 1, ..., n$) represents a city and each edge $e_{ij} \in E$ represents a road from v_i to v_j. The goal is to find a path $P = e_{ij}, e_{jk}, ...$ such that each vertex from V is visited only once. The computation cost of finding a solution to a HPP instance quickly becomes prohibitive as the number of cities grow.

Adleman made use of the natural parallelism in chemistry to address the HPP problem in a brute-force, yet novel and elegant way. First of all, each city (vertex) is assigned a unique single-stranded random DNA sequence v_i to identify it. Each sequence v_i is divided into two halves: the first half ($v_{i,in}$) for the incoming links, and the second half ($v_{i,out}$) for the outgoing links. The roads (edges) are represented as the concatenation of the single-strand complements of the corresponding half of the identifiers of the cities that they connect: for example, a road from v_i to v_j is represented as $e_{ij} = \bar{v}_{i,in}\bar{v}_{j,out}$, where $\bar{v}_i$ notes the complementary strand. Remember that a DNA strand has an orientation (from the 5' to the 3' end), so directed edges can be represented as required.

It is easy to see that when throwing "cities" and "roads" in a well-mixed "DNA soup", each half-strand v_{ik} is likely to encounter its complement $\bar{v}_{ik}$ and hybridize with it, forming double-strand sequences containing chunks of a HPP solution, as shown in figure 19.12. In a "soup" containing a large enough number of vertices and edges, it is likely that some of the double strands formed will contain the correct solution corresponding to a complete tour. Hence, the algorithm "discovers a solution" by itself and in a parallel way, only due to the property of DNA complementary base-pair matching.

The difficulty of Adleman's approach was to find a way to "extract" the correct results from the "soup." This was accomplished in several steps: first of all, the paths that begin with initial city v_0 and end with final city v_f were amplified via PCR. From those, the paths containing exactly n vertices were extracted using an agarose gel (which causes DNA strands to be segregated by length). After further amplification and purification, the paths in which each vertex appears only once were extracted (using a labor-intensive procedure consisting of successive magnetic bead separation steps that must be repeated for every vertex). These remaining strands are the DNA-encoded solutions to the HPP problem, and were "delivered" via a final step of amplification and purification in gel. The whole wet-lab "computation" procedure took about a week. The details of the original procedure can be found in [24], and good tutorials to DNA computing (including Adleman's HPP algorithm) can be found in [30, 31].

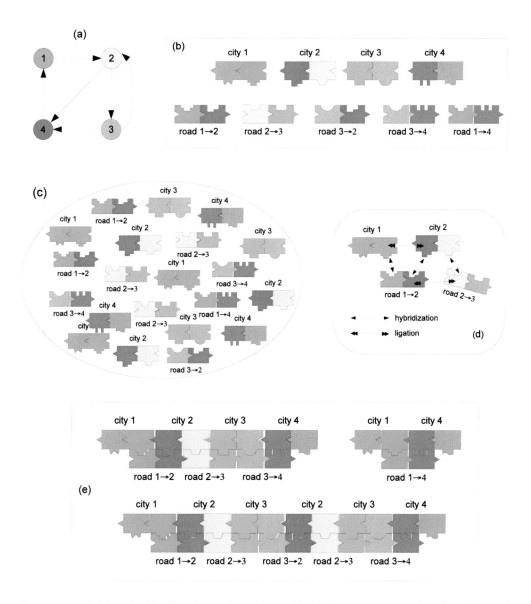

Figure 19.12 Solving the Hamiltonian path problem with DNA computing, as done by Adleman [24]. (a) A simple graph representing four cities with roads (arrows) between them. (b) Representation of cities and roads as single-stranded DNA sequences: the road sequences complement the sequences of the outgoing and incoming cities that they connect. (c) Initialization: Several road and city sequences are thrown in a "DNA soup." (d) Path construction process: roads match cities by complementary base-pairing; the sequences of roads and cities get ligated as hybridization proceeds. (e) Examples of candidate solutions: a correct solution (top left), and two incorrect ones.

Although a landmark in the field of DNA computing, the work of Adleman had serious limitations: the laboratory procedures involved were slow, labor-intensive and error-prone; moreover,

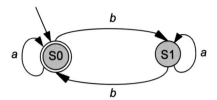

Figure 19.13 A two-state finite state machine with a simple task: to determine whether the input string contains an even number of b's.

the amount of DNA strands needed to obtain a solution increased exponentially with the size of the problem (in number of cities).

Following Adleman's work, many other DNA computing algorithms were proposed, mostly to solve other NP-hard problems. In 1995, Richard Lipton [514] extended Adleman's procedure to a more generic context, and showed how to solve the satisfiability problem (SAT) using DNA. Several other DNA-based algorithms were proposed around that time, tackling various NP-hard problems, such as graph coloring, maximal clique, or the Knight problem (a variety of SAT); see [30, 31] for an overview.

Eventually, however, the research community realized that the "killer application" for DNA computing did not lie in solving NP-hard problems: although a theoretical reduction in computation time could be achieved due to the massive parallelism of DNA computing, the (heavy) price to pay was a combinatorial explosion of DNA strands needed to search for solutions. In a famous article [370], Hartmanis showed that solving the HPP problem for 200 cities using Adleman's method would require more than the weight of the Earth in DNA strands.

Several promising applications for DNA computing exist however, such as self-assembly of nanostructures, disease diagnosis and drug administration. These applications rely on different approaches whereby DNA computation steps can proceed autonomously, in contrast with Adleman's elaborated procedure, which relied on human intervention at many steps. One example of such autonomous DNA computing is shown in the following.

DNA Automaton

In 1998 Shapiro [767] realized that the basic operations on DNA molecules, already present in any living cell, could be used to build a molecular version of the universal Turing machine. Such a hypothetical machine would work as follows: a molecule such as an enzyme would recognize or "read" a symbol from a polymer "tape" (such as a DNA strand), modify the information on the tape by cleaving the corresponding polymer region and ligating a new segment to replace the old one. The enzyme would then move on to the left or to the right, in order to process the next symbol. In practice however, building such a machine turned out to be too difficult due to the lack of sufficiently specialized enzymes capable of implementing highly specific recognition, cleavage and ligation operations in an accurate yet flexible way as required to build such a molecular Turing machine.

As a starting point, Shapiro and Benenson [768] then decided to build the simplest possible DNA machine based on these basic operations: a finite state automaton with two states. In contrast to Adleman's procedure, this DNA automaton would be able to run autonomously, without human intervention.

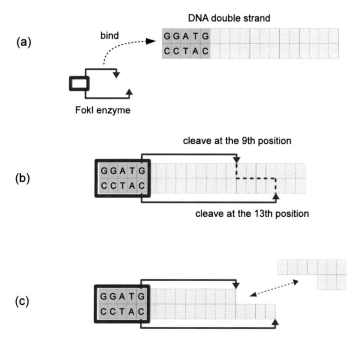

Figure 19.14 Schematic representation of the FokI enzyme operation [100, 768]: (a) recognizing and binding the DNA sequence 'GGATG'; (b) cleaving the DNA sequence at 9 and 13 nucleotides downstream from recognition site; (c) separation of the cleaved fragments.

Given an input string of symbols from the alphabet $\{a, b\}$ the automaton should accept the input if it contained an even number of b's. Figure 19.13 depicts the abstract finite state machine (FSM) that solves this simple problem: it consists of two states, S_0 and S_1, with state S_0 being both the initial and the accept state. Initially, there is an even number of b's (zero) so the automaton starts at S_0. When a b is read, it transitions to state S_1, representing an odd number of b's. The next b causes it to return to S_0, the even state. Any incoming a's are ignored, and the automaton remains in the same state. As a result, the automaton will remain in state S_0 (meaning that the input is accepted) as long as the input contains an even number of b's, as desired.

Although conceptually simple, the implementation of such an automaton in DNA turns out to be quite complicated. In this section we show a short summary of how it works. A detailed description of the DNA automaton can be found in [100], and an accessible introduction to the topic can be found in [768].

The DNA automaton relies on two enzymes to manipulate the DNA strands: FokI, an enzyme that recognizes a specific DNA segment (the recognition site) and cleaves the DNA a few nucleotides downstream from the recognition site; and ligase, an enzyme able to glue two DNA fragments together. The mechanism for input symbol recognition and the corresponding state transition relies on the special shape and action of FokI: the enzyme looks like a pair of asymmetric scissors, cutting one strand of the DNA duplex at 9 nucleotides away from the recognition site, and the other strand at 13 nucleotides away (see figure 19.14). Therefore, each FokI cleaving operation exposes a four-nucleotide sticky end that will be useful at a later stage in the computation.

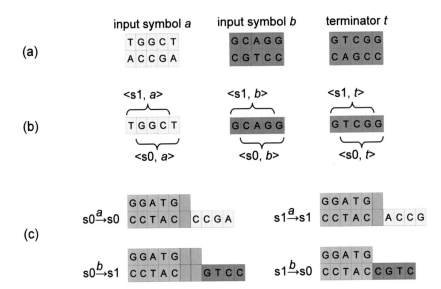

Figure 19.15 Representation of input symbols and state transitions in the DNA FSM [100]. **(a)** unprocessed input symbols in the input string; **(b)** cleaving of the input string by FokI reveals a sticky end to be interpreted as a tuple $< s_i, \sigma >$, where s_i is the current state of the FSM, and $\sigma \in \{a, b, t\}$ is the next input symbol to be read. **(c)** representation of the state transitions for the FSM in 19.13 as DNA fragments to be "trapped" within FokI: each state transition is represented as a DNA fragment consisting of three parts: the first part is the target sequence that binds to FokI; the second part is a space filling; the third part is the base-pair complement of the current state; the amount of space filling controls the FokI cleaving position to expose the next state as a sticky end according to (b).

Each input symbol is represented as a distinct 5-base pair (bp) DNA duplex, as shown in figure 19.15. Apart from the alphabet symbols a and b, a terminator symbol t is added in order to facilitate retrieving the output of the computation. The input string is represented as a sequence of input symbols separated by three-basepair space fillers. The first symbol of the input string is already cleaved in a way that indicates that the system starts in state S_0.

The FSM itself is implemented as a set of FokI enzymes with DNA fragments "trapped" inside. Each FokI-DNA complex is configured to implement one state transition of the FSM. Figure 19.15 shows the four FokI configurations needed to implement each of the state transitions of the FSM shown in figure 19.13. Each building block shown in figure 19.15(c) then gets trapped in a FokI enzyme to form a FokI-DNA complex able to perform the corresponding state transition. A number of these FokI-DNA complexes is then mixed as a "soup" with the input string, where they compete for binding to the input string. The binding and subsequent state transition are illustrated in figure 19.16. Only a FokI-DNA unit that matches the exposed sticky end (and therefore matches the corresponding current state and symbol there encoded) can successfully bind to the input strand (figure 19.16 (a)(b)). The FokI-DNA unit is designed such that when this binding occurs, it cleaves the input strand so that it exposes the next state and input symbol combination (figure 19.16 (b)(c)). A recycling process allows the enzyme to be reused after the cleaved parts have been eliminated and "digested". The sticky end representing the now current state and input symbol is then ready to be recognized and cleaved by another matching FokI-

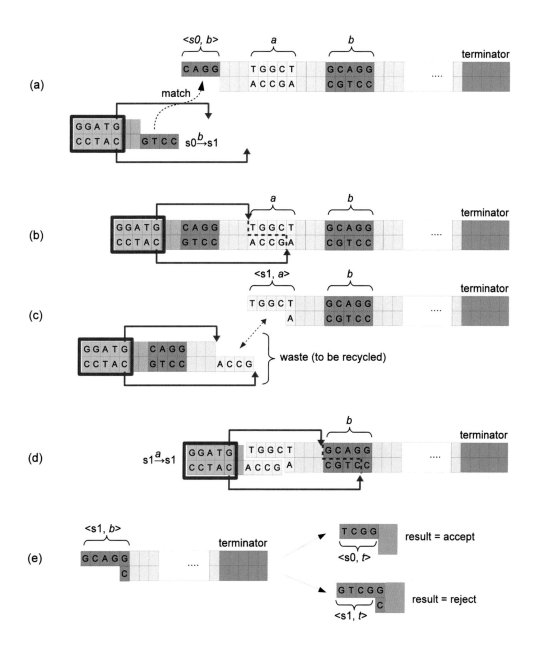

Figure 19.16 DNA FSM at work [100, 768]: at each stage of the computation, a matching FokI-DNA state transition complex binds to the input strand (a) and cleaves it (b), exposing the next state and input symbol (c). The computation proceeds (d) until the terminator symbol is exposed (e), revealing the final state of the FSM: if it is S_0, the input has been accepted (here, indicating that an even number of b's was present); otherwise, it has been rejected (due to an odd number of b's).

DNA complex, in the next step of the computation (figure 19.16 (d)). This process repeats as successive enzymes recognize the exposed sticky end, bind to it and cleave the input strand. It stops when a sticky end that is recognized as a terminal symbol is exposed. This final DNA fragment is then interpreted as the result of the computation, and can be read according to the state that is also exposed at the final sticky end: following the example FSM in figure 19.13, the input is accepted when the FSM ends up at state S_0, and rejected otherwise.

By composing the building blocks of the trapped DNA strands in different ways (beyond the examples of figure 19.15(c)), it is possible to implement any of the eight possible state transitions for a two-state two-symbol FSM [100]. Therefore, this DNA automaton is not restricted to the example FSM of figure 19.13. However, the implementation of large FSMs with several states and transitions would hit the physical limit of what can be packed inside the FokI structure.

By building this DNA FSM the authors realized that it actually had some very promising future applications, even in its apparently restricted form, far from the Universal Turing machine originally foreseen. By interpreting the states of the machine as "healthy" or "sick," a molecular device able to diagnose diseases could be envisaged: a "DNA doctor in a cell" [768]. Such a device would operate in a probabilistic manner, due to the competition between alternative biochemical pathways and the influence of the concentrations of the molecules used as cues to trigger the transitions. For this purpose, the DNA FSM was extended to deal with probabilistic transitions [23], and to respond to mRNA levels of disease-related genes [101,325].

19.3.2 Computing with Gene Regulatory Networks

Recall that chapter 5 briefly introduced gene regulatory networks (GRNs), and chapter 18 provided an overview of computational approaches to model them. Reengineering natural GRNs is a major topic in synthetic biology, and an important tool in bacterial computing, as we will see in section 19.4.1. In this section we introduce a simple and well-known engineered GRN, the repressilator, and briefly discuss its wet-lab implementation.

GRN Example: The Repressilator

The Repressilator [256] is one of the most well-known examples of how GRNs can be programmed to perform a particular function. It implements an oscillator with three genes that repress each other in a circular fashion, forming a cyclic negative feedback loop. The Repressilator was the first synthetic oscillator to be built in the wet lab using bacterial genes.

The conceptual design of the repressilator is illustrated in figure 19.17, and can be expressed by the following set of chemical reactions:

- Each protein P_i represses the expression of the next gene G_{i+1}, forming the inactive complex C_{i+1}, in a circular way:

$$n\,P_1 + G_2 \underset{k_r}{\overset{k_e}{\rightleftharpoons}} C_2 \qquad (19.1)$$

$$n\,P_2 + G_3 \underset{k_r}{\overset{k_e}{\rightleftharpoons}} C_3 \qquad (19.2)$$

$$n\,P_3 + G_1 \underset{k_r}{\overset{k_e}{\rightleftharpoons}} C_1. \qquad (19.3)$$

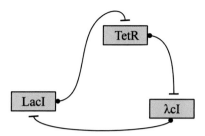

Figure 19.17 Schematic representation of the repressilator GRN: Three genes repress each other in a cyclic negative feedback loop: the repressor protein *LacI* (found in the Lac Operon from *E. coli*) represses the production of *TetR* (involved in tetracycline resistance), which in turn represses λcI (from the lambda phage virus). The λcI protein then represses *LacI*, closing the cycle. See [256] for details.

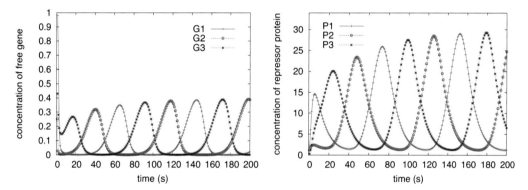

Figure 19.18 Idealized behavior of the repressilator, obtained by ODE integration of reactions 19.1 to 19.5. Oscillations in the concentrations of the various substances involved are observed. Here: concentration of free gene (left), and of repressor proteins (right). Parameters used: $n = 2$, $k_e = k_r = k_p = 1$, $k_m = 5$, $\mu_m = 0.5$, $\mu_p = 0.1$. Initial conditions: one gene is active ($g_1(t = 0) = 1$), while the other two are repressed ($c_2(t = 0) = c_3(t = 0) = 1$); the concentrations of all the other substances are set to zero.

- Each free gene G_i is transcribed to mRNA M_i:

$$G_i \xrightarrow{k_m} G_i + M_i \quad , \quad i = 1, 2, 3 \tag{19.4}$$

- Each mRNA M_i is translated to protein P_i:

$$M_i \xrightarrow{k_p} M_i + P_i \quad , \quad i = 1, 2, 3, \tag{19.5}$$

where P_i are repressor proteins, G_i are genes, C_i are gene-protein complexes, and n is the Hill coefficient for the expression of each gene. Furthermore, mRNAs and proteins decay at rates μ_m and μ_p, respectively (not shown).

Figure 19.18 shows a numerical integration of the ODEs corresponding to the Repressilator reactions 19.1 to 19.5, for a parameter choice that leads to an oscillatory behavior. The observed

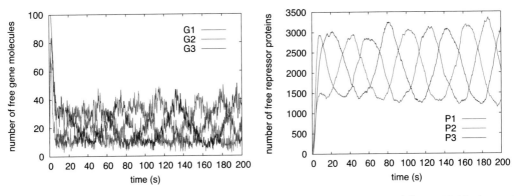

Figure 19.19 Stochastic simulation of the repressilator, using the parameters of figure 19.18 observed for a small number of active gene molecules $m = 100$ in a small volume $V = m/N_A$.

oscillations show the typical behavior of this circuit, however such oscillations only occur within a well-defined region of parameter space, which can be determined by a stability analysis of the system [256]. For instance, no oscillations occur for $n = 1$, indicating that the repressilator relies on a switchlike sigmoid response, which is characteristic of Hill kinetics (see chapter 18).

The Repressilator has also been demonstrated under more realistic conditions. In fact, it has been successfully implemented with real bacterial genes, despite the noise present in a real cellular GRN, indicating that this circuit is robust to noise up to a certain level. The extent of such noise tolerance has been more thoroughly evaluated in [518] using stochastic simulations.

Figure 19.19 shows a simulation of the repressilator circuit using the Gillespie SSA, with the same parameters as those of figure 19.18, however observed at a very small scale; individual molecule counts are now displayed instead of concentration values, macroscopic rate coefficients are translated to stochastic constants as explained in [926], and the volume of the system is set to a sufficiently small value ($V = m/N_A$), such that an amount m of molecules of each type initially present in the system leads to the initial concentrations $g_1 = c_2 = c_3 = 1$ of the ODE case. The stochastic noise looks more prominent in the gene plot (left side of figure 19.19) than in the protein plot (right), due to the lower amount of gene molecules when compared to the protein counts. However, the oscillations are sustained in spite of the inherent noise.

19.3.3 Reaction-Diffusion Computers

A reaction-diffusion (RD) computer [9, 13] is a chemical device capable of information processing via the propagation of waves in a fluid. Information is encoded in the concentrations of substances at any given point in the fluid space, and operations are executed by a combination of chemical reactions and the diffusion of substances in space. Typically, the device computes by propagating diffusive or excitation wave fronts in the medium.

Most of the reaction-diffusion computing approaches today make use of an excitable chemical medium. An excitable medium is a nonlinear dynamical system that is able to propagate waves in a self-sustaining way. Information can be encoded in such traveling waves for computational purposes. The computation proceeds as waves meet and propagate further. When they meet they cause state changes which can be considered as steps in the computational process. Propagating waves carry signals for subsequent steps in the computation, to be carried out in other regions of the fluid space. From a conceptual point of view, computation with reaction-diffusion resembles computation with cellular automata. Indeed, cellular automata have been used to simulate various aspects of reaction-diffusion computing. From another angle, a reaction-diffusion computer can be regarded as a possible physical realization of computation with cellular automata.

Excitable behavior can be obtained in chemistry through oscillating chemical reactions, such as the famous Belousov-Zhabotinsky (BZ) reaction. The BZ reaction is actually not a single chemical reaction but a family of catalyzed redox reactions leading to oscillatory behavior. The oscillations come from the coupling of three stages:

1. During the first stage, a metal ion (such as Cerium(III), or Ce^{3+}) is oxidized in an autocatalytic reaction to Ce^{4+}. The autocatalyst (bromous acid) is destroyed in another reaction, compensating for its otherwise exponential growth due to the autocatalysis.

2. In the second stage, the oxidized metal ion (Cerium(IV), Ce^{4+}) is reduced back to its original form Ce^{3+}, and an inhibitor (bromide ion, Br^-) of the first stage is produced.

3. The third stage consumes the inhibitor, allowing the first stage to resume, thereby closing the cycle.

The BZ reaction occurs in a sulfuric acid medium. In a homogeneous (well-mixed) solution, the successive oxidation-reduction steps from Ce^{3+} to Ce^{4+} and back cause visible changes in the color of the solution, from transparent (due to Ce^{3+}) to yellowish (due to Ce^{4+}) and back.

In a nonhomogeneous medium dynamic spatial patterns such as concentric rings expanding outwards, and swirling spirals can be observed, as waves of chemicals propagate in space. Figure 19.20 shows some wave patterns obtained with a simulation of a BZ reaction-diffusion system, using the Oregonator [276], an idealized model that captures the main qualitative aspects of the complex BZ reaction.

The Oregonator model [276] is described by the following set of reactions:

$$A + Y \xrightarrow{k_1} X \tag{19.6}$$

$$X + Y \xrightarrow{k_2} P \tag{19.7}$$

$$B + X \xrightarrow{k_3} 2X + Z \tag{19.8}$$

$$2X \xrightarrow{k_4} Q \tag{19.9}$$

$$Z \xrightarrow{k_5} fY. \tag{19.10}$$

The patterns in figure 19.20 were obtained via integration of the differential equations obtained from the Oregonator reactions 19.6 to 19.10, using the law of mass action, with $k_1 = k_5 = 1$, $k_2 = k_3 = 10$, and $k_4 = 2.5$.

Substance Z represents the oxidized metal ion (such as Cerium(IV)), whose color change can be easily observed. Substance X is the autocatalyst, whereas Y stands for the inhibitor Br^-. The three stages of the BZ reaction can be mapped to the Oregonator model as follows:

1. The first stage of the BZ reaction (the autocatalytic oxidation of the metal ion, with the simultaneous destruction of the autocatalyst) is captured by reactions 19.8 and 19.9. Note that the reduced form (such as Cerium(III)) that leads to Z is not explicitly represented. The autocatalyst X is properly destroyed in reaction 19.9.

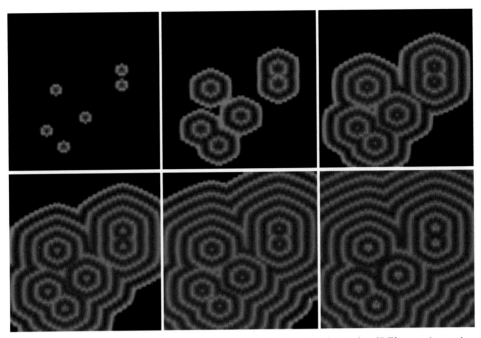

Figure 19.20 Wave patterns in our simulation of the Belousov-Zhabotinsky (BZ) reaction using the Oregonator model [276] on a hexagonal grid. The intensity of the orange color represents the concentration of substance Z (an oxidized metal ion such as Cerium(IV)). Wave patterns form starting from random drops of Z. As the waves spread outward, and the Petri Dish becomes filled with them, the positions of the original drops can still be seen. Simulation implemented using an early variant of the PyCellChemistry package (see Appendix 21.4) running on top of the Breve artificial life simulator (www.spiderland.org).

2. The second stage is subsumed in reaction 19.10, where Z is reduced and the inhibitor Y is produced.

3. The third stage is represented by reactions 19.6 and 19.7, where the inhibitor Y is consumed, allowing the cycle to resume.

Substances A and B are continuously replenished such that their concentrations remain constant in spite of their consumption in reactions 19.6 and 19.8. P and Q are waste products. The stoichiometric factor f is meant to make the Oregonator scheme more generic, and can be set to one in its simplest case [276].

Other metal ions can be used in the redox stages of the BZ reaction. One of them is $Ru(bpy)_3^{2+}$ (ruthenium bipyridyl complex), which leads to a photosensitive version of the BZ reaction: illumination accelerates the production of the inhibitor Br^-, forcing the system back to its stationary state [436]. In this way, the amount of Br^- can be steered by changing the light intensity, which can be used as a means to control the BZ medium to perform a desired task.

A pioneering application of reaction-diffusion computation in the wet lab was the demonstration of image processing with the photosensitive BZ reaction [477]. The projected picture is "stored" in the medium, and typical image processing operations such as contrast adjustment,

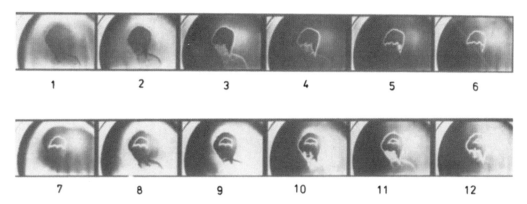

Figure 19.21 Demonstration of image processing using the photosensitive BZ reaction in a Petri dish [477]: as we sweep the successive frames (from 1 to 12) the contour of the image becomes visible, then gets blurred, then a new contour is formed. Reprinted by permission from Macmillan Publishers Ltd: *Nature*, vol. 337, pages 244-247 [477], © 1989.

contour detection, and smoothing can be demonstrated. Figure 19.21 shows an example, where the contour of the image becomes visible, then erodes, when a new contour is formed. Such BZ-based image processing method can be regarded as an inherently parallel computation, as various chemicals diffuse and react simultaneously. Other image processing applications using the BZ reaction include [687, 688].

Logic gates in the BZ medium can be constructed by using space dependent excitability [600, 813, 857]. In such systems, excitability depends on location: The system is excitable at some points and nonexcitable at others. A message can be coded in pulses of excitation: a peak of excitation can be interpreted as the logical *true* state, whereas the stationary state can be interpreted as the logical *false*. One bit of information can travel in space through a moving pulse of excitation. A train of pulses then carries a signal containing a sequence of bits. Logic gates can then be implemented by the collision of signals traveling in different directions.

Several other algorithms using the BZ reaction have also been shown, including algorithms for finding shortest paths in a labyrinth, computing Voronoi diagrams [12, 39], and controlling robots [14]. Other examples of RD computation with BZ include a chemical memory [600, 601], a chemical neuron [601], chemical sensors [341, 605], and more recently, neural networks implemented with BZ droplets that communicate via the propagation of wave fronts [15].

Although innovative and promising, RD computing also has some shortcomings [9, 12]: the computation speed is rather slow, restricting it to niche domains where conventional computers cannot be used; besides that, most of the current RD computing approaches are based on the BZ reaction, which runs in a sulfuric acid medium, that is not biocompatible.

19.4   *In Vivo* Computing with Bacteria and Other Living Organisms

So far we have mostly covered chemical computation *in vitro*, that is, within a test tube or a Petri dish. We now turn to computation *in vivo*, that is, using actual living cells. Typically, single cells or colonies of cells are used, such as bacteria, yeast, or slime molds. *In vivo* computing approaches aim at harnessing the computational properties of biological organisms and to reengineer them for various purposes. Ultimately, the target applications for such future wet computers would be new disease treatments, vaccines, biofuels, food, biosynthetic organs for replacement, environment depollution, and so on. In this context, a living organism is regarded a computer, with cells as processors, DNA as memory for data and program storage, and proteins as means of program execution.

We start our tour with bacterial computing and communication, and then move on to more complex organisms.

19.4.1   Bacterial Computing

Bacterial computing [32] is the application of synthetic biology to program bacterial GRNs to perform desired tasks.

Escherichia coli (*E. coli*) is a rod-shaped bacterium commonly found in the gut flora of many organisms, including humans. It is cheap and easy to grow *E. coli* in the laboratory, and as such it has been widely studied as a model organism for prokaryotic cells. For bacterial computing as well, *E. coli* is the organism of choice, due to the ready availability of genetic components that can be recombined in many ways to engineer versions of *E. coli* that can perform programmed operations.

In section 19.3.2 we looked at the repressilator by Elowitz and Leibler [256] as an example of a GRN built within *E. coli*. Another example is the genetic toggle switch by Gardner, Cantor and Collins [312]. Both the repressilator and the toggle switch were published in 2000 in the same issue of the *Nature* magazine. Since then, several research groups have worked on synthetic GRNs built within bacterial cells. A notable example is the implementation of logic gates using genetic circuits. Recall from chapter 5 that the lac operon is a well-known example of a naturally occurring logic gate that implements the boolean function "Lactose AND NOT Glucose." Numerous other GRNs behaving as logic gates have been identified, inspiring researchers to engineer GRNs to implement various boolean circuits *in vivo* using bacterial cells [35, 908].

The implementation of complex functions within bacterial GRNs suffers from scaling limitations similar to the ones faced by biochemical circuits in well-mixed media: A huge number of different molecules would be required in order to avoid cross-talk between calculation pathways. Although the amount of transcription factors that can be encoded with DNA is virtually unlimited, it is still cumbersome to engineer a large amount of them such that the resulting GRN works in a precise and reliable way. This is an obstacle for the implementation of large genetic circuits within single cells. An obvious solution is to use multiple communicating cells: each cell could implement one or a few functional modules, and the circuit connections would be implemented via cell signaling. Recent trends in bacterial computing therefore investigate distributed algorithms implemented by colonies of bacteria from single or multiple strains [133, 207, 307, 336, 340, 649].

Most of the proposed population-based bacterial algorithms so far rely on quorum sensing as a communication mechanism among bacteria [82, 207, 307, 340, 649]. Quorum sensing is a cell signaling mechanism used by some bacterial species to coordinate their behavior. For instance, the bacterium *Vibrio fischeri* lives in symbiosis with squids, providing them with bioluminescence. A colony of *V. fischeri* only emits light when in colonies inside the light organs of the squid, never in isolation. Another example of quorum sensing occurs in certain groups of pathogenic bacteria, which may decide to inject a toxin into the host's cell only when their population is high enough such that the attack has a high chance of defeating the host's immune responses.

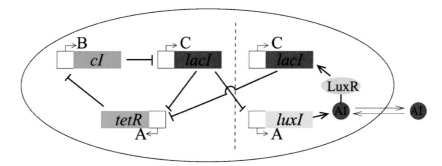

Figure 19.22 The multicellular clock by [307]: each bacterium contains a repressilator circuit (left side) coupled to a quorum sensing circuit (right side). The coupling is done by duplicating the *lacI* gene from the repressilator, which is now expressed upon binding of the *LuxR*-AI complex from the quorum sensing circuit. From [307], © 2004 National Academy of Sciences, U.S.A., reprinted with permission.

In [307] the Repressilator circuits of several bacteria were coupled together via quorum sensing. Figure 19.22 depicts the corresponding GRN solution. The Repressilator part consists of the three genes *cI*, *tetR*, and *lacI*, as depicted in figure 19.17. The quorum sensing circuit comprises the *luxI* gene, which produces the quorum sensing signaling molecule AI (or autoinducer). Another gene (not shown) produces the protein *LuxR* that forms a complex with AI. The *LuxR*-AI complex acts as a transcription factor that causes the expression of bioluminescence enzymes and also accelerates the rate of expression of both *LuxR* and AI, generating a positive feedback loop that considerably amplifies the amount of signaling AI molecules, so that they can be sensed by the other surrounding bacteria. In the coupled repressilator circuit, *lacI* represses *LuxI*. Moreover, the *lacI* gene is duplicated such that it can be activated by the *LuxR*-AI complex. Using mathematical and computational modeling, the authors show that this coupled architecture causes the bacteria to synchronize the oscillation frequencies of their internal repressilators, resulting in a global clock that is much more accurate than the individual noisy clocks without coupling. So the population-based algorithm is able to smooth the effects of the stochastic GRN dynamics that had been shown in figure 19.19 for the single-cell case.

Other examples of distributed bacterial oscillators based on quorum sensing include [82,207, 340], and a review of synthetic mechanisms for bacterial communication with focus on quorum sensing can be found in [649].

A recent approach based on conjugation (as opposed to quorum sensing) has been proposed in [336], and applied to the distributed implementation of logic circuits using a bacterial population. Conjugation is a mechanism by which two bacteria exchange genetic material directly, by injecting plasmids (DNA rings) into each other. Conjugation offers a one-to-one communication channel between two bacteria, as opposed to the "broadcast" method represented by quorum sensing. Through computational modeling and simulations, the authors show that the distributed implementation of logic gates is facilitated by this communication method.

So far most bacterial computing approaches use a population of bacteria from the same species, or from related strains. In [133] a vision of future bacterial computing is presented in which multiple species interact in a synthetic ecosystem. In such systems, interspecies interactions such as symbiosis and predator-prey relationships could be exploited for engineering purposes.

19.4.2 Biomolecular Communication Between Natural and Synthetic Cells

Besides the bacterial communication mechanisms, some researchers also envisage communication mechanisms between natural and synthetic cells.

Cronin and colleagues [202] proposed an analogy of the Turing test for synthetic cells, that could make use of cell signaling to "interrogate" artificial cells in order to measure how well they would imitate their natural counterparts. When the artificial cell becomes indistinguishable from a natural one from a signaling perspective, both natural and artificial cells would be able to communicate seamlessly, which is a desirable property for medical applications in which the intelligent drugs should be able to mimic the signaling language that the natural cells can understand. Later, an approach based on P systems was proposed for modeling the flow of substances through pores in the membrane for cellular communication [781].

Nakano, Suda, and coworkers [607] have proposed a number of engineered molecular communication solutions, including: a "cellular switch" made of cells interconnected by gap junctions; a micro-scale "cellular Internet" in which a "wire" made of connected cells propagates and amplifies chemical signals; and an engineered active transport system for the guided transport of selected substances to desired locations, using molecular motors that move along a network of filaments.

Stano and colleagues [803] argue that chemical communication is an important feature of a synthetic cell. Inspired by [202,607], they propose to endow their SSMCs with engineered molecular communication mechanisms similar to those in [607], in order to obtain a "minimal communication system" between natural and synthetic cells, as envisaged in [202]. For instance, SSMCs could be programmed to express a variety of proteins that would act as cell signals and receptors similar to those evolved in bacterial quorum sensing, establishing a two-way communication channel between SSMCs and natural cells.

19.4.3 Computation with Slime Molds and Other Living Organisms

Other examples of wet computation approaches use various living organisms for computation, such as yeasts and slime molds. In [706], genetically engineered yeast cells are combined to compose logic functions. A small number of cell types can be composed in flexible ways to form complex circuits. Each cell type is programmed to respond to a specific type of stimulus. By combining different cells, the authors show that all basic logic gates can be implemented. These gates can be further recombined into more complex circuits, which would have been difficult to obtain (and perhaps incompatible with an *in vivo* implementation) by engineering a single GRN inside a single cell.

Another approach to facilitate the construction of logic circuits with GRNs is presented in [43], this time with the use of mammalian cells. In their approach, the implementation of a logic XOR gate becomes possible with a GRN in a single cell, thanks to existing mammalian genes that behave similar to an XOR gate, whereas similar genes have not been found in bacteria.

A slime mold is a large unicellular organism whose cell may contain many nuclei (when in plasmodium stage). Slime molds have been used to control robots in hybrid biological-electronic devices [865,866], to navigate in mazes [11,606], and to compute spanning trees [9].

In [9], Adamatzky argues that the plasmodium of the slime mold *Physarum polycephalum* behaves like a reaction-diffusion (RD) computer encapsulated in a membrane. He calls the resulting biological computer a *Physarum machine* and shows that it is able to compute spanning trees in order to find shortest paths in a natural way, a problem that is difficult to solve with RD

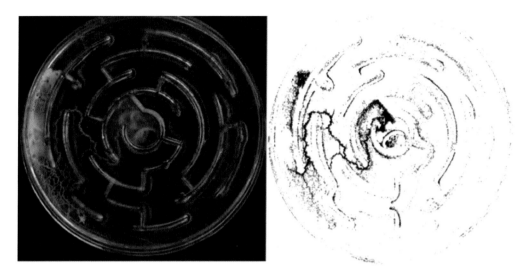

Figure 19.23 *In vivo* maze solving in a Petri Dish with the plasmodium of the slime mold *Physarum polycephalum* [11]. The plasmodium is initially placed at the center of the maze, with a food source at the periphery. After 72 hours, the slime mold has reached the food source, finding its way through the maze in the process. Left: photo of the experimental maze; right: corresponding filtered image. © 2012 IEEE. Reprinted, with permission, from [11].

computing alone due to its fluid nature. A spanning tree is an acyclic graph that connects a set of points such that there is always a path between any two points. Since the body of the slime mold is at the same time flexible and membrane-bound, it is possible to steer its shape by placing nutrients at target places. The plasmodium will then form pseudopodia (tubelike membrane extensions) that grow by chemotaxis toward the nutrient sources, producing a tree-shaped structure that can be regarded as a solution to the spanning tree problem. In this context, computing a spanning tree means growing a treelike structure in the amoeboid shape of the slime mold.

His experiments were recently expanded to solve complex mazes [11], following initial work by Nakagaki and coworkers [606]. An example of maze-solving from [11] can be visualized in figure 19.23. Even a book [10] is already published on the subject, with an overview of experimental work on Physarum machines and their various potential applications. It is important to emphasize, however, that, just like RD "computers," computing with slime molds is slow and not meant to compete with traditional computers. For instance, the computation of the spanning tree in [9] took three days. The main point of these unconventional computer architectures is to provide new ways to solve problems that can be useful in the "wetware" domains where traditional computers are not well positioned.

19.5 Ethical Issues

Humankind has since its beginning recruited and adapted the life around it. So far, we were restricted to technologies that did not blur the boundary between living and nonliving. But now by trying to cross this boundary, we might find whole new powers to exert on life. Programming living organisms and controlling animals on a chemical level: Should we do this just because we *can* do it? Where to draw the line? As soon as science and technology start using living beings for purposes, ethical questions arise and have to be answered. From the new design of molecular building blocks for integration in organisms to the use of organisms, e.g., slime molds, for computational purposes, the main question is whether humans have sufficient knowledge and wisdom to do this in a safe, responsible and ethical way. There is a lively debate among philosophers and bioethicists about these questions. In particular, the attempt to create life in the test tube via protocells is subject to scrutiny [88,96], though similar ethical discussions apply to many other aspects of research in synthetic biology and wet artificial chemistries.

Since its inception, protocell research has faced numerous public concerns, to the point that Venter and his team had to temporarily halt their research before receiving a green light from a commission of ethicists and religious leaders [96]. Some members of the popular press regard Venter's artificial cells, Rasmussen's Los Alamos Bug and similar protocell efforts as "Frankenstein microbes" or "Frankencells." The main concern about protocells is their potential ability to self-replicate and evolve: Self-replication would exponentially amplify any harmful effects of protocells; and evolution, especially if open-ended as would likely happen in the wild, could lead to unpredictable consequences. One obvious way to deal with these risks is to design protocells that could not survive or evolve outside the laboratory, or that would commit apoptosis (cellular suicide) after task completion. However, would these "crippled" protocells still be useful? After all it is exactly the lifelike character of protocells that make them attractive as new form of technology, including their biocompatibility, and their ability to replicate, adapt, and evolve.

Heavily cited in such ethical debates is the *precautionary principle*, according to which even in the absence of clear scientific evidence of harm, parties should take measures to protect public health and the environment. As a consequence, parties should (1) refrain from performing actions that might harm the environment, even in the face of scientific uncertainties; (2) accept that the burden of proof for the safety of an action lies with the proposer of that action (see [646]). We believe that the precautionary principle, applicable to human influence on Earth's climate, should be all the more applicable to the realm of biological systems. The reason is that the memory of Earth's climate system is minute compared to the memory of biological evolution: Whereas the climate has changed multiple times in the history of our planet, there has been only one evolution since the beginning of life. It has accumulated everything in the genomes that collectively are the memory of the biosphere. Hence human-made changes to the course of biological evolution might have even bigger and longer-lasting effects than the changes that impact our global climate, which can already have disastrous consequences. However, a discussion needs to be held about what the precautionary principle advises in regard to specific wet AC technology: What are the actual risks, and how to avoid them? How to exert caution without hampering research and its positive prospects? For the moment, most of the researchers in this area agree that the risks are still low, since they are not even able yet to produce entirely novel cells capable of surviving autonomously. The situation may change rapidly, though. In any case, it is necessary to accumulate more knowledge and arguments that can assist in risk assessment.

The ethical debate is only beginning in this area. Whatever ethical standpoint we may adopt, however, one thing we know for sure: we humans now rule the planet and have great power to change the course not only of our existence but also of all other species with whom we share this small spot in the universe. We can only hope that humans develop sufficient wisdom to exert their power with humbleness, caution and responsibility. The Earth is our village, and the biosphere is our only home to care for and to protect. Our intellect and our technology are powerful tools at our disposal, and we must learn to use them wisely.

19.6 Summary

In this chapter, we have seen a few examples of systems that can be legitimately counted as artificial chemistries, realized in matter. These ranged from very basic building blocks of life, like nucleic acids to the setting up of living systems that, due to the constraints in their interaction with the environment carry the label "artificial." We have, of course, looked at these systems from the particular vantage point argued in this book. That does not mean that this is the only way of looking at these systems, and in many cases it might not be the most important. However, we believe that the AC perspective allows us to understand these systems from a very natural viewpoint and to provide insights that otherwise might elude the observer.

20

BEYOND CHEMISTRY AND BIOLOGY

This chapter is devoted to the discussion of some of the variety of models of artificial chemistries employed beyond chemistry and biology. First, we shall look at self-assembly and discover how it refers not only to molecular interactions but also to the interaction of macroscopic bodies. The laws governing self-assembly can be analogously applied and again show the prevalence of chemical kinetics. This is followed by a discussion of nuclear and particle physics reactions. Again, the collision of entities leads to changes in their collective constitution and this can be modeled using ACs. Even in economic and social systems, similar phenomena abound, which are the subjects of sections 20.3 and 20.4. There, interactions through product transformation and communication provide the ground on which phenomena emerge that we all too well know from other ACs. It is thus natural to attempt to model production of goods and the communication of individuals by ACs.

20.1 Mechanical Self-Assembly

Perhaps the best known work in the area of macroscopic physical (mechanical) systems that self-assemble akin to the self-assembly of molecules in chemical reactions is the work by Hosokawa et al. [404]. The authors' motivation is to elucidate the methods that could be used to predict yield in a self-assembling system, what they call the yield problem.

To shed some light on this question, they prepare a set of magnetic plastic bodies in triangular shape (see figure 20.1(a)). The bodies are constructed such that a permanent magnet is attached inside two of its three faces (as in figure 20.1(b)). A large number of units is placed within a flat box that allows two-dimensional movement of bodies. Attachment and detachment happen if the box is agitated, which is made possible by a device able to turn the box (figure 20.1(c) and 20.2(a)).

The modeling of this system now proceeds in the following way: Individual encounters of boxes are interpreted in analogy to chemical reactions, with a successful attachment as an association reaction. The newly produced body now consists of some more triangle components as

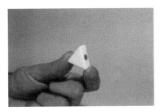

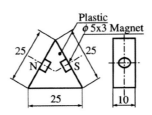

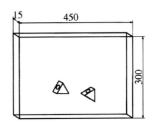

(a) Triangular-shaped bodies used as components

(b) Inside two sides of the triangle is a permanent magnet mounted

(c) A transparent box allows 2-dimensional movement of bodies

Figure 20.1 Self-assembly model experiments of Hosokawa et al. [404], © MIT Press 1994, with permission.

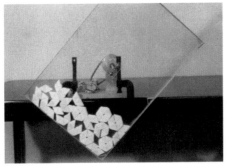

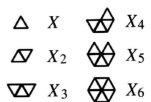

(a) Mechanism for agitating the triangular bodies to encounter each other

(b) Description of raw material and (intermediate) products

Figure 20.2 Self-assembly model experiments and description of Hosokawa et al. [404], © MIT Press 1994, with permission.

in figure 20.2(b), in which intermediate products are labeled with a variable name x_2 to x_5, with the raw material being symbolized by x and the final product x_6.

The reader can easily derive the artificial "chemical" reactions (bi-molecular only):

$$x + x \rightarrow x_2 \qquad x + x_2 \rightarrow x_3 \qquad x + x_3 \rightarrow x_4$$
$$x + x_4 \rightarrow x_5 \qquad x + x_5 \rightarrow x_6 \qquad x_2 + x_2 \rightarrow x_4$$
$$x_2 + x_3 \rightarrow x_5 \qquad x_2 + x_4 \rightarrow x_6 \qquad x_3 + x_3 \rightarrow x_6. \qquad (20.1)$$

We can now proceed to describe the reactions and the dynamics of the system,

$$\vec{x}(t+1) = \vec{x}(t) + \vec{F}(\vec{x}(t)), \qquad (20.2)$$

where we have chosen a discrete time description and $\vec{x}(t)$ symbolizes the vector of all concentration components. $\vec{F}$ describes the reaction probabilities, which the authors have divided into two parts, a collision probability P^c of encountering another body and a bonding probability P^b, which describes the dynamics of the actual bonding process. The collision probability P^c is proportional to the "concentration" of bodies formed, similar to concentration of chemical

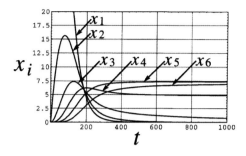

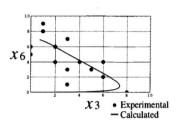

(a) Dynamics of the simulated system, starting from a count of 100 elementary components x. Intermediate products x_4, x_5 stabilize, together with final product x_6

(b) Comparison of model and experiment in the $x_3 - x_6$ plane.

Figure 20.3 Dynamics of self-assembly, comparison of model and experiments [404], © MIT Press 1994, with permission.

elements in mass action kinetics, whereas the bonding probability P^b depends on the geometry of the bodies encountering each other. Under reasonable assumptions for these parameters, the system can be simulated (see figure 20.3(a)) and compared to experimental results (see figure 20.3(b)). The statistics in this experiment didn't allow a more accurate comparison, yet we can get the idea. The authors perform other experiments with more sophisticated experimental equipment, but the fundamental argument remains the same: Self-assembly of macroscopic (mechanical) structures can be described in a similar way as molecular chemical reactions, using equations from chemical kinetics. This example thus implements an artificial chemistry with mechanical bodies.

While the system we discussed comes into equilibrium quickly, and is experimentally difficult to prepare as an open system, there are other macroscopic systems that lend themselves to non-equilibrium formation of patterns in a similar fashion [352]. In those examples, dissipative structures can be formed from small magnetic discs.

These are but examples of an emerging area of science and technology. A number of authors have pointed out that self-assembly of components, while considered first at the molecular level, can be applied at different scales [918, 919]. Thus, mesoscopic and macroscopic components might self-assemble through a process of interaction, with attractive and repulsive influences at work, often mediated by shape recognition. The definition of self-assembly contains the notion that some sort of more ordered state is being formed from a less ordered state. A key concept is mobility of components without which the correct ordered state would not be reachable.

Self-assembly can be generalized to hierarchical self-assembly, sometimes called self-formation, where sequential steps are taken to the self-construction of a complex object [69]. A host of phenomena have been considered as self-assembling structures, and it seems that the processes governing chemical self-assembly provide viable lessons for phenomena at larger scales.

Self-organizing assembly systems (SOAS) have been proposed in [296] as a practical application of ACs for self-assembly. In this work, the chemical abstract machine (CHAM, see section 9.3) is used to specify chemical rules according to which industrial robots can self-assemble into a customized assembly line, that is optimized for a given manufacturing order. The system is

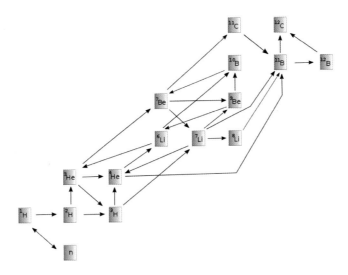

Figure 20.4 Network of nuclear reactions in big bang nucleosynthesis. Adapted from [187]. n is a neutron. Reaction arrows involve other participating particles, like protons, neutrons, photons, electrons, etc. (left out for clarity).

simulated with the help of the Maude specification language, and the authors show that it is able to find feasible configurations that can be proved correct. In [297] the potential for evolvability of such self-configuring manufacturing lines is discussed. Although this work is still at a prospective research stage and not ready for real industrial deployment, it indicates the potential of ACs in the manufacturing sector.

20.2   Nuclear and Particle Physics

Nuclear and particle physics are primary examples of reaction systems that could be formulated from an AC point of view. In fact, there even is a branch of the submolecular science called "nuclear chemistry." Speaking generally, nuclear physics concerns itself with the reaction and transformation of atomic nuclei. The most widespread process considered is the collision of two nuclei or of a nucleus with an atom or with an elementary particle like an electron. As in chemistry, nuclear reactions through collisions of two partners are much more probable than collisions of three and more partners.

A typical reaction is the production of deuterium from hydrogen, which can be described by the following equation:

$$^{1}H + {}^{1}H \rightarrow {}^{2}H + e^{+} + \nu_{e}. \tag{20.3}$$

This symbolizes the reaction of two protons (^{1}H) into a deuterium (^{2}H), a positron (e^{+}) and an electron neutrino (ν_{e}). Similar to the case in chemistry, there are constraints for reactions, such as for mass, energy, charge, momentum, etc., in nuclear reactions that must be obeyed by an actual reaction. Some reactions are more probable than others, and different pathways in nuclear reactions thus have different efficiencies.

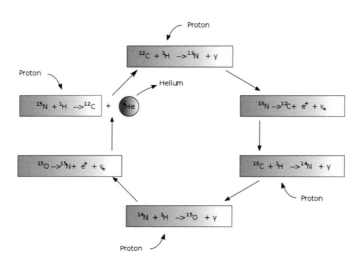

Figure 20.5 Cycle of catalytic nuclear reactions for production of helium in stars with hydrogen as fuel and carbon as catalyst.

For instance, one can visualize the reactions of big-bang nucleosynthesis in which primordial hydrogen produces heavier nuclei as a network of reactions. These reactions have taken place in the first few minutes of the existence of the universe, leading to the observed abundances of light elements in the universe (see figure 20.4).

As we have seen in previous reaction systems, "organizations" can emerge, characterized by the features of self-maintenance and closeness. In the case of nuclear physics, for instance, one reaction pathway is the Bethe-Weizsaecker cycle of helium production, which dominates as a reaction pathway in stars with larger masses than that of the sun. This cycle, also known as CNO cycle due to C, N and O nuclei acting as catalysts, consists of the following reactions:

$$^{12}C + {}^1H \rightarrow {}^{13}N + \gamma$$
$$^{13}N \rightarrow {}^{13}C + e^+ + v_e$$
$$^{13}C + {}^1H \rightarrow {}^{14}N + \gamma$$
$$^{14}N + {}^1H \rightarrow {}^{15}O + \gamma \tag{20.4}$$
$$^{15}O \rightarrow {}^{15}N + e^+ + v_e$$
$$^{15}N + {}^1H \rightarrow {}^{12}C + {}^4He,$$

which together produce helium, positrons, neutrinos, and energy from hydrogen in catalysis. The cycle becomes visible by graphing the reactions as in figure 20.5. This structure is open, with ^1H flowing in and $^4He, e^+, v_e$ flowing out.

In the realm of particle physics, again there are reactions brought about by the collision of particles, for instance

$$d + v_e \rightarrow p + n + v_e, \tag{20.5}$$

where d: deuterium, ν_e electron neutrino, p proton and n neutron.

One of the stranger phenomena of particle physics is that a network of particle reactions are the deepest level of how we can describe reality of matter. If we try (experimentally) to destroy elementary particles to see what they are made of, they will form new particles or simply scatter. Thus, at the lowest level, a network of reactions connects all matter.

20.3   Economic Systems

Economic systems also possess features that lend themselves to interpretation in the framework of artificial chemistries. One of the authors has collaborated on a proposal for a model of an economy that contains an artificial chemistry [816]. The idea is to consider the production of goods in analogy to chemical reactions. Of course, economies are very complex systems, so an abstract model of an economy can only provide some intuitions. We decided for an agent-based model that is divided into two parts, a production system part and a consumption system part. The agents of the former are firms or producers, while the agents of the latter are households or consumers. The connection between them is the flow of goods and services, as well as a market that establishes the value of goods and services.

In this section we would like to explain the AC aspect of the model in some more detail. It is based on the idea of a "natural number economy," published a few years ago as a webpage [377]. The idea there is to take natural numbers as stand-ins for different types of materials and products of an economy. Production agents (firms) act by transforming raw materials and previously produced products (certain natural numbers) required as input into output products (literally the products of these numbers). So what we require in a $\mathbb{N}$ economy is multisets of natural numbers, considered not as a set S of molecules, but as a set P of commodities or products. For instance, the product set P could consist of the following goods:

$$P = \{2, 3, 5, 13, 104, 130, 260\}$$

Primes stand for elementary or raw materials, and products of integers for composites that come about by production processes. It is clear that every product commodity can be factorized into its prime factors to see what raw materials are involved in making that product type. Labor is treated explicitly as a commodity in our model (the number 2). To simplify further, there is a special product, which serves as money, carrying the number 3. We have chosen the $\mathbb{N}$ economy as a simple simulation tool, however, each of the natural numbers could be translated into an actual raw material/product if a more realistic scenario were sought.

Typically, a product is the result of the application of a technology or skill, possessed by the production agent. So in addition to commodities, an agent needs to possess skills to be able to perform the transformation of inputs to outputs. In classical economy, this is the core of the so-called von Neumann technology matrix. This matrix connects commodities (input) with products (output), using technologies as a means. Von Neumann's model is an equilibrium model of an economy, which finds its balance through the assignment of appropriate prices to the commodities and products involved. Here we are not so much interested in the equilibrium features of this model as we are in its similarity with an artificial chemistry.

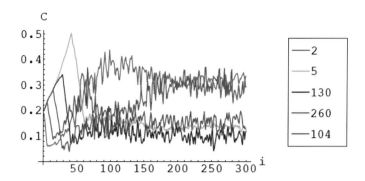

Figure 20.6 Typical run of a simulation of the production system of equation 20.6. From [816], with permission from authors.

An example of a reaction system model that would describe a production system that can be brought into equilibrium reads like the following:

$$2 \times 2 \rightarrow 2 \times 5$$

$$2 \times 2 + 2 \times 5 + 2 \times 13 \rightarrow 3 \times 130$$

$$3 \times 2 + 3 \times 130 \rightarrow 6 \times 260 \qquad (20.6)$$

$$6 \times 2 + 6 \times 260 \rightarrow 6 \times 104$$

$$6 \times 104 \rightarrow 13 \times 2$$

The second line of this production system, for instance, requires two units of commodity 2 (labor), two units of 5 and two units of 13, a free resource (e.g., sunlight) to produce three units of product 130. This example was constructed such that it is in balance.

Based on this production system, and with a proper organization of production agents into a cellular-automata like grid, simulations can be run which typically look like those in figure 20.6. We can see the noisy rise and fall of certain products, which seem to equilibrate around iteration 300.

The system we have proposed is not an artificial chemistry in pure form, but rather a mixture with a market and an agent-based model that takes care of the production and consumption agents. However, a perhaps simpler artificial economy could be designed that is more akin to an AC, though a complex one. In such a model, the agents themselves would be the equivalent of catalysts, possessing features determined by their skill set. Each such agent would then either be involved in the production or consumption of some of the goods of the economy, where consumption could be defined in a similar way as production, with the output being some sort of waste.

So, for example, general reactions of the type

$$\alpha_1 G_1 + \alpha_2 G_2 + \ldots \alpha_m G_m \rightarrow \beta_1 G_1 + \beta_2 G_2 + \cdots + \beta_m G_m + \beta_{m+1} G_{m+1} + \beta_{m+2} G_{m+2} + \beta_{m+n} G_{m+n} \quad (20.7)$$

with $\alpha_i, \beta_i \geq 0$ could model an economy, where $\beta_i \leq \alpha_i$ with the additional goods $G_{m+1} \ldots G_{m+n}$ being the result of the production process. This reaction does not contain the catalyst agent

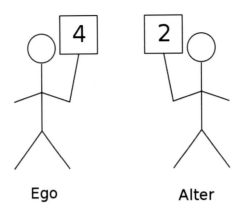

Figure 20.7 Two agents interact by showing signs symbolizing messages. Adapted from [234].

itself, though we could include it by adding it on both sides. A special feature of economies, however, is that the catalyst should be able to assume different states, e.g., based on skills and energy.

20.4   Social Systems

In the realm of social systems science, artificial chemistry models could also be able to success-fully capture phenomena. One of the most important functions a social system is based on is that of communication between social agents. Without communication, the interaction between so-cial agents would be severely limited. At the same time, communication introduces the com-plexity of language or other complex sign systems, which makes analysis difficult.

How can social order arise in a system with many agents that communicate with each other? This is the question the science of sociology has considered for many years. We tried to approach it from the angle of modeling and simulation, studying a simple artificial chemistry model for communication [234]. While this is not the only model that can be successfully applied to the question of the emergence of social order, it certainly is a legitimate approach, given the similar-ity between communication and chemical reaction.

The model consists of agents that exchange messages. In the most simple situation, we have just two agents, "ego" and "alter," that exchange messages from a set of possible messages $1 \ldots N$ consecutively (see figure 20.7). One agent shows a sign, upon which the other agent reacts by showing another sign. This latter behavior is called an activity and will make up the content of our observations.

Suppose now we have a population of M agents that all react (i.e., show activity) towards a sign shown by some other agent. We can build a network of these activities, with the nodes of the network being the activities (messages shown) and edges connecting to another mes-sage/activity if such a reaction exists among the behavior of all M agents. Think of this as a round table discussion, where each participant gets to react to opinions raised by another par-ticipant. It is interesting to note that in such a setting, sooner or later, the discussion which might

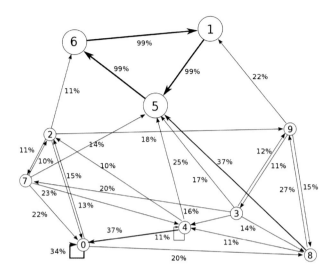

Figure 20.8 Activity graph with nodes symbolizing activities/messages and edges symbolizing the drawing of messages based on other messages. Edges are directed. Ten agents and ten different activities were used in this simulation. Only the more frequently used edges/activities are shown. Size of nodes indicates frequency of message used. In the scenario of a discussion, subgraph of nodes 1, 5, 6 can be interpreted as recurring topics. Adapted from [234].

be initially chaotic, settles on particular issues. Further along, participants cannot escape the impression that the discussion returns to the same issues again and again, giving the impression of a cycle: Social order has emerged.

Similarly, the network of activities has the tendency to close in on circles of activities, whose transition between each other is highly probable. figure 20.8 shows a typical simulation result. This example was generated by simulating 10 agents having 10 messages available. Activities are drawn from the set of messages based on the calculation of expectations and with a certain degree of probability. Agents are equipped with memory that allows them to steady their choice of messages as reaction to others. The interested reader should consult the paper for details [234]. It can be clearly seen that not all messages are equal, and that order emerges out of the interactions, leading to particular messages being sent most frequently.

Dittrich and Winter [235] have continued to work along these lines by studying a model for a political system, based on the formation of organizations in an appropriately set up AC. In another set of works, Adamatzky has examined collective belief systems and formulated a model based on an AC [8]. He considers beliefs ("doxa," as he calls them) as atomic states of individual minds. He then formulates doxastic states, based on these doxa and reactions between doxastic states. As a result, typical time-developments of the concentration of beliefs in a population of individual agents can be observed.

20.5 Summary

In this chapter again, we were only able to discuss a few examples of applications outside of chemistry and biology. What should have transpired from those examples, however, is the broad reach of ACs and their potential to find application in unsuspected areas. We think this is a consequence of the way our world is built from entities and their interactions. The first two examples were from physics, whereas the latter two examples were from the realm of complex systems. There is no doubt that complex systems in particular are a fertile ground for AC modeling efforts, and this is where we expect substantial development over the next decades.

Part VI

Conclusions

The false idea we have to put behind us is the idea that what is bound in time is an illusion and what is timeless is real.

LEE SMOLIN, TIME REBORN, 2013

CHAPTER

21

SUMMARY AND PERSPECTIVES

The field of artificial chemistry has developed substantially over the past several years. With the growth in the prowess of computer technology so grew the ability to perform simulations of fine-grained systems. From bulk matter to molecules to elementary particles, a whole range of material systems is now open for numerical studies in which to employ artificial chemistries. And not only the micro- and meso-scale of material entities is accessible to AC simulation, we have seen macro-scale examples, too, ranging from organisms to mechanical devices to economic and social entities.

What all of these approaches have in common is the attempt to gain understanding from the interaction of parts, rather than from their particular individual features. It is thus an antireductionist program of research that is promoted by artificial chemistries. Perhaps this is too much for scientists that were educated in a thoroughly reductionist atmosphere, so artificial chemistry is not without its castigators. In this last chapter, we therefore discuss a few of the main criticisms of artificial chemistries and try to provide some rebuttals to the most common arguments.

21.1 Some Common Criticisms of the Artificial Chemistry Approach

Some common criticisms of artificial chemistries are:

1. ACs would be only toy models that allow otherwise unoccupied scientific minds to "play" with entertaining systems of arbitrary symbols.

2. The goal of this play (beyond entertainment) would remain unspecified and, even if specified, unclear.

3. Moreover, if interesting results could be gleaned from simulation experiments of artificial chemistries, they could be only qualitative in nature, which would question their value for informing us what happens in the "real world."

4. Further, artificial chemistries would be such radical abstractions from real matter that they would be rendered without connection to reality.

5. Even if efforts were made to reduce the degree of abstraction, AC models would never really come close to real chemistry.

6. Finally, even if the models came closer to chemistry, they would never be readily simulable, due to the numerical effort involved.

The attentive reader might have noticed that most of these arguments have to do with the role artificial chemistries play in relation to the scientific chain of explanation from theory to computer model to simulation to experiment. We have discussed in chapter 2 how computer modeling and simulation have taken center stage in the scientific endeavor in recent years.

Artificial Chemistries are not the only subject of criticisms of this kind. Generally, what we call the sciences of the artificial [777] are subject to similar arguments. In particular, Artificial Life and Artificial Intelligence (AI) belong to the same class of methods. What all these methods have in common is that they are part of theoretical science methodology. They cannot and will never replace "reality." In the words of Froese and Gallagher, who comment on Artificial Life simulations: "To be sure, many researchers have unfortunately succumbed to the functionalist temptation to view their artificial systems as actual empirical instances of the phenomena they are investigating" [298]. While confusing model behavior with behavior of the actual system to be observed and explained is a common error, that does not say that model behavior is useless when trying to explain the behavior of a real system. In fact, the previously mentioned authors note that "this clarification does not diminish the value of the synthetic approach, but shifts its emphasis toward creating 'opaque thought experiments' by which it is possible to systematically explore the consequences of a theoretical position" [298, p. 96].

The term *opaque thought experiment* deserves some more discussion in this context. It was introduced by Di Paolo et al [230] to draw a distinction both in expectations and procedure between posing a classical ("clear") thought experiment versus an "opaque" thought experiment (using a complex simulation model). The role of a thought experiment in scientific inquiry is tightly connected to the concept of how the scientific process works in general. Paraphrased, science is a question-and-answer process in order to understand natural phenomena. The scientist poses questions to nature, which is expected to provide answers that can be built into a framework of explanation of the phenomena under consideration. This question and answer "game" is guided by current theories or models of the phenomena and operationalized in experiments, but the answers returned by nature — the results of the experiments — are sometimes surprising and require a revision of the explanatory concepts of the phenomenon. Thought experiments are not "real" experiments in that they directly pose questions to nature, so what could be learned from them? According to Kuhn [476], the understanding derivable from thought experiments is related to removing confusion among the conceptual apparatus of the researcher. It also reveals some previously marginal questions as probably essential for the explanation of the phenomenon, allowing a shift in attention as to what questions should be posed to nature in "real" experiments.

Opaque thought experiments have a similar function. They allow to sort out possible contradictory assumptions in models and to guide researchers to the important questions in the context of a simulation. If the simulation has a natural system used as the source of the model (regardless of the degree of abstraction), the simulation might have something to say about the essential questions to pose to nature. The complication for opaque thought experiments resides in the fact that interpreting outcomes from complex system simulation experiments is difficult and fraught with its own dangers.

Di Paolo propose three different phases for the use of simulation models as a way to understanding natural phenomena:

1. The exploratory phase: After an initial simulation model is built, different cases of interest should be explored and relevant observables should be defined. If necessary, other observables need to be sought, or even the entire model needs to be changed.

2. The experimental phase: The simulation model should be run. Observations should be used to formulate hypotheses, and crucial experiments should be undertaken to test the hypotheses in order to provide satisfactory explanations of the simulation's behavior.

3. Explanatory phase: The organization of observations into valid (tested) hypotheses needs to be related to the theories and ideas about the natural phenomenon under consideration that motivated the construction of the simulation model. The goal should be to tease out theoretical consequences from the explanation of the model.

The overall goal of the process is to be able to pose further questions to nature, the answers to which might shed more light on the phenomenon studied. In this way, the simulation of complex systems can help the scientific process. Artificial chemistry is thus working with the same motivations that drive other areas of science, and there is nothing particular about its approaches as far as the general methodology is concerned.

Having emphasized the similarity with other fields of inquiry, we now need to consider the borders of the field and how AC approaches can be delimited from other methods. This will be the subject of the next section.

21.2   Delimiting the Borders of the Field

Artificial chemistries deal with systems. Ever since the general theory of systems was established through Bertalanffy [115] it was clear the the main thrust of the systems approach was a holistic view the system objects under study, with an emphasis on the interaction of system components. Questions like how to measure the contribution of a particular component to the overall behavior of the system, or in other words, to the importance of certain components could be addressed in this conceptual framework. Systems science also holds the idea of a hierarchy of "systems within systems," i.e., that the components of a system could themselves be systems of a different kind, better considered subsystems of the whole. ACs could therefore be considered a part of systems science.

Some of the qualifying terms that often accompany systems can also be applied to artificial chemistries. For instance, they generally are dynamic systems, many of them having nonlinear interactions, making this subgroup nonlinear systems. Most ACs belong to the class of complex systems, some to the class of constructive systems. Not surprisingly, there can be chaotic ACs as well as adaptive ones, two other subclasses of systems belonging to the nonlinear system category.

The essence of systems science is to study and understand their behaviors. ACs, by virtue of them being systems themselves, can be used as tools whose behavior can become similar to the behavior of other systems in the world, provided the right interactions and the right features for its components are chosen. It definitely is a modeling tool, however, not to be mixed with the systems in the real world it is to explain.

In the tool set of computational modeling there are other tools that have a similar flavor to artificial chemistries. The most prominent of these are agent-based simulation systems, individual-based modeling approaches, and multiagent systems. What are the similarities and differences between these approaches? The most general of these terms is without a doubt *agent-based systems.*

The notion of agent-based systems is mainly used the social sciences and in computer science. It serves to describe systems whose components are able to behave in autonomous, often complex ways, exhibiting activities that possibly lead to emergent collective effects. Thus, system behavior is produced by individual activities and the interactions of agents. Bankes [60] lists three reasons why agent-based modeling (ABM) techniques are used in the social sciences: (1) the unsuitability of other modeling techniques; (2) agents as a natural ontology for many social problems; and (3) emergence. Similar arguments could be made for ACs in the context of social science. However, the activities of agents in an agent-based simulation are generally more complex and allow more powerful behaviors than the "reactions" allowed for ACs. Thus, ACs are a class of ABMs where the agents are severely restricted in their ability to behave, namely, they "behave chemically." There are generally no internal drives or elaborate activities (while atoms possess energy, they don't possess motivation, etc.). The tendency of ACs to simple features/activities of their atoms/molecules means that the emphasis of those systems is on the interaction events and their consequences. It also means that a simulation of an AC is substantially less complicated than the simulation of a general agent-based model. Bonabeau [127] points out this to be a potential Achilles heel of ABM systems. In particular for large systems, Bonabeau remains skeptical about the ability to model them.

The simplicity of ACs also lends itself more easily to mathematical analysis and the use of tools from dynamical systems or systems biology. Yet complex behavior is possible in ACs, as we have seen in many examples throughout this book, such as robots controlled by ACs, ACs that model immune systems or the emergence of language. It would be interesting to investigate how elaborate cognitive functions, such as "motivation" mentioned earlier, could emerge out of interactions supported by ACs. Hybrid models such as the brain model of grammar construction discussed in [272], which includes an AC component but is not exclusively an AC, may be the most promising approach.

In computer science, particularly in the field of software engineering, agents-based systems have also been applied. In this case, the notion of a software agent can be contrasted to that of a software object. The advantages of achieving system behavior using agents are listed in [770] as (1) autonomy; (2) complexity; (3) adaptability; (4) concurrency; (5) distribution, and (6) communication richness. These advantages can be put to use in complex software systems to help in the engineering process. Mostly in the Artificial Intelligence community, the term multiagent systems came into use. Multiagent systems [931] are a subset of agent-based systems, with an emphasis on the collective behavior of the system, as a result of the interactions among agents and the level of intelligence being assigned to each of the agents. Thus, the individual entity is able to adapt its activities/behaviors using some techniques, like neural networks, in order to better fit the overall purpose of the system. Multiagent systems have been used as an approach to distributed intelligence [915]. Again, in contrast to ACs, one would be hard pressed to assign intelligence to atoms or molecules, or even to larger complexes synthesized from macromolecules. The area of intelligence has not been the target of ACs, although the emergence of intelligence would be an extremely interesting outcome of AC models. We are not there yet though. We are still at the level of trying to obtain some sort of bacterial intelligence at best. It can be expected, however, that as ACs get more sophisticated and powerful, capturing higher levels of organization of living beings, such an outcome does not seem that far-fetched. Consider a morphogenetic AC that would enable the formation of tissues and organs directed by the chemistry of macromolecules, for instance, in the form of a genetic regulatory network; such an AC could lead to the formation of neural structures, or even to other higher-level information processing structures that do not exist naturally, but that would also be able to implement sophisticated cognitive functions. If such an AC were ever achieved, we would learn much from it, including what kind of alternative cognitive substrates might be possible beyond what we know from the biological world. It would provide an enormous contribution to our understanding of intelligence and of how real AI could look like. There is, however, still a lot that must be accomplished before such a chemistry can be built, notably concerning the way multiple nested levels of complexity and organization can be constructed and maintained in an emergent way.

The final term that merits mentioning here is *individual-based modeling*. This term is mainly used in ecology and biology more generally. The philosophy is quite similar to agent-based modeling techniques: "Phenomena occurring at higher levels of observation (populations, communities) emerge from the traits of individuals and characteristics of the environment that together determine how individuals interact with each other and the environment" [347, p. 59]. Thus, again emergence is in the center of interest for individual-based models (IBMs). Grimm and Railsback define the development cycle for IBMs into the following phases:

1. Define the trait(s) and ecological contexts of interest;

2. Propose alternative models how individuals decide in specific situations (trait(s));

3. Identify test patterns that are expected to emerge in an IBM;

4. Implement the IBM;

5. Analyze the IBM to test the proposed models.

Based on the results, one might go back to phase 1 and refine models and tests until a satisfactory model has been found. As is typical for biological applications, an emphasis on the individual will allow for variation of individual behavior. In contrast, atoms of the same type are by necessity identical in their reactions with other atoms. Thus, ACs are a much simpler approach, only interested in the effects of the species to which a particular atom/molecule belongs. Nevertheless, we have seen that ACs can also model ecosystems, and that ecosystems and collectively autocatalytic networks of chemical reactions share some similar dynamical aspects.

We have seen in this book that the areas of application for ACs are broad and cover many of the application areas of the techniques mentioned earlier. Artificial Life models are a primary target for ACs, as are the fields of chemistry and biology. Beyond those, we have seen ACs applied in computer science, physics, economics, social sciences, ecology, and even the arts. In some way, the simplicity of the approach works to its advantage. After applying ACs in so many different areas, common underlying phenomena that link seemingly disparate fields become apparent, such as ecosystems and autocatalytic networks, or economies and reaction chemistries. Another positive outcome of the AC mindset is that some systems that at first sight look daunting to us can be amenable to simple models by looking at them from a chemical perspective: by expressing the system as a set of chemical reactions that convert educts into products instead of a complex system of differential equations, for instance, the flow of objects through the system becomes more transparent. System behavior can be more intuitively grasped, at least for all the nonmathematicians among us. It is straightforward to derive the corresponding equations from the set of chemical reactions, and even to update these equations automatically as novel species emerge or extant species go extinct. Another advantage of ACs is the ability to switch from one realization of the model to another: Once the set of molecules and reaction rules are defined, it is easy to switch from a deterministic to a stochastic implementation of the system, or to spread the reactions out in space to investigate possible localized effects such as the formation of chemical patterns.

21.3   Main Features of Artificial Chemistries

After having read this book, if we ask ourselves now what the essential features of an Artificial Chemistry are, would we all agree on the answer? Let us make an attempt at a summary here. The reader might disagree with us on the importance of these aspects, but probably everyone would have them listed as aspects that are at least very typical for ACs. We see five essential features:

1. Emergent behavior at the nonelementary level

2. Room for novelty in the interactions and collective outcomes

3. Unpredictability of events and of the collective course the system takes during a particular run

4. Constructivity of the system in terms of levels, entities on levels and dynamics

5. A clear dependence of AC systems on time

Here we are going to discuss each of these separately. If the topic has been addressed sufficiently elsewhere in the book, we will simply point to that discussion.

Emergent behavior at the nonelementary level: Emergent behavior is a hallmark of artificial chemistries. As has been pointed out in chapter 14, the concept of emergence actually developed in the context of chemical reactions in the 19th century. This is not by chance, since chemistry has prime examples of collective phenomena of material entities, even at different scales. No wonder then, that a similar type of phenomenon should be present in artificial chemistries. In the light of the first section of this chapter, would that mean "real" or only simulated emergence? We believe "real," which is, however, not to state that an identical emergent phenomenon can be observed as in a real system. Thus, there has to remain at all times a distinction between the model and the system to be modeled, and a clear mapping from one to the other. Both should not be mixed. As for the mechanisms of emergence, however, we can safely assume that similar abstract relations are at work. Many call them (and we have subscribed to this point of view) "top-down causation."

The added value of artificial chemistries is the ability to experiment with interactions and features of entities that do not necessarily have a correspondence in nature. This allows to freely explore spaces of virtual phenomena that one would have great difficulty to realize in experimental settings, the point of rendering them impossible. The virtual phenomena from AC experiments could then be used to pose pertinent questions to natural systems, which might in turn lead to unexpected discoveries.

Room for novelty in the interactions and collective outcomes: Tightly connected to emergence is the fact that novelty can enter artificial chemistry systems. In chapter 12 we discussed this in more detail. It is worth repeating that the treatment of novelty and innovation in formal systems is difficult, and mathematics in particular has not made much progress in allowing novelty as a phenomenon that can be modeled by mathematical formalisms.

Again, chemistry is the prime example of novelty constantly being a fact of life. Every day, new chemical compounds are created by chemists around the world, by combining previously existing molecules in new and sometimes surprising ways. Even new elements are created occasionally, again through a combination (and subsequent transformation) of previously existing elements.

Besides emergence, the second driver of novelty and innovation is combinatorics. In spaces that are huge, due the possibility of combining elements to produce new elements, but with a limited supply of instantiations of combinations, we are bound to have a situation akin to the universe as a whole: mostly empty space, with a couple of places being occupied. The combinatorial spaces of chemistry and biology are even more extreme in regard to the relation between what is realized versus what could be potentially realized. Even in computer systems based on binary logic, the amount of combinatorics and thus the vastness of spaces (program and data spaces) is breathtaking. In the extent of their spaces, artificial chemistries are standing between real chemical and biological combinatorial spaces on the one side, and computer system combinatorial spaces on the other side.

The question should be posed how one would travel a space like the combinatorial universe of an AC. Think of space travel. Stuart Kauffman has an answer that closely resembles what could be done in the space of the real universe: Moving from one populated place to the other it is necessary to consider the immediate neighborhood. In his words, what is important is to consider the adjacent possible. But what does it mean to be restricted to immediate neighborhoods? This depends on the notion of operators that allow to move around in this space. In real space, we are bound by the dimensionality of space which forces us to move from a point A to a point B through the intermediate points. In our combinatorial spaces the equivalent of this kind of move is small variation, e.g., mutation in biological systems. However, our combinatorial spaces have another way of allowing movements: recombination of existing entities. This kind of move is the equivalent of a hypothetical hyperspace jump in real space. In combinatorial spaces, adjacency encompasses a great variety of moves, that essentially percolate the entire space.

Unpredictability of events and of the collective course the system takes: Another key feature of ACs is the unpredictability in detail of the outcomes. This shouldn't be mistaken for emergence. Predictability is not an ontological notion, it belongs to the realm of epistemology: What can we know and how? Nevertheless, given the enormous spaces, and a substantial choice from the "adjacent possible" contingency enters considerations of how AC systems will develop over time. Some events just cannot happen without other events having happened previously (contingency).

Thus, we frequently have the problem not to be able to predict what is going to happen in a particular simulation of an AC. We often have to contend with "typical" runs and qualitative characterizations of what was happening. Single encounters of molecules might produce novelty in a system that completely dominates the development in subsequent steps. Randomness enters on various levels, one being which of the molecules reacts with which other.

We observe similar phenomena in the realm of highly nonlinear social systems, which explains why ACs are a tool suitable to model social systems. Encounters between individuals have the potential to change the course of subsequent events radically.

It is only when we have access to an ensemble of systems (i.e., different runs) that we can have a better idea what "typical behavior" means in the context of that system.

Constructivity of the system in terms of levels, entities on levels and dynamics: Perhaps the most typical sign of the construction of a system that allows us to discern constructive processes from the appearance of novelty alone is the fact the constructive systems usually require scaffolding, that is structures (or processes) that are auxiliary in the construction, but later become unnecessary when the structure is built. Chapter 15 discusses examples of constructive systems, again focusing on the novelty at various levels.

Here, however, we want to emphasize the scaffolding aspect of constructive systems. In a widely cited paper from 1999, Stoltzfus proposes the possibility of "constructive neutral evolution" as an explanation for nonselective evolutionary innovations [815]. "Constructive" in the author's sense is synonym with growing complexity. The argument is that a complex sequence of steps is sometimes required to build something we want to explain. Often an intermediate step might not convey any functional advantage, and thus be neutral on the overall phenomenon yet necessary to increase complexity to the point where it can be "recruited" into a useful function. Thus, the neutrality of the step ultimately gives way to a functional step which is seized by selection and conserved.

We often see similar phenomena in ACs, where intermediate molecules are formed on the way from initial conditions of a population of molecules to an organization that is self-maintaining and closed. Ultimately, these intermediate molecules go away as they don't have the survival capacity to stay in the organization (i.e., they are not part of the organization). Yet, without these intermediate molecules, the organization could never have developed out of the initial conditions. One could argue that these intermediate molecules are part of the scaffolding that goes away once a stable organization has been found by the system.

A clear dependence of AC systems on time: Like all dynamical systems, artificial chemistries depend on time. The development of concentrations of molecules, the sequence of encounters between molecules, the set of interactions, the level of complexity attained by an AC, all depend on time. This places the notion of *time* squarely in the middle of AC models.

In the natural sciences, there is a lively discussion about time, recently invigorated by books like Smolin's [784] and others. The question posed is, whether time is "real," i.e., something deeply engrained as an indispensable component of our universe, or whether it is a convenient modeling tool, a parametric quantity that we use when describing systems, without being an essential part of reality. The answer to this metaphysical question has serious consequences for how we look at systems, ACs among them.

The Newtonian view of the world was one in which space and time were homogeneous, and time in particular was eclipsed in importance by "laws" that had eternal, i.e., time-independent validity. Time wasn't really required to make the universe work, which led scholars to suggest that the experience of time is an illusion [79]. In his book, Barbour argues that "the appearance of time can arise from utter timelessness" [79, p. 35], in the form of snapshots in configuration space. So there is no continuity, and moving forward or backward in time means to move to other snapshots in configuration space. This would allow to explain the existence of "eternal" laws of the universe, as well as some features of these laws, like reversibility (at the quantum level), and other strange phenomena built into physics.

On the other hand, there are other voices, like that of Smolin, who strongly argue that the universe is fundamentally open to the future, based on the fact that time is real and cannot be dispensed with. While Smolin rests his argument on very fundamental features of the universe, encapsulated in its quantum-gravity nature, others, like Ulanowicz [872] take a more macroscopic perspective, which comes, however, to the same conclusion: Ecosystems are nonNewtonian and open to the future.

The same can be said about artificial chemistries. They are similar to ecosystems, with their populations of species encountering each other in a spatial context, and possibly leading to new patterns of resource use and dynamic constitution over time. While limited in the diversity of their phenomena, due to their constraints, ACs of sufficient complexity have the potential to be open to the future.

21.4 Conclusion

How did artificial chemistries develop over the last two decades? We think the most striking aspect of their development is the surprising diversity and breadth of applicability of this concept. Whether we consider elementary particles or social systems, an AC model could be formulated to model and simulate those systems, with some degree of verisimilitude.

The appearance of organizations on all level of human activity, whether it be the exchange of arguments in a round-table discussion, which could at some point end up in an exchange of the same arguments repeated again (an organization!) or price-building mechanisms on the stock exchange, complex systems are approachable by formalizing some of their aspects as an artificial chemistry.

The other development was the adoption of a sort of standardized conceptual way to think about AC models. After it became clear how to formalize an AC [236], it was easier to understand which components of a model would be defined, and one could go about the modeling process more systematically. In chapter 2 we further drew a parallel between this standardized concept to the procedure when trying to set up any model for simulation, which on a fundamental level has to follow the same steps.

The final key development over the last decades was the progress in computational speed of our tools which allow much more complex simulations to happen today compared to a decade or two ago. This is, of course, unsurprising given the unrelenting progress in CPU technology as well as parallel hardware at our disposal, such as GPU coprocessors. Yet, there is a time when quantitative progress of this type can suddenly switch to qualitative differences both in expectations and results.

What to expect from the future? We think that this is largely in the hands of the reader of these lines. ACs are a research concept open to the future. For now, the number of real surprises seems still rather limited, and this might well remain so for the foreseeable future. But by writing this book, we hope to have paved the way for many more surprises, being brought about by creative new ways to use these ideas in the quest to understand our universe and our place within it.

And what are ideas other than the encounter of thoughts of creative minds?

FURTHER READING

Here at the end of the book, we would like to suggest a list of further useful readings that we consider especially inspiring. The intention is to provide readers with a few selected references that allow them to dig deeper in various directions and broaden their horizons.

1. M. Amos, *Genesis Machines: The New Science of Biocomputing*, Atlantic Books London, 2006.

2. R. Axelrod, *The Evolution of Cooperation*, Basic Books, 1984.

3. D. Bray, *Wetware: A Computer in Every Cell*, Yale University Press, 2009.

4. S.B. Carroll, J.K. Grenier and S.D. Weatherbee, *From DNA to Diversity: Molecular Genetics and the Evolution of Animal Design*, Blackwell Publishing, 2nd ed, 2005.

5. I.R. Cohen, *Tending Adam's Garden: Evolving the Cognitive Immune Self*, Elsevier Academic Press, 2000.

6. B.J. Copeland, *The Essential Turing*, Oxford University Press, 2004.

7. M. Eigen and P. Schuster, *The Hypercycle: A Principle of Natural Self-Organization*, Springer, 1979.

8. T. Gánti, *The Principles of Life*, Oxford University Press, 2003.

9. J.F. Haught, *Deeper Than Darwin*, Westview Press, 2003.

10. P. Herdewijn, M.V. Kisakürek, *Origin of Life — Chemical Approach*, Wiley VCH, 2008.

11. P. Herdina and U. Jessner, *A Dynamic Model of Multilingualism: Perspectives on Change in Psycholinguistics*, Multilingual Matters, 2002.

12. D.R. Hofstadter, *Gödel, Escher, Bach: An Eternal Golden Braid*, Penguin Books, 1979.

13. J.H. Holland, *Adaptation in Natural and Artificial Systems*, University of Michigan Press, 1975.

14. J.H. Holland, *Emergence: From Chaos to Order*, Basic Books, 1999.

15. S.A. Kauffman, *Investigations*, Oxford University Press, 2000.

16. S.A. Kauffman, *The origins of order: Self-organization and selection in evolution*, Oxford University Press, 1993.

17. M.W. Kirschner, J.C. Gerhart, *The Plausibility of Life*, Yale University Press, 2005.

18. P.L. Luisi, *The Emergence of Life: From Chemical Origins to Synthetic Biology*, Cambridge University Press, 2006.

19. J. Maynard Smith, E. Szathmáry, *The Major Transitions in Evolution*, Oxford University Press, 1997.

20. D.W. McShea, R.N. Brandon, *Biology's First Law*, University of Chicago Press, 2010.

21. H. Meinhardt, *The Algorithmic Beauty of Sea Shells*, Springer, 2003.

22. M. Mitchell, *Complexity: A Guided Tour*, Oxford University Press, 2009.

23. D. Noble, *The Music of Life: Biology beyond Genes*, Oxford University Press, 2006.

24. M. Nowak, *Evolutionary Dynamics: Exploring the Equations of Life*, Belknap Press, 2006.

25. S. Okasha, *Evolution and Levels of Selection*, Oxford University Press, 2006.

26. E. Regis, *What Is Life? Investigating the Nature of Life in the Age of Synthetic Biology*, Farrar, Straus & Giroux, 2008.

27. R. Rosen, *Life Itself*, Columbia University Press, 1991.

28. L. Smolin, *Time Reborn*, Houghton Mifflin Harcourt, 2013.

29. R.E. Ulanowicz, *A Third Window: Natural Life Beyond Newton and Darwin*, Templeton Press, 2009.

30. A. Wagner, *The Origins of Evolutionary Innovations*, Oxford University Press, 2011.

Appendix

Appendix: Setting up Your Own Artificial Chemistry System

This book has an associated software package that illustrates how to implement artificial chemistries through simple examples. Using this package, a rich set of chemistries can be constructed and tested in a very short time. The package is written in the Python language due to its popularity and ease of use. This appendix provides a short introduction to the package, focusing on its main functionalities and suggesting guidelines for writing your own artificial chemistry using it.

This appendix refers to the state of the package at the time of this writing (May 2014). We expect the package to remain in active development, so by the time you read these lines, more functionalities and features might be available; however, the basic core of the system should remain the same. The package can be downloaded from www.artificial-chemistries.org, where the latest updates and an online documentation can also be found.

The package is offered as Open Source, and the users are very welcome to contribute to it. The instructions about how to do so can also be found on the aforementioned web page.

The PyCellChemistry Package

The package has been designed with a minimalistic philosophy in mind. Our main concerns were an easy installation and programming, enabling a quick start for new users. To this end, we have tried to reduce the amount of external libraries and packages required to a minimum. This means that the basic package does not offer fancy graphics and user interfaces, in order to avoid dependency on elaborated graphics support, which can be time-consuming to install, depending on the configuration of the machine platform.

However, since graphics and visualization are also important in artificial chemistries, toward the end of this chapter a collection of resources including graphics support for Python are included.

Disclaimer: The package is provided as is, without any guarantees. Although we have done our best to offer a functional package, it is still possible that bugs persist to date. Beware that the package has been designed for maximum ease of use and didactic content, and not necessarily for maximum efficiency. We believe that the main usage of this package is as a rapid prototyping tool to test and compare a variety of ACs.

Installing and Running

First unpack the installation file. This should create a folder named `pycellchem`. In this folder you will find two categories of files: basic system files and application files. The basic system contains the core functions needed to implement artificial chemistries, such as multisets, chemical reactions, and simulation algorithms. It's content is explained in the next subsection. The application files provided with the package implement different examples of artificial chemistries. The subsequent subsection gives a list of these examples.

In order to keep the dependencies to a minimum we require only the use of the Python NumPy package, which contains the numerical support needed to implement ACs.

Most of the applications can be run from the command line by just typing:

```
YourMachinePrompt\% python [example.py] > output1.txt 2> output2.txt
```

Depending on the example chosen, the output file(s) will contain execution traces or data to be plotted, the latter usually in tab-separated format.

For more details about each core function and application, we refer the reader to the on-line reference manual in `www.artificial-chemistries.org`.

The Core Library of Artificial Chemistry Functionality

The basic system files are:

- `Multiset.py`: a bag of chemicals represented as a multiset of objects

- `KeyMultiset.py`: a multiset indexed by a given key; a key can be seen as a kind of binding site; this class is currently used to implement the reaction algorithm for Fraglets, but is also useful as a generic way to separate objects by category (for instance, all strings starting with a given pattern, or more generally, all objects satisfying a given condition), while still keeping them in the same bag.

- `Reaction.py`: implements explicit chemical reactions composed of an educt and a product multiset.

- `ReactionParser.py`: parses a text of the form `"A + B -> C"` representing an explicit chemical reaction, and converts it to a Reaction object.

- `ReactionVessel.py`: a reaction vessel for well-stirred explicit artificial chemistries (S, R, A) in which the molecules in S and the reactions in R are represented explicitly, and for which the set of possible molecular species S and the set of reactions R does not change. Two algorithms A are implemented as specializations of the base class `ReactionVessel`:

  - `WellStirredVessel`: a simple ODE integration algorithm based on Euler's method
  - `GillespieVessel`: a simple implementation of Gillespie's stochastic simulation algorithm (SSA) using the direct method

- `BinaryStrings.py`: routines to manipulate binary strings in the form of arbitrarily long integers

- `Cell.py`: a cell that can grow and divide; cells are hierarchical compartments akin to Membrane or P Systems; they may contain multisets of chemicals and other cell objects inside.

Examples of Artificial Chemistry Implementations

The following examples are included in the package:

- `Dimer.py`: An implementation of the simple reversible dimerization reaction used to illustrate equilibrium in section 2.2.4. It was used to obtain the plot of figure 2.3.

- `Chameleons.py`: The colored chameleon chemistry from section 2.5.1.

- `NumberChem.py`: The prime number chemistry from section 2.5.2.

- `MatrixChem.py`: An implementation of the matrix chemistry from chapter 3.

- `Logistic.py`: A chemical implementation of the logistic equation from section 7.2.4, including a comparison between ODE and SSA shown in figure 7.2.

- `Lotka.py`: A chemical implementation of the Lotka-Volterra equation, as explained in section 7.2.8. Here, the deterministic (ODE) and stochastic (SSA) implementations can also be compared.

- `Repressilator.py`: A stochastic simulation of the Repressilator circuit as discussed in section 19.3.2. It was used to obtain figure 19.19.[1]

- `Evolution.py`: Base class for the demonstration of concepts from evolutionary dynamics (chapter 7). The `Quasispecies` and `Tournament` classes below use this base class.

- `Quasispecies.py`: Stochastic version of the quasispecies evolutionary dynamics of section 7.2.7, used to produce figure 7.4.

- `Tournament.py`: A genetic algorithm with tournament selection in a chemistry, similar to the algorithm presented in [937, 938].

- `NKlandscape.py`: A simple implementation of Kauffman's NK model of rugged fitness landscapes [447] (see section 18.4.1); it constructs the epistatic matrix and generates the corresponding fitness values. Caution: This program uses a naive implementation of the epistatic matrix; therefore, it does not scale to large values of N and K.

- `Fraglets.py`: A simplified interpreter for the fraglets language [862], as described in section 16.2.1.

- `NetFraglets.py`: Networked fraglets where network nodes are represented by cells from the `Cell` class in the core library. It includes the CDP example from [862] shown in figure 16.10).

- `Disperser.py`: The Disperser load-balancing example from section 17.3.1.

- `HighOrderChem.py`: A simplified high-order chemistry in which the reaction rules are also molecules in a multiset.

---

[1] The ODE integration of the Repressilator shown in figure 19.18 was obtained with the ODE integrator of the Python SciPy library prior to the development of the `pycellchem` package.

Writing Your Own Artificial Chemistry in Python

Perhaps the easiest way to set up your own chemistry is by modifying an existing example. Let us walk through an example step by step in order to see how this can be done. Since constructive and nonconstructive chemistries are specified in very different ways, we select one example of each for modification.

Building a Nonconstructive Chemistry

Let us take the simplest program `Dimer.py`, shown in figure 1. It implements the following reversible dimerization reaction:

$$A + B \underset{k_r}{\overset{k_f}{\rightleftharpoons}} C \tag{1}$$

where C is a dimer of monomers A and B. This is an example of nonconstructive chemistry where the set of molecules S and the set of reactions R are defined explicitly. In `pycellchem` such chemistries are specified simply by the set R and the initial amount of chemicals present (S_0). It is not necessary to enumerate all possible molecules in S, since molecules that do not react do not participate in the dynamics of the system, and therefore do not need to be explicitly included in the simulation. The choice of the algorithm A determines the level of resolution at which we observe the system: The same chemistry (R, S_0) can be observed at a smaller scale by choosing A as a stochastic simulator, and at a coarser scale by choosing A as an ODE integrator.

We start by defining the set of reactions R, together with their kinetic coefficients k (lines 4-7 of figure 1):

```
reactionstrs = [
    "A + B --> C , k=1",
    "C --> A + B , k=1"
]
```

Note that each direction of a reversible reaction must be specified separately. If a kinetic coefficient is not specified for a reaction, its default value is one. Hence, this reaction set can also be written simply as:

```
reactionstrs = [
    "A + B --> C",
    "C --> A + B"
]
```

This set of reactions is specified as a human-readable list of strings `reactionstrs` that must be parsed so that they can be more efficiently processed by the chosen reaction algorithm. For this example we have chosen an ODE integrator implemented by the `WellStirredVessel` class (within `ReactionVessel.py`). Therefore, we now create a reaction algorithm object `reactor`, which is an instance of the `WellStirredVessel` class, and use it to parse the reaction list `reactionstrs` (lines 8–9 of figure 1):

```
self.reactor = WellStirredVessel()
self.reactor.parse(reactionstrs)
```

```
 1:   from ReactionVessel import *

 2:   class Dimerization:

 3:       def __init__( self ):
 4:           reactionstrs = [
 5:               "A + B --> C , k=1",
 6:               "C --> A + B , k=1"
 7:           ]
 8:           self.reactor = WellStirredVessel()
 9:           self.reactor.parse(reactionstrs)
10:           self.reactor.deposit('A', 2.0)
11:           self.reactor.deposit('B', 1.4)

12:       def run( self ):
13:           finalvt = 10.0
14:           dt = 0.1
15:           self.reactor.trace_title()
16:           while (self.reactor.vtime() <= finalvt):
17:               self.reactor.trace_conc()
18:               self.reactor.integrate(dt)

19:   dimer = Dimerization()
20:   dimer.run()
```

Figure 1 Source code for Dimer.py, an implementation of a reversible dimerization reaction.

The parse method calls the parser in ReactionParser.py to parse the list reactionstrs, adds the resulting Reaction objects to the reactor, and closes the system of reactions such that it is ready to be used for ODE integration. From this point on, no further reactions may be added to the system (or at least it is not straightforward to do so).

Before we start the integration, we deposit some initial amount of substances in the reactor (lines 10–11 of figure 1):

```
self.reactor.deposit('A', 2.0)
self.reactor.deposit('B', 1.4)
```

The method deposit(m, c) deposits a concentration c (in mols per unit of volume) of molecule m in the vessel. The amount of substance 'C' is not specified; therefore, it is considered as zero.

We now run the algorithm by invoking the integrate method at regular intervals dt, until the final simulation time finalvt is reached (lines 15–18 of figure 1):

```
self.reactor.trace_title()
while (self.reactor.vtime() <= finalvt):
```

```
self.reactor.trace_conc()
self.reactor.integrate(dt)
```

In this case, we would like to plot the concentrations of substances over time, so we invoke trace methods for that: `trace_title` prints a title line containing the names of the substances (in our case, 'A', 'B', and 'C') separated by tabs; `trace_conc` prints a line with the concentrations of each substance in the system, also separated by tabs. The output is written to `stdout` by default, so it will appear on the screen, but it can also be redirected to a file such that it can be plotted with a plotting tool such as `gnuplot` or other.

You can run this program by invoking it from Python. From a Unix command line shell:

```
python Dimer.py > dimer.gr
```

This will produce a tab-separated file `dimer.gr` with the concentrations of A, B, and C over time.

Modifying the Dimer Program

In order to build your own chemistry, you can start by modifying the dimer example, for instance, changing the values of k, or the initial concentrations of substances, and see what happens. You can also add, remove, or modify the reactions from the list `reactionstrs`.

For instance, if we want the equilibrium to move toward a higher yield of C, we can make the formation of C much faster with respect to its dissociation, for example, $k_f = 2.0$ and $k_r = 0.2$:

```
5:              "A + B --> C , k=2",
6:              "C --> A + B , k=0.2"
```

An interesting and easy modification is to change the reaction algorithm. Instead of ODE integration, let us now use Gillespie SSA. For this purpose it suffices to replace the class used to create the `reactor` object, and to invoke the `iterate` method to run one iteration of SSA, resulting in the following modifications with respect to figure 1:

```
8:              self.reactor = GillespieVessel()
      ...
16:             while (self.reactor.vtime() <= finalvt):
      ...
18:                 self.reactor.iterate()
```

If we run this new program, we will see that the concentrations now fluctuate a bit, but essentially we get the same equilibrium values as with the ODE case. So it is easy to simulate the same chemistry with different algorithms.

Such simplicity actually hides important differences between ODE and SSA, differences that can nevertheless be exposed when needed. One of the major differences between both is the explicit representation of molecule counts and reactor volume in `GillespieVessel`, as opposed to concentration values in `WellStirredVessel`. Another major difference concerns the kinetic coefficients k used by ODEs versus their mesoscopic counterparts c required by stochastic simulations. Understanding these differences is crucial to setting up stochastic simulations, so they will be discussed in more detail next.

Dealing with Discrete Molecule Counts and Volume in Stochastic Simulations

An ODE system can be seen as model for a system containing an arbitrarily large number of molecules (tending to infinity) in a sufficiently large volume. An ODE integration can also be interpreted as the average behavior of a sufficiently large number of repetitions of stochastic simulation experiments. Since it manipulates concentrations directly, the `WellStirredVessel` class has no concept of volume, and no concept of number of molecules.

As opposed to the ODE integrator, the SSA implementation in the `GillespieVessel` class works with integer numbers of molecules that are placed in a multiset. In order to facilitate programming, it is convenient to keep compatible sets of parameters between ODE and SSA. For this purpose, a reactor of type `GillespieVessel` has a finite volume V, such that concentration values can be obtained from molecule counts.

Recall from section 2.2 that the concentration $[S]$ of a molecular species S is measured as:

$$[S] = \frac{n_S}{N_A V} \tag{2}$$

where:

- n_S is the number of molecules of S in the vessel;

- $N_A = 6.02214 \times 10^{23}$ molecules is the Avogadro constant;

- V is the volume of the vessel.

Conversely, the number of molecules can be easily obtained from the concentration value, for given volume V:

$$[S] = \frac{n_S}{N_A V} \Rightarrow n_s = [S] N_A V \tag{3}$$

Since it is both a reactor vessel and a multiset, the `GillespieVessel` class inherits its structure from both `ReactionVessel` and `Multiset` classes. Individual molecules can be explicitly injected into the vessel using the `inject(m, n)` method from the `Multiset` class, where m is the name of the molecular species and n is the number of molecules of species m to be injected.

For compatibility purposes, `GillespieVessel` also offers a `deposit` method that allows programmers to deposit concentration amounts like in the ODE case. In the SSA case however, the concentrations values are first converted to molecule counts according to eq. 3, which are stored in the multiset. Similarly, the `trace_conc` method of `GillespieVessel` converts the internal molecule counts to concentration values according to eq. 2 before printing them for plotting. In this way, the interface of the `GillespieVessel` class is kept as similar as possible to that of the `WellStirredVessel` class, in spite of a different internal implementation.

For convenience, `GillespieVessel` keeps track of the quantity NAV = $N_A V$ which is the number of molecules of a given species present in the vessel when its concentration is one. NAV gives us an idea of the size of the system that we are simulating: a small NAV leads to a smaller vessel with less molecules, with a more prominent effect of stochastic fluctuations. The larger we make NAV, the closer we get to the ODE behavior. The default value of NAV is 1000, meaning that when we deposit a concentration of 1.0 of a given molecule in the vessel (using the `deposit` method), we are actually injecting 1000 units of this molecule into the vessel. This default value can be changed through the method `set_nav`. By adjusting NAV we can choose the desired size of the vessel without having to calculate its volume explicitly. In any case, the implementation of `GillespieVessel` makes sure that the values of `NAV` and V are kept consistent at all times (that is, when `NAV` changes via `set_nav`, the volume V is updated accordingly; conversely, `NAV` is updated when V changes).

Converting Macroscopic Kinetic Rate Coefficients to the Mesoscopic Scale

The kinetic rate coefficients k provided with each reaction refer to the macroscopic values frequently encountered in the chemistry literature. Before using a stochastic simulation, it is essential to convert each k coefficient to a volume-dependent mesoscopic coefficient c as explained in [328] and in chapter 4. This is done using the Wolkenhauer relation [926]:

$$c = \frac{k}{(N_A V)^{m-1}} \prod_{i=1}^{L} l_i! \tag{4}$$

where:

- c is the resulting mesoscopic coefficient

- k is the macroscopic kinetic coefficient of the reaction

- N_A is the Avogadro constant

- V is the volume of the vessel

- m is the total number of colliding molecules (which is equivalent to the size of the educt multiset of the reaction)

- L is the number of species in the educt multiset

- l_i is the educt stoichiometric coefficient of substance i (in other words, the number of molecules of species i in the educt multiset for the reaction)

Equation 4 says roughly that the more educt molecules the reaction has, the smaller c will become, because there will be a smaller chance that they collide together. Moreover the smaller the volume, the higher the probability that molecules will collide within the vessel.

The `parse` method from the `GillespieVessel` class performs the k to c conversion automatically for all the reactions in the system, by invoking the `close` method after all the reactions have been parsed. Any future changes in kinetic coefficients (through the `set_coefficient` method), changes in volume (`set_volume`) or in `NAV` value (`set_nav`) will automatically update the c values to keep them consistent.

Actually, collisions involving more than two molecules simultaneously are largely improbable. Nevertheless, multimolecular reactions are very common in the chemistry literature because they summarize more complex multistep reaction processes in a compact way. This is why it is important to support them through the generic relationship provided by eq. 4. Note that eq. 4 provides only part of the rule to calculate the reaction propensities for a generic reaction involving an arbitrary number of molecules. The implementation of `GillespieVessel` follows the propensity calculation method proposed in [926], where the full details can be found.

It is easy to use the `ReactionVessel` module to simulate any system of chemical reactions with minimum effort, and to compare its stochastic vs. deterministic simulations. Such a simple approach is very useful to model a wide variety of systems. See, for instance, the other examples of nonconstructive chemistries such as `Repressilator.py`, `Logistic.py`, and `Lotka.py`, included in the `pycellchem` package. However, this approach remains limited to cases where the full set of species and reactions can be enumerated explicitly. Often it is interesting to model the effect of the molecular structure on the kind of reactions that are possible in the system, and to see what kind of new molecules can be produced as a result. A constructive chemistry is more adequate for this purpose, as we shall see below.

Building a Constructive Chemistry

Constructive chemistries have no fixed recipe for their design, therefore the support from the `pycellchem` package can only be very lightweight in this respect. Since molecules and reactions may be defined in an implicit way, each constructive chemistry may have its own way to decide when and how reactions take place among molecules, the rules to apply in each case, and the algorithm to simulate their dynamics. It goes without saying that setting up a constructive chemistry is typically more complex than setting up a nonconstructive one. Such initial extra work will be compensated by a much greater flexibility and the open-ended possibilities of molecule types and dynamic behaviors.

Three modules from `pycellchem` might be useful to implement constructive chemistries: the `Multiset` class (together with its variant `KeyMultiset`) to deal with multisets of objects, the `Cell` class for compartmental chemistries, and the collection of binary string manipulation functions in the `BinaryStrings` module.

We start with the prime number chemistry implemented in `NumberChem.py`, then look at how to extend it.

The Prime Number Chemistry

The prime number chemistry from section 2.5.2 is implemented in `pycellchem` as `NumberChem.py`. A partial source code for this chemistry is shown in figure 2. In contrast to the Dimer example of figure 1, where the full source code could be shown, here we only show the most important parts that are relevant for our discussion.

We start by creating a multiset `mset`, and inserting some random numbers in it (lines 4–9 in figure 2). We draw 100 random integers picked from the interval `minn=2` to `maxn=1,000` and inject them into `mset`. After that, we run 10,000 iterations of a variant of algorithm 2.5.2: two random molecules m_1 and m_2 are extracted from the multiset (lines 15–16). Lines 17–20 make sure that $m_2 \geq m_1$ by exchanging the values of the two variables if needed. The collision is effective only when $m_2 > m_1$ and the division m_2/m_1 has no remainder (line 22). In this case (lines 23–25) the reaction is performed by dividing m_2 by m_1; the result replaces m_2. The counter for effective collisions is then incremented (line 24), and the product of the reaction is recorded (line 25). Lines 26–27 reinject m_1 and m_2 into the reactions, oblivious to the fact that one of them may have been modified or not.

The data for plotting are printed in the form of tab-separated columns, using the `trace_title` (line 12) and `trace` (lines 13 and 28): `trace_title` prints the header line with the names of the traced parameters separated by tabs, and `trace` is invoked at the end of every iteration to print a data line with the values of each parameter separated by tabs. The following parameters are traced: iteration number (column 1), total number of species (column 2), cumulative number of effective collisions up to the present iteration (column 3), total number of molecules representing prime numbers (column 4), and the event that happened in the present

```
1:   from Multiset import *

2:   class NumberChem:

3:       def __init__( self ):
4:           minn = 2
5:           maxn = 1000
6:           self.mset = Multiset()
7:           for i in range(100):
8:               dice = np.random.randint(minn, maxn+1)
9:               self.mset.inject(dice)

10:      def run( self ):
11:          self.neffect = 0
12:          self.trace_title()
13:          self.trace(0, '')
14:          for i in range(10000):
15:              m1 = self.mset.expelrnd()
16:              m2 = self.mset.expelrnd()
17:              if m1 > m2:
18:                  tmp = m1
19:                  m1 = m2
20:                  m2 = tmp
21:              reaction = ("%d / %d" % (m2, m1))
22:              if (m2 > m1 and m2 % m1 == 0):
23:                  m2 = m2 / m1
24:                  self.neffect += 1
25:                  reaction += (" = %d" % m2)
26:              self.mset.inject(m1)
27:              self.mset.inject(m2)
28:              self.trace(i+1, reaction)

29:  numchem = NumberChem()
30:  numchem.run()
```

Figure 2 Partial source code for `NumberChem.py`, an implementation of the prime number chemistry of chapter 2.

iteration (column 5: in case of an elastic collision, only the colliding molecules are shown; in case of an effective collision, the generated products are also shown).

A High-Order Chemistry

One way to write a new constructive chemistry would be to simply modify the reaction scheme in lines 15–27 of figure 2: we could replace the division rule by other rules performing other desired operations on the molecules. For instance, by interpreting each number as a binary string, and implementing reaction rules that perform string folding into a matrix and subsequent matrix multiplication, we could obtain the matrix chemistry from chapter 3. Indeed, the matrix chemistry is included in `pycellchem` as `MatrixChemistry.py`. The type of molecules handled could also be modified, for instance, by storing molecules as text strings instead of numbers. The `Fraglets.py` chemistry is an example of this case. Other constructive chemistries included in the package are `Quasispecies.py` and `Tournament.py`, which implement two variants of evolutionary dynamics.

Another way to write new constructive chemistries is through a generic high-order chemistry able to handle externally defined reaction rules. In this way, we do not need to reprogram everything in order to create a new chemistry. Instead, we can reuse the underlying algorithm for various different chemistries. For this purpose, we have implemented `HighOrderChem.py`, a simplified high-order chemistry in which the reaction rules are also molecules in a multiset. In this way, we build a chemistry able to run constructive chemistries: we can now change the chemistry by simply changing the set of reaction rules in the multiset, without changing the underlying algorithm implementation. With such a chemistry, the reactor algorithm *A* can execute a variety of chemical reactions that are specified simply as strings containing Python commands. Thanks to Python's `exec` command, it is possible to execute these strings as if they were part of the program's source code.

Figure 3 shows the core of the implementation of HighOrderChem.py. It makes use of Python's regular expression package (line 1) in order to parse a reaction rule given as a string. First of all, it creates two multisets, one for the data molecules (`self.mset`) and one for the rule molecules (`self.rset`). Ideally, all molecules should be in the same multiset such that they can be manipulated in the same way, but this would make our implementation unnecessarily complex at this stage. Method `iterate` (line 7) executes one iteration of the reactor algorithm, which proceeds as follows: First of all, one random rule is drawn from the rule multiset (line 8). The algorithm stops if no rules are present (line 9). We then parse the selected rule string (line 10) to identify its corresponding function (`funct` variable in line 12) and parameters (akin to binding sites, `bsite` in line 13). An example of rule would be `"bimolecular(m1, m2),"` which would call a global procedure `bimolecular` (which must be defined by the programmer) with two molecules m1 and m2 as parameters. Another example would be `"self.fold(m),"` which would call a method `fold` from a subclass derived from `HighOrderChem` with only one molecule (m) as a parameter. In these examples the parameters m1, m2, and m would be replaced by the values of the actual molecules drawn from the data multiset, as explained below.

Since each rule may require a different amount of educt molecules, the number of comma-separated parameters in `bsite` is counted (line 14) and stored in variable `nsites`, which will then contain the amount of educt molecules needed for the selected rule. After that, we draw `nsites` molecules at random from the data multiset (lines 18–20), and use them to compose a command line cmd (line 22) of the form `"function(m1, m2, ...)"` where m1, m2 are the actual contents of the molecules drawn. The command cmd is then executed (line 23), and its returned results are interpreted as a list of product molecules that is inserted in the data multiset (lines 24–25). Finally (line 26), the rule is reinjected into the rule multiset for later reuse. The algorithm returns a triple `(rule, educts, products)` that informs the caller about which reaction took place and which were the molecules involved.

Using `HighOrderChem` it is easy to implement any chemistry based on a set of rules that manipulate molecules based on their content. For instance, the prime number chemistry can be easily reimplemented on top of `HighOrderChem` by defining the following reaction rule:

```
 1:  import re
 2:  from Multiset import *

 3:  class HighOrderChem:
 4:      def __init__( self ):
 5:          self.mset = Multiset()
 6:          self.rset = Multiset()

 7:      def iterate( self ):
 8:          rule = self.rset.expelrnd()
 9:          if rule == '': return ('', [], [])
10:          match = re.match('([\.\w]+)\((.*)\)', rule)
11:          if (match == None): return (rule, [], [])
12:          funct = match.group(1)
13:          bsite = match.group(2)
14:          nsites = len(bsite.split(','))
15:          educts = []
16:          products = []
17:          if self.mset.mult() >= nsites:
18:              for j in range(nsites):
19:                  mol = self.mset.expelrnd()
20:                  educts.append(mol)
21:              edstr = str(educts).strip('[ ]')
22:              cmd = ("products = %s(%s)" % (funct, edstr))
23:              exec(cmd)
24:              for mol in products:
25:                  self.mset.inject(mol)
26:          self.rset.inject(rule)
27:          return (rule, educts, products)
```

Figure 3 A high-order chemistry able to execute rules containing Python commands. Like the data molecules, rule molecules are strings that float in a multiset.

```
def divrule(m1, m2):
    p = [m1, m2]
    if m1 > m2 and m1 % m2 == 0:
        p[0] = m1 / m2
    elif m2 > m1 and m2 % m1 == 0:
        p[1] = m2 / m1
    return p
```

We then inject at least one molecule of "divrule(m1, m2)" into the rule multiset of HighOrderChem, together with a few number molecules in its data multiset, and call its iterate method to simulate the system for a few time steps. The results should be the same as the original NumberChem implementation from figure 2.

Note, however, that we now have the extra flexibility of injecting as many different rules as we want into the multiset. We could also look at how rules compete, and so on. For instance, we could add multiplication, addition, subtraction, or square root or other operations in order to implement an algebraic chemistry similar to the ACGP explained in section 16.6. Or we could add machine molecules that operate on strings representing city tours in order to implement a version of the molecular TSP from section 17.2.1. The possibilities are endless, and we leave them open for the reader to explore.

In order to make `HighOrderChem` truly high-order, one would need the facility to operate on the rules themselves. One way to implement this is to specify types in the rule parameters, such as `"divrule(%d, %d)"` as a rule that takes two integer parameters, and `"ruleop(%s)"` as a rule that takes a text string as a parameter (and the text string could potentially contain another rule inside). During the parsing of arguments, instead of just counting the number of arguments as we are doing so far, we would also check their types: If the rule requires a string, we would draw the molecule from the rule multiset; in case the rule requires an integer parameter, it would be drawn from the data multiset. This is still not generic enough, since we could also have data molecules that have a string type. However, it is a starting point for further refinement.

So, turning `HighOrderChem` to truly High-Order is not a problem per se. The real difficulty is to design a rule set that is smart enough such that rules can rewrite other rules and still make sure that the rewritten rules are functional, or that nonfunctional rules (akin to mutation errors) can be detected and destroyed. This would constitute not only an interesting exercise but also an interesting research subject.

Using Compartments

So far we have focused on examples of well-mixed chemistries. As we have seen throughout this book, well-mixed chemistries have many limitations, and many chemistries rely on the notion of space, either in the form of tessellations of slots containing single molecules, or in the form of sets of potentially nested compartments, each compartment being essentially regarded as a well-mixed reactor.

In `pycellchem`, the `Cell` class implements the notion of nested compartments akin to P systems. It can be used to implement chemistries that rely on such nesting, or chemistries that require communication between compartments, such as Fraglets (see section 16.2.1). A `Cell` object contains a cytoplasm that is essentially a well-mixed vessel, and a list of zero or more internal subcells. Subcells may be added to an existing cell using the `add` method. They cannot be directly removed, though. Instead, as in P systems, the membrane of inner compartments can be dissolved through the `dissolve` method, which releases the content of the compartment to the outer compartment: the cytoplasm's multiset is merged with the outer compartment's cytoplasm, and the inner cell's list of subcells is merged with the list of subcells of the outer cell. Similarly, cells can divide through the `divide` method, which splits their contents randomly through the two daughter cells. A multicompartmental version of Gillespie's SSA is also included, such that reactions can be selected for simulation in a recursive way through the cell hierarchy.

At the time of this writing, the only application currently making use of the `Cell` class is `Fraglets.py`, but one can easily imagine an implementation of the ARMS system (section 9.4) or any other variant of P system on top of the base `Cell` class.

Further Resources

In this section we list some resources related to Python and artificial chemistries. Since most of the items point to existing web pages at the time of this writing, we apologize if they change or become unavailable by the time you read these lines.

Python

Introduction to Python:

- Online Tutorial: `http://docs.python.org/tut/`

- Tutorial on Object-Oriented Programming in Python: `https://docs.python.org/2/tutorial/classes.html`

For biologists:

- Python programming course for biologists, by the Pasteur Institute: `http://www.pasteur.fr/formation/infobio/python/`

- *Practical Computing for Biologists*, book by Haddock and Dunn [355]: Unix command line, regular expressions, python programming.

Extensive collections of Python learning resources:

- Beginner's Guide for Non-Programmers: `http://wiki.python.org/moin/BeginnersGuide/NonProgrammers`

- Beginner's Guide for Programmers: `http://wiki.python.org/moin/BeginnersGuide/Programmers`

Implementing binary strings in Python:

- Bitwise operators: `https://wiki.python.org/moin/BitwiseOperators`

- Bit manipulation tutorial: `https://wiki.python.org/moin/BitManipulation`

Python regular expressions:

- Online documentation: `http://docs.python.org/library/re.html`, or `https://docs.python.org/2/library/re.html`

- "Cheatsheet" summary: `http://cloud.github.com/downloads/tartley/python-regex-cheatsheet/cheatsheet.pdf`

Graphics and Mathematics

Graphics support:

- `gnuplot`, a plotting tool: `http://www.gnuplot.info/`

- Python Matplotlib, a plotting package for Python: `http://matplotlib.org/`

- Overview of Graphical User Interface (GUI) support in Python:
  `https://wiki.python.org/moin/GuiProgramming`
  `https://docs.python.org/2/faq/gui.html`

- A 2D physics engine in Python: http://www.pymunk.org/en/latest/

- 3D graphics and animations in Python: http://vpython.org/

Mathematical support:

- Python NumPy package, `http://www.numpy.org/`:
  a numeric package including vectors, matrices, random number generators, and many other functions.

- Python SciPy package, `http://www.scipy.org/`:
  a comprehensive collection of mathematical packages for Python, including NumPy and Matplotlib, among others.

Other AC Software Packages

Some available AC software packages (nonexhaustive list, alphabetically ordered):

- AlChemy lambda-chemistry:
  `http://tuvalu.santafe.edu/~walter/AlChemy/software.html`

- Avida Digital Evolution software: `http://devolab.msu.edu/software/`

- Chemical Casting Model (CCM) examples as Java applets:
  `http://www.kanadas.com/ccm/examples.html`

- Chemical Organization Theory software: `http://www.biosys.uni-jena.de/`
  `Research/Topics/Chemical+Organization+Theory.html`

- DiSCUS, a simulator of bacterial computing in Python:
  `https://code.google.com/p/discus/`

- Ecolab: `http://ecolab.sourceforge.net/`

- Fraglets: `http://www.fraglets.net/`

- MGS: `http://mgs.spatial-computing.org/`

- NK-NKC software: `http://www.cs.unibo.it/fioretti/CODE/NK/index.html`

- P systems software: `http://ppage.psystems.eu/index.php/Software`

- Organic Builder: `http://organicbuilder.sourceforge.net/`

- SimSoup: `http://www.simsoup.info/SimSoup_Download_Page.html`

- StochSim:
  `http://www.pdn.cam.ac.uk/groups/comp-cell/StochSim.html`
  `http://sourceforge.net/projects/stochsim/`

- SwarmChemistry: `http://bingweb.binghamton.edu/~sayama/SwarmChemistry/`

- Tierra: `http://life.ou.edu/tierra/`

- ToyChemistry: `http://www.bioinf.uni-leipzig.de/~gil/ToyChem/`

- Typogenetics in Python: `https://www.bamsoftware.com/hacks/geb/index.html`

Bibliography

[1]   The human brain project - website. https://www.humanbrainproject.eu.

[2]   http://www.tomilab.net/alife/music/, Accessed May 14, 2013.

[3]   IST PACE Final Report. http://www.istpace.org/Web_Final_Report/WP_1_artificial_
      cells_conception/the_los_alamos_bug/index.html, ©2004-2008, Last accessed Sep 29, 2014.

[4]   D. M. Abrams and S. H. Strogatz. Modelling the dynamics of language death. *Nature*, 424, Aug. 2003.

[5]   P. A. Abrams. The evolution of predator-prey interactions: Theory and evidence. *Annual Review of
      Ecology and Systematics*, 31:79–105, 2000.

[6]   K. Adamala and J. W. Szostak. Competition between model protocells driven by an encapsulated
      catalyst. *Nature Chemistry*, 5(6):495–501, 2013.

[7]   K. Adamala and J. W. Szostak. Nonenzymatic template-directed RNA synthesis inside model proto-
      cells. *Science*, 342(6162):1098–1100, 2013.

[8]   A. Adamatzky. Chemistry of beliefs: Experiments with doxastic solutions. *Kybernetes*, 30:1199–1208,
      2001.

[9]   A. Adamatzky. Physarum machines: Encapsulating reaction-diffusion to compute spanning tree.
      Naturwissenschaften, 94(12):975–980, Dec. 2007.

[10] A. Adamatzky. *Physarum Machines: Computers from Slime Mould*. World Scientific Series on Non-
      linear Science Series A, 2010.

[11] A. Adamatzky. Slime mold solves maze in one pass, assisted by gradient of chemo-attractants. *IEEE
      Transactions on NanoBioscience*, 11(2):131–134, June 2012.

[12] A. Adamatzky and B. D. L. Costello. On some limitations of reaction-diffusion chemical computers in
      relation to Voronoi diagram and its inversion. *Physics Letters A*, 309(5-6):397–406, 2003.

[13] A. Adamatzky, B. D. L. Costello, and T. Asai. *Reaction-Diffusion Computers*. Elsevier Science, 2005.

[14] A. Adamatzky, O. Holland, N. Rambidi, and A. Winfield. Wet artificial brains: Towards the chemical
      control of robot motion by reaction-diffusion and excitable media. In D. Floreano, J.-D. Nicoud, and
      F. Mondada, editors, *Proc. Fifth European Conference on Artificial Life*, pages 304–313. Springer, 1999.

[15] A. Adamatzky, J. Holley, P. Dittrich, J. Gorecki, B. D. L. Costello, K.-P. Zauner, and L. Bull. On architec-
      tures of circuits implemented in simulated Belousov-Zhabotinsky droplets. *Biosystems*, 109(1):72–77,
      2012.

[16] C. Adami. *Introduction to Artificial Life*. Springer, 1998.

[17] C. Adami. Ab initio modeling of ecosystems with artificial life. *Natural Resource Modeling*, 15(1):133–
      145, 2002.

[18] C. Adami. Digital genetics: Unravelling the genetic basis of evolution. *Nature Review Genetics*, 7(2):109–118, 2006.

[19] C. Adami and C. T. Brown. Evolutionary learning in the 2D artificial life system "Avida". In R. A. Brooks and P. Maes, editors, *Proc. Artificial Life IV*, pages 377–381. MIT Press, 1994.

[20] C. Adami, D. M. Bryson, C. Ofria, and R. T. Pennock, editors. *Proceedings of Artificial Life 13, the Thirteenth International Conference on the Simulation and Synthesis of Living Systems*. MIT Press, July 2012.

[21] C. Adami, C. Ofria, and J. D. Collier. Evolution of biological complexity. *Proceedings of the National Academy of Sciences*, 97:4463–4468, 2000.

[22] C. Adami and C. O. Wilke. Experiments in digital evolution (introduction to special issue). *Artificial Life*, 10:117–122, 2004.

[23] R. Adar, Y. Benenson, G. Linshiz, A. Rosner, N. Tishby, and E. Shapiro. Stochastic computing with biomolecular automata. *Proceedings of the National Academy of Sciences*, 101(27):9960–9965, 2004.

[24] L. M. Adleman. Molecular computation of solutions to combinatorical problems. *Science*, 266:1021, 1994.

[25] J. Adler and W. Tso. "Decision"-making in bacteria: Chemotactic response of Escherichia coli to conflicting stimuli. *Science*, 184:1292–1294, 1974.

[26] V. Ahl and T. F. H. Allen. *Hierarchy Theory: A Vision, Vocabulary, and Epistemology*. Columbia University Press, 1996.

[27] B. Alberts, D. Bray, K. Hopkin, A. Johnson, J. Lewis, M. Raff, K. Roberts, and P. Walter. *Essential Cell Biology*. Garland Science, Taylor & Francis Group, 3rd edition, 2010.

[28] L. Altenberg. The evolution of evolvability in genetic programming. In K. Kinnear, editor, *Advances in Genetic Programming*, chapter 3, pages 47–74. MIT Press, 1994.

[29] L. Altenberg. Genome growth and the evolution of the genotype-phenotype map. In W. Banzhaf and F. H. Eeckman, editors, *Evolution as a Computational Process*, pages 205–259. Springer, 1995.

[30] M. Amos. *Theoretical and Experimental DNA Computation*. Springer, 2005.

[31] M. Amos. DNA Computing. In R. A. Meyers, editor, *Encyclopedia of Complexity and Systems Science*, volume 4, pages 2089–2104. Springer, 2009.

[32] M. Amos. Bacterial Computing. In R. A. Meyers, editor, *Computational Complexity: Theory, Techniques and Applications*, pages 228–237. Springer, 2012.

[33] Z. An, Q. Pan, G. Yu, and Z. Wang. The spatial distribution of clusters and the formation of mixed languages in bilingual competition. *Physica A: Statistical Mechanics and its Applications*, 391(20):4943–4952, 2012.

[34] T. Andersen, R. Newman, and T. Otter. Shape homeostasis in virtual embryos. *Artificial Life*, 15(2):161–183, 2009.

[35] J. C. Anderson, C. A. Voigt, and A. P. Arkin. Environmental signal integration by a modular AND gate. *Molecular Systems Biology*, 3, Aug. 2007.

[36] O. Andrei and H. Kirchner. A port graph calculus for autonomic computing and invariant verification. *Electronic Notes in Theoretical Computer Science*, 253(4):17–38, 2009.

[37] S. S. Andrews and D. Bray. Stochastic simulation of chemical reactions with spatial resolution and single molecule detail. *Physical Biology*, 1(3):137–151, 2004.

[38] M. Andronescu, V. Bereg, H. H. Hoos, and A. Condon. RNA STRAND: The RNA secondary structure and statistical analysis database. *BMC Bioinformatics*, 9(1):340, 2008.

[39] T. Asai, B. D. L. Costello, and A. Adamatzky. Silicon implementation of a chemical reaction-diffusion processor for computation of Voronoi diagram. *International Journal of Bifurcation and Chaos*, 15(10):3307–3320, 2005.

[40] U. M. Ascher and L. R. Petzold. *Computer Methods for Ordinary Differential Equations and Differential Algebraic Equations*. SIAM, 1998.

[41] J. C. Astor and C. Adami. A developmental model for the evolution of artificial neural networks. *Artificial Life*, 6(3):189–218, 2000.

[42] P. Atkins and J. de Paula. *Physical Chemistry*. Oxford University Press, 2002.

[43] S. Ausländer, D. Ausländer, M. Mlüller, M. Wieland, and M. Fussenegger. Programmable single-cell mammalian biocomputers. *Nature*, 487:123–127, July 2012.

[44] H. Axelsen, R. Glück, and T. Yokoyama. Reversible machine code and its abstract processor architecture. In V. Diekert, M. V. Volkov, and A. Voronkov, editors, *Computer Science: Theory and Applications*, volume 4649 of *Lecture Notes in Computer Science*, pages 56–69. Springer, 2007.

[45] N. Baas. Emergence, hierarchies, and hyperstructures. In C. Langton, editor, *Artificial Life III, Proceedings of the 2nd Workshop on Artificial Life*, pages 515–537. Addison Wesley, 1994.

[46] N. Baas. Hyperstructures as abstract matter. *Advances in Complex Systems (ACS)*, 9:157–182, 2006.

[47] K. Bache and M. Lichman. UCI machine learning repository, 2013.

[48] J. Bada. How life began on earth: A status report. *Earth and Planetary Science Letters*, 226(1):1–15, 2004.

[49] R. J. Bagley and J. D. Farmer. Spontaneous emergence of a metabolism. In C. G. Langton, C. Taylor, J. D. Farmer, and S. Rasmussen, editors, *Artificial Life II*, pages 93–140. Addison-Wesley, 1992.

[50] R. J. Bagley, J. D. Farmer, and W. Fontana. Evolution of a metabolism. In C. G. Langton, C. Taylor, J. D. Farmer, and S. Rasmussen, editors, *Artificial Life II*, pages 141–158. Addison-Wesley, 1992.

[51] P. Bak and K. Sneppen. Punctuated equilibrium and criticality in a simple model of evolution. *Physical Review Letters*, 71:4083–4086, Dec. 1993.

[52] P. Bak, C. Tang, and K. Wiesenfeld. Self-organized criticality. *Physical Review*, A38:364–374, 1988.

[53] T. Ban, K. Tani, H. Nakata, and Y. Okano. Self-propelled droplets for extracting rare-earth metal ions. *Soft Matter*, 10(33):6316–6320, 2014.

[54] T. Ban, T. Yamagami, H. Nakata, and Y. Okano. pH-dependent motion of self-propelled droplets due to Marangoni effect at neutral pH. *Langmuir*, 29(8):2554–2561, 2013.

[55] J.-P. Banâtre, P. Fradet, J. L. Giavitto, and O. Michel. Higher-order chemical programming style. In J. P. Banâtre, P. Pascal Fradet, J. P. Giavitto, and O. Michel, editors, *Proceedings of the Workshop on Unconventional Programming Paradigms (UPP 2004)*, volume 3566 of *Lecture Notes in Computer Science*. Springer, 2005.

[56] J.-P. Banâtre, P. Fradet, and D. Le Métayer. Gamma and the chemical reaction model: Fifteen years after. *Lecture Notes in Computer Science*, 2235:17–44, 2001.

[57] J.-P. Banâtre, P. Fradet, and Y. Radenac. Chemical specification of autonomic systems. In *13th International Conference on Intelligent and Adaptive Systems and Software Engineering*, Nice, France, July 2004.

[58] J.-P. Banâtre and D. Le Métayer. The gamma model and its discipline of programming. *Science of Computer Programming*, 15(1):55–77, Nov. 1990.

[59] S. Bandini, G. Mauri, G. Pavesi, and C. Simone. Computing with a distributed reaction-diffusion model. In *Machines, Computations, and Universality*, volume 3354 of *Lecture Notes in Computer Science*, pages 93–103. Springer, 2005.

[60] S. Bankes. Agent-based modelling: A revolution? *Proceedings of the National Academy of Sciences*, 99, Suppl 3:7199–7200, 2002.

[61] Y. Bansho, N. Ichihashi, Y. Kazuta, T. Matsuura, H. Suzuki, and T. Yomo. Importance of parasite RNA species repression for prolonged translation-coupled RNA self-replication. *Chemistry & Biology*, 19(4):478–487, 2012.

[62] W. Banzhaf. The "molecular" traveling salesman. *Biol. Cybern.*, 64:7–14, 1990.

[63] W. Banzhaf. Self-replicating sequences of binary numbers. *Computers and Mathematics with Applications*, 26:1–8, 1993.

[64] W. Banzhaf. Self-replicating sequences of binary numbers. Foundations I: General. *Biological Cybernetics*, 69(4):269–274, 1993.

[65] W. Banzhaf. Self-replicating sequences of binary numbers. Foundations II: Strings of length N = 4. *Biol. Cybern.*, 69:275–281, 1993.

[66] W. Banzhaf. Self-replicating sequences of binary numbers: The build-up of complexity. *Complex Systems*, 8:215–225, 1994.

[67] W. Banzhaf. Artificial regulatory networks and genetic programming. In R. Riolo and B. Worzel, editors, *Genetic Programming Theory and Practice*, volume 6 of *Genetic Programming Series*, chapter 4, pages 43–61. Kluwer Academic, 2003.

[68] W. Banzhaf. On the dynamics of an artificial regulatory network. In *Advances in Artificial Life, Proceedings of ECAL 2003*, pages 217–227. Springer, 2003.

[69] W. Banzhaf. Artificial chemistries — Towards constructive dynamical systems. *Solid State Phenomena*, 97-98:43–50, 2004.

[70] W. Banzhaf. Self-organizing systems. In R. Meyer, editor, *Encyclopedia of Complexity and Systems Science*, pages 8040–8050. Springer, 2010.

[71] W. Banzhaf. Genetic programming and emergence. *Genetic Programming and Evolvable Machines*, 15:63–73, 2014.

[72] W. Banzhaf, P. Dittrich, and H. Rauhe. Emergent computation by catalytic reactions. *Nanotechnology*, 7:307–314, 1996.

[73] W. Banzhaf and P. D. Kuo. Network motifs in natural and artificial transcriptional regulatory networks. *Journal of Biological Physics and Chemistry*, 4:85–92, 2004.

[74] W. Banzhaf and C. Lasarczyk. Genetic programming of an algorithmic chemistry. In U.-M. O'Reilly, T. Yu, R. Riolo, and B. Worzel, editors, *Genetic Programming Theory and Practice II*, volume 8 of *Genetic Programming*, chapter 11, pages 175–190. Springer, 2005.

[75] W. Banzhaf and C. Lasarczyk. A new programming paradigm inspired by artificial chemistries. In J.-P. Banâtre, P. Fradet, J.-L. Giavitto, and O. Michel, editors, *Unconventional Programming Paradigms*, volume 3566 of *Lecture Notes in Computer Science*, pages 73–83. Springer, 2005.

[76] W. Banzhaf, P. Nordin, R. Keller, and F. Francone. *Genetic Programming: An Introduction*. Morgan Kaufmann, 1998.

[77]   Y. Bar-Yam. A mathematical theory of strong emergence using multiscale variety. *Complexity*, 9:15–24, 2004.

[78]   Y. Bar-Yam. Multiscale variety in complex systems. *Complexity*, 9:37–45, 2004.

[79]   J. Barbour. *The End of Time*. Oxford University Press, 1999.

[80]   H. P. Barendregt. *The Lambda Calculus, its Syntax and Semantics*. North-Holland, 1984.

[81]   M. Barrio, A. Leier, and T. T. Marquez-Lago. Reduction of chemical reaction networks through delay distributions. *The Journal of chemical physics*, 138:104114, 2013.

[82]   S. Basu, Y. Gerchman, C. H. Collins, F. H. Arnold, and R. Weiss. A synthetic multicellular system for programmed pattern formation. *Nature*, 434(7037):1130–1134, Apr. 2005.

[83]   A. L. Bauer, C. A. Beauchemin, and A. S. Perelson. Agent-based modeling of host-pathogen systems: The successes and challenges. *Information Sciences*, 179(10):1379–1389, 2009.

[84]   V. Y. Bazhenov, M. S. Soskin, V. B. Taranenko, and M. V. Vasnetsov. Biopolymers for real-time optical processing. In H. H. Arsenault, editor, *Optical Processing and Computing*, pages 103–44. Academic Press, 1989.

[85]   D. Beaver. A universal molecular computer. Technical report CSE-95-001, Penn State University, 1995.

[86]   A. L. Beberg, D. L. Ensign, G. Jayachandran, S. Khaliq, and V. S. Pande. Folding@home: Lessons from eight years of volunteer distributed computing. In *International Parallel and Distributed Processing Symposium*, pages 1–8. IEEE Computer Society, 2009.

[87]   M. Bedau, G. Church, S. Rasmussen, A. Caplan, S. Benner, M. Fussenegger, J. Collins, and D. Deamer. Life after the synthetic cell. *Nature*, 465(7297):422–424, May 2010.

[88]   M. Bedau and E. Parke, editors. *The Ethics of Protocells: Moral and Social Implications of Creating Life in the Laboratory*. MIT Press, 2009.

[89]   M. A. Bedau. Downward causation and the autonomy of weak emergence. *Principia: An International Journal of Epistemology*, 6:5–50, 2002.

[90]   M. A. Bedau. The evolution of complexity. In A. Barberousse, M. Morange, and T. Pradeu, editors, *Mapping the Future of Biology*, volume 266 of *Boston Studies in the Philosophy of Science*, pages 111–130. Springer, 2009.

[91]   M. A. Bedau and C. T. Brown. Visualizing evolutionary activity of genotypes. *Artificial Life*, 5(1):17–35, 1999.

[92]   M. A. Bedau, A. Buchanan, G. Gazzola, M. M. Hanczyc, T. Maeke, J. S. McCaskill, I. Poli, and N. H. Packard. Evolutionary design of a DDPD model of ligation. In E.-G. Talbi, P. Liardet, P. Collet, E. Lutton, and M. Schoenauer, editors, *Artificial Evolution*, volume 3871 of *Lecture Notes in Computer Science*, pages 201–212. Springer, 2006.

[93]   M. A. Bedau, J. S. McCaskill, N. H. Packard, S. Rasmussen, C. Adami, D. G. Green, T. Ikegami, K. Kaneko, and T. S. Ray. Open problems in artificial life. *Artificial Life*, 6(4):363–376, 2000.

[94]   M. A. Bedau and N. H. Packard. Measurement of evolutionary activity, teleology, and life. In C. G. Langton, C. Taylor, J. D. Farmer, and S. Rasmussen, editors, *Artificial Life II, SFI Studies in the Sciences of Complexity, vol. X*. Addison-Wesley, 1991.

[95]   M. A. Bedau and N. H. Packard. Evolution of evolvability via adaptation of mutation rates. *Biosystems*, 69(2-3):143–162, 2003.

[96] M. A. Bedau and E. C. Parke. Social and ethical issues concerning protocells. In Rasmussen et al. [691], chapter 28, pages 641–653.

[97] M. A. Bedau, E. Snyder, and N. H. Packard. A classification of long-term evolutionary dynamics. In *Proceedings of the Sixth International Conference on Artificial Life*, pages 228–237. MIT Press, 1998.

[98] Y. Benenson. Biocomputers: From test tubes to live cells. *Molecular BioSystems*, 5(7):675–685, 2009.

[99] Y. Benenson. Biomolecular computing systems: Principles, progress and potential. *Nature Reviews Genetics*, 13:455–468, July 2012.

[100] Y. Benenson, R. Adar, T. Paz-Elizur, Z. Livneh, and E. Shapiro. DNA molecule provides a computing machine with both data and fuel. *Proceedings of the National Academy of Sciences*, 100(5):2191–2196, 2003.

[101] Y. Benenson, B. Gil, U. Ben-Dor, R. Adar, and E. Shapiro. An autonomous molecular computer for logical control of gene expression. *Nature*, 2004.

[102] G. Benkö, C. Flamm, and P. F. Stadler. Generic properties of chemical networks: Artificial chemistry based on graph rewriting. In W. Banzhaf, J. Ziegler, T. Christaller, P. Dittrich, and J. T. Kim, editors, *Advances in Artificial Life*, volume 2801 of *Lecture Notes in Computer Science*, pages 10–19. Springer, 2003.

[103] G. Benkö, C. Flamm, and P. F. Stadler. A graph-based toy model of chemistry. *J Chem Inf Comput Sci*, 43(4):1085–93, 2003.

[104] G. Benkö, C. Flamm, and P. F. Stadler. Explicit collision simulation of chemical reactions in a graph based artificial chemistry. In *Advances in Artificial Life: Proceedings of the 8th European Conference on Artificial Life (ECAL)*, pages 725–733. Springer, 2005.

[105] G. Benkö, C. Flamm, and P. F. Stadler. The ToyChem package: A computational toolkit implementing a realistic artificial chemistry model. Technical Report, available electronically from http://www.bioinf.uni-leipzig.de/Publications/PREPRINTS/05-002.pdf, 2005.

[106] S. A. Benner and A. M. Sismour. Synthetic biology. *Nature Reviews Genetics*, 6:533–543, 2005.

[107] C. H. Bennett. Logical reversibility of computation. *IBM Journal of Research and Development*, 17(6):525–532, Nov. 1973.

[108] C. H. Bennett. The thermodynamics of computation — A review. *International Journal of Theoretical Physics*, 21(12):905–940, Dec. 1982.

[109] C. H. Bennett and R. Landauer. The fundamental physical limits of computation. *Scientific American*, 253(1):48–56, 1985.

[110] B. Berger and T. Leighton. Protein folding in the hydrophobic-hydrophilic (HP) model is NP-complete. *Journal of Computational Biology*, 5(1):27–40, 1998.

[111] E. R. Berlekamp, J. H. Conway, and R. K. Guy. What is life? In *Winning Ways for your Mathematical Plays*, volume 2, chapter 25, pages 817–850. Academic Press, 1982.

[112] F. Bernardini, M. Gheorghe, and N. Krasnogor. Quorum Sensing P Systems. *Theoretical Computer Science*, 371:20–33, Feb. 2007.

[113] G. Berry and G. Boudol. The chemical abstract machine. *Theoretical Computer Science*, 96(1):217–248, 1992.

[114] H. Bersini and F. J. Varela. Hints for adaptive problem solving gleaned from immune networks. In *Parallel Problem Solving from Nature, First Workshop PPSW 1*, Dortmund, Germany, Oct. 1990.

[115] L. V. Bertalanffy. *General System Theory: Foundations, Development, Applications.* George Braziller, 1976.

[116] K. Beuls and L. Steels. Agent-based models of strategies for the emergence and evolution of grammatical agreement. *PLoS ONE,* 8(3):e58960, 2013.

[117] E. Bigan, J.-M. Steyaert, and S. Douady. Properties of random complex chemical reaction networks and their relevance to biological toy models. In *Proceedings of JournÃĺes Ouvertes Biologie Informatique MathÃĺmatiques (JOBIM), Toulouse, July 1-4, 2013,* 2013.

[118] R. R. Birge. Protein-based optical computing and memories. *Computer,* 25:56–67, 1992.

[119] R. Bishop. Fluid convection, constraint and causation. *Interface Focus,* 2:4–12, 2012.

[120] R. Bishop and H. Atmanspacher. Contextual emergence in the description of properties. *Foundations of Physics,* 36:1753–1777, 2006.

[121] J. C. Blain and J. W. Szostak. Progress toward synthetic cells. *Annual Review of Biochemistry,* 83(1), 2014.

[122] C. Blilie. *The Promise and Limits of Computer Modeling.* World Scientific, 2007.

[123] D. Blitz. *Emergent Evolution.* Kluwer Academic Publishers, 1992.

[124] A. L. Blumenthal. A Wundt Primer: The operating characteristics of consciousness. In R. W. Rieber and D. K. Robinson, editors, *Wilhelm Wundt in History: The Making of a Scientific Psychology,* Path in Psychology, pages 121–144. Kluwer Academic, 2001.

[125] M. C. Boerlijst and P. Hogeweg. Spiral wave structures in prebiotic evolution: Hypercycle stable against parasites. *Physica D,* 48:17–28, 1991.

[126] H. Bolouri. *Computational Modeling of Gene Regulatory Networks — A Primer.* Imperial College Press, 2008.

[127] E. Bonabeau. Agent-based modelling: Methods and techniques for simluating human systems. *Proceedings of the National Academy of Sciences,* 99, Suppl 3:7280–7287, 2002.

[128] D. Boneh, C. Dunworth, and R. J. Lipton. Breaking DES using a molecular computer. Technical Report CS-TR-489-95, Princton University, 1995.

[129] D. Boneh, C. Dunworth, R. J. Lipton, and J. Sgall. On the computational power of DNA. *Discret Appl. Math.,* 71(1-3):79–94, 1996.

[130] J. Bongard. Evolving modular genetic regulatory networks. In *Proceedings of the 2002Congress on Evolutionary Computation (CEC),* volume 2, pages 1872–1877, 2002.

[131] R. Book and F. Otto. *String-Rewriting Systems.* Springer, 1993.

[132] D. Bray. *Wetware: A Computer in Every Living Cell.* Yale University Press, 2009.

[133] K. Brenner, L. You, and F. H. Arnold. Engineering microbial consortia: A new frontier in synthetic biology. *Trends in Biotechnology,* 26(9):483–489, 2008.

[134] J. J. Brewster and M. Conrad. EVOLVE IV: A metabolically-based artificial ecosystem model. In V. Porto, N. Saravannan, D. Waagen, and A. Eiben, editors, *Evolutionary Programming VII,* pages 473–82. Springer, 1998.

[135] J. J. Brewster and M. Conrad. Computer experiments on the development of niche specialization in an artificial ecosystem. In *Proc. Congress on Evolutionary Computation (CEC),* pages 439–44. IEEE, 1999.

[136]   J. Breyer, J. Ackermann, and J. S. McCaskill. Evolving reaction-diffusion ecosystems with self-assembling structure in thin films. *Artificial Life*, 4(1):25–40, 1999.

[137]   C. Briones, M. Stich, and S. C. Manrubia. The dawn of the RNA world: Toward functional complexity through ligation of random RNA oligomers. *RNA*, 15:743–749, 2009.

[138]   A. B. Brown. *A Recovery-oriented Approach to Dependable Services: Repairing Past Errors with System-wide Undo*. PhD thesis, University of California, Berkeley, 2003.

[139]   D. M. Bryson and C. Ofria. Understanding evolutionary potential in virtual CPU instruction set architectures. *PLoS ONE*, 8:e83242, 2013.

[140]   I. Budin and J. W. Szostak. Expanding roles for diverse physical phenomena during the origin of life. *Annu Rev Biophys*, 9(39):245–263, June 2010.

[141]   H. J. Buisman, H. M. M. ten Eikelder, P. A. J. Hilbers, and A. M. L. Liekens. Computing algebraic functions with biochemical reaction networks. *Artificial Life*, 15(1):5–19, 2009.

[142]   S. Bullock and M. A. Bedau. Exploring the dynamics of adaptation with evolutionary activity plots. *Artificial Life*, 12:1–5, 2006.

[143]   S. Bullock, J. Noble, R. Watson, and M. A. Bedau, editors. *Artificial Life XI: Proceedings of the Eleventh International Conference on the Simulation and Synthesis of Living Systems*. MIT Press, Aug. 2008.

[144]   M. Bunge. *Method, Model and Matter*. Reidel, 1973.

[145]   M. Bunge. *Emergence and Convergence: Qualitative Novelty and the Unity of Knowledge*. University of Toronto Press, 2003.

[146]   K. Burrage. *Parallel and Sequential Methods for Ordinary Differential Equations*. Oxford University Press, 1995.

[147]   J. Busch. *RESAC: Eine resolutionsbasierte Kuenstliche Chemie und deren Anwendungen*. PhD thesis, Department of Computer Science, Technical University of Dortmund, 2004.

[148]   J. Busch and W. Banzhaf. How to program artificial chemistries. In *Advances in Artificial Life, Proceedings of the 7th European Conference on Artificial Life (ECAL 2003)*, pages 20–30. Springer, 2003.

[149]   J. C. Butcher. *Numerical Methods for Ordinary Differential Equations*. John Wiley and Sons, 2003.

[150]   C. Calude and G. Păun. *Computing with Cells and Atoms*. Taylor and Francis, 2000.

[151]   M. Campbell-Kelly, W. Aspray, N. Ensmenger, and J. R. Yost. *Computer: A History of the Information Machine*. Westview Press, 3rd edition, 2013.

[152]   G. Canright, A. Deutsch, and T. Urnes. Chemotaxis-inspired load balancing. In *Proc. European Conference on Complex Systems*, Nov. 2005.

[153]   Y. Cao, D. T. Gillespie, and L. R. Petzold. Avoiding negative populations in explicit poisson tau-leaping. *Journal of Chemical Physics*, 123(5):054104–1–8, 2005.

[154]   Y. Cao, D. T. Gillespie, and L. R. Petzold. The slow-scale stochastic simulation algorithm. *Journal of Chemical Physics*, 122(1):014116–1–18, 2005.

[155]   Y. Cao, D. T. Gillespie, and L. R. Petzold. Efficient step size selection for the tau-leaping simulation method. *Journal of Chemical Physics*, 124(4):044109–1–11, 2006.

[156]   P. Capdepuy and M. Kaltenbach. Autonomous systems: Computational autopoiesis through artificial chemistries. In *Workshop on Artificial Chemistry and its Application, European Conference on Artificial Life*, 2005.

[157] L. Cardelli. Brane calculi. In V. Danos and V. Schachter, editors, *Computational Methods in Systems Biology*, volume 3082 of *Lecture Notes in Computer Science*, pages 257–278. Springer, 2005.

[158] L. Cardelli. Artificial biochemistry. In A. Condon, D. Harel, J. N. Kok, A. Salomaa, and E. Winfree, editors, *Algorithmic Bioprocesses*, Natural Computing Series, pages 429–462. Springer, 2009.

[159] T. Carletti and D. Fanelli. From chemical reactions to evolution: Emergence of species. *EPL (Europhysics Letters)*, 77(1):18005, 2007.

[160] C. D. Carothers, K. S. Perumalla, and R. M. Fujimoto. Efficient optimistic parallel simulations using reverse computation. *ACM Trans. Model. Comput. Simul.*, 9(3):224–253, July 1999.

[161] S. B. Carroll. Chance and necessity: The evolution of morphological complexity and diversity. *Nature*, 409:1102–1109, Feb. 2001.

[162] F. Caschera, S. Rasmussen, and M. M. Hanczyc. An oil droplet division-fusion cycle. *ChemPlusChem*, 78(1):52–54, Jan. 2013.

[163] F. Caschera, T. Sunami, T. Matsuura, H. Suzuki, M. M. Hanczyc, and T. Yomo. Programmed vesicle fusion triggers gene expression. *Langmuir*, 27(21):13082–13090, 2011.

[164] J. Castellanos, G. Păun, and A. Rodriguez-Paton. Computing with membranes: P systems with worm-objects. In *Proceedings of the Seventh International Symposium on String Processing and Information Retrieval (SPIRE)*, pages 65–74, Sept. 2000.

[165] X. Castelló, L. Loureiro-Porto, V. Eguíluz, and M. San Miguel. The fate of bilingualism in a model of language competition. In S. Takahashi, D. Sallach, and J. Rouchier, editors, *Advancing Social Simulation: The First World Congress*, pages 83–94. Springer, 2007.

[166] X. Castelló, F. Vazquez, V. Eguíluz, L. Loureiro-Porto, M. San Miguel, L. Chapel, and G. Deffuant. Viability and resilience in the dynamics of language competition. In G. Deffuant and N. Gilbert, editors, *Viability and Resilience of Complex Systems, Understanding Complex Systems*, pages 39–73. Springer, 2011.

[167] T. Cech, A. Zaug, P. J. Grabowski, et al. In vitro splicing of the ribosomal RNA precursor of Tetrahymena: involvement of a guanosine nucleotide in the excision of the intervening sequence. *Cell*, 27(3 Pt 2):487–496, 1981.

[168] F. Centler, C. Kaleta, P. S. di Fenizio, and P. Dittrich. Computing chemical organizations in biological networks. *Bioinformatics*, Dec 2008.

[169] H. Chandran, S. Garg, N. Gopalkrishnan, and J. H. Reif. Biomolecular computing systems. In Katz [444], chapter 11, pages 199–223.

[170] A. D. Channon. Passing the ALife test: Activity statistics classify evolution in Geb as unbounded. In J. Kelemen and P. Sosík, editors, *Advances in Artificial Life*, volume 2159 of *Lecture Notes in Computer Science*, pages 417–426. Springer, 2001.

[171] A. D. Channon. Unbounded evolutionary dynamics in a system of agents that actively process and transform their environment. *Genetic Programming and Evolvable Machines*, 7(3):253–281, 2006.

[172] A. Chavoya and Y. Duthen. A cell pattern generation model based on an extended artificial regulatory network. *Biosystems*, 94(1-2):95–101, 2008.

[173] I. A. Chen, R. W. Roberts, and J. W. Szostak. The emergence of competition between model protocells. *Science*, 305(5689):1474–1476, 2004.

[174] S.-K. Chen, W. K. Fuchs, and J.-Y. Chung. Reversible debugging using program instrumentation. *IEEE Transactions on Software Engineering*, 27(8):715–727, Aug. 2001.

[175] X. Chen and A. D. Ellington. Shaping up nucleic acid computation. *Current Opinion in Biotechnology*, 21(4):392–400, 2010.

[176] C. Chiarabelli, P. Stano, F. Anella, P. Carrara, and P. L. Luisi. Approaches to chemical synthetic biology. *FEBS Letters*, 586(15):2138–2145, 2012.

[177] C. Chiarabelli, P. Stano, and P. L. Luisi. Chemical approaches to synthetic biology. *Current Opinion in Biotechnology*, 20(4):492–497, 2009.

[178] D. Chowdhury and D. Stauffer. Evolutionary ecology *in silico*: Does mathematical modelling help in understanding 'generic' trends? *Journal of Biosciences*, 30(2):277–287, 2005.

[179] K. Christensen, S. A. D. Collobiano, M. Hall, and H. J. Jensen. Tangled nature: A model of evolutionary ecology. *Journal of Theoretical Biology*, 216(1):73–84, 2002.

[180] D. Chu and W. Ho. A category theoretical argument against the possibility of artificial life: Robert Rosen's central proof revisited. *Artificial Life*, 12(1):117–134, 2006.

[181] D. Chu and A. Zomaya. Parallel ant colony optimization for 3D protein structure prediction using the HP lattice model. In N. Nedjah, L. M. Mourelle, and E. Alba, editors, *Parallel Evolutionary Computations*, volume 22 of *Studies in Computational Intelligence*, pages 177–198. Springer, 2006.

[182] T. Clark. *A Handbook of Computational Chemistry: A Practical Guide to Chemical Structure and Energy Calculations*. Wiley, 1985.

[183] A. Clauset, C. Moore, and M. Newman. Hierarchical structure and the prediction of missing links in networks. *Nature*, 453:98–101, 2008.

[184] H. J. Cleaves II. Prebiotic chemistry, the primordial replicator, and modern protocells. In Rasmussen et al. [691], chapter 26, pages 583–613.

[185] C. E. Cleland and C. F. Chyba. Defining 'Life'. *Origins of Life and Evolution of Biospheres*, 32(4):387–393, 2002.

[186] J. Clune, J.-B. Mouret, and H. Lipson. The evolutionary origins of modularity. *Proc. R. Soc. B*, 280(1755), Mar. 2013.

[187] A. Coc, S. Goriely, Y. Xu, M. Saimpert, and E. Vangioni. Standard big bang nucleosynthesis up to CNO with an improved extended nuclear network. *The Astrophysical Journal*, 744:158, 2012.

[188] I. R. Cohen. *Tending Adam's Garden: Evolving the Cognitive Immune Self*. Elsevier Academic Press, 2000.

[189] P. Collet. Why GPGPUs for Evolutionary Computation? In S. Tsutsui and P. Collet, editors, *Massively Parallel Evolutionary Computation on GPGPUs*, Natural Computing Series, pages 3–14. Springer, 2013.

[190] M. B. Comarmond, R. Giege, J. C. Thierry, D. Moras, and J. Fischer. Three-dimensional structure of yeast T-RNA-ASP. I. structure determination. *Acta Crystallogr. Sect. B*, 42:272–280, 1986.

[191] M. Conrad. *Computer Experiments on the Evolution of Co-adaptation in a Primitive Ecosystem*. PhD thesis, Stanford University, 1969.

[192] M. Conrad. Information processing in molecular systems. *Currents in Modern Biology*, 5:1–14, 1972.

[193] M. Conrad. On design principles for a molecular computer. *Communications of the ACM*, 28(5):464–480, May 1985.

[194] M. Conrad. Molecular computing: The lock-key paradigm. *Computer*, 25:11–22, Nov. 1992.

[195] M. Conrad and M. M. Pattee. Evolution experiments with an artificial ecosystem. *Journal of Theoretical Biology*, 28:293–409, 1970.

[196] M. Conrad and K.-P. Zauner. Conformation-based computing: A rationale and a recipe. In Sienko et al. [776], chapter 1, pages 1–31.

[197] S. Cooper, F. Khatib, A. Treuille, J. Barbero, J. Lee, M. Beenen, A. Leaver-Fay, D. Baker, Z. Popovic, and F. players. Predicting protein structures with a multiplayer online game. *Nature*, 466(7307):756–760, Aug. 2010.

[198] T. F. Cooper and C. Ofria. Evolution of stable ecosystems in populations of digital organisms. In R. K. Standish, M. A. Bedau, and H. A. Abbass, editors, *Proceedings of Artificial Life VIII, the 8th International Conference on the Simulation and Synthesis of Living Systems*, pages 227–232. MIT Press, 2002.

[199] B. J. Copeland, editor. *The Essential Turing*. Oxford University Press, 2004.

[200] V. Cornish, D. Mendel, and P. G. Schultz. Probing protein structure and function with an expanded genetic code. *Angewandte Chemie, International Edition*, 34(6):621–633, Mar. 1995.

[201] R. Cottam, W. Ranson, and R. Vounckx. Re-mapping Robert Rosen's (M,R)-systems. *Chemical Biodiversity*, 4:2352– 2368, 2007.

[202] L. Cronin, N. Krasnogor, B. G. Davis, C. Alexander, N. Robertson, J. H. G. Steinke, S. L. M. Schroeder, A. N. Khlobystov, G. Cooper, P. M. Gardner, P. Siepmann, B. J. Whitaker, and D. Marsh. The imitation game: A computational chemical approach to recognizing life. *Nature Biotechnology*, 24(10):1203–1206, Oct. 2006.

[203] M. Cross and P. Hohenberg. Pattern formation outside of equilibrium. *Reviews of Modern Physics*, 65:851–1112, 1993.

[204] T. Csendes. A simulation study on the chemoton. *Kybernetes*, 13(2):79–85, 1984.

[205] G. Cybenko. Dynamic load balancing for distributed memory multiprocessors. *Journal of Parallel and Distributed Computing*, 7:279–301, 1989.

[206] K. Dale and P. Husbands. The evolution of reaction-diffusion controllers for minimally cognitive agents. *Artificial Life*, 16(1):1–19, 2010.

[207] T. Danino, O. Mondragon-Palomino, L. Tsimring, and J. Hasty. A synchronized quorum of genetic clocks. *Nature*, 463(7279):326–330, Jan. 2010.

[208] V. Danos and C. Laneve. Formal molecular biology. *Theoretical Computer Science*, 325(1):69–110, 2004.

[209] F. Dardel and F. Képès. *Bioinformatics: Genomics and post-genomics*. John Wiley & Sons, 2006.

[210] R. Das and D. Baker. Macromolecular modeling with Rosetta. *Annual review of biochemistry*, 77:363–382, 2008.

[211] D. Dasgupta, editor. *Artificial Immune Systems and Their Applications*. Springer, 1999.

[212] J. De Beule, E. Hovig, and M. Benson. Introducing dynamics into the field of biosemiotics. *Biosemiotics*, 4(1):5–24, Apr. 2011.

[213] F. K. de Boer and P. Hogeweg. Co-evolution and ecosystem based problem solving. *Ecological Informatics*, 9:47–58, 2012.

[214] L. N. de Castro and J. Timmis. *Artificial Immune Systems: A New Computational Intelligence Approach*. Springer, 2002.

[215] L. N. De Castro and F. J. Von Zuben. Artificial immune systems part I: Basic theory and applications. Technical Report TR-DCA 01/99, Universidade Estadual de Campinas, Brazil, 1999.

[216] C. de Duve. The other revolution in the life sciences. *Science*, 339:1148, Mar. 2013.

[217] J. de Haan. How emergence arises. *Ecological Complexity*, 3:293–201, 2006.

[218] A. De Vos. *Reversible Computing: Fundamentals, Quantum Computing, and Applications*. John Wiley & Sons, Oct. 2010.

[219] T. De Wolf and T. Holvoet. Emergence versus self-organisation: Different concepts but promising when combined. In S. A. Brueckner, G. Di Marzo Serugendo, A. Karageorgos, and R. Nagpal, editors, *Engineering Self-Organising Systems*, volume 3464 of *Lecture Notes in Computer Science*, pages 1–15. Springer, 2005.

[220] A. Deckard and H. M. Sauro. Preliminary studies on the in silico evolution of biochemical networks. *ChemBioChem*, 5(10):1423–1431, 2004.

[221] J. Decraene and B. McMullin. The evolution of complexity in self-maintaining cellular information processing networks. *Advances in Complex Systems*, 14(01):55–75, 2011.

[222] J. Decraene, G. G. Mitchell, and B. McMullin. Evolving artificial cell signaling networks using molecular classifier systems. In *Proceedings of Bio-Inspired Models of Network, Information and Computing Systems (BIONETICS)*, Cavalese, Italy, Dec. 2006. IEEE.

[223] J. Decraene, G. G. Mitchell, and B. McMullin. Exploring evolutionary stability in a concurrent artificial chemistry. In *5th European Conference on Complex Systems (ECCS)*, Jerusalem, Israel, Sept. 2008.

[224] J. Decraene, G. G. Mitchell, and B. McMullin. Unexpected evolutionary dynamics in a string based artificial chemistry. In Bullock et al. [143], pages 158–165.

[225] J. Decraene, G. G. Mitchell, and B. McMullin. Crosstalk and the cooperation of collectively autocatalytic reaction networks. In *IEEE Congress on Evolutionary Computation (CEC)*, pages 2249–2256. IEEE, 2009.

[226] J. Demongeot, E. Golès, and M. Tchuente. *Dynamical systems and cellular automata*. Academic Press, 1985.

[227] I. Derényi and I. Lagzi. Fatty acid droplet self-division driven by a chemical reaction. *Physical Chemistry Chemical Physics*, 16:4639–4641, 2014.

[228] N. Dershowitz and J. Jouannaud. *Handbook of Theoretical Computer Science, Volume B, Formal Methods and Semantics*, chapter Rewrite Systems, pages 243–320. North-Holland, 1990.

[229] A. K. Dewdney. In the game called Core War hostile programs engage in a battle of bits. *Scientific American*, 250:15–19, 1984.

[230] E. Di Paolo, J. Noble, and S. Bullock. Simulation models as opaque thought experiments. In M. Bedau and et al, editors, *Artificial Life VII: Proceedings of the 7th International Conference on Artificial Life*. MIT Press, 2000.

[231] P. Dittrich. Chemical computing. In J.-P. Banâtre, P. Fradet, J.-L. Giavitto, and O. Michel, editors, *Unconventional Programming Paradigms*, volume 3566 of *Lecture Notes in Computer Science*, pages 19–32. Springer, 2005.

[232] P. Dittrich and W. Banzhaf. A topological structure based on hashing — Emergence of a "spatial" organization. In *Fourth European Conference on Artificial Life (ECAL'97)*, Brighton, UK, 1997.

[233] P. Dittrich and W. Banzhaf. Self-evolution in a constructive binary string system. *Artificial Life*, 4(2):203–220, 1998.

[234] P. Dittrich, T. Kron, and W. Banzhaf. On the scalability of social order: Modeling the problem of double and multi contingency following luhmann. *Journal of Artificial Societies and Social Simulation*, 6(1), 2003.

[235] P. Dittrich and L. Winter. Chemical organizations in a toy model of the political system. *Advances in Complex Systems*, 11(04):609–627, 2008.

[236] P. Dittrich, J. Ziegler, and W. Banzhaf. Artificial chemistries — A review. *Artificial Life*, 7(3):225–275, 2001.

[237] C. B. Do, D. A. Woods, and S. Batzoglou. CONTRAfold: RNA secondary structure prediction without physics-based models. *Bioinformatics*, 22(14):e90–e98, 2006.

[238] U. Dobramysl and U. C. Täuber. Spatial variability enhances species fitness in stochastic predator-prey interactions. *Physical Review Letteers*, 101:258102, Dec. 2008.

[239] M. Dorigo and T. Stützle. *Ant Colony Optimization*. MIT Press, 2004.

[240] A. Dorin and K. Korb. Building virtual ecosystems from artificial chemistry. In F. Almeida e Costa, L. Rocha, E. Costa, I. Harvey, and A. Coutinho, editors, *Advances in Artificial Life*, volume 4648 of *Lecture Notes in Computer Science*, pages 103–112. Springer, 2007.

[241] R. M. Downing. On population size and neutrality: Facilitating the evolution of evolvability. In Ebner et al., editors, *Proceedings of the 10th European Conference on Genetic Programming (EuroGP)*, volume 4445 of *LNCS*, pages 181–192, Apr. 2007.

[242] A. D. Doyle, R. J. Petrie, M. L. Kutys, and K. M. Yamada. Dimensions in cell migration. *Current Opinions in Cell Biology*, 25:642–649, 2013.

[243] B. Drossel. Random boolean networks. In H. G. Schuster, editor, *Reviews of Nonlinear Dynamics and Complexity*, volume 1. Wiley, 2008.

[244] M. Durvy and P. Thiran. Reaction-diffusion based transmission patterns for ad hoc networks. In *Proc. of IEEE INFOCOM*, pages 2195–2205. IEEE, 2005.

[245] A. Eddington. *The Nature of the Physical World, Gifford Lectures*. Cambridge University Press, 1928.

[246] L. Edwards and Y. Peng. Computational models for the formation of protocell structures. *Artificial Life*, 4(1):61–77, 1998.

[247] R. Ehricht, T. Ellinger, and J. S. McCaskill. Cooperative amplification of templates by cross-hybridization (CATCH). *European Journal of Biochemistry*, 243(1/2):358–364, 1997.

[248] A. E. Eiben and J. Smith. *Introduction to Evolutionary Computing*. Springer, 2010.

[249] M. Eigen and P. Schuster. The hypercycle: A principle of natural self-organization, Part A: Emergence of the hypercycle. *Naturwissenschaften*, 64(11):541–565, Nov. 1977.

[250] M. Eigen and P. Schuster. The hypercycle: A principle of natural self-organization, Part B: The abstract hypercycle. *Naturwissenschaften*, 65(1):7–41, Jan. 1978.

[251] M. Eigen and P. Schuster. The hypercycle: A principle of natural self-organization, Part C: The realistic hypercycle. *Naturwissenschaften*, 65(7):341–369, July 1978.

[252] M. Eigen and P. Schuster. *The Hypercycle: A Principle of Natural Self-Organization*. Springer, 1979.

[253] J. Elf and M. Ehrenberg. Spontaneous separation of bi-stable biochemical systems into spatial domains of opposite phases, supplementary material: Next subvolume method. *Proceedings of IEE Systems Biology*, 1(2):230–236, Dec. 2004.

[254] G. Ellis. Physics and the real world. *Physics Today*, 58:49, 2005.

[255] G. Ellis. Top-down causation and emergence: Some comments on mechanisms. *Interface Focus*, 2:126–140, 2012.

[256] M. B. Elowitz and S. Leibler. A synthetic oscillatory network of transcriptional regulators. *Nature*, 403:335–338, Jan. 2000.

[257] R. Elsässer and B. Monien. Diffusion load balancing in static and dynamic networks. In *Proc. International Workshop on Ambient Intelligence Computing*, pages 49–62, Dec. 2003.

[258] I. R. Epstein and J. A. Pojman. *An Introduction to Nonlinear Chemical Dynamics: Oscillations, Waves, Patterns, and Chaos*. Oxford University Press, 1998.

[259] A. Erskine and J. M. Herrmann. Cell division behaviour in a heterogeneous swarm environment. In *Advances in Artificial Life, Proceedings ECAL 2013*, pages 35–42. MIT Press, 2013.

[260] J. D. Farmer, S. A. Kauffman, and N. H. Packard. Autocatalytic replication of polymers. *Physica D*, 2(1-3):50–67, Oct.-Nov. 1986.

[261] J. D. Farmer, S. A. Kauffman, N. H. Packard, and A. S. Perelson. Adaptive dynamic networks as models for the immune system and autocatalytic sets. *Annals of the New York Academy of Sciences*, 504(1):118–131, 1987.

[262] J. D. Farmer and N. H. Packard. The immune system, adaptation, and machine learning. *Physica D*, pages 187–204, 1986.

[263] A. Faulconbridge, S. Stepney, J. F. Miller, and L. S. Caves. RBN-World: The hunt for a rich AChem. In Fellermann et al. [268], pages 261–268.

[264] A. Faulconbridge, S. Stepney, J. F. Miller, and L. S. Caves. RBN-World: A sub-symbolic artificial chemistry. In G. Kampis, I. Karsai, and E. Szathmáry, editors, *Advances in Artificial Life. Darwin Meets von Neumann*, volume 5777 of *Lecture Notes in Computer Science*, pages 377–384. Springer, 2011.

[265] D. Federici and K. Downing. Evolution and development of a multicellular organism: Scalability, resilience, and neutral complexification. *Artificial Life*, 12(3):381–409, 2006.

[266] L. Felix, F. Rosselló, and G. Valiente. Optimal artificial chemistries and metabolic pathways. In *Sixth Mexican International Conference on Computer Science*, pages 298–305. IEEE, Sept. 2005.

[267] L. Félix and G. Valiente. Validation of metabolic pathway databases based on chemical substructure search. *Biomolecular Engineering*, 24(3):327–335, 2007. 6th Atlantic Symposium on Computational Biology.

[268] H. Fellermann, M. Dörr, M. M. Hanczyc, L. L. Laursen, S. Maurer, D. Merkle, P.-A. Monnard, K. Støy, and S. Rasmussen, editors. *Artificial Life XII*. MIT Press, Aug. 2010.

[269] H. Fellermann, M. Hadorn, E. Bönzli, and S. Rasmussen. Integrating molecular computation and material production in an artificial subcellular matrix. In T. Prescott, N. Lepora, A. Mura, and P. Verschure, editors, *Biomimetic and Biohybrid Systems*, volume 7375 of *Lecture Notes in Computer Science*, pages 343–344. Springer, 2012.

[270] H. Fellermann, S. Rasmussen, H.-J. Ziock, and R. V. Solé. Life cycle of a minimal protocell — a dissipative particle dynamics study. *Artificial Life*, 13(4):319–345, 2007.

[271] C. Fernando. *The Evolution of the Chemoton*. PhD thesis, University of Sussex, UK, 2005.

[272] C. Fernando. Fluid construction grammar in the brain. In L. Steels, editor, *Computational Issues in Fluid Construction Grammar*, volume 7249 of *Lecture Notes in Computer Science*, pages 312–330. Springer, 2012.

[273] C. Fernando and E. D. Paolo. The chemoton: A model for the origin of long RNA templates. In *Proceedings of the Ninth International Conference on the Simulation and Synthesis of Living Systems (ALIFE9)*. MIT Press, Sept. 2004.

[274] C. Fernando and E. Szathmáry. Natural selection in the brain. In B. M. Glatzeder et al., editors, *Towards a Theory of Thinking, On Thinking*, pages 291–322. Springer, 2010.

[275] C. Fernando, G. von Kiedrowski, and E. Szathmáry. A stochastic model of nonenzymatic nucleic acid replication: "Elongators" sequester replicators. *Journal of Molecular Evolution*, 64(5):572–585, May 2007.

[276] R. J. Field and R. M. Noyes. Oscillations in chemical systems. IV. limit cycle behavior in a model of a real chemical reaction. *Journal of Chemical Physics*, 60(5):1877–1884, Mar. 1974.

[277] A. Filisetti, A. Graudenzi, R. Serra, M. Villani, D. D. Lucrezia, R. M. Füchslin, S. A. Kauffman, N. H. Packard, and I. Poli. A stochastic model of the emergence of autocatalytic cycles. *Journal of Systems Chemistry*, 2(1), 2011.

[278] A. Filisetti, R. Serra, M. Villani, R. M. Füchslin, N. H. Packard, S. A. Kauffman, and I. Poli. A stochastic model of autocatalytic reaction networks. In *Proceedings of the European Conference on Complex Systems (ECCS)*, Sept. 2010.

[279] J. Fisher and T. A. Henzinger. Executable cell biology. *Nature Biotechnology*, 25(11):1239–1249, Nov. 2007.

[280] C. Flamm, A. Ullrich, H. Ekker, M. Mann, D. Högerl, M. Rohrschneider, S. Sauer, G. Scheuermann, K. Klemm, I. L. Hofacker, and P. F. Stadler. Evolution of metabolic networks: A computational framework. *Journal of Systems Chemistry*, 1(1):1–14, 2010.

[281] W. Fontana. Algorithmic chemistry. In C. G. Langton, C. Taylor, J. D. Farmer, and S. Rasmussen, editors, *Artificial Life II*, pages 159–210. Westview Press, 1991.

[282] W. Fontana and L. Buss. "The arrival of the fittest": Toward a theory of biological organization. *Bulletin of Mathematical Biology*, 56(1):1–64, 1994.

[283] W. Fontana and L. Buss. What would be conserved if "the tape were played twice"? *Proceedings of the National Academy of Sciences*, 91(2):757–761, 1994.

[284] W. Fontana and L. Buss. The barrier of objects: From dynamical systems to bounded organizations. In J. Casti and A. Karlqvist, editors, *Boundaries and Barriers*, pages 56–116. Addison Wesley, 1996.

[285] W. Fontana and P. Schuster. A computer model of evolutionary optimization. *Biophysical Chemistry*, 26:123–147, 1987.

[286] S. Forrest. Emergent computation: Self-organizing, collective, and cooperative phenomena in natural and artificial computing networks (introduction to the proceedings of the ninth annual CNLS conference). *Physica D: Nonlinear Phenomena*, 42(1):1–11, 1990.

[287] S. Forrest, B. Javornik, R. E. Smith, and A. S. Perelson. Using genetic algorithms to explore pattern recognition in the immune system. *Evolutionary Computation*, 1(3):191–211, 1993.

[288] S. Forrest, A. S. Perelson, L. Allen, and R. Cherukuri. Self-nonself discrimination in a computer. In *Proceeding of the IEEE Computer Society Symposium on Research in Security and Privacy*, pages 202–212, May 1994.

[289] S. Fortunato. Community detection in graphs. *Physics Reports*, 486:75–174, 2010.

[290] D. V. Foster, M. M. Rorick, T. Gesell, L. M. Feeney, and J. G. Foster. Dynamic landscapes: A model of context and contingency in evolution. *Journal of Theoretical Biology*, 334:162–172, 2013.

[291] J. J. Fox and C. C. Hill. From topology to dynamics in biochemical networks. *Chaos: An Interdisciplinary Journal of Nonlinear Science*, 11(4):809–815, 2001.

[292] L. Frachebourg, P. L. Krapivsky, and E. Ben-Naim. Spatial organization in cyclic Lotka-Volterra systems. *Physical Review E*, 54:6186–6200, Dec. 1996.

[293] P. Fradet and D. L. Métayer. Structured gamma. *Science of Computer Programming*, 31(2-3):263–289, 1998.

[294] A. Francez-Charlot, J. Frunzke, C. Reichen, J. Z. Ebneter, B. Gourion, and J. Vorholt. Sigma factor mimicry involved in regulation of general stress response. *Proceedings of the National Academy of Sciences*, 106:3467–3472, 2009.

[295] E. Fredkin and T. Toffoli. Conservative logic. *International Journal of Theoretical Physics*, 21(3-4):219–253, 1982.

[296] R. Frei, G. Di Marzo Serugendo, and T. Serbanuta. Ambient intelligence in self-organising assembly systems using the chemical reaction model. *Journal of Ambient Intelligence and Humanized Computing*, 1:163–184, 2010. 10.1007/s12652-010-0016-0.

[297] R. Frei and J. Whitacre. Degeneracy and networked buffering: Principles for supporting emergent evolvability in agile manufacturing systems. *Natural Computing*, 11(3):417–430, 2012.

[298] T. Froese and S. Gallagher. Phenomenology and artificial life: Toward a technological supplementation of phenomenological methodology. *Husserl Studies*, 26:83–106, 2010.

[299] I. Fry. The origins of research into the origins of life. *Endeavour*, 30(1):24–28, 2006.

[300] S. Fujita. Description of organic reactions based on imaginary transition structures. *J. Chem. Inf. Comput. Sci.*, 26(4):205–230, 1986.

[301] C. Furusawa and K. Kaneko. Emergence of multicellular organisms with dynamic differentiation and spatial pattern. *Artificial Life*, 4(1):79–93, 1998.

[302] C. Furusawa and K. Kaneko. Origin of multicellular organisms as an inevitable consequence of dynamical systems. *The Anatomical Record*, 268(3):327–342, 2002.

[303] C. Furusawa and K. Kaneko. Morphogenesis, plasticity and irreversibility. *International Journal of Developmental Biology*, 50:223–232, 2006.

[304] C. Furusawa and K. Kaneko. Chaotic expression dynamics implies pluripotency: When theory and experiment meet. *Biology Direct*, 4(1):17, 2009.

[305] T. Gánti. *Chemoton Theory, Volume 1: Theoretical Foundations of Fluid Machineries*. Kluwer Academic, 2003.

[306] T. Gánti. *The Principles of Life*. Oxford University Press, 2003.

[307] J. Garcia-Ojalvo, M. B. Elowitz, and S. H. Strogatz. Modeling a synthetic multicellular clock: Repressilators coupled by quorum sensing. *Proceedings of the National Academy of Sciences*, 101(30):10955–10960, 2004.

[308] M. García-Quismondo, R. Gutiérrez-Escudero, M. A. Martínez-del-Amor, E. Orejuela-Pinedo, and I. Pérez-Hurtado. P-Lingua 2.0: A software framework for cell-like P systems. *International Journal of Computers, Communications & Control*, IV(3):234–243, 2009.

[309] M. Gardner. Mathematical games: The fantastic combinations of John Conway's new solitaire game "Life". *Scientific American*, 223(4):120–123, 1970.

[310] M. Gardner. Mathematical games: On cellular automata, self-reproduction, the Garden of Eden and the game "Life". *Scientific American*, 224(2):112–117, 1971.

[311] P. P. Gardner and R. Giegerich. A comprehensive comparison of comparative RNA structure prediction approaches. *BMC Bioinformatics*, 5(140), 2004.

[312] T. S. Gardner, C. R. Cantor, and J. J. Collins. Construction of a genetic toggle switch in Escherichia coli. *Nature*, 403:339–342, Jan. 2000.

[313] M. Gargaud, P. López-García, and H. Martin, editors. *Origins and Evolution of Life: An Astrobiological Perspective*. Cambridge University Press, 2011.

[314] P. Gerlee and T. Lundh. Productivity and diversity in a cross-feeding population of artificial organisms. *Evolution*, 64(9):2716–2730, 2010.

[315] P. Gerlee and T. Lundh. Rock-paper-scissors dynamics in a digital ecology. In Fellermann et al. [268], pages 283–292.

[316] C. R. Geyer, T. R. Battersby, and S. A. Benner. Nucleobase pairing in expanded Watson-Crick-like genetic information systems. *Structure*, 11(12):1485–1498, 2003.

[317] J.-L. Giavitto. Topological collections, transformations and their application to the modeling and the simulation of dynamical systems. In R. Nieuwenhuis, editor, *Rewriting Techniques and Applications*, volume 2706 of *Lecture Notes in Computer Science*, pages 208–233. Springer, 2003.

[318] J. L. Giavitto, G. Malcolm, and O. Michel. Rewriting systems and the modelling of biological systems. *Comparative and Functional Genomics*, 5(1):95–99, 2004.

[319] J.-L. Giavitto and O. Michel. MGS: A programming language for the transformations of topological collections. Technical Report 61-2001, Laboratoire de Méthodes Informatiques (LaMI), Université d'Evry Val d'Essonne, Evry, France, May 2001.

[320] J.-L. Giavitto and O. Michel. MGS: A rule-based programming language for complex objects and collections. *Electronic Notes in Theoretical Computer Science*, 59(4):286–304, 2001. RULE 2001, Second International Workshop on Rule-Based Programming (Satellite Event of PLI 2001).

[321] J. L. Giavitto and O. Michel. Modeling the topological organization of cellular processes. *BioSystems*, 70(2):149–163, 2003.

[322] J.-L. Giavitto, O. Michel, and A. Spicher. Unconventional and nested computations in spatial computing. *IJUC*, 9(1-2):71–95, 2013.

[323] D. G. Gibson, J. I. Glass, C. Lartigue, V. N. Noskov, R.-Y. Chuang, M. A. Algire, G. A. Benders, M. G. Montague, L. Ma, M. M. Moodie, C. Merryman, S. Vashee, R. Krishnakumar, N. Assad-Garcia, C. Andrews-Pfannkoch, E. A. Denisova, L. Young, Z.-Q. Qi, T. H. Segall-Shapiro, C. H. Calvey, P. P. Parmar, C. A. Hutchison, H. O. Smith, and J. C. Venter. Creation of a bacterial cell controlled by a chemically synthesized genome. *Science*, 329(5987):52–56, 2010.

[324] M. Gibson and J. Bruck. Efficient exact stochastic simulation of chemical systems with many species and many channels. *Journal of Chemical Physics*, 104:1876–1889, 2000.

[325] B. Gil, M. Kahan-Hanum, N. Skirtenko, R. Adar, and E. Shapiro. Detection of multiple disease indicators by an autonomous biomolecular computer. *Nano Letters*, 11(7):2989–2996, 2011.

[326] S. F. Gilbert. *Developmental Biology*. Sinauer Associates, 6th edition, 2000.

[327] W. Gilbert. The RNA world. *Nature*, 319(6055):618, 1986.

[328] D. T. Gillespie. A general method for numerically simulating the stochastic time evolution of coupled chemical reacting systems. *Journal of Computational Physics*, 22:403–434, 1976.

[329] D. T. Gillespie. Exact stochastic simulation of coupled chemical reactions. *Journal of Physical Chemistry*, 81:2340–2361, 1977.

[330] D. T. Gillespie. A rigorous derivation of the chemical master equation. *Physica A*, 188:404–425, 1992.

[331] D. T. Gillespie. Approximate accelerated stochastic simulation of chemically reacting systems. *Journal of Chemical Physics*, 115:1716–1733, 2001.

[332] D. T. Gillespie and L. R. Petzold. Improved leap-size selection for accelerated stochastic simulation method. *Journal of Chemical Physics*, 119:8229–8234, 2003.

[333] R. Gillespie and R. Nyholm. Inorganic stereochemistry. *Quarterly Reviews, Chemical Society*, 11:339–380, 1957.

[334] J. I. Glass, N. Assad-Garcia, N. Alperovich, S. Yooseph, M. R. Lewis, M. Maruf, C. A. Hutchison, H. O. Smith, and J. C. Venter. Essential genes of a minimal bacterium. 103(2):425–430, 2006.

[335] J. A. Glazier and F. Graner. Simulation of the differential adhesion driven rearrangement of biological cells. *Physics Review E*, 47(3):2128–2154, 1993.

[336] A. Goñi Moreno, M. Amos, and F. de la Cruz. Multicellular computing using conjugation for wiring. *PLoS ONE*, 8:e65986, 06 2013.

[337] S. Goings, H. Goldsby, B. H. Cheng, and C. Ofria. An ecology-based evolutionary algorithm to evolve solutions to complex problems. In Adami et al. [20], pages 171–177.

[338] S. Goings and C. Ofria. Ecological approaches to diversity maintenance in evolutionary algorithms. In *IEEE Symposium on Artificial Life*, pages 124–130, 2009.

[339] D. E. Goldberg. *Genetic Algorithms in Search, Optimization, and Machine Learning*. Addison-Wesley, 1989.

[340] A. Goñi-Moreno and M. Amos. Model for a population-based microbial oscillator. *Biosystems*, 105(3):286–294, 2011.

[341] J. Gorecki, J. N. Gorecka, K. Yoshikawa, Y. Igarashi, and H. Nagahara. Sensing the distance to a source of periodic oscillations in a nonlinear chemical medium with the output information coded in frequency of excitation pulses. *Physical Review E*, 72:046201 1–7, Oct. 2005.

[342] D. Görlich, S. Artmann, and P. Dittrich. Cells as semantic systems. *Biochimica et Biophysica Acta (BBA) - General Subjects*, 1810(10):914–923, 2011.

[343] A. Goudsmit. Some reflections on Rosen's conception of semantics and finality. *Chemical Diversity*, 4:2427–2435, 2007.

[344] S. Gould and E. Vrba. Exaptation: A missing term in the science of form. *Paleobiology*, 8:4–15, 1982.

[345] P. Gray and S. Scott. *Chemical Oscillations and Instabilities: Nonlinear Chemical Kinetics*. Oxford Science Publications, 1990.

[346] R. Grima and S. Schnell. Modelling reaction kinetics inside cells. *Essays in Biochemistry*, 4541-56, 2008.

[347] V. Grimm and S. Railsback. *Individual-Based Modeling and Ecology*. Princeton University Press, 2005.

[348] C. Gros. Random Boolean Networks. In *Complex and Adaptive Dynamical Systems*, pages 109–143. Springer, 2011.

[349] A. R. Gruber, R. Lorenz, S. H. Bernhart, R. Neuböck, and I. L. Hofacker. The Vienna RNA websuite. *Nucleic Acids Research*, 36(suppl 2):W70–W74, 2008.

[350] T. Gruber and C. Gross. Multiple sigma subunits and the partitioning of bacterial transcription space. *Annual Review of Microbiology*, 57:441–466, 2003.

[351] G. Gruenert, B. Ibrahim, T. Lenser, M. Lohel, T. Hinze, and P. Dittrich. Rule-based spatial modeling with diffusing, geometrically constrained molecules. *BMC Bioinformatics*, 11(307), 2010.

[352] B. A. Grzybowski, H. A. Stone, and G. M. Whitesides. Dynamic self-assembly of magnetized, millimetre-sized objects rotating at a liquid-air interface. *Nature*, 405(6790):1033–1036, 2000.

[353] R. R. Gutell. Comparative analysis of the higher-order structure of RNA. In *Biophysics of RNA Folding* [731], chapter 2, pages 11–22.

[354] J. Haas, S. Roth, K. Arnold, F. Kiefer, T. Schmidt, L. Bordoli, and T. Schwede. The protein model portal: A comprehensive resource for protein structure and model information. *Database*, 2013, 2013.

[355] S. H. D. Haddock and C. W. Dunn. *Practical Computing for Biologists*. Sinauer Associates, 2011.

[356] R. C. Haddon and A. A. Lamola. The molecular electronic device and the biochip computer: present status. *Proceedings of the National Academy of Sciences*, 82(7):1874–1878, Apr. 1985.

[357] M. Hadorn, E. Boenzli, K. T. Sørensen, H. Fellermann, P. Eggenberger Hotz, and M. M. Hanczyc. Specific and reversible DNA-directed self-assembly of oil-in-water emulsion droplets. *Proceedings of the National Academy of Sciences*, 109(50):20320–20325, 2012.

[358] H. Haken. *Synergetics: An Introduction*. Springer, 3rd edition, 1983.

[359] H. Haken. *Synergetics: Introduction and Advanced Topics*. Springer, 2004.

[360] M. Hall, K. Christensen, S. A. di Collobiano, and H. Jeldtoft Jensen. Time-dependent extinction rate and species abundance in a tangled-nature model of biological evolution. *Physical Review E*, 66:011904, July 2002.

[361] N. Hampp, C. Bräuchle, and D. Oesterhelt. Bacteriorhodopsin wildtype and variant aspartate-96 → aspargine as reversible holographic media. *Biophysical Journal*, 58:83–93, 1990.

[362] N. Hampp, C. Bräuchle, and D. Oesterhelt. Mutated bacteriorhodopsins: Competitive materials for optical information processing? *Materials Research Society Bulletin*, 17:56–60, 1992.

[363] M. M. Hanczyc. Droplets: Unconventional protocell model with life-like dynamics and room to grow. *Life*, 4(4):1038–1049, 2014.

[364] M. M. Hanczyc and T. Ikegami. Chemical basis for minimal cognition. *Artificial Life*, 16(3):233–243, 2010.

[365] M. M. Hanczyc, T. Toyota, T. Ikegami, N. Packard, and T. Sugawara. Fatty acid chemistry at the oil-water interface: Self-propelled oil droplets. *Journal of the American Chemical Society*, 129(30):9386–9391, 2007.

[366] S. Harding and W. Banzhaf. Artificial development. In R. Wuertz, editor, *Organic Computing*, Complex Systems Series, pages 201–220. Springer, 2008.

[367] D. Haronian and A. Lewis. Elements of a unique bacteriorhodopsin neural network architecture. *Applied Optics*, 30:597–608, 1991.

[368] E. Hart and J. Timmis. Application areas of AIS: The past, the present and the future. *Applied Soft Computing*, 8(1):191–201, 2008.

[369] F. U. Hartl, A. Bracher, and M. Hayer-Hartl. Molecular chaperones in protein folding and proteostasis. *Nature*, 475(7356):324–332, July 2011.

[370] J. Hartmanis. On the weight of computations. *Bulletin of the European Association for Theoretical Computer Science*, 55:136–138, 1995.

[371] G. Hasegawa and M. Murata. TCP symbiosis: congestion control mechanisms of TCP based on Lotka-Volterra competition model. In *Proc. Workshop on Interdisciplinary systems approach in performance evaluation and design of computer & communications sytems (Interperf)*. ACM, 2006.

[372] M. Hatcher. Unpublished, 2013.

[373] M. Hatcher, W. Banzhaf, and T. Yu. Bondable cellular automata. In T. Lenaerts, M. Giacobini, H. Bersini, P. Bourgine, M. Dorigo, and R. Doursat, editors, *Proceedings of the 11th European Conference on Artificial Life ECAL-2011*, pages 326–333. MIT Press, 2001.

[374] B. Hayes. Reverse engineering. *American Scientist*, 94(3), May-June 2006.

[375] P. Herdina and U. Jessner. *A Dynamic Model of Multilingualism: Perspectives on Change in Psycholinguistics*. Multilingual Matters, 2002.

[376] G. Herman, A. Lindenmayer, and G. Rozenberg. Description of developmental languages using recurrence systems. *Mathematical Systems Theory*, 8:316–341, 1975.

[377] J. Herriott and B. Sawmill. Web-economy as a self-organizing systems. http://www.redfish.com/research/webEconomyIdeas/, (Last accessed Oct 11, 2013).

[378] D. Herschlag. RNA chaperones and the RNA folding problem. *Journal of Biological Chemistry*, 270:20871–20874, 1995.

[379] S. Hickinbotham, E. Clark, A. Nellis, M. Pay, S. Stepney, T. Clarke, and P. Young. Molecular microprograms. In G. Kampis, I. Karsai, and E. Szathmáry, editors, *Advances in Artificial Life. Darwin Meets von Neumann*, volume 5777 of *Lecture Notes in Computer Science*, pages 291–298. Springer, 2011.

[380] S. Hickinbotham, E. Clark, S. Stepney, T. Clarke, A. Nellis, M. Pay, and P. Young. Diversity from a monoculture: Effects of mutation-on-copy in a string-based artificial chemistry. In S. Rasmussen, editor, *Proceedings of ALIFE XII, Odense Denmark*, pages 24–31. MIT Press, 2010.

[381] S. Hickinbotham, E. Clark, S. Stepney, T. Clarke, A. Nellis, M. Pay, and P. Young. Specification of the stringmol chemical programming language version 0.2. Technical Report YCS-2010-458, Department of Computer Science, University of York, 2010.

[382] W. Hillis. *The Connection Machine*. MIT Press, 1989.

[383] J. Hindley and J. Seldin. *Introduction to Combinators and λ-calculus*. Cambridge University Press, 1986.

[384] A. Hintze and C. Adami. Evolution of complex modular biological networks. *PLoS Computational Biology*, 4(2):e23, 2008.

[385] N. Hirokawa. Kinesin and dynein superfamily proteins and the mechanism of organelle transport. *Science*, 279(5350):519–526, 1998.

[386] A. Hjelmfelt, E. D. Weinberger, and J. Ross. Chemical implementation of neural networks and Turing machines. *Proceedings of the National Academy of Sciences*, 88:10983–10987, 1991.

[387] A. Hjelmfelt, E. D. Weinberger, and J. Ross. Chemical implementation of finite-state machines. *Proceedings of the National Academy of Sciences*, 89(1):383–387, 1992.

[388] W. S. Hlavacek, J. R. Faeder, M. L. Blinov, R. G. Posner, M. Hucka, and W. Fontana. Rules for modeling signal-transduction systems. *Science's STKE*, 2006/344/re6, July 2006.

[389] I. L. Hofacker, W. Fontana, P. F. Stadler, S. Bonhoeffer, M. Tacker, and P. Schuster. Fast folding and comparison of RNA secondary structures. *Monatshefte für Chemie / Chemical Monthly*, 125(2):167–188, 1994.

[390] I. L. Hofacker and P. F. Stadler. Memory efficient folding algorithms for circular RNA secondary structures. *Bioinformatics*, 2006.

[391] D. R. Hofstadter. *Gödel, Escher, Bach: An Eternal Golden Braid*. Penguin Books, 1979.

[392] P. Hogeweg. Shapes in the shadow: Evolutionary dynamics of morphogenesis. *Artificial Life*, 6(1):85–101, 2000.

[393] T. Hohsaka, Y. Ashizuka, H. Murakami, and M. Sisido. Five-base codons for incorporation of nonnatural amino acids into proteins. *Nucleic acids research*, 29(17):3646–3651, 2001.

[394] T. Hohsaka, Y. Ashizuka, H. Taira, H. Murakami, and M. Sisido. Incorporation of nonnatural amino acids into proteins by using various four-base codons in an Escherichia coli in vitro translation system. *Biochemistry*, 40(37):11060–11064, 2001.

[395] J. H. Holland. *Adaptation in Natural and Artificial Systems*. University of Michigan Press, 1975.

[396] J. H. Holland and J. Reitman. Cognitive systems based on adaptive algorithms. In D. Waterman and F. Hayes-Roth, editors, *Pattern Directed Inference Systems*, pages 313–329. Academic Press, 1978.

[397] J. Hopcroft and J. Ullman. *Introduction to Automata Theory, Languages and Computation*. Addison Wesley, 1979.

[398] W. Hordijk, J. P. Crutchfield, and M. Mitchell. Embedded-particle computation in evolved cellular automata. In T. Toffoli, M. Biafore, and J. Leäo, editors, *Physics and Computation*, pages 153–158. New England Systems Institute, 1996.

[399] W. Hordijk, J. P. Crutchfield, and M. Mitchell. Mechanisms of emergent computation in cellular automata. In *Parallel Problem Solving from Nature (PPSN V)*, pages 613–622. Springer, 1998.

[400] W. Hordijk, J. Hein, and M. Steel. Autocatalytic sets and the origin of life. *Entropy*, (12):1733–1742, 2010.

[401] W. Hordijk and S. A. Kauffman. Correlation analysis of coupled fitness landscapes. *Complexity*, 10(6):41–49, 2005.

[402] W. Hordijk and M. Steel. Detecting autocatalytic, self-sustaining sets in chemical reaction systems. *Journal of Theoretical Biology*, 227(4):451–461, 2004.

[403] N. Horibe, M. M. Hanczyc, and T. Ikegami. Mode switching and collective behavior in chemical oil droplets. *Entropy*, 13(3):709–719, 2011.

[404] K. Hosokawa, I. Shimoyama, and H. Miura. Dynamics of self-assembling systems - analogy with chemical kinetics. In R. Brooks and P. Maes, editors, *Artificial Life IV*, pages 172–180. MIT Press, 1994.

[405] T. Hoverd and S. Stepney. Energy as a driver of diversity in open-ended evolution. In T. Lenaerts, M. Giacobini, H. Bersini, P. Bourgine, M. Dorigo, and R. Doursat, editors, *Advances in Artificial Life, ECAL 2011: Proceedings of the Eleventh European Conference on the Synthesis and Simulation of Living Systems*, pages 356–363. MIT Press, 2011.

[406] W. A. Howard. The formulae-as-types notion of construction. In *To HB Curry: Essays on Combinatory Logic, Lambda Calculus and Formalism*, pages 479–490. Academic Press, 1980.

[407] T. Hu and W. Banzhaf. Evolvability and speed of evolutionary algorithms in light of recent developments in biology. *Journal of Artificial Evolution and Applications*, 2010. Article ID 568375.

[408] T. Hu, J. L. Payne, W. Banzhaf, and J. H. Moore. Robustness, evolvability, and accessibility in linear genetic programming. In S. Silva, J. A. Foster, M. Nicolau, P. Machado, and M. Giacobini, editors, *Genetic
Programming*, volume 6621 of *Lecture Notes in Computer Science*, pages 13–24. Springer, 2011.

[409] C. Huber, F. Kraus, M. Hanzlik, W. Eisenreich, and G. Wächtershäuser. Elements of metabolic evolution. *Chemistry: A European Journal*, 18(7):2063–2080, 2012.

[410] P. Huneman. Determinism, predictability and open-ended evolution: Lessons from computational
emergence. *Synthese*, 185(2):195–214, 2012.

[411] T. J. Hutton. Evolvable self-replicating molecules in an artificial chemistry. *Artificial Life*, 8(4):341–
356, 2002.

[412] T. J. Hutton. Evolvable self-reproducing cells in a two-dimensional artificial chemistry. *Artificial Life*,
13(1):11–30, 2007.

[413] T. J. Hutton. The organic builder: A public experiment in artificial chemistries and sel-replication.
Artificial Life, 15(1):21–28, 2009.

[414] K. Hyodo, N. Wakamiya, and M. Murata. Reaction-diffusion based autonomous control of camera
sensor networks. In *Proceedings of the 2nd International Conference on Bio-Inspired Models of Network, Information, and Computing Systems (BIONETICS)*. ICST, 2007.

[415] B. Ibrahim, R. Henze, G. Gruenert, M. Egbert, J. Huwald, and P. Dittrich. Spatial rule-based modeling:
A method and its application to the human mitotic kinetochore. *Cells*, 2(3):506–544, 2013.

[416] N. Ichihashi, T. Matsuura, H. Kita, T. Sunami, H. Suzuki, and T. Yomo. Constructing partial models of
cells. *Cold Spring Harbor Perspectives in Biology*, 2(6):a004945, 2010.

[417] N. Ichihashi, K. Usui, Y. Kazuta, T. Sunami, T. Matsuura, and T. Yomo. Darwinian evolution in a
translation-coupled RNA replication system within a cell-like compartment. *Nature Communications*, 4:2494, 2013.

[418] Z. Ignatova, I. Martínez-Pérez, and K.-H. Zimmermann. *DNA Computing Models*. Springer, 2008.

[419] T. Ikegami and T. Hashimoto. Active mutation in self-reproducing networks of machines and tapes.
Artificial Life, 2(3):305–318, 1995.

[420] T. Ikegami and T. Hashimoto. Replication and diversity in machine-tape coevolutionary systems. In
C. G. Langton and K. Shimohara, editors, *Artificial Life V*, pages 426–433. MIT Press, 1997.

[421] T. Ikegami and K. Suzuki. From a homeostatic to a homeodynamic self. *BioSystems*, 91:388–400, 2008.

[422] P. Inverardi and A. Wolf. Formal specification and analysis of software architectures using the chemical abstract machine model. *IEEE Transactions on Software Engineering*, 21(4):373–386, 1995.

[423] E. J. Izquierdo and C. T. Fernando. The evolution of evolvability in gene transcription networks. In
Bullock et al. [143], pages 265–273.

[424] Z. Izri, M. N. van der Linden, S. Michelin, and O. Dauchot. Self-propulsion of pure water droplets by
spontaneous Marangoni-stress-driven motion. *Physical Review Letters*, 113:248302, Dec 2014.

[425] S. Jain and S. Krishna. Autocatalytic sets and the growth of complexity in an evolutionary model.
Physical Review Letters, 81(25):5684–5687, 1998.

[426] S. Jain and S. Krishna. A model for the emergence of cooperation, interdependence, and structure in
evolving networks. *Proceedings of the National Academy of Sciences of the United States of America*,
98(2):543, 2001.

[427] S. Jain and S. Krishna. Graph theory and the evolution of autocatalytic networks. In S. Bornholdt and H. G. Schuster, editors, *Handbook of Graphs and Networks: From the Genome to the Internet*, chapter 16, pages 355–395. Wiley, 2003.

[428] S. Jain and S. Krishna. Can we recognize an innovation? perspective from an evolving network model. In H. Meyer-Ortmanns and S. Thurner, editors, *Principles of Evolution*, pages 145–172. Springer, 2011.

[429] F. Jensen. *Introduction to Computational Chemistry*. Wiley, 2nd ed., 2007.

[430] H. Jeong, B. Tombor, R. Albert, Z. N. Oltvai, and A.-L. Barabasi. The large-scale organization of metabolic networks. *Nature*, 407:651–654, Oct. 2000.

[431] M. Joachimczak and B. Wróbel. Open ended evolution of 3D multicellular development controlled by gene regulatory networks. In Adami et al. [20], pages 67–74.

[432] D. G. Jones and A. K. Dewdney. Core War guidelines. http://users.obs.carnegiescience.edu/birk/COREWAR/DOCS/guide2red.txt (Last accessed Dec 14, 2013), 1994.

[433] G. F. Joyce. RNA evolution and the origin of life. *Nature*, 338:217–24, Mar. 1989.

[434] G. F. Joyce. The RNA world: Life before DNA and protein. In *Extraterrestrials: Where are they?* Cambridge University Press, 1994.

[435] G. F. Joyce. Toward an alternative biology. *Science*, 336:307–308, Apr. 2012.

[436] S. Kádár, T. Amemiya, and K. Showalter. Reaction mechanism for light sensitivity of the $ru(bpy)_3^{2+}$-catalyzed Belousov-Zhabotinsky reaction. *Journal of Physical Chemistry A*, 101(44):8200–8206, 1997.

[437] G. Kampis. *Self-Modifying Systems in Biology and Cognitive Science*. Pergamon Press, 1991.

[438] Y. Kanada. Combinatorial problem solving using randomized dynamic tunneling on a production system. In *1995 IEEE International Conference on Systems, Man and Cybernetics. Intelligent Systems for the 21st Century*, volume 4, pages 3784–3789. IEEE, 1995.

[439] Y. Kanada and M. Hirokawa. Stochastic problem solving by local computation based on self-organization paradigm. In *27th Hawaii International Conference on System Science*, pages 82–91, 1994.

[440] I. Karonen. The beginner's guide to Redcode. http://vyznev.net/corewar/guide.html (Last accessed Dec 14, 2013), 2004.

[441] N. Kashtan and U. Alon. Spontaneous evolution of modularity and network motifs. *Proceedings of the National Academy of Sciences*, 102(39):13773–13778, 2005.

[442] N. Kashtan, M. Parter, E. Dekel, A. E. Mayo, and U. Alon. Extinctions in heterogeneous environments and the evolution of modularity. *Evolution*, 63(8):1964–1975, 2009.

[443] N. Kato, T. Okuno, A. Okano, H. Kanoh, and S. Nishihara. An ALife approach to modelling virtual cities. In *IEEE Conference on Systems, Man and Cybernetics*, volume 2, pages 1168–1173. IEEE Press, 1998.

[444] E. Katz, editor. *Biomolecular Information Processing: From Logic Systems to Smart Sensors and Actuators*. Wiley, 2012.

[445] S. A. Kauffman. Metabolic stability and epigenesis in randomly constructed genetic nets. *Journal of Theoretical Biology*, 22:437–467, 1969.

[446] S. A. Kauffman. Autocatalytic sets of proteins. *Journal of Theoretical Biology*, 119:1–24, 1986.

[447] S. A. Kauffman. *The origins of order: Self-organization and selection in evolution*. Oxford University Press, 1993.

[448] S. A. Kauffman. *Investigations*. Oxford University Press, 2000.

[449] S. A. Kauffman. Approaches to the origin of life on earth. *Life*, 1:34–38, 2011.

[450] S. A. Kauffman and S. Johnsen. Coevolution to the edge of chaos: Coupled fitness landscapes, poised states, and coevolutionary avalanches. *Journal of Theoretical Biology*, 149(4):467–505, 1991.

[451] A. V. Kazantsev, A. A. Krivenko, D. J. Harrington, S. R. Holbrook, P. D. Adams, and N. R. Pace. Crystal structure of bacterial ribonuclease P RNA. *Proceedings of the National Academy of Sciences*, 102:13392–13397, 2005.

[452] C. Kelly, B. McMullin, and D. O'Brien. Enrichment of interaction rules in a string-based artificial chemistry. In *Proc Artificial Life XI*, 2008.

[453] J. G. Kemeny. Man viewed as a machine. *Scientific American*, 192(4):58–67, Apr. 1955.

[454] J. Kennedy and R. Eberhart. Particle swarm optimization. In *Proceedings of the IEEE International Conference on Neural Networks*, volume 4, pages 1942–1948, Nov.-Dec. 1995.

[455] B. Kerner. Experimental features of self-organization in traffic flow. *Physical Review Letters*, 81:3797–3800, 1998.

[456] M. Kerszberg and L. Wolpert. Specifying positional information in the embryo: Looking beyond morphogens. *Cell*, 130(2):205–209, 2007.

[457] F. Khatib, F. DiMaio, S. Cooper, M. Kazmierczyk, M. Gilski, S. Krzywda, H. Zabranska, I. Pichova, J. Thompson, Z. Popovic, M. Jaskolski, and D. Baker. Crystal structure of a monomeric retroviral protease solved by protein folding game players. *Nature Structural & Molecular Biology*, 18(10):1175–1177, Oct. 2011.

[458] D.-E. Kim and G. F. Joyce. Cross-catalytic replication of an RNA ligase ribozyme. *Chemistry & Biology*, 11(11):1505–1512, 2004.

[459] J. Kim. Making sense of emergence. *Philosophical Studies*, 95:3–36, 1999.

[460] K. Kim and S. Cho. A comprehensive overview of the applications of artificial life. *Artificial Life*, 12(1):153–182, 2006.

[461] M. W. Kirschner and J. C. Gerhart. Evolvability. *Proceedings of the National Academy of Sciences*, 95(15):8420–8427, July 1998.

[462] M. W. Kirschner and J. C. Gerhart. *The plausibility of life: Resolving Darwin's dilemma*. Yale University Press, 2005.

[463] H. Kita, T. Matsuura, T. Sunami, K. Hosoda, N. Ichihashi, K. Tsukada, I. Urabe, and T. Yomo. Replication of genetic information with self-encoded replicase in liposomes. *ChemBioChem*, 9(15):2403–2410, 2008.

[464] H. Kitano. Evolution of metabolism for morphogenesis. In R. Brooks and P. Maes, editors, *Artificial Life IV*, pages 49–58. MIT Press, 1994.

[465] H. Kitano. Systems biology: Toward system-level understanding of biological systems. In H. Kitano, editor, *Foundations of Systems Biology*, chapter 1. MIT Press, 2001.

[466] J. F. Knabe, M. J. Schilstra, and C. Nehaniv. Evolution and morphogenesis of differentiated multicellular organisms: Autonomously generated diffusion gradients for positional information. In Bullock et al. [143], pages 321–328.

[467] D. B. Knoester and P. K. McKinley. Evolution of synchronization and desynchronization in digital organisms. *Artificial Life*, 17(1):1–20, 2011.

[468] D. B. Knoester, P. K. McKinley, and C. A. Ofria. Using group selection to evolve leadership in populations of self-replicating digital organisms. In *Proceedings of the Genetic and Evolutionary Computation Conference (GECCO)*, pages 293–300, 2007.

[469] A. J. Koch and H. Meinhardt. Biological pattern formation: From basic mechanisms to complex structures. *Reviews of Modern Physics*, 66(4), 1994.

[470] M. Komosiński and Ádám Rotaru-Varga. From directed to open-ended evolution in a complex simulation model. In M. A. Bedau, J. S. McCaskill, N. H. Packard, and S. Rasmussen, editors, *Artificial Life VII - Proc. Seventh International Conference on Artificial Life*, pages 293–299. MIT Press, 2000.

[471] K. Korb and A. Dorin. Evolution unbound: Releasing the arrow of complexity. *Biology & Philosophy*, 26(3):317–338, 2011.

[472] J. R. Koza. Artificial life: Spontaneous emergence of self-replicating and evolutionary self-improving computer programs. In C. Langton, editor, *Artificial Life III*, pages 225–262. Addison-Wesley, 1994.

[473] N. Krasnogor, W. E. Hart, J. Smith, and D. A. Pelta. Protein structure prediction with evolutionary algorithms. In *Proceedings of the Genetic and Evolutionary Computation Conference*, volume 2, pages 1596–1601, July 1999.

[474] P. Kreyssig and P. Dittrich. Reaction flow artificial chemistries. In T. Lenaerts, M. Giacobini, H. Bersini, P. Bourgine, M. Dorigo, and R. Doursat, editors, *Advances in Artificial Life, ECAL 2011: Proceedings of the Eleventh European Conference on the Synthesis and Simulation of Living Systems*, pages 431–437. MIT Press, 2011.

[475] J. Kuby. *Immunology*. W. H. Freeman & Co., 3rd edition, 1997.

[476] T. Kuhn. A function for thought experiments. *Ontario Journal of Educational Research*, 10:211–231, 1968.

[477] L. Kuhnert, K. Agladze, and V. I. Krinsky. Image processing using light-sensitive chemical waves. *Nature*, 337:244–247, 1989.

[478] P. D. Kuo and W. Banzhaf. Small world and scale-free network topologies in an artificial regulatory network model. In J. Pollack, M. Bedau, P. Husbands, T. Ikegami, and R. A. Watson, editors, *Artificial Life IX: Proceedings of the Ninth International Conference on the Simulation and Synthesis of Living Systems*, pages 404–409. MIT Press, 2004.

[479] P. D. Kuo, W. Banzhaf, and A. Leier. Network topology and the evolution of dynamics in an artificial genetic regulatory network model created by whole genome duplication and divergence. *Biosystems*, 85:177–200, 2006.

[480] P. D. Kuo, A. Leier, and W. Banzhaf. Evolving dynamics in an artificial regulatory network model. In X. Yao, E. K. Burke, J. A. Lozano, J. Smith, J. J. Merelo-Guervós, J. A. Bullinaria, J. E. Rowe, P. Tino, A. Kabán, and H.-P. Schwefel, editors, *Parallel Problem Solving from Nature - PPSN VIII*, volume 3242 of *Lecture Notes in Computer Science*, pages 571–580. Springer, 2004.

[481] V. Kvasnička and J. Pospíchal. Artificial chemistry and molecular Darwinian evolution in silico. *Collection of Czechoslovak Chemical Communications*, 68(1):139–177, 2003.

[482] V. Kvasnička, J. Pospíchal, and T. Kaláb. A study of replicators and hypercycles by typogenetics. In *Advances in Artificial Life: Proceedings of ECAL 2001*, pages 37–54. Springer, 2001.

[483] I. Lagzi, S. Soh, P. J. Wesson, K. P. Browne, and B. A. Grzybowski. Maze solving by chemotactic droplets. *Journal of the American Chemical Society*, 132(4):1198–1199, 2010.

[484] R. Laing. Artificial organisms and autonomous cell rules. *Journal of Cybernetics*, 2(1):38–49, 1972.

[485] R. Laing. The capabilities of some species of artificial organism. *Journal of Cybernetics*, 3(2):16–25, 1973.

[486] R. Laing. Some alternative reproductive strategies in artificial molecular machines. *Journal of Theoretical Biology*, 54:63–84, 1975.

[487] R. Laing. Automaton introspection. *Journal of Computer and System Sciences*, 13(2):172–83, 1976.

[488] R. Laing. Automaton models of reproduction by self-inspection. *Journal of Theoretical Biology*, 66:437–56, 1977.

[489] A. Lam and V.-K. Li. Chemical-reaction-inspired metaheuristic for optimization. *IEEE Transactions on Evolutionary Computation*, 14(3):381–399, June 2010.

[490] D. Lancet, E. Sadovsky, and E. Seidemann. Probability model for molecular recognition in biological receptor repertoires: Significance to the olfactory system. *Proceedings of the National Academy of Sciences*, 90:3715–3719, 1993.

[491] R. Landauer. Irreversibility and heat generation in the computing process. *IBM Journal of Research and Development*, 5(3):183–191, July 1961.

[492] T. J. Lane, D. Shukla, K. A. Beauchamp, and V. S. Pande. To milliseconds and beyond: Challenges in the simulation of protein folding. *Current Opinion in Structural Biology*, 23(1):58–65, 2013.

[493] W. Langdon and W. Banzhaf. Repeated patterns in genetic programming. *Natural Computing*, 7(4):589–613, 2008.

[494] R. Lange and R. Hengge-Aronis. Identification of a central regulator of stationary-phase gene expression in Escherichia coli. *Molecular Microbiology*, 5:49–59, 1991.

[495] C. Langton. Self-reproduction in cellular automata. *Physica D*, 10:135–144, 1984.

[496] C. Langton. Life at the edge of chaos. In C. Langton, C. Taylor, J. Farmer, and S. Rasmussen, editors, *Artificial Life II, Proceedings of the 2nd workshop on Artificial Life*. Addison-Wesley, 1992.

[497] C. Lasarczyk and W. Banzhaf. An algorithmic chemistry for genetic programming. In *Proceedings of the 8th European Conference on Genetic Programming*, volume 3447, pages 1–12. Springer, 2005.

[498] C. Lasarczyk and W. Banzhaf. Total synthesis of algorithmic chemistries. In *Proceedings of the 2005 Conference on Genetic and Evolutionary Computation*, pages 1635–1640, 2005.

[499] A. Leach. *Molecular Modelling: Principles and Applications*. Prentice Hall, 2nd edition, 2001.

[500] J. Lebowitz, W. Bortz, and M. Kalos. A new algorithm for Monte Carlo simulation of ising spin systems. *Journal of Computational Physics*, 17:10–18, 1975.

[501] A. M. Leconte, G. T. Hwang, S. Matsuda, P. Capek, Y. Hari, and F. E. Romesberg. Discovery, characterization, and optimization of an unnatural base pair for expansion of the genetic alphabet. *Journal of the American Chemical Society*, 130:2336–2343, 2008.

[502] J. Lehman and K. O. Stanley. Exploiting open-endedness to solve problems through the search for novelty. In Bullock et al. [143], pages 329–336.

[503] A. Leier, M. Barrio, and T. T. Marquez-Lago. Exact model reduction with delays: closed-form distributions and extensions to fully bi-directional monomolecular reactions. *Journal of The Royal Society Interface*, 11:20140108, 2014.

[504] A. Leier, P. D. Kuo, and W. Banzhaf. Analysis of preferential network motif generation in an artificial regulatory network model created by duplication and divergence. *Advances in Complex Systems*, 10:155–172, 2007.

[505] T. Lenaerts and H. Bersini. A synthon approach to artificial chemistry. *Artificial Life*, 15(1):89–103, 2009.

[506] T. Lenser, N. Matsumaru, T. Hinze, and P. Dittrich. Tracking the evolution of chemical computing networks. In Bullock et al. [143], pages 343–350.

[507] R. E. Lenski, C. Ofria, R. Pennock, and C. Adami. The evolutionary origin of complex features. *Nature*, 423(6745):139–144, supplementary material, 2003.

[508] T. Lenton. Gaia and natural selection. *Nature*, 394:439–447, 1998.

[509] J. Letelier, G. Marin, and J. Mpodozis. Autopoietic and (M,R) systems. *Journal of Theoretical Biology*, 222:261–272, 2003.

[510] G. Lewes. *Problems of Life and Mind*. Trubner and Company, 1874.

[511] A. M. L. Liekens, t. H. M. M. Eikelder, M. N. Steijaert, and P. A. J. Hilbers. Simulated evolution of mass conserving reaction networks. In *Artificial Life XI*, pages 351–358. MIT Press, 2008.

[512] A. Lindenmayer. Mathematical models for cellular interaction in development, Parts I and II. *Journal of Theoretical Biology*, 18:280–315, 1968.

[513] H. Lipson, J. Pollack, and N. Suh. On the origin of modular variation. *Evolution*, Dec. 2002.

[514] R. J. Lipton. DNA solution of hard computational problems. *Science*, 268(5210):542–545, Apr. 1995.

[515] R. J. Lipton. Speeding up computation via molecular biology. In R. J. Lipton and E. B. Baum, editors, *DNA Based Computers, Proceedings of a DIMACS Workshop, Apr. 1995*, volume 27 of *DIMACS Series in Discrete Mathematics and Theoretical Computer Science*. 1996.

[516] J. E. Lisman. A mechanism for memory storage insensitive to molecular turnover: A bistable autophosphorylating kinase. *Proceedings of the National Academy of Sciences*, 82:3055–3057, May 1985.

[517] H. Lodish, A. Berk, P. Matsudaira, C. Kaiser, M. Krieger, M. Scott, L. Zipursky, and J. Darnell. *Molecular Cell Biology*. W. H. Freeman & Co., 7th edition, 2013.

[518] A. Loinger and O. Biham. Stochastic simulations of the repressilator circuit. *Physical Review E*, 76(5):051917, Nov. 2007.

[519] M. A. Lones, L. A. Fuente, A. P. Turner, L. S. Caves, S. Stepney, S. L. Smith, and A. M. Tyrrell. Artificial biochemical networks: Evolving dynamical systems to control dynamical systems. *IEEE Transactions on Evolutionary Computation*, 18(2):145–166, Apr. 2014.

[520] M. A. Lones, A. M. Tyrrell, S. Stepney, and L. S. Caves. Controlling complex dynamics with artificial biochemical networks. In A. Esparcia-Alcázar, A. Ekárt, S. Silva, S. Dignum, and A. Uyar, editors, *Genetic Programming*, volume 6021 of *Lecture Notes in Computer Science*, pages 159–170. Springer, 2010.

[521] R. L. Lopes and E. Costa. Rencode: A regulatory network computational device. In S. Silva, J. A. Foster, M. Nicolau, P. Machado, and M. Giacobini, editors, *Genetic Programming*, volume 6621 of *Lecture Notes in Computer Science*, pages 142–153. Springer, 2011.

[522] R. L. Lopes and E. Costa. The regulatory network computational device. *Genetic Programming and Evolvable Machines*, 13(3):339–375, 2012.

[523] E. Lorenz. Deterministic nonperiodic flow. *Journal of Atmospheric Science*, 20:130–141, 1963.

[524] J. Lovelock. Geophysiology, the science of Gaia. *Reviews of Geophysics*, 27(2):215–222, May 1989.

[525] J. Lovelock and L. Margulis. Atmospheric homoeostasis by and for the biosphere: The gaia hypothesis. *Tellus*, 26:2–9, 1974.

[526] D. Lowe and D. Miorandi. All roads lead to Rome: Data highways for dense wireless sensor networks. In *Proc. 1st International Conference on Sensor Systems and Software (S-Cube)*. ICST, 2009.

[527] D. Lowe, D. Miorandi, and K. Gomez. Activation-inhibition-based data highways for wireless sensor networks. In *Proceedings of the 4th International Conference on Bio-Inspired Models of Network, Information, and Computing Systems (BIONETICS)*. ICST, 2009.

[528] T. Lu, D. Volfson, L. Tsimring, and J. Hasty. Cellular growth and division in the Gillespie algorithm. *Systems Biology*, 1(1):121–128, June 2004.

[529] M. W. Lucht. Size selection and adaptive evolution in an artificial chemistry. *Artificial Life*, 18(2):143–163, Spring 2012.

[530] M. W. Lugowski. Computational metabolism: Towards biological geometries for computing. In C. G. Langton, editor, *Artificial Life*, pages 341–368. Addison-Wesley, 1989.

[531] P. L. Luisi. The chemical implementation of autopoisis. In G. R. Fleischaker, S. Colonna, and P. L. Luisi, editors, *Self-Production of Supramolecular Structures*, pages 179–197. Kluwer, 1991.

[532] P. L. Luisi. About various definitions of life. *Origins of Life and Evolution of Biospheres*, Dec. 1998.

[533] P. L. Luisi. Autopoiesis: A review and a reappraisal. *Naturwissenschaften*, 90:49–59, 2003.

[534] P. L. Luisi. *The Emergence of Life: From Chemical Origins to Synthetic Biology*. Cambridge University Press, 2006.

[535] P. L. Luisi, F. Ferri, and P. Stano. Approaches to semi-synthetic minimal cells: A review. *Naturwissenschaften*, Dec. 2006.

[536] P. L. Luisi and F. J. Varela. Self replicating micelles — A chemical version of a minimal autopoietic system. *Origins of Life and Evolution of the Biosphere*, 19:633–643, 1989.

[537] T. Lundh and P. Gerlee. Cross-feeding dynamics described by a series expansion of the replicator equation. *Bulletin of Mathematical Biology*, 75(5):709–724, 2013.

[538] D. Madina, N. Ono, and T. Ikegami. Cellular evolution in a 3D lattice artificial chemistry. In W. Banzhaf, T. Christaller, P. Dittrich, J. T. Kim, and J. Ziegler, editors, *Advances in Artificial Life: Proceedings of the 7th European Conference on Artificial Life (ECAL)*, volume 2801 of *Lecture Notes in Computer Science*, pages 59–68. Springer, 2003.

[539] M. Magnasco. Chemical kinetics is Turing universal. *Physical Review Letters*, 78(6):1190–1193, 1997.

[540] C. C. Maley. Four steps toward open-ended evolution. In *Proceedings of the Genetic and Evolutionary Computation Conference (GECCO)*, pages 1336–1343, July 1999.

[541] S. S. Mansy, J. P. Schrum, M. Krishnamurthy, S. Tobe, D. A. Treco, and J. W. Szostak. Template-directed synthesis of a genetic polymer in a model protocell. *Nature*, 454(7200):122–125, July 2008.

[542] S. S. Mansy and J. W. Szostak. Reconstructing the emergence of cellular life through the synthesis of model protocells. *Cold Spring Harbor Symposia on Quantitative Biology Quantitative Biology*, 74:47–54, Sept. 2009.

[543] D. L. Marcy, B. W. Vought, and R. R. Birge. Bioelectronics and protein-based optical memories and processors. In Sienko et al. [776], chapter 1, pages 1–31.

[544] T. T. Marquez-Lago and K. Burrage. Binomial tau-leap spatial stochastic simulation algorithm for applications in chemical kinetics. *Journal of Chemical Physics*, 127(10):104101–1–9, 2007.

[545] J. D. Marth. A unified vision of the building blocks of life. *Nature Cell Biology*, 10(9):1015–1016, Sept. 2008.

[546] D. Martinelli. Introduction (to the issue and to zoomusicology). *Trans. Revista Transcultural de Música*, (12), 2008.

[547] J. Masel and M. V. Trotter. Robustness and evolvability. *Trends in Genetics*, 26(9):406–414, 2010.

[548] W. H. Mather, J. Hasty, and L. S. Tsimring. Fast stochastic algorithm for simulating evolutionary population dynamics. *Bioinformatics*, 28(9):1230–1238, 2012.

[549] D. H. Mathews and D. H. Turner. Prediction of rna secondary structure by free energy minimization. *Current opinion in structural biology*, 16:270–278, 2006.

[550] N. Matsumaru, F. Centler, P. Speroni di Fenizio, and P. Dittrich. Chemical organization theory as a theoretical base for chemical computing. *International Journal of Unconventional Computing*, 3(4):285–309, 2007.

[551] N. Matsumaru and P. Dittrich. Organization-oriented chemical programming for the organic design of distributed computing systems. In *Proceedings of Bio-Inspired Models of Network, Information and Computing Systems (BIONETICS)*, Cavalese, Italy, Dec. 2006. IEEE.

[552] N. Matsumaru, P. Kreyssig, and P. Dittrich. Organisation-oriented chemical programming. In C. Müller-Schloer, H. Schmeck, and T. Ungerer, editors, *Organic Computing — A Paradigm Shift for Complex Systems*, volume 1 of *Autonomic Systems*, pages 207–220. Springer, 2011.

[553] N. Matsumaru, T. Lenser, T. Hinze, and P. Dittrich. Designing a chemical program using chemical organization theory. *BMC Systems Biology*, 1(Suppl 1):1–2, May 2007.

[554] N. Matsumaru, T. Lenser, T. Hinze, and P. Dittrich. Toward organization-oriented chemical programming: A case study with the maximal independent set problem. In F. Dressler and I. Carreras, editors, *Advances in Biologically Inspired Information Systems*, volume 69 of *Studies in Computational Intelligence*, pages 147–163. Springer, 2007.

[555] N. Matsumaru, P. Speroni di Fenizio, F. Centler, and P. Dittrich. On the evolution of chemical organizations. In S. Artmann and P. Dittrich, editors, *Proc. 7th German Workshop on Artificial Life*, pages 135–146, 2006.

[556] H. Matsuno, M. M. Hanczyc, and T. Ikegami. Self-maintained movements of droplets with convection flow. In *Progress in Artificial Life*, volume 4828 of *Lecture Notes in Computer Science*, pages 179–188. 2007.

[557] H. R. Maturana. The organization of the living: A theory of the living organization. *International Journal of Man-Machine Studies*, 7(3):313–332, May 1975.

[558] H. R. Maturana and F. J. Varela. Autopoiesis: The organization of the living. In *Autopoiesis and Cognition: The Realization of the Living* [559], pages 59–138.

[559] H. R. Maturana and F. J. Varela. *Autopoiesis and Cognition: The Realization of the Living*, volume 42 of *Series: Boston Studies in the Philosophy of Science*. D. Reidel, 1980.

[560] F. Mavelli and K. Ruiz-Mirazo. Stochastic simulations of minimal self-reproducing cellular systems. *Philosophical Transactions of the Royal Society B: Biological Sciences*, 362(1486):1789–1802, Oct. 2007.

[561] R. May. Simple mathematical models with very complicated dynamics. *Nature*, 261:459–467, 1976.

[562] B. Mayer and S. Rasmussen. The lattice molecular automaton (LMA): A simulation system for constructive molecular dynamics. *Int. J. of Modern Physics C*, 9(1):157–177, 1998.

[563] J. Maynard Smith. *Evolution and the Theory of Games*. Cambridge University Press, 1982.

[564] J. Maynard Smith. *Evolutionary Genetics*. Oxford University Press, 2nd edition, 1998.

[565] J. Maynard Smith and E. Szathmáry. *The Major Transitions in Evolution.* Oxford University Press, 1997.

[566] J. S. McCaskill. Polymer chemistry on tape: A computational model for emergent genetics. Technical report, MPI for Biophysical Chemistry, 1988.

[567] P. K. McKinley, B. H. C. Cheng, C. Ofria, D. B. Knoester, B. Beckmann, and H. Goldsby. Harnessing digital evolution. *IEEE Computer,* 41(1):54–63, Jan. 2008.

[568] B. McMullin. Remarks on autocatalysis and autopoiesis. *Annals of the New York Academy of Sciences,* 901(1):163–174, 2000.

[569] B. McMullin. 30 years of computational autopoiesis: A review. *Artificial Life,* 10(3):277–296, 2004.

[570] B. McMullin, C. Kelly, and D. O'Brien. Multi-level selectional stalemate in a simple artificial chemistry. In F. Almeida e Costa, L. M. Rocha, E. Costa, I. Harvey, and A. Coutinho, editors, *Advances in Artificial Life,* volume 4648 of *Lecture Notes in Computer Science,* pages 22–31. Springer, 2007.

[571] D. W. McShea. Metazoan complexity and evolution: Is there a trend? *International Journal of Organic Evolution,* 50(2):477–492, Apr. 1996.

[572] H. Meinhardt. *Models of Biological Pattern Formation.* Academic Press, 1982.

[573] H. Meinhardt. *The Algorithmic Beauty of Sea Shells.* Springer, 2003.

[574] N. Metropolis and S. Ulam. The Monte Carlo method. *Journal of the American Statistical Association,* 44:335–341, 1949.

[575] T. Meyer, D. Schreckling, C. Tschudin, and L. Yamamoto. Robustness to code and data deletion in autocatalytic quines. In C. Priami, F. Dressler, O. B. Akan, and A. Ngom, editors, *Transactions on Computational Systems Biology (TCSB) X,* volume 5410 of *Lecture Notes in Bioinformatics (LNBI),* pages 20–40. Springer, 2008.

[576] T. Meyer and C. Tschudin. Chemical networking protocols. In *Proceedings of the 8th ACM Workshop on Hot Topics in Networks (HotNets-VIII),* Oct. 2009.

[577] T. Meyer and C. Tschudin. Robust network services with distributed code rewriting. In P. Lio and D. Verma, editors, *Biologically Inspired Networking and Sensing: Algorithms and Architectures,* chapter 3, pages 36–57. IGI Global, 2012.

[578] T. Meyer and C. Tschudin. A theory of packet flows based on law-of-mass-action scheduling. In *31st IEEE International Symposium on Reliable Distributed Systems (SRDS,* 2012.

[579] T. Meyer, L. Yamamoto, W. Banzhaf, and C. Tschudin. Elongation control in an algorithmic chemistry. In G. Kampis, I. Karsai, and E. Szathmáry, editors, *Advances in Artificial Life. Darwin Meets von Neumann,* volume 5777 of *Lecture Notes in Computer Science,* pages 267–274. Springer, 2011.

[580] T. Meyer, L. Yamamoto, and C. Tschudin. An artificial chemistry for networking. In P. Liò, E. Yoneki, J. Crowcroft, and D. C. Verma, editors, *Bio-Inspired Computing and Communication,* volume 5151 of *LNCS,* pages 45–57. Springer, 2008.

[581] T. Meyer, L. Yamamoto, and C. Tschudin. A self-healing multipath routing protocol. In *Proceedings 3rd International Conference on Bio-Inspired Models of Network, Information, and Computing Systems (BIONETICS),* pages 1–8. ICST, Nov. 2008.

[582] R. Michod. *Darwinian Dynamics.* Princeton University Press, 1999.

[583] T. Miconi. Evolution and complexity: The double-edged sword. *Artificial Life,* 14:325–344, 2008.

[584] J. Mill. *System of Logic.* John W. Parker, 1843.

[585] J. F. Miller. Evolving a self-repairing, self-regulating, French flag organism. In K. Deb, editor, *Genetic and Evolutionary Computation (GECCO)*, volume 3102 of *Lecture Notes in Computer Science*, pages 129–139. Springer, 2004.

[586] J. F. Miller and W. Banzhaf. Evolving the program for a cell: From French flags to Boolean circuits. In *On Growth, Form and Computers*. 2003.

[587] S. L. Miller. Production of amino acids under possible primitive Earth conditions. *Science*, 117(3046):528–529, May 1953.

[588] S. L. Miller and H. C. Urey. Organic compound synthesis on the primitive Earth. *Science*, 130(3370):245–251, July 1959.

[589] R. Milner. *Communication and Concurrency*. Prentice Hall, 1989.

[590] M. Mitchell. *An Introduction to Genetic Algorithms*. Bradford, 1998.

[591] M. Mitchell. *Complexity: A Guided Tour*. Oxford University Press, 2009.

[592] T. Miura and K. Tominaga. An approach to algorithmic music composition with an artificial chemistry. In S. Artmann and P. Dittrich, editors, *Explorations in the Complexity of Possible Life, Proceedings of the 7th German Workshop on Artificial Life (GWAL-2006)*, pages 21–30. Aka and IOS Press, 2006.

[593] M. Mobilia, I. T. Georgiev, and U. C. Täuber. Phase transitions and spatio-temporal fluctuations in stochastic lattice Lotka-Volterra models. *Journal of Statistical Physics*, 128(1-2):447–483, 2007.

[594] M. Monti, T. Meyer, C. Tschudin, and M. Luise. Stability and sensitivity analysis of traffic-shaping algorithms inspired by chemical engineering. *IEEE Journal on Selected Areas in Communication (JSAC), Special Issue on Network Science*, 31(6), 2013.

[595] M. Monti, M. Sifalakis, T. Meyer, C. Tschudin, and M. Luise. A chemical-inspired approach to design distributed rate controllers for packet networks. In *IFIP/IEEE International Symposium on Integrated Network Management (IM), 6th workshop on Distributed Autonomous Network Management Systems (DANMS)*, May 2013.

[596] L. A. Moran, H. R. Horton, K. G. Scrimgeour, and M. D. Perry. *Principles of Biochemistry*. Pearson, 5th edition, 2014.

[597] H. C. Morris. Typogenetics: A logic for artificial life. In C. G. Langton, editor, *Artificial Life*, pages 341–368. Addison-Wesley, 1989.

[598] H. C. Morris. *Typogenetics: A logic of artificial propagating entities*. PhD thesis, University of British Columbia, 1989.

[599] E. Mossel and M. Steel. Random biochemical networks: The probability of self-sustaining autocatalysis. *Journal of Theoretical Biology*, 233:327–336, 2005.

[600] I. Motoike and K. Yoshikawa. Information operations with an excitable field. *Physical Review E*, 59:5354–5360, May 1999.

[601] I. N. Motoike, K. Yoshikawa, Y. Iguchi, and S. Nakata. Real-time memory on an excitable field. *Physical Review E*, 63:036220 1–4, Feb. 2001.

[602] J. Moult, K. Fidelis, A. Kryshtafovych, and A. Tramontano. Critical assessment of methods of protein structure prediction (CASP)-round IX. *Proteins: Structure, Function, and Bioinformatics*, 79(S10):1–5, 2011.

[603] G. E. Mueller and G. P. Wagner. Novelty in evolution: Restructuring the concept. *Annual Review of Ecology and Systematics*, 22:229–256, 1991.

[604] A. Munteanu and R. V. Solé. Phenotypic diversity and chaos in a minimal cell model. *Journal of Theoretical Biology*, 240(3):434–442, 2006.

[605] H. Nagahara, T. Ichino, and K. Yoshikawa. Direction detector on an excitable field: Field computation with coincidence detection. *Physical Review E*, 70:036221 1–5, Sept. 2004.

[606] T. Nakagaki, H. Yamada, and A. Toth. Intelligence: Maze-solving by an amoeboid organism. *Nature*, 407(6803):470, Sept. 2000.

[607] T. Nakano, M. Moore, A. Enomoto, and T. Suda. Molecular communication technology as a biological ICT. In *Biological Functions for Information and Communication Technologies*, volume 320 of *Studies in Computational Intelligence*, pages 49–86. Springer, 2011.

[608] G. Neglia and G. Reina. Evaluating activator-inhibitor mechanisms for sensors coordination. In *Proceedings of the 2nd International Conference on Bio-Inspired Models of Network, Information, and Computing Systems (BIONETICS)*. ICST, 2007.

[609] A. Nellis and S. Stepney. Automatically moving between levels in artificial chemistries. In Fellermann et al. [268], pages 269–276.

[610] H. Neumann, K. Wang, L. Davis, M. Garcia-Alai, and J. W. Chin. Encoding multiple unnatural amino acids via evolution of a quadruplet-decoding ribosome. *Nature*, 464(7287):441–444, 2010.

[611] G. Nicolis and I. Prigogine. *Self-Organization in Nonequilibrium Systems*. Wiley, 1977.

[612] G. Nierhaus. *Algorithmic composition*. Springer, 2009.

[613] T. Y. Nishida. Simulations of photosynthesis by a K-subset transforming system with membranes. *Fundamenta Informaticae*, 49(1-3):249–259, 2002.

[614] T. Y. Nishida. Membrane algorithms. In R. Freund, G. Păun, G. Rozenberg, and A. Salomaa, editors, *Membrane Computing*, volume 3850 of *Lecture Notes in Computer Science*, pages 55–66. Springer, 2006.

[615] T. Nishikawa, T. Sunami, T. Matsuura, and T. Yomo. Directed evolution of proteins through in vitro protein synthesis in liposomes. *Journal of Nucleic Acids*, 2012(Article ID 923214), 2012.

[616] W. Nitschke and M. J. Russell. Beating the acetyl coenzyme A-pathway to the origin of life. *Philosophical Transactions of the Royal Society B: Biological Sciences*, 368(1622):20120258, 2013.

[617] W. Niu, P. G. Schultz, and J. Guo. An expanded genetic code in mammalian cells with a functional quadruplet codon. *ACS Chemical Biology*, 2013.

[618] D. Noble. A theory of biological relativity. *Interface Focus*, 2:55–64, 2012.

[619] V. Noireaux and A. Libchaber. A vesicle bioreactor as a step toward an artificial cell assembly. *Proceedings of the National Academy of Sciences*, 101(51):17669–17674, 2004.

[620] P. Nordin and W. Banzhaf. Genetic reasoning evolving proofs with genetic search. Technical report, University of Dortmund, Department of Computer Science, SysReport, Dortmund, Germany, 1996.

[621] M. Nowak. *Evolutionary Dynamics: Exploring the Equations of Life*. Belknap Press, 2006.

[622] M. Nowak and N. L. Komarova. Towards an evolutionary theory of language. *Trends in Cognitive Sciences*, 5(7):288–295, 2001.

[623] M. Nowak, N. L. Komarova, and P. Niyogi. Evolution of universal grammar. *Science*, 291(5501):114–118, 2001.

[624] M. Nowak and D. C. Krakauer. The evolution of language. *Proceedings of the National Academy of Sciences*, 96(14):8028–8033, 1999.

[625] M. Nowak and H. Ohtsuki. Prevolutionary dynamics and the origin of evolution. *Proceedings of the National Academy of Sciences*, 105(39), Sept. 2008.

[626] M. Nowak, J. B. Plotkin, and V. A. A. Jansen. The evolution of syntactic communication. *Nature*, 404:495–498, Mar. 2000.

[627] M. Nowak and P. Schuster. Error thresholds of replication in finite populations mutation frequencies and the onset of Muller's ratchet. *Journal of Theoretical Biology*, 137(4):375–395, 1989.

[628] C. Ofria, C. Adami, and T. C. Collier. Design of evolvable computer languages. *IEEE Transactions on Evolutionary Computation*, 6(4):420–424, Aug. 2002.

[629] C. Ofria, D. M. Bryson, and C. O. Wilke. Avida: A software platform for research in computational evolutionary biology. In M. Komosinksi and A. Adamatzky, editors, *Artificial Life Models in Software*, chapter 1, pages 3–34. Springer, 2009.

[630] C. Ofria, W. Huang, and E. Torng. On the gradual evolution of complexity and the sudden emergence of complex features. *Artificial Life*, 14(3):255–263, 2008.

[631] M. Okamoto and K. Hayashi. Network study of integrated biochemical switching system I: Connection of basic elements. *Biosystems*, 24(1):39–52, 1990.

[632] M. Okamoto, A. Katsurayama, M. Tsukiji, Y. Aso, and K. Hayashi. Dynamic behavior of enzymatic system realizing two-factor model. *Journal of Theoretical Biology*, 83(1):1–16, Mar. 1980.

[633] M. Okamoto, T. Sakai, and K. Hayashi. Switching mechanism of a cyclic enzyme system: role as a "chemical diode". *Biosystems*, 21(1):1–11, 1987.

[634] M. Okamoto, T. Sakai, and K. Hayashi. Biochemical switching device realizing McCulloch-Pitts type equation. *Biological Cybernetics*, 58(5):295–299, Apr. 1988.

[635] M. Okamoto, T. Sakai, and K. Hayashi. Biochemical switching device: How to turn on (off) the switch. *Biosystems*, 22(2):155–162, 1989.

[636] S. Okasha. Multilevel selection and the major transitions in evolution. *Philosophy of Science*, 72:1013–1025, 2005.

[637] N. Ono. Computational studies on conditions of the emergence of autopoietic protocells. *BioSystems*, 81(3):223–233, 2005.

[638] N. Ono, Y. Fujiwara, and K. Yuta. Artificial metabolic system: An evolutionary model for community organization in metabolic networks. In M. Capcarrere et al., editors, *Proceedings of the European Artificial Life Conference (ECAL)*, LNAI 3630, pages 716–724. Springer, 2005.

[639] N. Ono and T. Ikegami. Model of self-replicating cell capable of self-maintenance. In D. Floreano, J.-D. Nicoud, and F. Mondana, editors, *Proceedings of Fifth European Conference on Artificial Life (ECAL'99)*, pages 399–406. Springer, 1999.

[640] N. Ono and T. Ikegami. Self-maintenance and self-reproduction in an abstract cell model. *Journal of Theoretical Biology*, 206(2):243–253, 2000.

[641] N. Ono and T. Ikegami. Artificial chemistry: Computational studies on the emergence of self-reproducing units. In J. Kelemen and P. Sosík, editors, *Advances in Artificial Life*, volume 2159 of *Lecture Notes in Computer Science*, pages 186–195. Springer, 2001.

[642] N. Ono and H. Suzuki. String-based artificial chemistry that allows maintenance of different types of self-replicators. *Journal of Three Dimensional Images*, 16:148–153, 2002.

[643] T. Oohashi, O. Ueno, T. Maekawa, N. Kawai, E. Nishina, and M. Honda. An effective hierarchical model for the biomolecular covalent bond: An approach integrating artificial chemistry and an actual terrestrial life system. *Artificial Life*, 15(1):29–58, 2009.

[644] A. I. Oparin. *The Origin of Life*. Dover, 2nd edition, 2003.

[645] L. E. Orgel. The origin of life: A review of facts and speculations. *Trends in Biochemical Sciences*, 23(12):491–495, 1998.

[646] T. O'Riordan and J. Cameron, editors. *Interpreting the Precautionary Principle*. Earthscan, 1994.

[647] M. Ormo, A. B. Cubitt, K. Kallio, L. A. Gross, R. Y. Tsien, and S. J. Remington. Crystal structure of the Aequorea victoria green fluorescent protein. *Science*, 273:1392–1395, 1996.

[648] L. Pagie and P. Hogeweg. Evolutionary consequences of coevolving targets. *Evolutionary Computation*, 5(4):401–418, 1997.

[649] A. Pai, Y. Tanouchi, C. H. Collins, and L. You. Engineering multicellular systems by cell-cell communication. *Current Opinion in Biotechnology*, 20(4):461–470, 2009.

[650] A. N. Pargellis. The evolution of self-replicating computer organisms. *Physica D*, 98(1):111–127, 1996.

[651] A. N. Pargellis. The spontaneous generation of digital "life". *Physica D*, 91(1-2):86–96, 1996.

[652] A. N. Pargellis. Digital life behavior in the amoeba world. *Artificial Life*, 7(1):63–75, 2001.

[653] J. Parkinson and D. Blair. Does E. coli have a nose? *Science*, 259:1701–1702, 1993.

[654] D. P. Parsons, C. Knibbe, and G. Beslon. Homologous and nonhomologous rearrangements: Interactions and effects on evolvability. In T. Lenaerts, M. Giacobini, H. Bersini, P. Bourgine, M. Dorigo, and R. Doursat, editors, *Advances in Artificial Life, ECAL 2011: Proceedings of the Eleventh European Conference on the Synthesis and Simulation of Living Systems*, pages 622–629. MIT Press, 2011.

[655] M. Parter, N. Kashtan, and U. Alon. Environmental variability and modularity of bacterial metabolic networks. *BMC Evolutionary Biology*, 7(169), 2007.

[656] M. Patriarca and T. Leppänen. Modeling language competition. *Physica A: Statistical Mechanics and its Applications*, 338(1-2):296–299, 2004.

[657] N. Paul and G. F. Joyce. A self-replicating ligase ribozyme. *Proceedings of the National Academy of Sciences*, 99(20):12733–12740, October 2002.

[658] J. E. Pearson. Complex patterns in a simple system. *Science*, 261(5118):189–192, July 1993.

[659] J. T. Pedersen and J. Moult. Protein folding simulations with genetic algorithms and a detailed molecular description. *Journal of Molecular Biology*, 269(2):240–259, 1997.

[660] J. Penner, R. Hoar, and C. Jacob. Modelling bacterial signal transduction pathways through evolving artificial chemistries. In *Proceedings of the First Australian Conference on Artificial Life, Camberra, ACT, Australia*, 2003.

[661] M. J. Pérez-Jiménez and F. J. Romero-Campero. P systems, a new computational modelling tool for systems biology. *Transactions on Computational Systems Biology VI, Lecture Notes in Bioinformatics, 4220*, pages 176–197, 2006.

[662] K. S. Perumalla. *Introduction to Reversible Computing*. Chapman and Hall/CRC, 2013.

[663] R. Pfeifer, F. Iida, and G. Gómez. Designing intelligent robots: On the implications of embodiment. *Journal of the Robotics Society of Japan*, 24:9–16, 2006.

[664] R. Pfeifer, M. Lungarella, and F. Iida. Self-organization, embodiment, and biologically inspired robotics. *Science*, 318(5853):1088–1093, 2007.

[665] W. Piaseczny, H. Suzuki, and H. Sawai. Chemical genetic programming: Evolution of amino acid rewriting rules used for genotype-phenotype translation. In *Congress on Evolutionary Computation (CEC)*, volume 2, pages 1639–1646, June 2004.

[666] V. B. Pinheiro, A. I. Taylor, C. Cozens, M. Abramov, M. Renders, S. Zhang, J. C. Chaput, J. Wengel, S.-Y. Peak-Chew, S. H. McLaughlin, P. Herdewijn, and P. Holliger. Synthetic genetic polymers capable of heredity and evolution. *Science*, 336(6079):341–344, 2012.

[667] K. W. Plaxco and M. Gross. *Astrobiology: A Brief Introduction*. Johns Hopkins University Press, 2nd edition, 2011.

[668] H. Poincaré. *Science et Méthode*. Flammarion, 1909.

[669] R. Poli, J. Kennedy, and T. Blackwell. Particle swarm optimization. *Swarm Intelligence*, 1(1):33–57, 2007.

[670] A. M. Poole, M. J. Phillips, and D. Penny. Prokaryote and eukaryote evolvability. *Biosystems*, 69(2-3):163–185, 2003.

[671] K. Popper. *Objective Knowledge: An Evolutionary Approach*. Oxford University Press, 1972.

[672] M. A. Potter and K. A. De Jong. Cooperative coevolution: An architecture for evolving coadapted subcomponents. *Evolutionary Computation*, 8(1):1–29, 2000.

[673] M. W. Powner, B. Gerland, and J. D. Sutherland. Synthesis of activated pyrimidine ribonucleotides in prebiotically plausible conditions. *Nature*, 459:239–242, May 2009.

[674] C. Priami, A. Regev, E. Shapiro, and W. Silverman. Application of a stochastic name-passing calculus to representation and simulation of molecular processes. *Information Processing Letters*, 80(1):25–31, 2001.

[675] I. Prigogine and R. Lefever. Symmetry breaking instabilities in dissipative systems. II. *Journal of Chemical Physics*, 48(4):1695–1700, 1968.

[676] M. Prokopenko, F. Boschetti, and A. Ryan. An information-theoretic primer on complexity, self-organization, and emergence. *Complexity*, 15:11–28, 2009.

[677] P. Prusinkiewicz and A. Lindenmayer. *The Algorithmic Beauty of Plants*. Springer, 1990.

[678] G. Păun. Computing with membranes. Technical report, Turku Center for Computer Science, TUCS Report 208, 1998.

[679] G. Păun. P systems with active membranes: Attacking NP-complete problems. *Journal of Automata, Languages and Combinatorics*, 6(1):75–90, Jan. 2001.

[680] G. Păun. *Membrane Computing: An Introduction*. Springer, 2002.

[681] G. Păun and G. Rozenberg. A guide to membrane computing. *Theoretical Computer Science*, 287:73–100, 2002.

[682] G. Păun, G. Rozenberg, and A. Salomaa. *DNA Computing: New Computing Paradigms*. Texts in Theoretical Computer Science. An EATCS Series. Springer, 1998.

[683] G. Păun, Y. Suzuki, H. Tanaka, and T. Yokomori. On the power of membrane division in P systems. *Theoretical Computer Science*, 324(1):61–85, 2004.

[684] A. A. Rabow and H. A. Scheraga. Improved genetic algorithm for the protein folding problem by use of a cartesian combination operator. *Protein Science*, 5(9):1800–1815, 1996.

[685] N. Ramakrishnan and U. S. Bhalla. Memory switches in chemical reaction space. *PLoS Computational Biology*, 4(7), July 2008.

[686] S. Raman, R. Vernon, J. Thompson, M. Tyka, R. Sadreyev, J. Pei, D. Kim, E. Kellogg, F. DiMaio, O. Lange, L. Kinch, W. Sheffler, B.-H. Kim, R. Das, N. V. Grishin, and D. Baker. Structure prediction for CASP8 with all-atom refinement using Rosetta. *Proteins*, 77(9):89–99, 2009.

[687] N. Rambidi and A. V. Maximychev. Molecular neural network devices based on non-linear dynamic media: Basic primitive information processing operations. *Biosystems*, 36(2):87–99, 1995.

[688] N. Rambidi and A. V. Maximychev. Molecular image-processing devices based on chemical reaction systems. 5. processing images with several levels of brightness and some application potentialities. *Advanced Materials for Optics and Electronics*, 7(4):161–170, 1997.

[689] S. Rasmussen, N. Baas, B. Mayer, M. Nilsson, and M. Olesen. Ansatz for dynamical hierarchies. *Artificial Life*, 7(4):329–353, 2002.

[690] S. Rasmussen and C. L. Barrett. Elements of a theory of simulation. In F. Morán, A. Moreno, J. J. M. Guervós, and P. Chacón, editors, *Advances in Artificial Life, Third European Conference on Artificial Life*, volume 929 of *Lecture Notes in Computer Science*, pages 515–529. Springer, 1995.

[691] S. Rasmussen, M. A. Bedau, L. Chen, D. Deamer, D. C. Krakauer, N. H. Packard, and P. F. Stadler, editors. *Protocells: Bridging Nonliving and Living Matter*. The MIT Press, 2008.

[692] S. Rasmussen, M. A. Bedau, J. S. McCaskill, and N. H. Packard. A roadmap to protocells. In Rasmussen et al. [691], chapter 4.

[693] S. Rasmussen, L. Chen, D. Deamer, D. C. Krakauer, N. H. Packard, P. F. Stadler, and M. A. Bedau. Transitions from nonliving to living matter. *Science*, 303:963–965, Oct. 2004.

[694] S. Rasmussen, L. Chen, M. Nilsson, and S. Abe. Bridging nonliving and living matter. *Artificial Life*, 9(3):269–316, 2003.

[695] S. Rasmussen, L. Chen, B. M. R. Stadler, and P. F. Stadler. Proto-organism kinetics: Evolutionary dynamics of lipid aggregates with genes and metabolism. *Origins of Life and Evolution of the Biosphere*, 34(1/2):171–180, 2004.

[696] S. Rasmussen, C. Knudsen, and R. Feldberg. Dynamics of programmable matter. In C. G. Langton, C. Taylor, J. D. Farmer, and S. Rasmussen, editors, *Artificial Life II*, pages 211–291. Addison-Wesley, Feb. 1992.

[697] S. Rasmussen, C. Knudsen, R. Feldberg, and M. Hindsholm. The Coreworld: Emergence and evolution of cooperative structures in a computational chemistry. *Physica*, 42D:111–134, 1990.

[698] M. Rathinam, L. R. Petzold, Y. Cao, and D. T. Gillespie. Stiffness in stochastic chemically reacting systems: The implicit tau-leaping method. *Journal of Chemical Physics*, 119:12784–12794, 2003.

[699] C. Ratze, F. Gillet, J. Mueller, and K. Stoffel. Simulation modelling of ecological hierarchies in constructive dynamical systems. *Ecological Complexity*, 4:13–25, 2007.

[700] E. Ravasz, A. L. Somera, D. A. Mongru, Z. N. Oltvai, and A.-L. Barabási. Hierarchical organization of modularity in metabolic networks. *Science*, 297(5586):1551–1555, 2002.

[701] T. S. Ray. An approach to the synthesis of life. In C. G. Langton, C. Taylor, J. D. Farmer, and S. Rasmussen, editors, *Artificial Life II*, pages 371–408. Addison-Wesley, 1992.

[702] T. S. Ray. A proposal to create two biodiversity reserves: One digital and one organic. Project proposal for Network Tierra, http://life.ou.edu/pubs/reserves/, 1995.

[703] I. Rechenberg. *Evolutionsstrategien*. Frommann-Holzboog Verlag, 1973.

[704] A. Regev, E. M. Panina, W. Silverman, L. Cardelli, and E. Shapiro. BioAmbients: An abstraction for biological compartments. *Theoretical Computer Science*, 325(1):141–167, 2004. Computational Systems Biology.

[705] J. A. Reggia, S. L. Armentrout, H. H. Chou, and Y. Peng. Simple systems that exhibit self-directed replication. *Science*, 259:1282–1287, 1993.

[706] S. Regot, J. Macia, N. Conde, K. Furukawa, J. Kjellen, T. Peeters, S. Hohmann, E. de Nadal, F. Posas, and R. Sole. Distributed biological computation with multicellular engineered networks. *Nature*, 469(7329):207–211, Jan. 2011.

[707] K. A. Reich. The search for essential genes. *Research in Microbiology*, 151(5):319–324, 2000.

[708] J. Reisinger, K. O. Stanley, and R. Miikkulainen. Towards an empirical measure of evolvability. In *Proceedings of the 2005 Workshops on Genetic and Evolutionary Computation*, pages 257–264. ACM, 2005.

[709] C. W. Reynolds. Flocks, herds and schools: A distributed behavioral model. In *Proceedings of the 14th Annual Conference on Computer Graphics and Interactive Techniques (ACM SIGGRAPH)*, volume 21, pages 25–34. ACM, 1987.

[710] A. Ricardo and J. W. Szostak. Origin of life on earth. *Scientific American*, pages 54–61, Sept. 2009.

[711] P. Richmond, D. Walker, S. Coakley, and D. Romano. High performance cellular level agent-based simulation with FLAME for the GPU. *Briefings in Bioinformatics*, 11(3):334–347, 2010.

[712] P. A. Rikvold. Self-optimization, community stability, and fluctuations in two individual-based models of biological coevolution. *Journal of Mathematical Biology*, 55(5-6):653–677, 2007.

[713] P. A. Rikvold and R. K. P. Zia. Punctuated equilibria and $1/f$ noise in a biological coevolution model with individual-based dynamics. *Physical Review E*, 68:031913, Sep 2003.

[714] M. Rizki and M. Conrad. Computing the theory of evolution. *Physica D*, 22:83–99, 1986.

[715] D. Roggen, D. Federici, and D. Floreano. Evolutionary morphogenesis for multi-cellular systems. *Genetic Programming and Evolvable Machines*, 8(1):61–96, Mar. 2007.

[716] A. Romer. *The Vertebrate Body*. W.B. Saunders, 1977.

[717] J. Romero and P. Machado. *The Art of Artificial Evolution: A Handbook of Evolutionary Art and Music*. Springer, 2007.

[718] F. J. Romero-Campero, J. Twycross, M. Camara, M. Bennett, M. Gheorghe, and N. Krasnogor. Modular assembly of cell systems biology models using P systems. *International Journal of Foundations of Computer Science*, 20(3):427–442, 2009.

[719] F. Roselló and G. Valiente. Graph transformation in molecular biology. In H.-J. Kreowski, U. Montanari, F. Orejas, G. Rozenberg, and G. Taentzer, editors, *Formal Methods in Software and Systems Modeling*, volume 3393 of *Lecture Notes in Computer Science*, pages 116–133. Springer, 2005.

[720] R. Rosen. A relational theory of biological systems. *Bulletin of Mathematical Biophysics*, 20:245–260, 1958.

[721] R. Rosen. Two-factor models, neural nets and biochemical automata. *Journal of Theoretical Biology*, 15(3):282–297, June 1967.

[722] R. Rosen. Recent developments in the theory of control and regulation of cellular processes. *International Review of Cytology*, 23:25–88, 1968.

[723] R. Rosen. Some realizations of (M, R)-systems and their interpretation. *Bulletin of Mathematical Biophysics*, 33(3):303–319, 1971.

[724] R. Rosen. Some relational cell models: The metabolism-repair systems. In R. Rosen, editor, *Foundations of Mathematical Biology*, volume II. Academic Press, 1972.

[725] R. Rosen. *Life Itself*. Columbia University Press, 1991.

[726] S. Rosenwald, R. Kafri, and D. Lancet. Test of a statistical model for molecular recognition in biological repertoires. *Journal of Theoretical Biology*, 216(3):327–336, 2002.

[727] O. E. Rössler. A multivibrating switching network in homogeneous kinetics. *Bulletin of Mathematical Biology*, 37:181–192, 1975.

[728] P. Rothemund. A DNA and restriction enzyme implementation of Turing machines. In *DNA Based Computers: Proceedings of the DIMACS Workshop*, pages 75–119. American Mathematical Society, Apr. 1995.

[729] K. Ruiz-Mirazo and F. Mavelli. Simulation model for functionalized vesicles: Lipid-peptide integration in minimal protocells. In F. Almeida e Costa, L. Rocha, E. Costa, I. Harvey, and A. Coutinho, editors, *Advances in Artificial Life*, volume 4648 of *Lecture Notes in Computer Science*, pages 32–41. Springer, 2007.

[730] K. Ruiz-Mirazo and F. Mavelli. On the way towards 'basic autonomous agents': Stochastic simulations of minimal lipid-peptide cells. *Biosystems*, 91(2):374–387, 2008.

[731] R. Russell. *Biophysics of RNA Folding*, volume 3 of *Biophysics for the Life Sciences*. Springer, 2013.

[732] D. Rutherford. *Introduction to Lattice Theory*. Oliver and Boyd, 1966.

[733] A. Ryan. Emergence is coupled to scope not level. *Complexity*, 13:67–77, 2007.

[734] H. Saibil. Chaperone machines for protein folding, unfolding and disaggregation. *Nature Reviews Molecular Cell Biology*, 14(10):630–642, Oct. 2013.

[735] Y. Sakae, T. Hiroyasu, M. Miki, and Y. Okamoto. Protein structure predictions by parallel simulated annealing molecular dynamics using genetic crossover. *Journal of Computational Chemistry*, 32(7):1353–1360, 2011.

[736] A. Salomaa. *Formal Languages*. Academic Press, 1973.

[737] C. Salzberg. A graph-based reflexive artificial chemistry. *BioSystems*, 87:1–12, 2007.

[738] C. Salzberg, A. Antony, and H. Sayama. Complex genetic evolution of self-replicating loops. In J. Pollack, M. Bedau, P. Husbands, T. Ikegami, and R. A. Watson, editors, *Artificial Life IX: Proceedings of the Ninth International Conference on the Simulation and Synthesis of Living Systems*, pages 262–267. MIT Press, 2004.

[739] C. Salzberg, A. Antony, and H. Sayama. Evolutionary dynamics of cellular automata-based self-replicators in hostile environments. *Biosystems*, 78(1-3):119–134, 2004.

[740] R. Santana, P. Larranaga, and J. Lozano. Protein folding in simplified models with estimation of distribution algorithms. *IEEE Transactions on Evolutionary Computation*, 12(4):418–438, 2008.

[741] M. Santillán. On the use of the Hill functions in mathematical models of gene regulatory networks. *Mathematical Modelling of Natural Phenomena*, 3:85–97, 2008.

[742] J. Sardanyés. *Dynamics, Evolution and Information in Nonlinear Dynamical Systems of Replicators*. PhD thesis, Universitat Pompeu Fabra, Barcelona, Spain, 2009.

[743] J. Sardanyés, J. Duarteb, C. Januáriob, and N. Martinsc. Topological entropy of catalytic sets: Hyper-cycles revisited. *Communications in Nonlinear Science and Numerical Simulation*, 17(2):795–803, Feb. 2012.

[744] H. Sayama. A new structurally dissolvable self-reproducing loop evolving in a simple cellular automata. *Artificial Life*, 5(4):343–65, 1999.

[745] H. Sayama. Toward the realization of an evolving ecosystem on cellular automata. In M. Sugisaka and H. Tanaka, editors, *Fourth International Symposium on Artificial Life and Robotics*, pages 254–257. Beppu, Oita, 1999.

[746] H. Sayama. Swarm chemistry. *Artificial Life*, 15(1):105–114, 2009.

[747] H. Sayama. Seeking open-ended evolution in swarm chemistry. In *Proceedings of the Third IEEE Symposium on Artificial Life*, pages 186–193. IEEE, 2011.

[748] H. Sayama. Morphologies of self-organizing swarms in 3D swarm chemistry. In *Proceedings of the Fourteenth International Conference on Genetic and Evolutionary Computation Conference*, pages 577–584. ACM, 2012.

[749] H. Sayama. Swarm-based morphogenetic artificial life. In R. Doursat, H. Sayama, and O. Michel, editors, *Morphogenetic Engineering*, pages 191–208. Springer, 2012.

[750] I. Scheuring, T. Czárán, P. Szabó, G. Károly, and Z. Toroczkai. Spatial models of prebiotic evolution: Soup before pizza? *Origins of Life and Evolution of Biospheres*, 33(4-5):319–355, Oct. 2003.

[751] T. Schlick. *Molecular Modeling and Simulation: An Interdisciplinary Guide*. Springer, 2nd edition, 2010.

[752] S. Schnell and T. E. Turner. Reaction kinetics in intracellular environments with macromolecular crowding: simulations and rate laws. *Progress in Biophysics and Molecular Biology*, 85(2-3):235–260, 2004.

[753] D. Schreckling and T. Marktscheffel. An artificial immune system approach for artificial chemistries based on set rewriting. In E. Hart, C. McEwan, J. Timmis, and A. Hone, editors, *Artificial Immune Systems*, volume 6209 of *Lecture Notes in Computer Science*, pages 250–263. Springer, 2010.

[754] J. P. Schrum, T. F. Zhu, and J. W. Szostak. The origins of cellular life. *Cold Spring Harbor Perspectives in Biology*, May 2010.

[755] P. Schuster. How does complexity arise in evolution. *Complexity*, 2(1):22–30, Sept.-Oct. 1996.

[756] S. Schuster and T. Höfer. Determining all extreme semi-positive conservation relations in chemical reaction systems: A test criterion for conservativity. *Journal of the Chemical Society, Faraday Transactions*, 87:2561–2566, 1991.

[757] R. Schwartz. *Biological Modeling and Simulation: A Survey of Practical Models, Algorithms, and Numerical Methods*. MIT Press, 2008.

[758] L. A. Seelig and O. E. Rössler. A chemical reaction flip-flop with one unique switching input. *Zeitschrift für Naturforschung*, 27b:1441–1444, 1972.

[759] M. G. Seetin and D. H. Mathews. RNA structure prediction: An overview of methods. In K. C. Keiler, editor, *Bacterial Regulatory RNA*, volume 905 of *Methods in Molecular Biology*, pages 99–122. Humana Press, 2012.

[760] D. Segrè, D. Ben-Eli, D. Deamer, and D. Lancet. The Lipid World. *Origins of Life and Evolution of Biospheres*, 31(1):119–145, 2001.

[761] D. Segrè, D. Ben-Eli, and D. Lancet. Compositional genomes: Prebiotic information transfer in mu-
       tually catalytic noncovalent assemblies. *Proceedings of the National Academy of Sciences*, 97(8):4112–
       4117, 2000.

[762] D. Segrè and D. Lancet. Composing life. *EMBO Reports*, 1(3):217–222, Sept. 2000.

[763] D. Segrè and D. Lancet. Theoretical and computational approaches to the study of the origins of life.
       In *Origins: Genesis, Evolution and Diversity of Life, COLE*, volume 6. Springer, 2004.

[764] D. Segrè, D. Lancet, O. Kedem, and Y. Pilpel. Graded autocatalysis replication domain (GARD): Kinetic
       analysis of self-replication in mutually catalytic sets. *Origins of Life and Evolution of Biospheres*, 28(4-
       6):501–514, 1998.

[765] D. Segrè, Y. Pilpel, and D. Lancet. Mutual catalysis in sets of prebiotic organic molecules: Evolution
       through computer simulated chemical kinetics. *Physica A*, 249(1-4):558–564, 1998.

[766] J. Setubal and J. Meidanis. *Introduction to Computational Molecular Biology*. Brooks/Cole Publish-
       ing, 1997.

[767] E. Shapiro. A mechanical Turing machine: Blueprint for a biomolecular computer. *Interface Focus*,
       2(4):497–503, Aug. 2012.

[768] E. Shapiro and Y. Benenson. Bringing DNA computers to life. *Scientific American*, May 2006.

[769] R. Shapiro. A simpler origin for life. *Scientific American*, 296:46–53, June 2007.

[770] O. Shehory and A. Sturm. Evaluation of modeling techniques for agent-based systems. In *Agents*,
       volume 2001, pages 624–631, 2001.

[771] B. Shenhav, A. Bar-Even, R. Kafri, and D. Lancet. Polymer GARD: Computer simulation of covalent
       bond formation in reproducing molecular assemblies. *Origins of Life and Evolution of Biospheres*,
       35(2):111–133, 2005.

[772] B. Shenhav, R. Kafri, and D. Lancet. Graded artificial chemistry in restricted boundaries. In *Pro-
       ceedings of 9th International Conference on the Simulation and Synthesis of Living Systems (ALIFE9)*,
       Boston, Massachusetts, USA, pages 501–506, 2004.

[773] B. Shenhav, A. Oz, and D. Lancet. Coevolution of compositional protocells and their environment.
       Philosophical Transactions of the Royal Society B: Biological Sciences, 362(1486):1813–1819, Oct. 2007.

[774] M. R. Shirts and V. S. Pande. Screen savers of the world, unite! *Science*, 290(5498):1903–1904, Dec.
       2000.

[775] A. Shmygelska and H. H. Hoos. An ant colony optimisation algorithm for the 2D and 3D hydrophobic
       polar protein folding problem. *BMC Bioinformatics*, 6, 2005.

[776] T. Sienko, A. Adamatzky, N. Rambidi, and M. Conrad, editors. *Molecular Computing*. MIT Press, 2003.

[777] H. Simon. *The Sciences of the Artificial*. MIT Press, 1999 (3rd Edition).

[778] K. T. Simons, R. Bonneau, I. Ruczinski, and D. Baker. Ab initio protein structure prediction of CASP
       III targets using ROSETTA. *Proteins: Structure, Function, and Genetics*, Suppl. 3:171–176, 1999.

[779] M. Sipper. Fifty years of research on self-replication: An overview. *Artificial Life*, 4(3):237–257, 1998.

[780] A. Slepoy, A. Thompson, and S. Plimpton. A constant-time kinetic Monte Carlo algorithm for simula-
       tion of large biochemical reaction networks. *Journal of Chemical Physics*, 128:205101–1 – 205101–8,
       2008.

[781] J. Smaldon, J. Blakes, N. Krasnogor, and D. Lancet. A multi-scaled approach to artificial life simulation with P systems and dissipative particle dynamics. In *Proceedings of the 10th Annual Conference on Genetic and evolutionary computation (GECCO)*, pages 249–256. ACM, 2008.

[782] H. O. Smith, C. A. Hutchison, C. Pfannkoch, and J. C. Venter. Generating a synthetic genome by whole genome assembly: φx174 bacteriophage from synthetic oligonucleotides. *Proceedings of the National Academy of Sciences*, 100(26):15440–15445, 2003.

[783] T. Smith and M. Waterman. Identification of common molecular subsequences. *Journal of Molecular Biology*, 147(1):195–197, Mar. 1981.

[784] L. Smolin. *Time Reborn*. Houghton Mifflin Hartcourt, 2013.

[785] A. Snare. Typogenetics. Master's thesis, School of Computer Science and Software Engineering, Monash University, Australia, 1999.

[786] R. Sole and J. Bascompte. *Self-Organization in Complex Ecosystems*. Princeton University Press, 2006.

[787] R. Sole, A. Munteanu, C. Rodriguez-Caso, and J. Macia. Synthetic protocell biology: From reproduction to computation. *Philosophical Transactions of the Royal Society B: Biological Sciences*, 362:1727–1739, 2007.

[788] L. Spector, J. Klein, and M. Feinstein. Division blocks and the open-ended evolution of development, form, and behavior. In *Proceedings of the Genetic and Evolutionary Computation Conference (GECCO)*, pages 316–323, July 2007.

[789] T. Sperl. Taking the redpill: Artificial evolution in native x86 systems. arXiv:1105.1534v1, 2011.

[790] P. Speroni di Fenizio. A less abstract artficial chemistry. In M. A. Bedau, J. S. McCaskill, N. H. Packard, and S. Rasmussen, editors, *Artificial Life VII*, pages 49–53. MIT Press, 2000.

[791] P. Speroni di Fenizio. *Chemical Organization Theory*. PhD thesis, Friedrich Schiller Univ. Jena, 2007.

[792] P. F. Stadler, W. Fontana, and J. H. Miller. Random catalytic reaction networks. *Physica D*, 63:378–392, 1993.

[793] R. Stadler. Molecular, chemical, and organic computing. *Communications of the ACM*, 50(9):43–45, Sept. 2007.

[794] M. Staley. Darwinian selection leads to Gaia. *Journal of Theoretical Biology*, 218:35–46, 2002.

[795] R. K. Standish. Population models with random embryologies as a paradigm for evolution. In R. J. Stonier and X. H. Yu, editors, *Complex Systems: Mechanism of Adaption*. IOS Press, 1994.

[796] R. K. Standish. Ecolab: Where to now? In R. Stocker, H. Jelinek, B. Durnota, and T. Bossomeier, editors, *Complex Systems: From Local Interaction to Global Phenomena*. IOS, 1996.

[797] R. K. Standish. An ecolab perspective on the Bedau evolutionary statistics. In M. A. Bedau, J. S. McCaskill, N. H. Packard, and S. Rasmussen, editors, *Artificial Life VII*, pages 238–242. MIT Press, 2000.

[798] R. K. Standish. Diversity evolution. In *Artificial Life VIII: Proceedings of the Eighth International Conference on Artificial Life*. MIT Press, 2003.

[799] R. K. Standish. Complexity of networks. In H. A. Abbass, T. Bossomaier, and J. Wiles, editors, *Recent Advances in Artificial Life*, volume 3 of *Advances in Natural Computation*, chapter 19, pages 253–263. World Scientific, Dec. 2005.

[800] R. K. Standish. Network complexity of foodwebs. In Fellermann et al. [268], pages 337–343.

[801] P. Stano and P. L. Luisi. Achievements and open questions in the self-reproduction of vesicles and synthetic minimal cells. *Chemical Communications*, 46:3639–3653, 2010.

[802] P. Stano and P. L. Luisi. Semi-synthetic minimal cells: Origin and recent developments. *Current Opinion in Biotechnology*, 24(4):633–638, 2013.

[803] P. Stano, G. Rampioni, P. Carrara, L. Damiano, L. Leoni, and P. L. Luisi. Semi-synthetic minimal cells as a tool for biochemical ICT. *Biosystems*, 109(1):24–34, 2012.

[804] L. Steels. The synthetic modeling of language origins. *Evolution of Communication*, 1(1):1–34, 1997.

[805] L. Steels. The origins of ontologies and communication conventions in multi-agent systems. *Journal of Agents and Multi-Agent Systems*, 1(2):169–194, Oct. 1998.

[806] L. Steels. The origins of syntax in visually grounded robotic agents. *Artificial Intelligence*, 103(1-2):133–156, 1998.

[807] L. Steels. Synthesising the origins of language and meaning using co-evolution, self-organisation and level formation. In J. Hurford, C. Knight, and M. Studdert-Kennedy, editors, *Approaches to the Evolution of Language: Social and Cognitive bases*, chapter 23, pages 384–404. Edinburgh University Press, 1998.

[808] L. Steels. The emergence of grammar in communicating autonomous robotic agents. In W. Horn, editor, *Proceedings of the 14th European Conference on Artificial Intelligence (ECAI)*. IOS Publishing, Aug. 2000.

[809] L. Steels. Language as a complex adaptive system. In M. Schoenauer, K. Deb, G. Rudolph, X. Yao, E. Lutton, J. Merelo, and H.-P. Schwefel, editors, *Parallel Problem Solving from Nature PPSN VI*, volume 1917 of *Lecture Notes in Computer Science*, pages 17–26. Springer, 2000.

[810] L. Steels. A first encounter with fluid construction grammar. In L. Steels, editor, *Design Patterns in Fluid Construction Grammar*, chapter 2, pages 31–68. John Benjamins, 2011.

[811] L. Steels. Introducing Fluid Construction Grammar. In L. Steels, editor, *Design Patterns in Fluid Construction Grammar*, chapter 1, pages 3–30. John Benjamins, 2011.

[812] L. Steels. Modeling the cultural evolution of language. *Physics of Life Reviews*, 8(4):339–356, 2011.

[813] O. Steinbock, P. Kettunen, and K. Showalter. Chemical wave logic gates. *Journal of Physical Chemistry*, 100(49):18970–18975, 1996.

[814] S. Stepney. Nonclassical computation: A dynamical systems perspective. In G. Rozenberg, T. Bäck, and J. N. Kok, editors, *Handbook of Natural Computing*, chapter 59, pages 1979–2025. Springer, 2012.

[815] A. Stoltzfus. On the possibility of contructive neutral evolution. *Journal of Molecular Evolution*, 49:169–181, 1999.

[816] B. Straatman, R. White, and W. Banzhaf. An artificial chemistry-based model of economies. In *Proceedings Artificial Life XI*, pages 592–602, 2008.

[817] A. B. Stundzia and C. J. Lumsden. Stochastic simulation of coupled reaction-diffusion processes. *Journal of Computational Physics*, 127(1):196–207, 1996.

[818] K. Sugiura, H. Suzuki, T. Shiose, H. Kawakami, and O. Katai. Evolution of rewriting rule sets using string-based Tierra. In W. Banzhaf, J. Ziegler, T. Christaller, P. Dittrich, and J. T. Kim, editors, *Advances in Artificial Life*, volume 2801 of *Lecture Notes in Computer Science*, pages 69–77. Springer, 2003.

[819] T. Sunami, K. Sato, K. Ishikawa, and T. Yomo. Population analysis of liposomes with protein synthesis and a cascading genetic network. In Rasmussen et al. [691], chapter 7, pages 157–168.

[820] H. Suzuki. Minimum density of functional proteins to make a system evolvable. *Artificial Life and Robotics*, Dec. 2001.

[821] H. Suzuki. String rewriting grammar optimized using an evolvability measure. In J. Kelemen and P. Sosík, editors, *Advances in Artificial Life*, volume 2159 of *Lecture Notes in Computer Science*, pages 458–468. Springer, 2001.

[822] H. Suzuki. An example of design optimization for high evolvability: String rewriting grammar. *Biosystems*, 69(2-3):211–221, 2003.

[823] H. Suzuki. Models for the conservation of genetic information with string-based artificial chemistry. In W. Banzhaf, J. Ziegler, T. Christaller, P. Dittrich, and J. T. Kim, editors, *Advances in Artificial Life*, volume 2801 of *Lecture Notes in Computer Science*, pages 78–88. Springer, 2003.

[824] H. Suzuki. Mathematical folding of node chains in a molecular network. *Biosystems*, 87(2-3):125–135, 2007.

[825] H. Suzuki. A network cell with molecular agents that divides from centrosome signals. *Biosystems*, 94(1-2):118–125, 2008.

[826] H. Suzuki and N. Ono. Universal replication in a string-based artificial chemistry system. *Journal of Three Dimensional Images*, 16(4):154–159, 2002.

[827] H. Suzuki and N. Ono. Statistical mechanical rewiring in network artificial chemistry. In *The 8th European Conference on Artificial Life (ECAL) Workshop Proceedings CD-ROM. Canterbury, UK*, 2005.

[828] Y. Suzuki, Y. Fujiwara, J. Takabayashi, and H. Tanaka. Artificial life applications of a class of P systems: Abstract rewriting systems on multisets. In C. S. Calude, G. Păun, G. Rozenberg, and A. Salomaa, editors, *Multiset Processing*, volume 2235 of *Lecture Notes in Computer Science*, pages 299–346. Springer, 2001.

[829] Y. Suzuki, J. Takabayashi, and H. Tanaka. Investigation of tritrophic interactions in an ecosystem using abstract chemistry. *Artificial Life and Robotics*, 6(3):129–132, 2002.

[830] Y. Suzuki and H. Tanaka. Symbolic chemical system based on abstract rewriting and its behavior pattern. *Artificial Life and Robotics*, 1:211–219, 1997.

[831] Y. Suzuki and H. Tanaka. Order parameter for a symbolic chemical system. In C. Adami, R. Belew, H. Kitano, and C. Taylor, editors, *Artificial Life VI*, pages 130–139. MIT Press, 1998.

[832] Y. Suzuki and H. Tanaka. Modeling p53 signaling pathways by using multiset processing. In G. Ciobanu, G. Păun, and M. J. Pérez-Jiménez, editors, *Applications of Membrane Computing*, Natural Computing Series, pages 203–214. Springer, 2006.

[833] H. L. Sweeney and A. Houdusse. Myosin VI rewrites the rules for myosin motors. *Cell*, 141:573–582, 2010.

[834] G. Szabó and T. Czárán. Phase transition in a spatial Lotka-Volterra model. *Physical Review E*, 63:061904–1–4, May 2001.

[835] E. Szathmáry. A classification of replicators and lambda-calculus models of biological organization. *Proceedings: Biological Sciences*, 260(1359):279–286, 1995.

[836] E. Szathmáry. The evolution of replicators. *Philosophical Transactions: Biological Sciences*, 355(1403):1669–1676, 2000.

[837] E. Szathmáry and J. Maynard Smith. The major evolutionary transitons. *Nature*, 374:227–232, Mar. 1994.

[838] E. Szathmáry and J. Maynard Smith. From replicators to reproducers: The first major transitions leading to life. *Journal of Theoretical Biology*, 187(4):555–571, Aug. 1997.

[839] J. W. Szostak, D. P. Bartel, and P. L. Luisi. Synthesizing life. *Nature*, 409:387–390, 2001.

[840] H. Takagi and K. Kaneko. Pattern dynamics of a multi-component reaction diffusion system: Differentiation of replicating spots. *International Journal of Bifurcation and Chaos*, 12(11):2579–2598, 2002.

[841] K. Takahashi, K. Kaizu, B. Hu, and M. Tomita. A multi-algorithm, multi-timescale method for cell simulation. *Bioinformatics*, 20(4):538–546, 2004.

[842] K. Takahashi and S. Yamanaka. Induction of pluripotent stem cells from mouse embryonic and adult fibroblast cultures by defined factors. *Cell*, 126(4):663–676, Aug. 2006.

[843] U. Tangen, L. Schulte, and J. S. McCaskill. A parallel hardware evolvable computer POLYP. In K. L. Pocek and J. Arnold, editors, *Proceedings of the 5th Annual IEEE Symposium on Field-Programmable Custom Computing Machines*, pages 238–239. IEEE Computer Society, Apr. 1997.

[844] N. K. Tanner. Ribozymes: The characteristics and properties of catalytic RNAs. *FEMS Microbiology Reviews*, 23(3):257–275, June 1999.

[845] T. Taylor. *From Artificial Evolution to Artificial Life*. PhD thesis, School of Informatics, University of Edinburgh, May 1999.

[846] T. Taylor. A genetic regulatory network-inspired real-time controller for a group of underwater robots. In F. Groen, N. Amato, A. Bonarini, E. Yoshida, and B. Kröse, editors, *Intelligent Autonomous Systems (Proceedings of IAS8)*, pages 403–412. IOS Press, 2004.

[847] T. Taylor and J. Hallam. Studying evolution with self-replicating computer programs. In P. Husbands and I. Harvey, editors, *Proc. Fourth European Conference on Artificial Life (ECAL'97)*, pages 550–559. MIT Press, 1997.

[848] H. Terasawa, K. Nishimura, H. Suzuki, T. Matsuura, and T. Yomo. Coupling of the fusion and budding of giant phospholipid vesicles containing macromolecules. *Proceedings of the National Academy of Sciences*, 109(16):5942–5947, 2012.

[849] C. Thachuk. Logically and physically reversible natural computing: A tutorial. In G. Dueck and D. Miller, editors, *Reversible Computation*, volume 7948 of *Lecture Notes in Computer Science*, pages 247–262. Springer, 2013.

[850] P. Tijerina and R. Russell. The roles of chaperones in RNA folding. [731], chapter 11, pages 205–230.

[851] J. Timmis, A. Hone, T. Stibor, and E. Clark. Theoretical advances in artificial immune systems. *Theoretical Computer Science*, 403(1):11–32, 2008.

[852] T. Toffoli. Reversible computing. In J. Bakker and J. Leeuwen, editors, *Automata, Languages and Programming*, volume 85 of *Lecture Notes in Computer Science*, pages 632–644. Springer, 1980.

[853] T. Toffoli and N. Margolus. *Cellular Automata Machines*. MIT Press, 1987.

[854] K. Tominaga and M. Setomoto. An artificial chemistry approach to generating polyphonic musical phrases. In M. Giacobini and et al, editors, *Proceedings of EvoWorkshops 2008*, LNCS 4974, pages 463–472. Springer, 2008.

[855] K. Tominaga, Y. Suzuki, K. Kobayashi, T. Watanabe, K. Koizumi, and K. Kishi. Modeling biochemical pathways using an artificial chemistry. *Artificial Life*, 15(1):115–129, 2009.

[856] K. Tominaga, T. Watanabe, K. Kobayashi, M. Nakamura, K. Kishi, and M. Kazuno. Modeling molecular computing systems by an artificial chemistry — Its expressive power and application. *Artificial Life*, 13(3):223–247, 2007.

[857] Á. Tóth and K. Showalter. Logic gates in excitable media. *Journal of Chemical Physics*, 103(6):2058–2066, 1995.

[858] T. Toyota, N. Maru, M. M. Hanczyc, T. Ikegami, and T. Sugawara. Self-propelled oil droplets consuming "fuel" surfactant. *Journal of the American Chemical Society*, 131(14):5012–5013, 2009.

[859] M. Trautz. Das Gesetz der Reaktionsgeschwindigkeit und der Gleichgewichte in Gasen. Bestätigung der Additivität von Cv-3/2R. Neue Bestimmung der Integrationskonstanten und der Molekäldurchmesser. *Zeitschrift für anorganische und allgemeine Chemie*, 96:1–28, 1916.

[860] M. Treiber and D. Helbing. Explanation of observed features of self-organization in traffic flow. Technical report, Arxiv preprint cond-mat/9901239, 1999.

[861] S. Tschernyschkow, S. Herda, G. Gruenert, V. Döring, D. Görlich, A. Hofmeister, C. Hoischen, P. Dittrich, S. Diekmann, and B. Ibrahim. Rule-based modeling and simulations of the inner kinetochore structure. *Progress in Biophysics and Molecular Biology*, 113(1):33–45, 2013.

[862] C. Tschudin. Fraglets: A metabolistic execution model for communication protocols. In *Proceedings of the 2nd Annual Symposium on Autonomous Intelligent Networks and Systems (AINS)*, July 2003.

[863] C. Tschudin and T. Meyer. Programming by equilibria. In *15. Kolloquium Programmiersprachen und Grundlagen der Programmierung (KPS)*, pages 37–46, Oct. 2009.

[864] C. Tschudin and L. Yamamoto. A metabolic approach to protocol resilience. In M. Smirnov, editor, *Autonomic Communication*, volume 3457 of *Lecture Notes in Computer Science*, pages 191–206. Springer, 2005.

[865] S. Tsuda, S. Artmann, and K.-P. Zauner. The Phi-Bot: A robot controlled by a slime mould. In A. Adamatzky and M. Komosinski, editors, *Artificial Life Models in Hardware*, pages 213–232. Springer, 2009.

[866] S. Tsuda, K. P. Zauner, and Y. P. Gunji. Robot control with biological cells. *Biosystems*, 87(2-3):215–223, 02 2007.

[867] A. M. Turing. On computable numbers, with an application to the Entscheidungsproblem. *Proceedings of the London Mathematical Society*, Series 2, Vol. 42:230–265, 1936.

[868] A. M. Turing. The chemical basis of morphogenesis. *Philosophical Transactions of the Royal Society (B)*, 237:37–72, 1952.

[869] A. M. Turing. Intelligent machinery. In C. Evans and A. Robertson, editors, *Reprinted in: Cybernetics: Key Papers*. University Park Press, Baltimore, 1968.

[870] T. E. Turner, S. Schnell, and K. Burrage. Stochastic approaches for modelling in vivo reactions. *Computational Biology and Chemistry*, 28(3):165–178, 2004.

[871] T. Tuschl, C. Gohlke, T. M. Jovin, E. Westhof, and F. Eckstein. A three-dimensional model for the hammerhead ribozyme based on fluorescence measurements. *Science*, 266:785–789, 1994.

[872] R. E. Ulanowicz. *The Third Window: A Natural Life beyond Newton and Darwin*. Templeton Foundation Press, 2009.

[873] A. Ullrich and C. Flamm. Functional evolution of ribozyme-catalyzed metabolisms in a graph-based toy-universe. In M. Heiner and A. Uhrmacher, editors, *Computational Methods in Systems Biology*, volume 5307 of *Lecture Notes in Computer Science*, pages 28–43. Springer, 2008.

[874] A. Ullrich and C. Flamm. A sequence-to-function map for ribozyme-catalyzed metabolisms. In G. Kampis, I. Karsai, and E. Szathmáry, editors, *Advances in Artificial Life. Darwin Meets von Neumann*, volume 5778 of *Lecture Notes in Computer Science*, pages 19–26. Springer, 2011.

[875] A. Ullrich, C. Flamm, M. Rohrschneider, and P. F. Stadler. *In silico* evolution of early metabolism. In Fellermann et al. [268], pages 57–64.

[876] A. Ullrich, M. Rohrschneider, G. Scheuermann, P. F. Stadler, and C. Flamm. In silico evolution of early metabolism. *Artificial Life*, 17:87–108, 2011.

[877] K. Usui, N. Ichihashi, Y. Kazuta, T. Matsuura, and T. Yomo. Effects of ribosomes on the kinetics of $Q\beta$ replication. *FEBS Letters*, 588(1):117–123, 2014.

[878] N. Vaidya, M. L. Manapat, I. A. Chen, R. Xulvi-Brunet, E. J. Hayden, and N. Lehman. Spontaneous network formation among cooperative RNA replicators. *Nature*, 491(7422):72–77, 2012.

[879] N. van Kampen. *Stochastic Processes in Physics and Chemistry*. Elsevier, 1992.

[880] S. Van Segbroeck, A. Nowé, and T. Lenaerts. Stochastic simulation of the chemoton. *Artificial Life*, 15(2):213–226, Apr. 2009.

[881] F. J. Varela, H. R. Maturana, and R. Uribe. Autopoiesis: The organization of living systems, its characterization and a model. *BioSystems*, 5(4):187–196, 1974.

[882] L. Varetto. Typogenetics: An artificial genetic system. *Journal of Theoretical Biology*, 160(2):185–205, 1993.

[883] L. Varetto. Studying artificial life with a molecular automaton. *Journal of Theoretical Biology*, 193(2):257–285, 1998.

[884] V. Vasas, E. Szathmáry, and M. Santos. Lack of evolvability in self-sustaining autocatalytic networks constraints metabolism-first scenarios for the origin of life. *Proceedings of the National Academy of Sciences*, 107(4), Jan. 2010.

[885] J. Čejková, M. Novák, F. Štěpánek, and M. M. Hanczyc. Dynamics of chemotactic droplets in salt concentration gradients. *Langmuir*, 30(40):11937–11944, 2014.

[886] S. Verel. *Étude et exploitation des réseaux de neutralité dans les paysages adaptatifs pour l'optimisation difficile*. PhD thesis, Université de Nice-Sophia Antipolis, France, Dec. 2005.

[887] L. Vinicius. *Modular evolution: How natural selection produces biological complexity*. Cambridge University Press, 2010.

[888] N. Virgo, C. Fernando, B. Bigge, and P. Husbands. The elongation catastrophe in physical self-replicators. In T. Lenaerts, M. Giacobini, H. Bersini, P. Bourgine, M. Dorigo, and R. Doursat, editors, *Advances in Artificial Life, ECAL 2011: Proceedings of the Eleventh European Conference on the Synthesis and Simulation of Living Systems*, pages 828–835. MIT Press, 2011.

[889] T. Viscek and A. Zafeiris. Collective motion. *Physics Reports*, 517:71–140, 2012.

[890] G. von Kiedrowski. A self-replicating hexadeoxynucleotide. *Angewandte Chemie International Edition in English*, 25(10):932–935, Oct. 1986.

[891] G. von Kiedrowski. Minimal replicator theory I: Parabolic versus exponential growth. *Bioorganic Chemistry Frontiers*, 3:114–146, 1993.

[892] G. von Kiedrowski and E. Szathmáry. Selection versus coexistence of parabolic replicators spreading on surfaces. *Selection*, Dec. 2001.

[893] J. Von Neumann and A. Burks. *Theory of self-reproducing automata*. University of Illinois Press Champaign, 1966.

[894] B. Vowk, A. S. Wait, and C. Schmidt. An evolutionary approach generates human competitive coreware programs. In *Workshop and Tutorial Proceedings Ninth International Conference on the Simulation and Synthesis of Living Systems (Alife XI)*, pages 33–36, 2004.

[895] G. Wächtershäuser. Before enzymes and templates: Theory of surface metabolism. *Microbiological Reviews*, 52(4):452–484, 1988.

[896] G. Wächtershäuser. Evolution of the first metabolic cycles. *Proceedings of the National Academy of Sciences*, 87:200–204, Jan. 1990.

[897] G. Wächtershäuser. Life as we don't know it. *Science*, 289(5483):1307–1308, 2000.

[898] G. Wächtershäuser. From volcanic origins of chemoautotrophic life to bacteria, archaea and eukarya. *Philosophical Transactions of the Royal Society B: Biological Sciences*, 361(1474):1787–1808, 2006.

[899] G. Wächtershäuser. 17 Origin of life by metabolic avalanche breakthrough in an iron-sulfur world. *Journal of Biomolecular Structure and Dynamics*, 31(sup1):9–10, 2013.

[900] A. Wagner. *Robustness and Evolvability in Living Systems*. Princeton University Press, 2005.

[901] A. Wagner. Robustness, evolvability, and neutrality. *FEBS Letters*, 579:1772–1778, 2005.

[902] A. Wagner. Robustness and evolvability: A paradox resolved. *Proceedings of the Royal Society B: Biological Sciences*, 275:91–100, 2008.

[903] A. Wagner. *The Origins of Evolutionary Innovations: A Theory of Transformative Change in Living Systems*. Oxford University Press, 2011.

[904] A. Wagner and D. A. Fell. The small world inside large metabolic networks. *Proceedings of the Royal Society of London. Series B: Biological Sciences*, 268(1478):1803–1810, 2001.

[905] G. P. Wagner and L. Altenberg. Perspective: Complex adaptations and the evolution of evolvability. *Evolution*, 50:967–976, 1996.

[906] G. P. Wagner, M. Pavlicev, and J. M. Cheverud. The road to modularity. *Nature Reviews Genetics*, 8:921–931, Dec. 2007.

[907] D. C. Walker, J. Southgate, G. Hill, M. Holcombe, D. R. Hose, S. Wood, S. M. Neil, and R. H. Smallwood. The epitheliome: Agent-based modelling of the social behaviour of cells. *Biosystems*, 76(1-3):89–100, 2004.

[908] B. Wang, R. I. Kitney, N. Joly, and M. Buck. Engineering modular and orthogonal genetic logic gates for robust digital-like synthetic biology. *Nature Communications*, 2:508, Oct. 2011.

[909] L. Wang, J. Xie, and P. G. Schultz. Expanding the genetic code. *Annual Review of Biophysics and Biomolecular Structure*, 35:225–249, 2006.

[910] H. Watson. The stereochemistry of the protein myoglobin. *Prog. Stereochem.*, 4:299, 1969.

[911] R. A. Watson. *Compositional Evolution: Interdisciplinary Investigations in Evolvability, Modularity, and Symbiosis*. PhD thesis, Department of Computer Science, Brandeis University, May 2002.

[912] A. Weeks and S. Stepney. Artificial catalysed reaction networks for search. In *ECAL Workshop on Artificial Chemistry*, 2005.

[913] E. Weinan. *Principles of Multiscale Modeling*. Cambridge University Press, 2011.

[914] T. Weise and K. Tang. Evolving distributed algorithms with genetic programming. *IEEE Transactions on Evolutionary Computation*, 16(2):242–265, Apr. 2012.

[915] G. Weiss. *Multiagent Systems: A Modern Approach to Distributed Artificial Intelligence*. MIT Press, 1999.

[916] J. N. Weiss. The Hill equation revisited: Uses and misuses. *The FASEB Journal*, 11(11):835–41, Sept. 1997.

[917] R. White and G. Engelen. Cellular automata and fractal urban form: A cellular modelling approach to the evolution of urban land-use patterns. *Environment and Planning A*, 25:1175–1175, 1993.

[918] G. M. Whitesides and M. Boncheva. Boyond molecules: Self-assembly of mesoscopic and macroscopic components. *Proceedings of the National Academy of Sciences*, 99:4769–4774, 2002.

[919] G. M. Whitesides and B. A. Grzybowski. Self-assembly at all scales. *Science*, 295:2418–2421, 2002.

[920] D. Wilson. A theory of group selection. *Proceedings of the National Academy of Sciences*, 72:143–146, 1975.

[921] P. Winkler. Puzzled – Solutions and sources. *Communcations of the ACM*, 52(6):103, 2009.

[922] P. Winkler. Puzzled – Understanding relationships among numbers. *Communcations of the ACM*, 52(5):112, 2009.

[923] S. Wolfram. Computation theory of cellular automata. *Communications in Mathematical Physics*, 96(1):15–57, 1984.

[924] S. Wolfram. *Theory and Applications of Cellular Automata*. World Scientific, 1986.

[925] S. Wolfram. *Cellular Automata and Complexity: Collected Papers*. Perseus, 1994.

[926] O. Wolkenhauer, M. Ullah, W. Kolch, and K.-H. Cho. Modelling and simulation of intracellular dynamics: Choosing an appropriate framework. *IEEE Transactions on Nano-Bioscience*, 3(3):200–207, 2004.

[927] L. Wolpert. Positional information and the spatial pattern of cellular differentiation. *Journal of Theoretical Biology*, 25(1):1–47, 1969.

[928] L. Wolpert. One hundred years of positional information. *Trends in Genetics*, 12(9):359–364, 1996.

[929] L. Wolpert. Positional information and patterning revisited. *Journal of Theoretical Biology*, 269(1):359–365, 2011.

[930] L. Wolpert, C. Tickle, et al. *Principles of Development*. Oxford University Press, 4th edition, 2011.

[931] M. Wooldridge. *An Introduction to Multiagent Systems*. John Wiley & Sons, 2009.

[932] S. X. Wu and W. Banzhaf. A hierarchical cooperative evolutionary algorithm. In *Proceedings of the 12th Annual Conference on Genetic and Evolutionary Computation*, pages 233–240. ACM, 2010.

[933] L. Yaeger, V. Griffith, and O. Sporns. Passive and driven trends in the evolution of complexity. In Bullock et al. [143], pages 725–732.

[934] L. Yaeger, O. Sporns, S. Williams, X. Shuai, and S. Dougherty. Evolutionary selection of network structure and function. In Fellermann et al. [268], pages 313–320.

[935] L. Yamamoto. Code regulation in open ended evolution. In Ebner et al., editors, *Proceedings of the 10th European Conference on Genetic Programming (EuroGP 2007)*, volume 4445 of *LNCS*, pages 271–280. Springer, Apr. 2007.

[936] L. Yamamoto. PlasmidPL: A plasmid-inspired language for genetic programming. In *Proc. 11th European Conference on Genetic Programming (EuroGP)*, volume 4971 of *LNCS*, pages 337–349. Springer, Mar. 2008.

[937] L. Yamamoto. Evaluation of a catalytic search algorithm. In J. R. González, D. A. Pelta, C. Cruz, G. Terrazas, and N. Krasnogor, editors, *Nature Inspired Cooperative Strategies for Optimization (NICSO 2010)*, volume 284 of *Studies in Computational Intelligence*, pages 75–87. Springer, 2010.

[938] L. Yamamoto and W. Banzhaf. Catalytic search in dynamic environments. In Fellermann et al. [268], pages 277–285.

[939] L. Yamamoto, W. Banzhaf, and P. Collet. Evolving reaction-diffusion systems on GPU. In L. Antunes and H. S. Pinto, editors, *Progress in Artificial Intelligence*, volume 7026 of *LNAI*, pages 208–223. Springer, 2011.

[940] L. Yamamoto, P. Collet, and W. Banzhaf. Artificial Chemistries on GPU. In S. Tsutsui and P. Collet, editors, *Massively Parallel Evolutionary Computation on GPGPUs*, Natural Computing Series, pages 389–419. Springer, 2013.

[941] L. Yamamoto, D. Miorandi, P. Collet, and W. Banzhaf. Recovery properties of distributed cluster head election using reaction-diffusion. *Swarm Intelligence, ANTS 2010 Special Issue, Part 1*, 5(3-4):225–255, 2011.

[942] L. Yamamoto, D. Schreckling, and T. Meyer. Self-replicating and self-modifying programs in Fraglets. In *Proc. 2nd International Conference on Bio-Inspired Models of Network, Information, and Computing Systems (BIONETICS)*, Dec. 2007.

[943] L. Yamamoto and C. Tschudin. Experiments on the automatic evolution of protocols using genetic programming. In *Proc. 2nd Workshop on Autonomic Communication (WAC)*, volume 3854 of *LNCS*, pages 13–28. Springer, Oct. 2005.

[944] T. Yokoyama. Reversible computation and reversible programming languages. *Electronic Notes in Theoretical Computer Science*, 253(6):71–81, 2010.

[945] A. Yoshida, K. Aoki, and S. Araki. Cooperative control based on reaction-diffusion equation for surveillance system. In *Knowledge-Based Intelligent Information and Engineering Systems (KES)*, volume 3683 of *Lecture Notes in Computer Science*, pages 533–539. Springer, 2005.

[946] A. Yoshida, T. Yamaguchi, N. Wakamiya, and M. Murata. Proposal of a reaction-diffusion based congestion control method for wireless mesh networks. In *Proc. 10th International Conference on Advanced Communication Technology (ICACT)*, pages 455–460. IEEE, 2008.

[947] W. Young and E. Elcock. Monte Carlo studies of vacancy migration in binary ordered alloys. *Proceedings of the Physical Society*, 89:735–747, 1966.

[948] W. Yu, K. Sato, M. Wakabayashi, T. Nakaishi, E. P. Ko-Mitamura, Y. Shima, I. Urabe, and T. Yomo. Synthesis of functional protein in liposome. *Journal of Bioscience and Bioengineering*, 92(6):590–593, 2001.

[949] I. Zachar, A. Fedor, and E. Szathmáry. Two different template replicators coexisting in the same protocell: Stochastic simulation of an extended chemoton model. *PLoS ONE*, 6(7):e21380, 07 2011.

[950] K. Zahnle, L. Schaefer, and B. Fegley. Earth's earliest atmospheres. *Cold Spring Harbor Perspectives in Biology*, Oct. 2010.

[951] K.-P. Zauner. Molecular Information Technology. *Critical Reviews in Solid State and Material Sciences*, 30(1):33–69, 2005.

[952] K.-P. Zauner and M. Conrad. Simulating the interplay of structure, kinetics, and dynamics in complex biochemical networks. In R. Hofestädt, T. Lengauer, M. Löffler, and D. Schomburg, editors, *Proc. German Conference on Bioinformatics (GCB'96)*, pages 336–338, 1996.

[953] K.-P. Zauner and M. Conrad. Conformation-driven computing: Simulating the context-conformation-action loop. *Supramolecular Science*, 5(5-6):791–794, 1998.

[954] K.-P. Zauner and M. Conrad. Enzymatic computing. *Biotechnology Progress*, 17(3):553–559, 2001.

[955] L. Zhang. The extended glider-eater machine in the spiral rule. In C. Calude, M. Hagiya, K. Morita, G. Rozenberg, and J. Timmis, editors, *Unconventional Computation*, volume 6079 of *Lecture Notes in Computer Science*, pages 175–186. Springer, 2010.

[956] H.-X. Zhou, G. Rivas, and A. P. Minton. Macromolecular crowding and confinement: Biochemical, biophysical, and potential physiological consequences. *Annual Review of Biophysics*, 37:375, 2008.

[957] T. Zhu, Y. Hu, Z.-M. Ma, D.-X. Zhang, T. Li, and Z. Yang. Efficient simulation under a population genetics model of carcinogenesis. *Bioinformatics*, 27(6):837–843, 2011.

[958] J. Ziegler and W. Banzhaf. Evolving control metabolisms for a robot. *Artificial Life*, 7(2):171–190, 2001.

[959] J. Ziegler, P. Dittrich, and W. Banzhaf. Towards a metabolic robot control system. In M. Holcombe and R. Paton, editors, *Information Processing in Cells and Tissues*, pages 305–318. Plenum Press, Dec. 1998.

Author Index

Subject Index

Printed in the United States
by Baker & Taylor Publisher Services